Electric Renewable Energy Systems

Electric Renewable Energy Systems

Muhammad H. Rashid, Ph.D.,
Fellow IET, Life Fellow IEEE
University of West Florida, U.S.A.

AMSTERDAM • BOSTON • HEIDELBERG • LONDON
NEW YORK • OXFORD • PARIS • SAN DIEGO
SAN FRANCISCO • SINGAPORE • SYDNEY • TOKYO
Academic Press is an Imprint of Elsevier

British Library Cataloguing-in-Publication Data
A catalogue record for this book is available from the British Library

Library of Congress Cataloging-in-Publication Data
A catalog record for this book is available from the Library of Congress

ISBN: 978-0-12-804448-3

For information on all Academic Press publications
visit our website at http://store.elsevier.com/

Working together
to grow libraries in
developing countries

www.elsevier.com • www.bookaid.org

Publisher: Joe Hayton
Acquisition Editor: Raquel Zanol
Editorial Project Manager: Mariana Kühl Leme
Editorial Project Manager Intern: Ana Claudia A. Garcia
Production Project Manager: Kiruthika Govindaraju
Marketing Manager: Louise Springthorpe
Cover Designer: Mark Rogers

Dedication

Dedicated to those who encourage and empower students
through the knowledge of renewable energy sources,
conversion, and applications.

Contents

List of Contributors

Abdullah, Tuan Ab Rashid Tuan College of Engineering, Universiti Tenaga Nasional, Jalan IKRAM-UNITEN, Selangor Darul Ehsan, Malaysia

Abidin, Izham Zainal College of Engineering, Universiti Tenaga Nasional, Jalan IKRAM-UNITEN, Selangor Darul Ehsan, Malaysia

Abu-aisheh, Akram Ahamd Department of Electrical and Computer Engineering, University of Hartford, West Hartford, CT, USA

Ali, Rosnazri School of Electrical System Engineering, Universiti Malaysia Perlis (UniMAP), Arau, Perlis, Malaysia

Baharudin, Nor Hanisah School of Electrical System Engineering, Universiti Malaysia Perlis (UniMAP), Arau, Perlis, Malaysia

Batarseh, Majd Ghazi Yousef Princess Sumaya University for Technology, King Abdullah II Faculty of Engineering, Electrical Engineering Department, University of Jordan, Al-Jubaiha, Amman, Jordan

Batarseh, Majd Ghazi Department of Electrical Engineering, Princess Sumaya University for Technology, Amman, Jordan

Beig, Abdul R. Department of Electrical Engineering, The Petroleum Institute, Abu Dhabi, UAE

Bentley, Edward Electrical Power Engineering at Northumbria University Newcastle upon Tyne, United Kingdom

Daud, Abdel Karim Khaled Department of Electrical Engineering, College of Engineering, Palestine Polytechnic University, Hebron–West Bank, Palestine

Gao, David (Zhiwei) Department of Physics and Electrical Engineering, Faculty of Engineering and Environment, University of Northumbria, Newcastle upon Tyne, UK

Haque, Ahteshamul Department of Electrical Engineering, Faculty of Engineering & Technology, Jamia Millia Islamia University, New Delhi, India

Hassan, Syed Idris Syed School of Electrical System Engineering, Universiti Malaysia Perlis (UniMAP), Arau, Perlis, Malaysia

Hegde, Sriram Department of Applied Mechanics, Indian Institute of Technology, Delhi, India

Khader, Sameer Hanna Department of Electrical Engineering, College of Engineering, Palestine Polytechnic University, Hebron–West Bank, Palestine

Mansur, Tunku Muhammad Nizar Tunku School of Electrical System Engineering, Universiti Malaysia Perlis (UniMAP), Arau, Perlis, Malaysia

Marsadek, Marayati College of Engineering, Universiti Tenaga Nasional, Jalan IKRAM-UNITEN, Selangor Darul Ehsan, Malaysia

Mishra, Sukumar Department of Electrical Engineering, Indian Institute of Technology Delhi, New Delhi, India

Murthy, Sreenivas S. Department of Electrical Engineering, Indian Institute of Technology, Delhi; CPRI, Bengaluru, India

Muyeen, S.M. Department of Electrical Engineering, The Petroleum Institute, Abu Dhabi, UAE

Nasir Ani, Farid Faculty of Mechanical Engineering, Universiti Teknologi Malaysia

Nasiri, Adel Electrical Engineering and Computer Science Department, College of Engineering and Applied Sciences, University of Wisconsin-Milwaukee, Milwaukee, WI, USA

Nehrir, M. Hashem Electrical & Computer Engineering Department, Montana State University, Bozeman, MT, USA

Novakovic, Bora Electrical Engineering and Computer Science Department, College of Engineering and Applied Sciences, University of Wisconsin-Milwaukee, Milwaukee, WI, USA

Osman, Miszaina College of Engineering, Universiti Tenaga Nasional, Jalan IKRAM-UNITEN, Selangor Darul Ehsan, Malaysia

Putrus, Ghanim Electrical Power Engineering at Northumbria University Newcastle upon Tyne, United Kingdom

Sekaran, Easwaran Chandira Associate Professor, Department of Electrical and Electronics Engineering, Coimbatore Institute of Technology, Coimbatore, INDIA

Sharma, Dushyant Department of Electrical Engineering, Indian Institute of Technology Delhi, New Delhi, India

Soelaiman, Tubagus Ahmad Fauzi Mechanical Engineering Department, Faculty of Mechanical and Aerospace Engineering, Thermodynamics Laboratory, Engineering Centre for Industry, Institut Teknologi Bandung, Bandung, Indonesia

Sun, Kai Department of Physics and Electrical Engineering, Faculty of Engineering and Environment, University of Northumbria, Newcastle upon Tyne, UK

Vasantharathna, S. Department of Electrical and Electronics Engineering, Coimbatore Institute of Technology, Coimbatore, Tamil Nadu, India

Wang, Caisheng Electrical & Computer Engineering Department, Wayne State University, Detroit, MI, USA

About the Editor-in-Chief

Muhammad H. Rashid is employed by the University of West Florida as a Professor of Electrical and Computer Engineering. Previously, he was employed by the University of Florida as Professor and Director of UF/UWF Joint Program. He received BSc degree in Electrical Engineering from the Bangladesh University of Engineering and Technology and MSc and PhD degrees from the University of Birmingham in United Kingdom. Previously, he worked as Professor of Electrical Engineering and the Chair of the Engineering Department at Indiana University – Purdue University Fort Wayne. Also, he worked as Visiting Assistant Professor of Electrical Engineering at the University of Connecticut, Associate Professor of Electrical Engineering at Concordia University (Montreal, Canada), Professor of Electrical Engineering at Purdue University Calumet, Visiting Professor of Electrical Engineering at King Fahd University of Petroleum and Minerals (Saudi Arabia), design and development engineer with Brush Electrical Machines Ltd (England, United Kingdom), Research Engineer with Lucas Group Research Centre (England, United Kingdom), and Lecturer and Head of Control Engineering Department at the Higher Institute of Electronics (Libya and Malta).

He is actively involved in teaching, researching, and lecturing in electronics, power electronics, and professional ethics. He has published 18 books listed in the US Library of Congress and more than 160 technical papers. His books are adopted as textbooks all over the world. His book, *Power Electronics* has translations in Spanish, Portuguese, Indonesian, Korean, Italian, Chinese, Persian, and Indian edition. His book, *Microelectronics* has translations in Spanish (in Mexico and Spain), Italian, and Chinese.

He has received many invitations from foreign governments and agencies to give keynote lectures and consultation; foreign universities to serve as an external examiner for undergraduate, master's, and PhD examinations; funding agencies to review research proposals; and US and foreign universities to evaluate promotion cases for professorship. He has worked as a regular employee or consultant in Canada, Korea, United Kingdom, Singapore, Malta, Libya, Malaysia, Saudi Arabia, Pakistan, and Bangladesh. He has traveled to almost all the States in USA and many other countries to give lectures and present papers (Japan, China, Hong Kong, Indonesia, Taiwan, Malaysia, Thailand, Singapore, India, Pakistan, Turkey, Saudi Arabia, United Arab Emirates, Qatar, Libya, Jordan, Egypt, Morocco, Malta, Italy, Greece, United Kingdom, Brazil, and Mexico).

He is a *Fellow* of the Institution of Engineering and Technology (IET, United Kingdom) and a *Life Fellow* of the Institute of Electrical and Electronics Engineers (IEEE, USA). He was elected as an IEEE Fellow with the citation "Leadership in power electronics education and contributions to the analysis and design methodologies of solid-state power converters." He is the recipient of the 1991 Outstanding Engineer Award from IEEE. He received the 2002 IEEE Educational Activity Award (EAB) and Meritorious Achievement Award in Continuing Education with the following citation "For contributions to the design and delivery of continuing education in power electronics and computer-aided-simulation." He is the recipient of the 2008 IEEE Undergraduate Teaching Award with citation "For his distinguished leadership and dedication to quality undergraduate electrical engineering education, motivating students and publication of outstanding textbooks." He is also the recipient of the IEEE 2013 Industry Applications Society Outstanding Achievement Award.

He is an ABET program evaluator for electrical and computer engineering (1995–2000) and was an engineering evaluator for the Southern Association of Colleges and Schools (SACS, USA). He is also an ABET program evaluator for (general) engineering program. He is the Series Editor of *Power Electronics and Applications* and *Nanotechnology and Applications* with the CRC Press. He serves as the Editorial Advisor of *Electric Power and Energy* with Elsevier Publishing. He gives lectures and conducts workshops on Outcome-Based Education (OBE) and its implementations including assessments.

He is a distinguished lecturer for the IEEE Education Society and a Regional Speaker (previously distinguished lecturer) for the IEEE Industrial Applications Society. He also authored a book *The Process of Outcome-Based Education – Implementation, Assessment and Evaluations* (2012, UiTM Press, Malaysia).

Preface

The demand for energy, particularly in electrical forms, is ever increasing in order to improve the standard of living. Electric power generated from renewable energy sources is getting increasing attention and supports for new initiatives and developments in order to meet the increased energy demands around the world while minimizing the environmental effects. The renewable energy sources include solar, wind, hydroelectric, fuel cells, geothermal, biomass, and ocean thermal. The conversion of these energy sources into electric form requires the knowledge of three-phase supply, magnetic circuit, power transformers, and electric generators and control. Power electronic converters convert the electrical energy from the energy sources into different forms: AC–DC, DC–DC, and DC–AC to meet the utility demand.

Power electronics is an integral part of renewable energy for interfacing between the renewable sources and the utility supply and/or customers. Semiconductor devices are used as switches for power conversion or processing, as are solid-state electronics for efficient control of the amount of power and energy flow. The utility interface requites the knowledge of electric transmission, power systems, and the integration of power grid.

This book covers the energy sources, the energy conversion into electric form, the power conversion or processing to meet the utility demand, and transmission and interface to the power system and power grid. This book is designed to cover the state-of-the-art development in electric renewable energy covering from the source to the power grid.

The book is organized into four parts: (1) energy sources, (2) conversion to electric energy, (3) electric power conversion (or processing), and (4) interfacing to the electric power grid. It uses the tutorial approach so that engineering and nonengineering students or engineers can follow and then develop clear understanding leading to the in-depth analysis and design. It is intended as an undergraduate textbook for regular students and/or continuing education.

The list of contributors to this book spans the globe. The contributors are leading authorities in their areas of expertise. All were chosen because of their intimate knowledge of their subjects, and their contributions make this a comprehensive state-of-the-art guide to the expanding field of renewable energy. As expected, writing style of each contributor is different although the editorial staff attempted to make it more uniform.

Muhammad H. Rashid
Editor-in-Chief

Acknowledgments

Thanks to all the contributors without whose dedication, commitment, and hard work, it would not be possible to complete and publish this book.

Thanks to Ms Mariana Kühl Leme (Editorial Project Manager), Ms Chelsea Johnston, and other editorial and production staff members.

Finally, thanks to my family for their patience while I was occupied with this book and other projects.

Any comments and suggestions regarding this book are welcome. They should be sent to

Dr Muhammad H. Rashid
Professor, Department of Electrical and Computer Engineering,
University of West Florida, Pensacola, Florida, USA
E-mail: mrashid@uwf.edu
Web: http://uwf.edu/mrashid

Introduction to electrical energy systems

1

Bora Novakovic, Adel Nasiri
Electrical Engineering and Computer Science Department, College of Engineering and
Applied Sciences, University of Wisconsin-Milwaukee, Milwaukee, WI, USA

Chapter Outline

1.1   Electrical energy systems

Electrical energy is one of the most commonly used forms of energy in the world. It can be easily converted into any other energy form and can be safely and efficiently transported over long distances. As a result, it is used in our daily lives more than any other energy source. It powers home appliances, cars, and trains; supplies the machines that pump water; and energizes the light bulbs lighting homes and cities.

Any system that deals with electrical energy on its way between energy sources and loads can be considered to be an electrical energy system. These systems vary greatly in size and complexity. For instance, consider a supply system of a computer, large data center, or power system of a country. All these systems include some sort of power conversion to generate electrical energy, power transmission, and distribution.

In all cases, some kind of power transformation is also included either to convert between direct current (DC) and alternating current (AC) systems or to adapt electric supply voltage to the load's requirements.

All electrical energy systems are characterized by the voltage waveform, rated voltage, power levels, and the number of lines or phases in the case of AC systems. Based on the voltage waveform, electrical energy systems can be divided in two main categories, AC systems and DC systems. AC systems transport and distribute energy using alternating voltages and currents while DC systems use direct currents and voltages for the same purpose. Another classification of low voltage (less than 600 V), medium voltage (600 V–69 kV), high voltage (69–230 kV), and extra-high voltage systems (more than 230 kV) can be done based on the voltage ratings of the system [1]. AC systems exist at all voltage levels. DC systems, on the other hand, are more common at extra-high and low voltage levels. AC systems can be further divided to single-phase and polyphase systems depending on the number of phases used for power transmission. Among polyphase systems, the three-phase system is most commonly used.

Another classification of electrical energy systems can be based on their purpose. Assuming that the electrical energy goes through a chain of systems on its path from sources to the loads, it is possible to identify system groups that have common or similar purposes. At the beginning and end of this chain, we generally have some types of energy conversion systems. These devices convert mechanical, thermal, chemical, or some other form of energy into electrical energy or vice versa. They are usually considered to be sources or loads, depending on whether they produce or consume electrical energy. In between we have transmission and distribution systems, which may include a number of electrical energy conversion (transformation) systems. The main purpose of transmission systems is to transport electrical energy in the most efficient manner. Distribution systems distribute electrical energy among loads and make sure that the form of the electrical energy fits the load requirements. Electrical energy conversion (transformation) systems are usually part of transmission and distribution systems. They change the form of electrical energy by modifying either the voltage levels, voltage waveforms, or the number of phases in the case of polyphase systems.

As an example, consider a large utility grid system. In this type of system, the end points are sources and loads. On the source side, we have energy conversion systems in the form of large power plants, converting the energy from fossil, renewable, or nuclear energy sources into electrical energy. These systems are highly complicated and may include several energy conversion stages. However, output is in most cases at the terminals of a medium voltage three-phase generator, which is an electromechanical energy conversion system on its own. In a large power system, generators are usually far from loads. As a consequence, medium voltage three-phase power from the generators first goes to the transmission system, which transports the energy over long distances. Before it is transported, electrical energy is transformed into a form that is suitable for low loss transportation over large distances, usually high voltage AC or DC. From the transmission system, the high voltage electrical energy is transferred to a distribution system where it is converted into low and medium voltage levels and distributed among loads.

1.2 Energy and power

In an effort to define the properties of a connection between two electrical energy systems in a manner that can be used for engineering and economical purposes, in addition voltages and currents, it is useful to take power and energy into consideration. This is especially true for multiphase AC systems. In this case, observation of multiple alternating voltages and currents provides little useful information, which means that additional analysis is required. Power and exchanged energy, on the other hand, together with the rated voltage, would be enough to quickly evaluate the connection properties and behavior.

In electrical energy systems, energy and power terms are easily defined by looking at the simple circuit shown in Figure 1.1. The equations for instantaneous electrical power (p) and energy (E) drawn by the load during a time period of Δt are given in (1.1) and (1.2) [2].

$$p(t) = v(t)i(t) \tag{1.1}$$

$$E = \int_{t}^{t+\Delta t} p(t)\, dt = \int_{t}^{t+\Delta t} v(t)i(t)\, dt \tag{1.2}$$

In the case where the voltage does not change over time (DC supply case) and the load is constant, the current and power are also constant, while the integral in (1.2) is reduced to $E = vi\Delta t$. These values are simple to calculate and are very informative.

If it is assumed that the supply voltage is an alternating waveform (AC supply), the equation for power (1.1) becomes a value that changes in time while the energy equation (1.2) becomes a bit more complicated to calculate. The simplicity of the DC case no longer exists. To make things similarly informative for the AC systems, the concepts of root main square (RMS) values, active power (P), reactive power (Q), and apparent power (S) have been introduced. Assuming that voltage and current in the system can be described by Equations (1.3) and (1.4), RMS values are defined by Equations (1.5) and (1.6), active power is given by Equation (1.7), reactive power by Equation (1.8), and apparent power by Equation (1.9) [3].

$$v(t) = V\,\cos(\omega t) \tag{1.3}$$

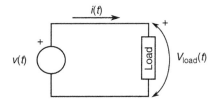

Figure 1.1 Voltage source with connected load.

$$i(t) = I \, \cos(\omega t - \theta) \hspace{4cm} (1.4)$$

$$V_{rms} = \sqrt{\frac{1}{T} \int_{t}^{t+T} v(t)^2 \, dt} = \frac{V}{\sqrt{2}} \hspace{3cm} (1.5)$$

$$I_{rms} = \sqrt{\frac{1}{T} \int_{t}^{t+T} i(t)^2 \, dt} = \frac{I}{\sqrt{2}} \hspace{3cm} (1.6)$$

$$P = V_{rms} I_{rms} \cos(\theta) \hspace{4cm} (1.7)$$

$$Q = V_{rms} I_{rms} \sin(\theta) \hspace{4cm} (1.8)$$

$$S = V_{rms} I_{rms} = \sqrt{P^2 + Q^2} \hspace{3.5cm} (1.9)$$

If the consumer notation convention is assumed, which states that the load draws positive active power, the source will draw negative active power (or produce positive power). Active power is given in watts (W) and describes how fast energy is converted from electrical into some other form (e.g., thermal in the case of a resistive load). Reactive power, on the other hand, characterizes the ability of reactive loads (capacitors and inductors) to return the energy stored in them back to the source. It actually describes the rate at which the energy circulates between sources and reactive loads.

Parameter $\cos(\theta)$ is defined as the system power factor. The load that has a power factor with positive angle θ is called lagging and by convention draws reactive power. In practice the inductive loads have current lagging behind the voltage waveform. The angle θ is positive in this case if expressed with Equation (1.4), which means that the inductor draws positive reactive power. A load that has a power factor with negative angle θ is called leading and draws negative reactive power (or supplies positive reactive power). These loads are mostly capacitive. Reactive power is usually expressed in volt-amperes-reactive (VAr).

In AC systems, all instantaneous values have an alternating nature. As a result, instantaneous power and energy also have alternating components. Active power is used to quantify the energy producing component of power and the rate at which energy flows, on average, in one direction and is converted into another form by the load. It can be proved that Equation (1.7) represents the average value of the instantaneous power expressed in Equation (1.1) for current and voltage waveforms given in Equations (1.3) and (1.4). Consequently, the useful amount of energy drawn by the load during the time period Δt can be expressed as an integral of active power,

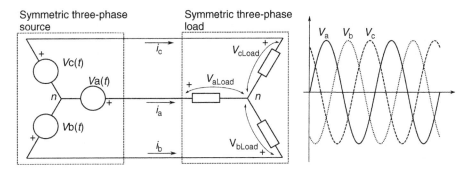

Figure 1.2 Symmetric three-phase system. Voltage waveforms have the same amplitude and have 120° phase difference.

given by Equation (1.10). If the power factor remains constant, Equation (1.10) is reduced to (1.11), which is simple to calculate and is very informative.

$$E = \int_t^{t+\Delta t} P \, dt = \int_t^{t+\Delta t} V_{rms} I_{rms} \cos(\theta) \, dt \qquad (1.10)$$

$$E = V_{rms} I_{rms} \cos(\theta) \Delta t \qquad (1.11)$$

It should be also noted that reactive power only circulates between sources and loads and is not converted into other energy forms (at least not in a useful manner), which means that it does not contribute to useful energy consumption by load. However, it increases losses in the transmission systems as the energy is moved back and forth between generators and loads. For this reason, reactive power must be taken into account and is equally as important as active power in large distribution systems.

For the three-phase system exemplified in Figure 1.2, voltage and current waveform expressions are given by Equations (1.12) and (1.13). In this case, power and energy equations are similar to a single-phase system. The difference is that there are three power carrying lines and three times more power can be transferred with the same voltage and current ratings. The expression for active and reactive power and energy consumed by the load during a time interval of Δt is given by Equations (1.14–1.16) [3].

$$v_a(t) = V_{rms} \sqrt{2} \, \cos(\omega t),$$

$$v_b(t) = V_{rms} \sqrt{2} \, \cos\left(\omega t - \frac{2\pi}{3}\right), \qquad (1.12)$$

$$v_c(t) = V_{rms} \sqrt{2} \, \cos\left(\omega t - \frac{4\pi}{3}\right)$$

$$i_a(t) = I_{rms}\sqrt{2}\,\cos(\omega t - \theta),$$
$$i_b(t) = I_{rms}\sqrt{2}\,\cos\left(\omega t - \theta - \frac{2\pi}{3}\right), \tag{1.13}$$
$$i_c(t) = I_{rms}\sqrt{2}\,\cos\left(\omega t - \theta - \frac{4\pi}{3}\right)$$

$$P_{3\phi} = 3V_{rms}I_{rms}\cos(\theta) \tag{1.14}$$

$$Q_{3\phi} = 3V_{rms}I_{rms}\sin(\theta) \tag{1.15}$$

$$E = \int_t^{t+\Delta t} P_{3\phi}\,dt = 3V_{rms}I_{rms}\cos(\theta)\Delta t \tag{1.16}$$

For the sake of simplicity, Equations (1.12–1.16) are given in terms of rated phase to neutral RMS voltage (V_{rms}). In power systems engineering, it is more common to express multiphase voltages in terms of phase-to-phase rated voltage (V_{rmsll}). The relationship between phase-to-phase and phase-to-neutral voltages in a symmetric three-phase system is described by Equation (1.17). Using this equation, it is easy to express energy and power in three-phase systems in terms of line-to-line voltages. This task is left to the reader as an exercise.

$$V_{rmsll} = V_{rms}\sqrt{3} \tag{1.17}$$

Another important point is related to the measurement units used for power systems calculations. Instead of joule (J) for energy, which is essentially watt-second, units like watt-hour (Wh) or the prefixed units kilowatt-hour (kWh), megawatt-hour (MWh), or terrawatt-hour (TWh) are commonly used. For volt and ampere units, due to the high amplitudes, prefixes kilo (k) and mega (M) are also commonly used, for example, kA, MV.

1.3   AC versus DC supply

It is worth mentioning that a great dispute took place in the late nineteenth century between the proponents of AC and DC distribution systems. AC systems were promoted by George Westinghouse and Nikola Tesla while Thomas Edison promoted his DC distribution system concept. AC systems offered many benefits at the time and prevailed in the end [4].

One of the main benefits of AC systems is easy voltage transformation. Relatively simple and robust transformers can be used for this purpose. The ability to transform voltage level is very important for the electrical energy distribution systems and it should be noted that DC voltage transformation was not an easy task in the late nineteenth century. Voltage transformation enables high voltage transmission, which reduces conduction losses. For the same amount of transmitted power, the conduction losses are reduced with the square of the voltage used for transmission. Another important benefit of easy voltage transformation is the ability to regulate voltage at load centers and to keep it nearly constant with, again, relatively simple and robust multiple tap transformers.

The use of three-phase systems and Tesla's rotating magnetic field is another advantage of AC systems. Three-phase transmission additionally reduces losses in the system while the principle of the rotating magnetic field enables easy electromechanical energy conversion with the help of robust high power synchronous generators and induction machines.

Nowadays DC and AC systems are intertwined at many levels. DC systems are more common for lower voltages while AC systems still dominate large-scale power distribution and power transmission. In recent decades, DC systems have become more popular mainly as a consequence of development of digital systems and power electronics converters. Some types of energy sources such as solar photovoltaic (PV) and fuel cells, which are becoming increasingly popular, also require a DC interface. Another field where high voltage DC systems have made significant progress is long distance energy transmission where AC systems can have stability issues.

1.4 Basic energy conversion processes

In most cases, when engineers discuss energy conversion systems, they refer to systems that convert various types of energy into electrical energy. The systems that convert electrical energy into other forms are generally called loads. In fact, energy conversion is a much broader term. Most existing natural and manufactured systems convert energy from one form into another all the time. Plants convert the energy of sunlight into chemical bonds. Animals eat plants and use the stored energy for processes in their bodies. Humans use fossil fuels from dead plants and animals to fuel engines that power cars, planes, trains, and eventually generators that produce electrical energy, which powers most of the things created in modern times.

If we limit ourselves to engineering science, one of the most important goals is to find an efficient, clean, and convenient way to convert the energy available in nature into a form that we can easily use. For electrical engineers, the most usable form of energy is electrical.

Arguably, the most common form of energy in nature is thermal; every object that has a temperature higher than its surroundings can be considered as a source of thermal energy. A device that converts electrical into thermal energy can be a simple resistor. The conversion process from thermal directly into electrical is a bit more complicated and requires the use of a special type of device called a *thermoelectric energy converter* based on Peltier and Seebeck effects. It should be noted that thermal energy can be converted into a mechanical form (steam pressure), which can be used for electromechanical energy conversion. This is in fact the most common way of energy conversion from thermal into electrical in two steps: thermal–mechanical–electrical.

Mechanical energy is also very common in everyday life. Wind, for example, carries large amounts of mechanical (kinetic) energy. Large accumulation lakes of hydropower systems also hold tremendous amounts of mechanical (potential) energy. This energy can be converted into electrical energy with the help of electromechanical energy conversion systems. Most of these systems, usually referred to as electrical generators, convert rotational kinetic energy into electrical energy. As a consequence, electromechanical energy conversion systems usually require devices that convert other mechanical energy types into rotational energy. These devices are called prime movers and are usually in the form of a turbine system.

Solar (light or electromagnetic radiation) energy can be considered as another form of energy commonly used to produce electricity. PV systems convert sunlight directly into electrical energy. These systems are becoming increasingly popular as they produce clean energy with very little environmental impact. The technology is still relatively expensive and new but is the focus of many researchers and is constantly developing.

Chemical energy is used by all life forms and is the earliest form of energy used by humankind. Most of the substances that hold chemical energy can be used for creating fire that gives off thermal energy (coal and wood burn very well). In fact, fire can be seen as a self-sustaining energy conversion process that converts chemical into thermal energy. Most of the modern power conversion systems still convert chemical energy into thermal and then thermal energy into electrical. Large coal power plants, for example, burn coal in order to heat the steam and then use the pressure of the heated steam for electromechanical energy conversion. Systems that perform direct conversion of chemical energy into electrical energy are called electrochemical energy conversion systems. The most common electrochemical energy conversion system is a common battery. These systems require special energy storage chemicals and specialized energy conversion structures, which are one of the most popular research areas today.

Nuclear energy can be considered as another form of energy, which comes directly from nuclear reactions at an atomic level inside a nuclear fuel. There are two forms of nuclear reactions, nuclear fusion and nuclear fission. Nuclear fission is a nuclear reaction where the atom of an element breaks apart into two smaller atoms. Nuclear fusion is the reaction where two smaller atoms form a larger one. In both cases, a part of the mass of the initial atoms is converted to energy. These reactions produce high-energy particles and intense heat. The produced heat can be used for the generation of electrical energy. All nuclear power plants are currently based on nuclear fission. Nuclear fusion is still in its early stages and a feasible fusion reactor with a positive net energy balance still does not exist. Research efforts are under way and some predictions state that fusion power generation could become a reality by 2030.

1.5   Review of the laws of thermodynamics

In order to understand the limitations and operation principles of any energy conversion system, one must fully understand the fundamental laws of thermodynamics. Thermodynamics is a branch of physics that deals with energy conversion processes in general. It describes the basic physical properties and postulates the fundamental laws that govern the behavior of an energy conversion system.

Thermodynamics describes a system as a whole and does not take individual particles or even parts of the system into account. A thermodynamic system is described with aggregate properties such as temperature, energy, pressure, and work. The relations between these properties in a thermodynamic system are governed by the laws of thermodynamics.

1.5.1   Zeroth law of thermodynamics

The zeroth law of thermodynamics introduces the concept of thermodynamic equilibrium. If a physical system or a set of systems is in thermodynamic equilibrium, any measurable macroscopic property of the system or the set of systems (e.g., temperature, pressure, volume) are in a steady state and do not change with time. The law itself states that if two systems are in equilibrium with a third system, they are also in equilibrium with each other.

1.5.2   First law of thermodynamics

The first law of thermodynamics reformulates one of the basic laws of physics – the law of conservation of energy, which states that *the total amount of energy of a closed system cannot change* (without external influence). The first law of thermodynamics is usually formulated with Equation (1.18), which states that the total change in the system energy (ΔE) equals the difference between the energy supplied to the system (Q) and the amount of work that the system performs on its surroundings (W).

$$\Delta E = Q - W \qquad (1.18)$$

Note that most natural systems consist of many thermodynamic systems and it may seem that input energy does more than just increase system energy and the work the system performs. Before applying the laws of thermodynamics, thermodynamic systems must be strictly defined.

As an example, consider starting a rotating electrical machine. This system is designed to convert electrical energy into mechanical work and can be considered to have three subsystems – electrical, electromagnetic, and mechanical. Part of the invested electrical energy increases machine temperature, its rotational kinetic energy, and energy in inductances and capacitances. These increases can be contributed to (ΔE) of all three subsystems. The supplied energy (Q) in this case would be the electrical power flowing into the machine terminals. The work of the forces created by the electromagnetic fields within the machine together with all the thermal losses and irradiated electromagnetic fields contribute to the work of the system (W) [5,6].

One important consequence of the first law of thermodynamics is that it is impossible to design a machine that performs work without an energy input (usually referred to as *perpetuum mobile of the first kind*).

1.5.3   Second law of thermodynamics

The first law of thermodynamics deals with total energy within a system, but it does not limit what can be done with that energy. These limitations are addressed in the second law of thermodynamics, which states that it is impossible to make a machine that extracts energy from its environment only to perform work (also known as *perpetuum mobile of the second kind*).

We know from experience that all devices produce heat, no matter what their main purpose is. This seems to be true for all physical systems. No matter what kind of system we have, in the process of performing work (or converting the energy into a useful form), part of the input energy is always lost in the form of heat. In order to quantify this property of nature, thermodynamics introduces the notion of entropy. For thermodynamic systems the entropy depends on the system state and a convenient way to define it is in its differential form given by Equation (1.19).

$$dS = \frac{dQ}{T} \tag{1.19}$$

where dS is the change in system entropy, Q is heat (internal energy) of the system and T is absolute temperature of the system. With the entropy defined in this way, the second law of thermodynamics states that *the entropy of a closed system never decreases and systems naturally tend to the state of maximum entropy*. The extreme cases are ideal reversible processes, where the system changes its state in such a way that the entropy does not change [5,6].

1.5.4   Third law of thermodynamics

The third law provides the referent value for entropy and the conditions that a system must satisfy in order to have the referent entropy. It states that *the entropy of a perfect crystal of any pure substance approaches zero (minimum value) as the temperature of the substance approaches absolute zero* [5,6]. One must note that substances, such as glass that do not have a uniform crystal structure, may have constant entropy greater than zero at absolute zero temperature.

1.6   Photovoltaic energy conversion systems

PV systems convert energy from light directly into electrical energy. The energy conversion principle is based on the special properties of certain semiconductor materials that allow local ionization of atoms in the crystal structure by visible light photons. The ionization process produces two particles, a negative charge electron and a positive charge particle known as a hole in semiconductor physics. A hole is just an empty space, previously occupied by the electron that has left the atom after absorbing the visible light photon. It can be shown that both particles can move freely across the crystal structure of a specific semiconductor.

A production of free positive and negative charged particles in the crystal is still not enough to create a useful macroscopic potential difference. If there is no other

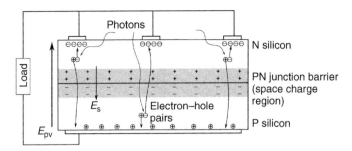

Figure 1.3 PN junction and a field distribution inside a PV cell.

influence, electrons and holes will be created randomly and will be recombined after they lose the energy acquired from the photons in thermal interactions with the atoms in the crystal. In order to create a useful potential difference, we need a barrier that will separate holes and electrons. A large PN junction similar to the one found in common semiconductor diodes can be used for this purpose.

A PN junction with its space charge region electric field serves as a barrier that will separate holes and electrons and create a useful potential difference across the junction. As soon as the generated electron (hole) enters the space charge region, the field accelerates it toward the N(P) side of the panel. Once the electrons reach the N region and the holes reach the P region, it is not likely that they will be recombined as a majority carrier, so we have a buildup of charges, as shown in Figure 1.3. The charge buildup creates a field (E_{pv}) and a useful potential difference across the panel terminals. Note that when the panel is not lit, charge buildup due to the diffusion of carriers on the electrode–panel junction, which is not explicitly shown in Figure 1.3, cancels out the field of the PN junction space charge so there is no useful net potential difference on the panel terminals.

In practice, silicon is most commonly used for PV energy conversion. Single-crystal ingots or polycrystalline silicon rods are cut into plates (usually several centimeters in diameter and about 0.5-mm thick). P-type silicon is used as a base and the other side is doped to create an N-type layer across the entire surface, creating one large PN junction [7]. Metal electrodes are then applied to both surfaces, which are then used to connect the plates in series and create panels as shown in Figure 1.4.

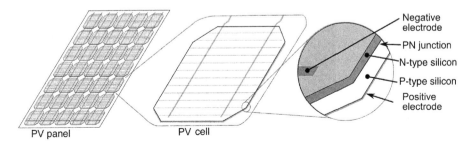

Figure 1.4 Structure of a PV panel.

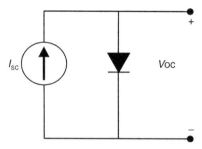

Figure 1.5 Model of a PV cell.

The simplest model of a PV cell can be represented with a circuit shown in Figure 1.5. When the cell is unloaded, the voltage across its terminal can be expressed by Equation (1.20), where k represents the Boltzmann constant, q is the charge of the electron in coulombs, T is the absolute temperature of the panel in degrees Kelvin, I_0 is reverse saturation current of the PN junction, and I_{sc} is the so-called short circuit current of the cell. Both I_0 and I_{sc} depend on the manufacturing parameters of the cell, while I_{sc} also depends on the incoming light irradiance [2].

$$V_{OC} = \frac{kT}{q} \ln\left(\frac{I_{sc}}{I_0} + 1\right)$$ (1.20)

1.7 Electrochemical energy conversion systems

Electrochemical energy conversion systems convert energy stored in chemical bonds directly into electrical energy and vice versa. These systems are characterized by their chemical properties (e.g., a reaction that creates electrical energy from the chemical bonds) and their electrical properties such as the voltage produced by the basic cell, internal resistances, capacitances, power (current) capabilities, and efficiency. Electrical properties are determined by the chemical reaction properties, materials the cell is made of, and the design geometry of the cell.

The chemical processes involved in the energy conversion can be reversible or irreversible. Devices that use irreversible processes are usually called nonrechargeable. For these devices, the electrolyte reacts with electrodes, creates an electric current flow and usually is chemically changed into by-products that cannot be reused in the cell. As an example, we can look at a hydrogen fuel cell. The cell uses the hydrogen gas as a fuel, special polymer electrolyte membrane, and atmosphere oxygen to create electricity. The by-product of the reaction is water, as shown in Figure 1.6. There is no way to separate water into oxygen and hydrogen for possible reuse by this fuel cell, which means that the reaction is irreversible and the device is not rechargeable. This particular reaction produces about 1 V across the electrodes and must be stacked to achieve higher voltages [8,9].

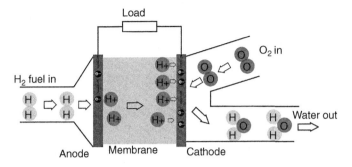

Figure 1.6 Simplified structure of a hydrogen fuel cell.

Reversible processes are used in devices such as rechargeable and flow batteries. These devices have two distinct operation cycles, charge and discharge, which are characterized by different chemical reactions and different electrical properties. During the discharge cycle, the electrolyte is transformed in order to release the energy from the chemical bonds and generate electrical energy. For the charge cycle, electrical energy flows into the device and transforms the electrolyte back into its original state, storing the electrical energy back into chemical bonds.

One example of a rechargeable electrochemical energy conversion device is the common lead–acid battery. The battery uses lead and lead dioxide plates submerged in a water solution of sulfuric acid as shown in Figure 1.7. Plates react with the solution and produce potential difference; the reaction is shown in Equation (1.21). The lead plate reacts with the water solution of sulfuric acid and forms lead sulfate. The reaction releases positive hydrogen atoms into the solution and two electrons into the plate making it negative. Lead dioxide plate also reacts with the water solution of sulfuric acid and also creates lead sulfate but it needs three hydrogen atoms and one electron for the reaction; the electron comes from the plate leaving it positive. The reaction produces the potential difference of about 2.1 V and it varies with the level of charge. The cells are usually stacked to achieve the voltage range of about 12–14 V.

$$\text{Negative plate: } Pb + HSO_4^- \leftrightarrows PbSO_4 + H^+ + 2e^-$$
$$\text{Positive plate: } PbO_2 + HSO_4^- + 3H^+ + e^- \rightleftarrows PbSO_4 + 2H_2O$$

(1.21)

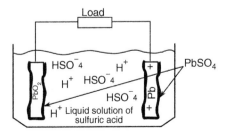

Figure 1.7 Simplified structure of a lead–acid battery.

It should be noted that the reactions are completely reversible if the opposite current flow is applied to the battery terminals [2,10].

1.8 Thermoelectric energy conversion systems

Thermoelectric devices convert thermal energy directly into electrical energy or vice versa. A simple resistor can convert electrical into thermal energy. Devices that convert thermal into electrical energy are more complicated and are based on the Seebeck effect, named after Thomas Seebeck, its discoverer. The Seebeck effect causes the voltage difference across a junction of two different conductors experiencing a thermal gradient. The voltage difference is a consequence of different reactions of two conductors on the applied temperature gradient. Figure 1.8 shows the simplified structure of a Seebeck generator where the conductor materials are usually implemented as doped P- and N-type semiconductors. This structure can produce only a couple of millivolts and must be stacked to achieve meaningful electromotive force. Electromotive force of the Seebeck generator is given in Equation (1.22) and is proportional to the temperature gradient ΔT. The coefficient of proportionality S is called the Seebeck coefficient, which depends both on temperature gradient across the conductors and conductor material properties.

$$V_L = -S\Delta T \tag{1.22}$$

It is worth mentioning that Peltier and Thomson effects are also linked to thermoelectric phenomena. The Peltier effect is opposite to the Seebeck effect. If we replace the load in Figure 1.8 with a voltage source, we can force the flow of thermal energy from the cold plate to the hot plate. This makes Peltier plates effective as cooling devices. The Thomson effect is similar to the Seebeck effect but explains the voltage gradient present across a single conductor experiencing a thermal gradient [11,12].

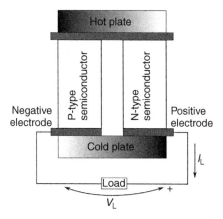

Figure 1.8 Simplified structure of a Seebeck generator.

1.9 Electromechanical energy conversion systems

Electromechanical energy conversion processes are used to generate the majority of electrical energy consumed today. In these processes, mechanical energy is converted to electrical energy using rotating machines commonly known as electric generators. On the load side of the power system, similar rotating machines – electric motors – are among the largest consumers of electrical power.

An electromechanical energy conversion system usually has three subsystems, electrical, mechanical, and a system of electromagnetic fields, which is used to couple mechanical forces and electric currents. The basic coupling principle is described by Lorentz force law, which gives the force on a particle carrying an electric change moving in an electromagnetic field, given by Equation (1.23).

$$\vec{F} = q\left(\vec{E} + \vec{v} \times \vec{B}\right)$$

(1.23)

$\vec{F}$ is the vector of the force acting on the charged particle, q is the electric charge of the particle, $\vec{E}$ is the electric field vector, $\vec{v}$ is the speed of the particle, and $\vec{B}$ is the magnetic induction vector. If we have multiple charged particles flowing in the wire in the stationary magnetic field, the forces acting on the short wire segment (actually on the electric charges in it) can be calculated using Equation 1.24, which is derived from Equation (1.23).

$$\overrightarrow{dF} = i\left(\overrightarrow{dl} \times \vec{B}\right)$$

(1.24)

Consider an example shown in Figure 1.9. The force vector $d\vec{F}$, given in Equation (1.24), acts on the very short straight wire segment $\overrightarrow{dl}$ in the magnetic field $\vec{B}$ along the current carrying conductor c with the current magnitude i [13].

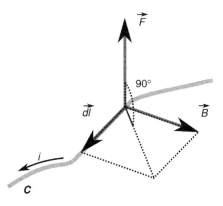

Figure 1.9 Illustration of vector orientation from Equation (1.24).

Wire segment vector $\vec{dl}$ has a magnitude equal to the length of a segment and a direction equal to the direction of the current in the wire segment. Equation (1.24) can be integrated to calculate forces on the entire wire.

For systems such as electric machines, where the magnetic field is not stationary and system geometry is not simple, it is not an easy task to calculate the forces acting in the system using Equations (1.23) and (1.24). In these cases, we can apply basic thermodynamic principles; we can turn to Equation (1.18) to evaluate the system behavior. We can separate any electromechanical conversion system into three subsystems, electrical, mechanical, and the system of coupling fields. We can now apply the first law of thermodynamic and Equation (1.18) on all three subsystems and treat them as if they are linked through their inputs and outputs.

Assuming that our system converts mechanical energy into electrical energy, for the mechanical subsystem the energy balance can be written in a form described by Equation (1.25). Mechanical work at the system input E_{m_input} increases the mechanical subsystem energy E_{m_sys} (e.g., kinetic energy of the moving parts) and creates useful work of mechanical forces E_{m_work}. Useful work of mechanical forces is converted to mechanical losses E_{m_losses} and to energy of the generated coupling fields E_{f_in}, as shown in Equation (1.26). Energy invested in the coupling fields E_{f_in} splits again into a couple of parts (1.27). Some of invested energy is stored in the coupling field energy E_{f_sys}. Energy denoted as E_{f_work} in Equation (1.27) splits between the losses in the field energy E_{f_losses} (e.g., electromagnetic radiation, iron core losses, dielectric losses, etc.) and the energy that is transferred to the induced electromotive force E_{e_in} as shown in Equation (1.28). Energy invested into induced electromotive forces E_{e_in} again splits into energy of the system E_{e_sys} (energy of the fields other than coupling fields), losses E_{e_losses}, and useful work E_{e_work} as given by Equations (1.29) and (1.30). Energy E_{e_work} is the actual electrical energy that is delivered from the energy conversion system and energy E_{e_losses} accounts for the losses in the electrical conductors of the system.

$$E_{m\_sys} = E_{m\_in} - E_{m\_work} \tag{1.25}$$

$$E_{f\_in} = E_{m\_work} - E_{m\_losses} \tag{1.26}$$

$$E_{f\_sys} = E_{f\_in} - E_{f\_work} \tag{1.27}$$

$$E_{e\_in} = E_{f\_work} - E_{f\_losses} \tag{1.28}$$

$$E_{e\_sys} = E_{e\_in} - E_{e\_work} \tag{1.29}$$

$$E_{e\_out} = E_{e\_work} - E_{e\_losses} \tag{1.30}$$

We can usually express all variables in Equations (1.25), (1.27), and (1.29) in terms of values that are known for a particular electromechanical system. Note that all the

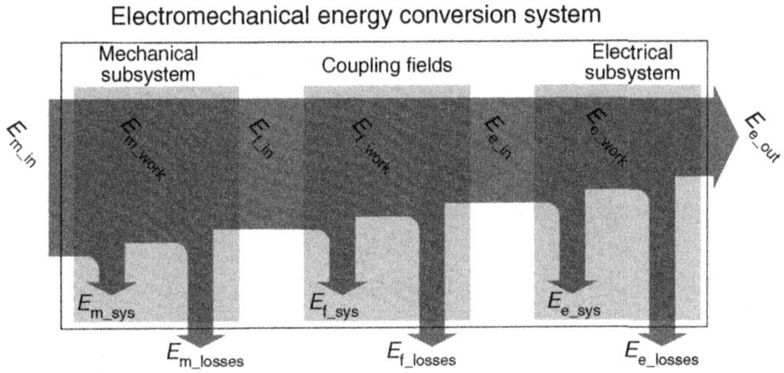

Electromechanical energy conversion system

Figure 1.10 Energy balance of an electromechanical energy conversion system (i.e., electric generator).

parameters in these equations are aggregate properties of the system and not related to geometry and detail field distributions in the actual machine. If we can determine system losses from the Equations (1.26), (1.28), and (1.30), we can easily link the mechanical input work and electrical energy output as described by Equation (1.31) and Figure 1.10. The machine power balance can be obtained by performing a time derivative of Equation (1.31) [14].

$$E_{m\_in} = E_{e\_out} - E_{m\_sys} - E_{m\_losses} - E_{f\_sys} - E_{f\_losses} - E_{e\_sys} - E_{e\_losses} \tag{1.31}$$

1.9.1 Prime movers for the electric generators

Rotating electric machines known as electric generators are most commonly used to produce electrical energy on a large scale today. Almost all power plants use rotating generators at some stage of power production. The energy resources, on the other hand, provide energy in various forms and they seldom produce work that can be seamlessly supplied along a rotating shaft, which is necessary to run electric generators. Systems that adapt the forces from the ones naturally produced by the energy resources to the form that can spin the electric generator are called prime movers.

Prime movers can take many forms and can vary in complexity. The form of prime mover depends on the type of the energy resource and form of energy it contains. For example, if we need to extract kinetic energy of a fluid, we would use a turbine of some kind. Good examples are hydro turbines and wind turbines. For chemical energy resources such as oil, a combustion engine is used as a prime mover. For natural gas, a gas turbine is used as a prime mover. Thermal energy is an interesting example as the conversion process requires an additional step in most cases. Thermal energy is converted to kinetic energy of steam and then a steam turbine is used to spin an electric generator.

Besides the main purpose to adapt forces, prime movers have an important role in system regulation. The precise rotating frequency and sharing of load among numerous generators in a power system can only be done with the help of controllable prime movers. In that sense, prime movers also serve as regulators, letting through only the needed amount of power to electric generators.

1.10   Energy storage

Energy storage systems are essential to the operation of power systems. They ensure continuity of energy supply and improve the reliability of the system. Energy storage systems can be in many forms and sizes. The size, cost, and scalability of an energy storage system highly depend on the form of the stored energy. Energy can be stored as potential, kinetic, chemical, electromagnetic, thermal, etc. Some energy storage forms are better suited for small-scale systems and some are used only for large-scale storage systems. For example, chemical batteries are well suited for small systems ranging from watches and computers to building backup systems but are still expensive when megawatt scales are considered. Pumped hydropower storage, on the other hand, which stores huge amounts of energy in the form of potential energy of water, can be found only in large power systems.

Examples of chemical energy storage systems include batteries, flow batteries, and fuel cells. Mechanical (kinetic and potential) energy storage systems include pumped storage hydropower, flywheels, and pressurized gas storage systems. Thermal energy can be stored as a molten salt and is also mainly used for large-scale systems. Magnetic energy can be stored in superconducting magnetic storage systems, which is still a relatively new and expensive technology [2].

1.11   Efficiency and losses

For an energy conversion system, we can usually distinguish input and output energy flows. The input energy is flowing into the system where it is converted into other types; in practice always more than one form of energy. Usually only one of the output energy forms (e.g., electrical energy or mechanical work) is considered useful for a particular system. This allows us to define the efficiency of a system for a particular time period, which can be defined as the ratio between the amount of useful output energy or work and the amount of invested energy during the time period, as shown by Equation (1.32). In the case of systems with constant output power, we can look at the power levels instead of energy at the system inputs and outputs, and efficiency can be expressed by Equation (1.33). The amount of input energy that is converted to any other form than the useful form is usually considered as loss and is unwanted in the system [2,3].

$$\eta = \frac{\text{Useful energy or work}}{\text{Input energy}} \tag{1.32}$$

$$\eta = \frac{\text{Useful output power}}{\text{Input power}} \tag{1.33}$$

System efficiency is usually denoted with the Greek letter eta (η). It is always less than one for practical systems, or less than 100% if expressed in percentage. The first law of thermodynamics excludes the possibility of a system that has efficiency greater than one. Furthermore, as a consequence of the second law of thermodynamics, for any practical system, there is always some amount of energy that is converted to other forms (other than useful form). This makes the efficiency of any practical system less than one. An extreme case of a system with 100% efficiency is theoretical and cannot be found in nature [5,6].

Consider an example of a rotating machine with electrical energy as an input and generated mechanical work at its shaft as useful output. Besides mechanical work, the machine performs other things. It radiates thermal energy, generates mechanical vibrations, creates electromagnetic interference, moves air around, etc. All of these effects require some amount of invested energy, which can be considered as a loss in the system. The ideal machine that converts all electrical energy into mechanical work at 100% efficiency would have to work at the ambient temperature; it would be completely silent, and would not radiate any electromagnetic signature, which is clearly impossible.

1.12 Energy resources

Energy resources are all forms of fuels used in the modern world, either for heating, generation of electrical energy, or for other forms of energy conversion processes. Energy resources can be roughly classified in three categories: renewable, fossil, and nuclear.

Fossil energy resources are obtained from dead plant and animal deposits created over the long history of the planet. These resources are vast, but limited, and are not renewable. Until recently fossil fuels have provided for the majority of humanity's energy demands. These resources mainly include coal, oil, and natural gas.

Renewable energy resources are forms of energy that are naturally replenished on our planet. Examples of traditional renewable resources are hydropower and biomass (e.g., plant fuels such as wood traditionally have been used throughout history, mostly for heating). Modern renewable resources include wind, wave, tidal, solar, and geothermal. Some forms of fuels created from biomass (plants and animals) also fall under this category.

Deposits of certain radioactive elements in Earth's crust can be classified as nuclear energy resources. These resources are used as fuel for nuclear fission-based power plants. The amount of these rare radioactive elements is limited on our planet and cannot be replenished. Over the years, there has been some research on fusion power but it is still not proven to be a feasible energy resource. This form of energy conversion aims to harvest the energy from sustained fusion of hydrogen atoms into helium [15].

1.13 Environmental considerations

Environmental considerations have become an important part of any energy system. Almost all of the energy production or conversion systems have some negative impact on the environment. For example, systems that use fossil fuels unavoidably produce greenhouse gases along with solid by-products. These products usually negatively affect the environment and wildlife. Renewable energy systems also can have a negative impact; fortunately this impact is on a much subtler and less damaging level. Even the cleanest power conversion systems unavoidably radiate thermal energy. The aggregate thermal energy produced by millions of such systems, in a city for example, can cause undesirable local climate changes. With the constant growth of energy demands and production in the last decades, the negative effects of various manufactured systems have been accumulated and are now creating serious concerns on a global level. As a consequence, one of the most important engineering tasks is not only to find new and cleaner ways to produce energy but also to make existing systems more efficient and cleaner [16].

References

[1] IEEE recommended practice for electric power distribution for industrial plants. IEEE Std 141-1993, p. 1, 768, April 29, 1994.
[2] Masters GM. Renewable and efficient electric power systems. NJ: John Wiley & Sons; 2004.
[3] Grainger J, Stevenson W Jr. Power system analysis. NY: McGraw-Hill; 1994.
[4] de Andrade L, de Leao TP. A brief history of direct current in electrical power systems. HISTory of ELectro-technology CONference (HISTELCON), 2012. Third IEEE, p.1, 6, September 5–7, 2012.
[5] Fermi E. Thermodynamics. NY: Dover Publications; 1956.
[6] Van Ness HC. Understanding thermodynamics. NY: Dover Publications; 1969.
[7] SERI. Basic photovoltaic principles and methods. Golden, CO: SERI; 1982. SP29-1448.
[8] Fuel cell systems. Office of Energy Efficiency & Renewable Energy. Available from: http://energy.gov/eere/fuelcells/fuel-cell-systems; 2014.
[9] Fuel cell basics. Smithsonian Institution. Available from: http://americanhistory.si.edu/fuelcells/basics.htm; 2008.
[10] Lead–acid battery. Wikipedia article. Available from: http://en.wikipedia.org/wiki/Lead%E2%80%93acid_battery; 2015.
[11] Thermoelectrics: the science of thermoelectric materials. Materials Science and Engineering, Northwestern University. Available from: http://thermoelectrics.matsci.northwestern.edu/thermoelectrics/index.html; 2015.
[12] Rowe D. CRC handbook of thermoelectrics. Boca Raton, FL: CRC Press; 1995.
[13] Magnetic field. Wikipedia article. Available from: http://en.wikipedia.org/wiki/Magnetic_field; 2015.
[14] Krause PC, Wasynczuk O, Sudhoff SD. Analysis of electric machinery and drive systems. NY: Wiley-IEEE Press; 2002.
[15] Sawin JL, Sverrisson F. Renewables: 2014 Global Status Report, REN21 Secretariat, Paris, France, 2014.
[16] Global warming. Natural Resources Defense Council. Available from: http://www.nrdc.org/globalwarming/; 2015.

Components of electric energy systems

2

Majd Ghazi Yousef Batarseh
Princess Sumaya University for Technology, King Abdullah II Faculty of Engineering, Electrical Engineering Department, University of Jordan, Al-Jubaiha, Amman, Jordan

Chapter Outline

2.1 Introduction

The three basic components of any electrical energy system are as follows:

1. Generation including conventional sources with various fuel types, along with new emerging renewable resources.
2. Transmission through traditional alternating current (AC) lines or the more recently developed high voltage DC transmission (HVDC).

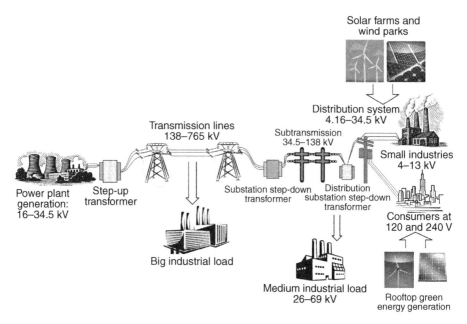

Figure 2.1 Typical radial electrical energy system [1–5].

3. Distribution comprising transformers at substations feeding consumer loads via distribution
lines or power cables.

The most commonly used radial diagram for an electrical energy system is shown
in Figure 2.1, where the generation of electricity is typically in a three-phase AC
at line-to-line RMS voltages of 16–34.5 kV. Step-up transformers at the remote
generation site increase the voltage to the appropriate transmission levels of nom-
inally 138–765 kV, and this power is then transmitted across long distances over
AC overhead TLs and underground cables. Substations reduce the voltage levels to
34.5–138 kV for subtransmission, which is further stepped down to the range of
4.16–34.5 kV for the distribution system. This is eventually reduced to the customer's
operating standards of 240 V predominantly in Europe or 120 V mainly in the United
States.

This flow and its elements will be discussed in subsequent sections starting with
power plants.

2.2   Power plants

Other than nuclear energy, the sun is the main provider of all electrical energy and the
core driver of power plants, whether directly from solar radiation via photovoltaics
and solar heating via thermal electrics, or indirectly via wind (due to atmospheric temperature difference), hydro (due to the water cycle that begins with evaporation), and

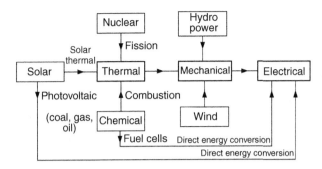

Figure 2.2 Methods for energy conversions into electricity [1].

biofuels (due to photosynthesis) or through solar energy stored over millions of years in the form of fossil fuels, which is still considered to be the main source of electricity worldwide.

Power plants are where energy conversion takes place from other forms of energy into electricity. They are diverse in size, energy conversion technology, fuel, and accordingly, environmental impact.

Figure 2.2 demonstrates the different forms of energy conversion into electricity; both direct and indirect [1], which comprises of various power plants.

Currently the mainstream path of generating electricity is from thermal energy into mechanical energy, as displayed through the bold path in Figure 2.2. The process starts with burning fossil fuel (primarily coal, oil, and natural gas) and converting it into thermal energy, which in turn is converted into mechanical energy through turbines. Then synchronous generators convert mechanical energy into electricity; this will be discussed later. Emissions of carbon dioxide and other toxic gases, thermal pollution, and the fact that consumption rate is millions of years faster than regeneration rate of this critical fuel, are serious issues concerning these types of power plants. The World Bank's statistics show that 67% of the total electricity production of 22,158.5 terawatt hours (TWh) in 2011 was generated from fossil fuel, out of which 41.2% was from coal, 21.9% was from natural gas, and 3.9% was from oil [6].

Hydroelectricity is one of the oldest generators of electricity and does not include thermal energy in the conversion process, as shown in Figure 2.2. Hydropower plants convert potential energy into mechanical energy through water turbines, which then generate electricity. Efficiencies of 90% and higher are achieved through this clean and renewable route. However, its availability is limited to sites with certain geographical and environmental suitability. Electricity production from hydroelectric sources contributed to 15.9% of the total electricity generation in 2011 [6].

Nuclear-based power plants generate thermal energy without emitting greenhouse gases. This thermal energy takes the same route as fossil fuels to generate electricity. However, radioactive waste disposal and storage is a serious challenge and a controversial topic as to how safe and clean nuclear energy is. It is worth mentioning the Fukoshima nuclear disaster in this regard.

Wind is another renewable source of electricity where kinetic energy causes wind turbines to rotate and the resultant mechanical energy is converted into electricity through a coupling magnetic field.

The sun can directly generate electricity using photovoltaics or indirectly through the use of thermal energy via solar thermal concentrators.

According to the World Bank, 11.7% of the total production of electricity was generated from nuclear plants in 2011, versus a modest contribution of 4.2% from renewable sources excluding hydroelectric (i.e., including geothermal, solar, tidal, wind, biomass, and biofuels). It is worth noting, however, that electrical energy generated from renewable resources is currently attracting more attention worldwide, with all EU countries making it a national target.

Figure 2.3 shows energy production in 2011 by different sources and different economies, where low-income economies are those in which the 2013 gross national income (GNI) per capita was $1,045 or less, upper and lower middle-income economies are those in which the GNI per capita was between $1,046 and $12,745 and high-income economies are those in which the GNI per capita was $12,746 or more [6].

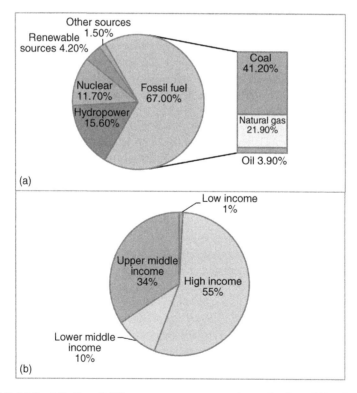

Figure 2.3 (a) Contribution of different energy sources to the production of 22,158.5 TWh in 2011 and (b) energy production by different economies [6].

2.3   Electric power generators

Other than photovoltaics, which generate direct current (DC) electricity directly using semiconductor materials, all other sources of electrical energy eventually convert mechanical energy from a spinning turbine (steam, gas, or hydro) into three-phase AC electricity, through an electromechanical synchronous generator, as shown in Figure 2.4.

In addition to delivering a total constant instantaneous power to three-phase load, these AC power systems allow the use of transformers, which make the long-distance transmission of high voltage levels through TLs feasible. The DC electricity from the photovoltaic array is thus integrated through a power electronics inverter into the existing AC grid.

Electric voltages can be star (Y)- or delta (Δ)-connected three-phase sources as shown in Figure 2.5.

A phase voltage is the voltage between a line terminal and neutral, whereas line voltage is that between two line terminals. For a balanced generation, the three-phase voltages, V_{an}, V_{bn}, and V_{cn}, have the same amplitude V_p and are 120° out of phase. A positive (a, b, c) sequence entails phase a leads phase b, and phase b leads phase c, all by 120°. Table 2.1 lists phase voltages V_{an}, V_{bn}, and V_{cn} and the three line voltages V_{ab}, V_{bc}, and V_{ca} for a balanced positive sequenced system.

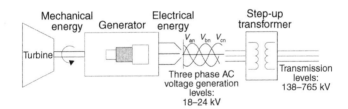

Figure 2.4 Three-phase electricity produced by a three-phase synchronous generator.

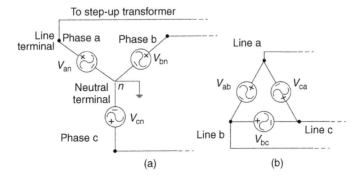

Figure 2.5 (a) Y-connected and (b) Δ-connected three-phase sources.

Table 2.1 Phase and line voltages for a balanced three-phase positive sequenced system

Phase voltages (volt)	Line voltages (volt)
$V_{an} = V_p \angle 0°$	$V_{ab} = \sqrt{3}\,V_p \angle 30°$
$V_{bn} = V_p \angle -120°$	$V_{bc} = \sqrt{3}\,V_p \angle -90°$
$V_{cn} = V_p \angle -240°$	$V_{ca} = \sqrt{3}\,V_p \angle -210°$

2.3.1 The synchronous generator

A synchronous generator, which is considered to be the major source of global electricity generation, consists of two magnetic structures separated by an air gap; one outer hollow, cylindrical-shaped stationary part called the stator and another moving part slotted inside the stator called the rotor.

Faraday's law of electromagnetic induction is the fundamental concept in electricity generation; an electromotive force (voltage, i.e., electrical energy) is induced and collected by an electrical conductor when moving it (mechanical energy) through a magnetic field (magnetism). This mechanical–magnetic–electrical trio structure is satisfied by a synchronous generator, where the source of magnetic field in the machine is produced by a DC current through the rotor windings (also known as field windings), and the mechanical rotation of the rotor is provided by the spinning turbine. Finally, this rotating magnetic field induces three-phase AC voltage within the stator windings (also known as armature windings) of the generator. In other words, two conditions required to convert energy are, an exciting field winding on the rotor to provide a coupling magnetic field inside the machine, and a spinning rotor with kinetic energy coming from mechanically coupled turbines. Output frequency of the AC voltage is synchronized with the input mechanical rotation rate of the generator, hence the name synchronous generator. As shown in Figure 2.6, the rotor can be round (nonsalient) with uniform air gap, as used in nuclear and gas power plants, or a generally lesser expensive and lesser stress tolerant salient rotor, as typically used in hydropower plants.

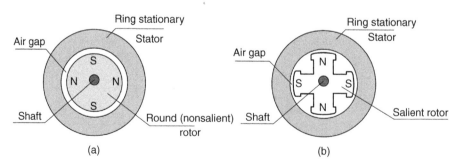

Figure 2.6 A synchronous generator with (a) a four-pole round (nonsalient) rotor and (b) a four-pole salient rotor [3].

2.4 Transformers

An inevitable path for the three-phase, 16–34.5 kV, AC voltages produced at the remote generation site is through a step-up transformer to increase the generated levels up to 138–765 kV suitable for long-distance transmission at reduced current and power losses. These high transmission levels are subsequently decreased to lower distribution levels, which are followed by a further reduction to safe nominal operating levels, for customers' use at various substations.

The core component of a substation is the power transformer. A transformer is a magnetic device with two or more coils, all wrapped around a common magnetic core material with different numbers of turns N. The input winding of the transformer (the primary side) has N_1 turns and is fed with an AC source operating at a certain voltage level. The output, which is magnetically coupled to the input by the same flux ϕ for ideal cores, powers the load at the secondary winding (N_2 turns) with an induced higher or lower level of AC voltage. The gain of the transformer is decided by the turns ratio a, which is the number of turns of the coils at the output N_2 to the input N_1 as expressed in Equation (2.1).

$$a = \frac{N_2}{N_1} \tag{2.1}$$

The right-hand rule is followed to determine both direction of flux at the primary side and polarity of induced voltage across the windings of secondary side. This rule states that if the right-hand fingers are curled around the coil with the direction of current, then the extended thumb points in the direction of the flux. Similarly, with the extended thumb in the direction of generated flux crossing the windings on the secondary side, then current will flow in the secondary windings in the direction of curled fingers around that coil.

Figure 2.7 shows an ideal transformer with two windings of N_1 turns at the input and N_2 turns at the output, coiled around a common magnetic core and the circuit symbol used to model it.

The circuit symbol of the ideal transformer in Figure 2.7b shows the windings at primary and secondary sides as inductive coils, and the magnetic core is represented by a pair of parallel bars in between. In this case, the dot notation is used instead of the right-hand rule to determine the polarity of the induced voltage, where a positive marked dot at the primary side induces a voltage with its positive terminal at the dot of the secondary windings.

The induced voltage at the secondary side can power a load or can be fed into another transformer. By Faraday's law and in accordance with the right-hand rule, one can express the gain with the following set of equations, assuming zero leakage flux:

$$v_1 = N_1 \frac{d\varphi}{dt} \quad \text{and} \quad v_2 = N_2 \frac{d\varphi}{dt} \tag{2.2}$$

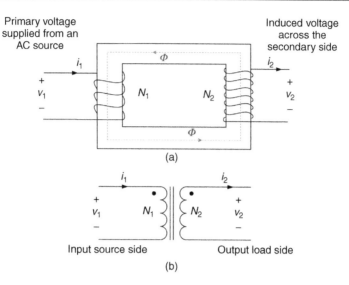

Primary voltage
supplied from an
AC source

Induced voltage
across the
secondary side

(a)

Input source side   Output load side

(b)

Figure 2.7 (a) An ideal two winding transformer and (b) its circuit symbol.

From which the gain can be expressed as:

$$\frac{v_2}{v_1} = \frac{N_2}{N_1} = \alpha \tag{2.3}$$

From Equation (2.3), a step-up transformer has more turns coiled at its secondary side compared to its primary side, that is, the turns ratio of a step-up transformer is $\alpha > 1$, whereas for a step-down transformer the ratio is $\alpha < 1$.

For a loss-less transformer the power conversion is 100% efficient, in other words, all the power generated at the primary side gets transferred to the secondary, as given in Equation (2.4):

$$P_{in} = P_{out} \tag{2.4}$$

$$v_1 \times i_1 = v_2 \times i_2$$

$$\frac{v_1}{v_2} = \frac{i_2}{i_1} = \frac{N_1}{N_2} = \frac{1}{\alpha} \tag{2.5}$$

Equations (2.3)–(2.5) govern the voltage, current, and accordingly impede levels on the secondary side of the transformers, which are carried via TLs, as discussed in Section 2.5.

2.5 Transmission lines

Since the launch of the first AC electrical system in 1886, the Westinghouse Electric Company won the battle over the DC electrical system. However, with development and advances in power electronics as an enabling technology, HVDC transmission systems are emerging and competing once again with the currently dominant AC system. Whether AC or DC, whether hanging overhead TLs or extended underground or underwater cable lines, electricity will be carried via power lines or cables; that is, electrical energy is transmitted to distribution stations and eventually to the consumer via power transmission or cable lines. This route extending from a sending to a receiving end, should be highly efficient, reliable, and loss-less. For this reason it is very important to analyze and model TLs.

TLs are effectively a number of relatively high conductance (low resistance), uninsulated strands of aluminum bundled together, carrying the generated three-phases, armored at the center with a steel core support to reduce sagging, and are shielded against weather derating and lightning strike. These conductors are referred to as aluminum cable steel reinforced (ACSR) and are shown in Figure 2.8. The parameters of the TLs, separation between conductors, and steel towers or wooden poles that they are mounted onto, vary with the power handling capability that they are designed to carry, the distance they stretch to, and accordingly their cost [3,4].

2.5.1 Transmission lines parameters

The currents through the TLs produce electric and magnetic fields within the vicinity of the lines. The effect of interaction of these fields can be modeled by uniformly distributed per unit length: series impedance of resistance and inductance, and parallel admittance of capacitance and conductance. However, the conductance parameter of electric TL is usually neglected due to its minimal effect [5]. Per unit length values will be applied and thus Ω/m, H/m, and F/m are the units that are used. TLs extending over 240 km are designated in long, less than 80 km short, and in between medium TLs.

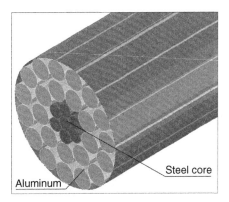

Figure 2.8 A cross-sectional view of ACSR TL conductor.

2.5.2 Transmission line resistance

Like any other conductor, TLs are modeled by their desirably small DC resistance per meter, $R_{UL} = \dfrac{\rho}{A}$, where the subscript UL denotes unit length. Increasing the diameter of TL decreases its resistance per meter and thus decreases the losses of $I^2 xR$,[1] in effect this improves the efficiency of power transmission at higher costs. As the length of the TL increases, so does the resistance and thus power losses in the form of heat. Furthermore, as the resistance increases linearly with temperature, this will cause the conductor to expand, which results in increased sagging. Therefore, in addition to increased losses, sagging may also lead to damages and even blackouts if the sagged lines touch underneath trees [3,5,6].

DC currents (at zero frequency) distribute uniformly across the cross-sectional area of conductors. As frequency increases, the AC tends to concentrate towards the outer surface more than at the center and thereby nonuniformly covering the area. This phenomenon, known as the skin effect, increases the AC resistance of the conductor compared to the DC resistance as the effective cross-sectional area becomes less [4]. Many manufactures provide tabulated values for the per-unit length resistance of the TL.

2.5.3 Transmission line inductance

Back to Faraday's law, and a sinusoidally varying magnetic field will be produced around the conductor carrying a sinusoidal AC current following the right-hand rule. This magnetic field will induce a voltage, which will oppose the flow of current. This resistance to AC current due to magnetic field is the inductive reactance effect of the TL. Figure 2.9 shows a diagram of a single conductor and a neutral line, and the magnetic field produced. Here, r_1 and r_2 are the radii of the phase and neutral lines, respectively, and D is the separation distance between them.

Many textbooks and references provide derivations for the equations calculating the parameter values. We will suffice in stating them here, the values of the per-unit length inductance for single- and three-phase lines are given in Equations (2.6) and (2.7):

$$L_{1\varnothing} = 2 \times 10^{-7} \ln \frac{D_{1\varnothing}}{D_s} \tag{2.6}$$

$$L_{3\varnothing} = 2 \times 10^{-7} \ln \frac{D_{3\varnothing}}{D_s} \tag{2.7}$$

where, $L_{1\varnothing}$ and $L_{3\varnothing}$ are the single- and three-phase per-unit length inductance in H/m, respectively, $D_{1\varnothing}$ is the geometrical mean diameter (GMD), which is the distance between the single-phase conductor and the neutral. $D_{3\varnothing}$ is the GMD between the three-phases, that is, $D_{3\varnothing} = \sqrt[3]{D_{12} \times D_{23} \times D_{31}}$, where D_{12} is the distance between conductors of the first and second phases, D_{23} is the distance between conductors of the second

---

[1] All AC currents are in effective (RMS) values.

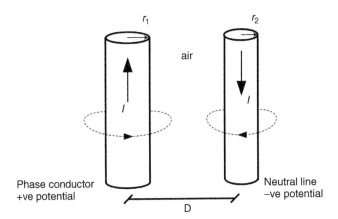

Figure 2.9 A single phase transmission line with the neutral return line.

and third phases, and D_{31} is that between the third and first phases, D_s is the geometrical mean radius of the conductors for the three-phases, and D_s for each phase can be calculated as $D_s = e^{-1/4} \times r_i = 0.778 \; r_i$ [5].

2.5.4 Transmission line capacitance

The last parameter to be considered for TL modeling is capacitance. As shown in Figure 2.9, having two conductors running parallel to each other, separated by a dielectric material of air, with a potential difference between them, is the definition of capacitance. Equation (2.8) gives the value of this capacitance for single- and three-phase TLs with equal spacing in F/m, the derivation can be found in [4].

$$C_n = \frac{2\pi \; k}{\ln(D/r)} \tag{2.8}$$

where C_n is the line to neutral per-unit length capacitance in F/m and k is the permittivity of air. For a TL with unequal spacing, D is replaced with $D_{3\emptyset} = \sqrt[3]{D_{12} \times D_{23} \times D_{31}}$.

2.5.5 Transmission line models

The three equivalent π models are listed in Table 2.2 according to TL length l (Figure 2.10). The sending end voltages and currents can be represented by the two port network equations as:

$$V_S = A V_R + B I_R \quad \text{and} \quad I_S = C V_R + D I_R$$

where the *ABCD* constants are expressed in Table 2.2 for each line length.

Table 2.2 **TL models**

TL lengths, l	Equivalent π circuits	Analysis
Short line: $l \le 80$ km	Figure 2.10a	$A = 1$, $B = Z$, $C = 0$, $D = 1$, S or $\mho$
Medium line: $80 \le l \le 240$ in km	Figure 2.10b	$A = D = \dfrac{Z\,Y}{2} + 1$ $B = Z$, $C = Y\left(1 + \dfrac{Z\,Y}{4}\right)$, S or $\mho$
Long line: $l \ge 240$ km	Figure 2.10c	$A = D = \dfrac{Z'\,Y'}{2} + 1$ $B = Z'$, $C = Y^{\left(1 + \frac{Z'Y'}{4}\right)}$, S or $\mho$

where $Z' = Z\dfrac{\sin h\gamma l}{\gamma l}$, $\dfrac{Y'}{2} = \dfrac{Y}{2}\dfrac{\tan h(\gamma l/2)}{\gamma l/2}$, and $\gamma = \sqrt{zy}$ and l line length, γ is the propagation constant with total impedance $Z = zl$, and total admittance $Y = yl$.

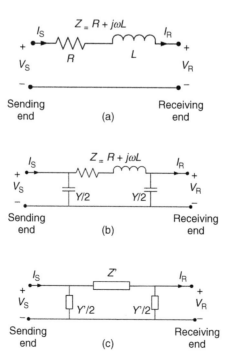

Figure 2.10 (a) Short TL model, (b) medium TL model, and (c) long TL model.

2.6   Relays and circuit breakers

Faults in electrical systems are of two main types; symmetrical (with three-phases) and unsymmetrical (with one or two phases) faults. Lightning, which is a form of excessive charge, is the number one cause of electrical faults, as it interrupts normal current flow at high level transmission, and provides a short circuit path for these charges through the ground rather than the normal line current flow. Permanent damage to TL insulators and/or transformers can be avoided by taking some detective measures (through relays), and protective actions (through circuit breakers) against these very short but big impact faults. Relays, which are considered to be the brain of protection systems, are designed to send an "OPEN" control signal to high–speed switching circuit breakers, in order to isolate the faulty part from the rest of the system, until the fault clears. A few cycles after the transient dies out, the relay sends a "CLOSE" control signal to the circuit breakers for normal steady-state operation to resume.

Fault analysis and overcurrent and voltage calculations are out of the scope of this chapter. However, this topic is highly important for relay and circuit breaker selection and ratings, and is covered in many references.

Protection equipment includes; instrument transformers as sensing devices, relays, and circuit breakers, which are installed along the electrical system in various locations and can be illustrated as in Figure 2.11.

Dependent on the sensed values of instrument transformers and compared to nominal values, output of the relay sent to circuit breakers can be one of the three possible control decisions:

1. Normal operation detected: keep flow (stay ON).
2. Fault detected: disconnect flow "OPEN" (open circuit breaker).
3. Fault cleared: transient died out "CLOSE" (reclose for normal flow).

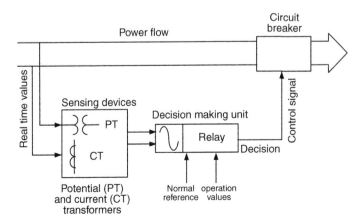

Figure 2.11 Main components of a protection system and how it operates.

2.7 Voltage regulators

Loads at the receiving side of electrical systems are fed through transformers. However, even if the input source was kept constant, the voltage at the secondary side of the transformer varies with changing the load and thus the load voltage at the receiving side (V_R) also varies. It is usually recommended to keep this absolute voltage variation within a small range, with the extreme load change from no load to full load. This is known as voltage or load regulation. Voltage regulation is often expressed as a percentage ($V_R\%$) and is given at full load as in Equation (2.9):

$$V_{R \text{ at full load}} \;\% = \left(\frac{\left| V_{R \text{ at no load}} \right| - \left| V_{R \text{ at full load}} \right|}{\left| V_{R \text{ at full load}} \right|} \right) \times 100 \qquad (2.9)$$

This voltage regulation percentage can be used as a measure to compare TLs in terms of their ability to keep output voltage as constant as possible, with load variation. The smallest the voltage regulation, the better the performance of the transmission system.

2.8 Subtransmission

Different industrial loads are powered at various operating levels and thus are accordingly connected to the remotely located generators via either high voltage long TLs in the voltage range of 138–765 kV, the upper medium (34.5–138 kV) subtransmission systems, or the lower medium (4.16–34.5 kV) shorter distribution lines, as shown in Figure 2.1. The diameter of the TL and the spacing between the lines decreases with voltage levels.

Whether to consider this intermediate subtransmission network part of the high transmission end or the lower distribution end is still ambiguous.

2.9 Distribution systems

Intercity, industrial, commercial, and residential loads are supplied by distribution substations through distribution lines at primary ranges, which are as high as 34.5 kV. They are then stepped down through distribution transformers along the way to various secondary levels until at the vicinity of the customer, where they reach their lowest effective values, at standard residential operating levels of 120/240 V through feeder lines. A simple radial diagram depicting a distribution system is shown in Figure 2.12.

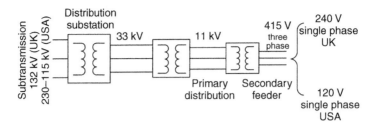

Figure 2.12 A radial distribution network [1,5,7].

With the increase in electrical demand, proper system planning, accurate current, future load forecasting at all times, and power transfer capability are essential factors in operating and running distribution systems [3,7].

The relatively recent modification on existing distribution systems, is the integration of distributed generators, where many renewable sources of energy feed the system at subtransmission and distribution voltage levels.

2.10 Loads

The final destination of generated electrical energy is at the consumer's site via the short distribution lines. Electric loads come in a variety of sizes, and similar to three-phase generated sources, loads can also be star (Y)- or delta (Δ)-connected, as shown in Figure 2.13.

A balanced three-phase Y-connected load indicates equal phase impedances: $Z_{L\_Y} = Z_A = Z_B = Z_C$. Similarly, for a balanced three-phased Δ-connected load $Z_{L\_\Delta} = Z_{AB} = Z_{BC} = Z_{CA}$.

Therefore, four possible electrical system configurations may be encountered:

1. A Y-connected three-phase source feeding a three-phase Y-connected load.
2. A Y-connected three-phase source feeding a three-phase Δ-connected load.
3. A Δ-connected three-phase source feeding a three-phase Y-connected load.
4. A Δ-connected three-phase source feeding a three-phase Δ-connected load.

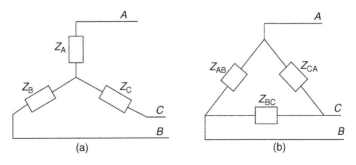

Figure 2.13 (a) Y- and (b) Δ-connected three-phase loads.

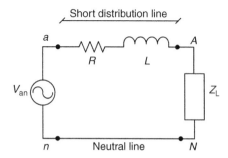

Figure 2.14 A per-phase equivalent circuit for a balanced three-phase system.

For a balanced system, it is sufficient to carry out analysis on a single per-phase circuit, in which delta (Δ) gets transformed into a star (Y), and thus the per-phase loop of the source and the load gets enclosed between the line and neutral return. The per-phase diagram used to solve balanced systems is shown in Figure 2.14.

2.11   Power capacitors

Loads are generally inductive in nature, and combined with the inductance parameter of the TL, it is quite common for electrical systems to operate at lagging power factors. However, this leads to more reactive power in VARs to be generated in order to satisfy the real power need in watts at the customer's load. This is translated to more current in TLs, thereby resulting in higher voltage drops, more power losses, reduced efficiencies, and additional costs. Therefore, for better system performance, it is always desirable to have a close to the unity power factor. This power factor correction technique aims at injecting reactive power VARs in order to compensate for and reduce the inductive lagging effect. This can be practically achieved by adding capacitor units or banks in parallel to the load. For this reason, capacitors are applied throughout the electrical system, mainly at feeders and substation buses.

2.12   Control centers

The changes in size, nature, and demand taking place over the electric grid, in addition to the rapid advances in information and communication technology sector, are also reflected on control centers. The need for distributed (decentralized), integrated, and intelligent control systems is proving to be essential for enhanced stability, reliability, efficiency, and improved energy management in terms of decision making. Control centers are moving from remote terminal units toward implementing supervisory control and data acquisition, energy management systems, and business management systems [8].

2.13   Worldwide standards for household voltage and frequency

The standards for household voltage and frequency differ among countries and even states within the same country. The enabling technology of power electronics in almost all handheld electrical appliances, in addition to adaptable plugs, which allow the safe use of these electric devices at different voltage and frequency standards. All of Europe and most other countries use the efficient 50 Hz frequency at typical effective voltage values between 220 V and 240 V. Voltage standards of 100–127 V at 60 Hz come second worldwide and dominate North America. Other combinations like 220–240 V at 60 Hz and 100–127 V at 50 Hz are used in other countries. Many websites and textbooks provide listings of different voltage and frequency standards used worldwide [9].

2.14   Representation of an electrical energy system

One-line diagrams with standard symbols rather than equivalent circuits are used to graphically represent electrical power systems. They give an overview of the whole system, however, the information conveyed by them and the purpose of this representation varies with the issue of concern. IEEE lists many of the standard symbols in its publication "The Graphic Symbols for Electrical and Electronics Diagrams" [10].

2.15   Equivalent circuits and reactance diagrams

In contrast to the one-line diagram, impedance and reactance diagrams are used for analysis and calculation purposes. Information derived from the one-line diagram assists in obtaining the impedance and reactance diagrams, which are used for performance examination. A TL that is merely a line connection in a single-line diagram is modeled by its parameters in the impedance and reactance diagrams. Figure 2.15 shows the single-line diagram of a simple electrical power system with the impedance and reactance diagrams, respectively.

2.16   Per-unit system

In order to avoid the confusion of different voltage levels related to transformers and the use of large units (kilo or mega) used in electrical energy systems, a per-unit representation is adopted, where all values are expressed as a ratio or a percentage from a base or reference value prespecified for each. Voltages of 110 V and 127 V are expressed, respectively as 0.917 and 1.06 per-unit for a chosen base voltage of 120 V. Due to the interconnected relationships between voltage (kV), current (A), apparent

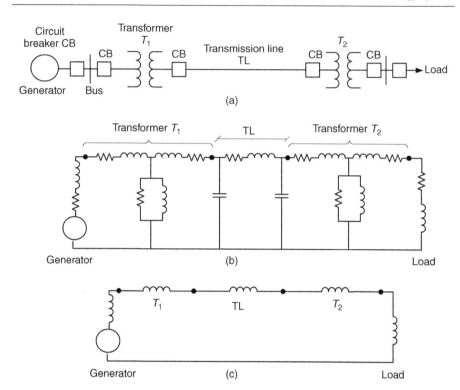

Figure 2.15 (a) One-line diagram of a power system, (b) impedance, and (c) reactance diagrams.

power (MVA), and impedance (Ω), it is a common practice to specify the two base units of kV and MVA and accordingly derive the rest.

2.17 Summary

This chapter reviewed the components that comprise an electrical energy system. Electricity generation, transmission, and distribution were presented along with complementary circuitries such as circuit breakers, TL parameters, and control centers. One-line, impedance, and reactance diagrams have been described and per-unit notation has been introduced.

Problems

1. Discuss the reason behind generating electricity in three-phases rather than any other number of phases.
2. Derive the instantaneous power for a balanced positive sequence three-phase generation.
3. In Figure 2.7b, a 240 V source at the primary side is stepped down to 120 V feeding a 50 Ω load at the secondary side. Find:
   a. The transformer turn ratio.
   b. The power generated by the source.

4. A 416 V balanced, Y-connected three-phase source (Figure 2.5a) is feeding a balanced three-phase, Y-connected load (Figure 2.13a) with phase impedance 12 + j8 Ω. The TLs have per phase impedances of 0.09 + j0 15 Ω. For this system:
  a. Draw the per phase equivalent circuit.
  b. Find the line current.
  c. Find the efficiency of power transmission.
5. List the differences between 50 Hz and 60 Hz systems.

References

[1] Freris L, Infield D. Renewable energy in power systems. West Sussex, UK: John Wiley & Sons; 2008.

[2] Masters GM. Renewable and efficient electric power systems. New Jersey: John Wiley & Sons; 2010.

[3] Mohan N. Electric power systems first course. New Jersey: John Wiley & Sons; 2012.

[4] Chapman S. Electric machinery and power system fundamentals. New York: McGraw-Hill; 2001.

[5] Grainger JJ, Stevenson WD. Power system analysis. New Jersey: McGraw-Hill; 1994.

[6] The Worldbank. Available from: wdi.worldbank.org/table/3.7#; 2014.

[7] Gonen T. Electric power distribution engineering. 3rd ed. Florida: CRC Press; 2014.

[8] Wu F, Moslehi K, Bose A. Power system control centers: past, present, and future. Proc IEEE 2005;93(11):1890–908.

[9] El-Sharkawi MA. Electric energy: an introduction. 3rd ed. Florida: CRC Press; 2012.

[10] 315-1975 – IEEE Standard, American National Standard, Canadian Standard, Graphic Symbols for Electrical and Electronic Diagrams.

Solar energy

3

Ahteshamul Haque
Department of Electrical Engineering, Faculty of Engineering & Technology,
Jamia Millia Islamia University, New Delhi, India

Chapter Outline

3.1 Introduction

Human beings have tried to harness solar energy in the past for their convenience. In the fifth century BC, passive solar systems were designed by the Greeks to utilize solar energy for heating their houses during the winter season. This invention was further improved with the advancement of mica and glass, which prevent the escape of solar heat during the daytime. Another invention was made in USA to use solar energy to heat water. The first commercial solar water heater was sold in the 1890s. In the nineteenth century scientists in Europe constructed the first solar powered steam engine [1].

In the 1950s scientists working at Bell Labs developed the first commercial photovoltaic (PV) cells. These PV cells were capable of converting sunlight into electrical energy to power electric equipment. These PV cells began to be used in space programs, that is, to power satellites, etc. Further advancement in the technology reduced the price of solar PV and it began to be used for household applications [2].

Currently, global demand for electricity is increasing [3]. The limited reservoirs of fossil fuels and emission of greenhouse gases have led to serious concerns regarding energy crises and climate threats. These concerns led researchers to look for alternative sources of energy, and solar energy is considered as the most acceptable source among all renewable energy sources. Solar energy is available in abundance and free of cost all across the globe. It is reported that Earth receives energy from the Sun, which is 10,000 times more than the total energy demand of the planet [4].

The conversion from solar energy to electrical energy is done by using solar PV. Solar PV has a nonlinear characteristic and its output varies with ambient conditions like solar irradiation, ambient temperatures, etc. [5].

In this chapter passive and active solar energy conversion is discussed. PV modeling, its operation, module, integration, and evaluation parameters are described. Finally, practical problems are given.

3.2 Passive solar energy system

The passive solar energy system (PSES) is used to utilize the Sun's energy for heating and cooling of a living space. PSES can be used for reducing heating and cooling energy bills and provides increased comfort. In this system, either the complete living space (building) or parts of it take advantage of natural energy characteristics. PSES requires few moving parts, low maintenance, and eliminates the need for mechanical heating and cooling systems.

The design of PSES is not very complex but knowledge of solar geometry, climate, and window technology is an essential requirement. PSES can be integrated into any building if a suitable site is available.

The following are the categories of passive solar heating technique [1].

1. Direct gain: In this type, solar irradiation directly penetrates into the building and is stored.
2. Indirect gain: In this type, solar irradiation is collected, stored, and distributed using thermal storage materials like a Trombe wall.

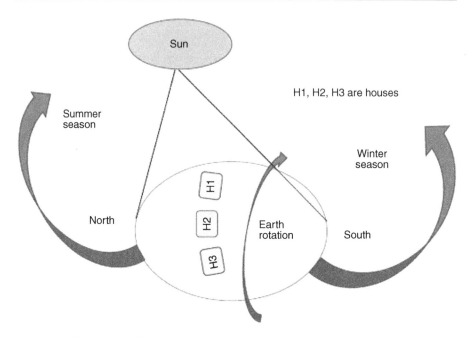

Figure 3.1 Orientation of a house for a solar passive energy system.

3. Isolated gain: In this system, solar irradiation is collected in an area of the building that can be open or closed off selectively.

The basic aim of PSES is to maximize solar heat gain in winter and minimize heat gain in summer. The following specific techniques are used for PSES implementation:

- Orientation of the building is kept with the long axis running east to west.
- Glass is used and its size and orientation are chosen to maximize solar heat gain in winter and minimize heat gain in summer.
- Overhangs are sized south facing to shade windows in summer and allow gains in winter (Figure 3.1).
- Thermal mass is stored either in walls or floors for heat storage.
- Daylight should be used to provide lighting.

With effective shading, window selection, and insulation, the natural air can be used to cool the building. In many PSES designs when considering climate conditions, the windows can be opened at night to flush the air inside and closed during the daytime to significantly reduce the need for supplemental cooling.

The drawback of PSES is that its effectiveness only lasts for 16–18 h daily. The rest of the time (morning hours for heating) heating/cooling is dependent on a supplemental heating/cooling system. However, heating/cooling by using PSES saves significant amount of money.

3.3   Active solar energy system (photovoltaic)

PV cells are used to convert solar energy into electrical energy. This concept was discovered in 1839 by French scientist Edmund Becquerel and is known as the photovoltaic effect. The PV effect was first studied in solid like selenium in 1870. The converting efficiency of selenium solar was 1–2% and it was very costly, which prevents engineers using it in energy converters. With expansion of work in this area a method was developed for producing highly pure crystalline silicon in the 1950s. In 1954, Bell Labs developed silicon PV cells whose efficiency was 4%, which was further improved to 11%. At this time a new era of power-producing cells began. In 1958, a US space satellite used a small array of cells to power its radio. The currently available PV cells are made of silicon and are also known as solar cells [2].

3.3.1   Principle

Sunlight is composed of photons. These photons contain different amounts of energies corresponding to various wavelengths of light. When photons strike a PV cell, they may be absorbed, reflected, or pass through these cells. Absorption of photons in solar cell results in the generation of an electron hole pair. This generation of electron hole pairs result in the generation of a voltage, which can drive the current in an external circuit. Figure 3.2 shows the effect of light on a silicon PV cell. Figure 3.3 shows the connection of a PV cell to an external load/circuit.

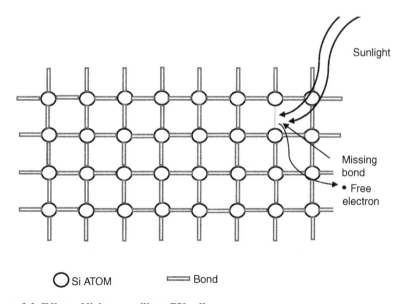

Figure 3.2 Effect of light on a silicon PV cell.

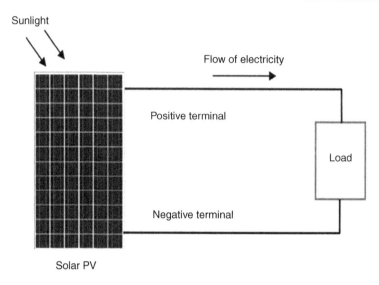

Figure 3.3 Connection of a PV cell to an external load circuit.

3.3.2 Types of solar PV cells

Solar PV cells are made of silicon, which is available in abundance. Solar PV cells are categorized into the following manufacturing technologies:

- Monocrystalline
- Polycrystalline
- Bar crystalline silicon
- Thin film technology

The conversion efficiency of monocrystalline cells ranges from 13% to 17%, whereas for polycrystalline the range is 10–14%. The polycrystalline cells are economical as compared to monocrystalline. The expected life of polycrystalline cells is between 20 and 25 years and for monocrystalline it is between 25 and 30 years. The conversion efficiency of bar crystalline is around 11%. The production cost of thin film technology is reduced but the efficiency is very low and ranges between 5% and 13% with a lifespan of around 15–20 years.

In addition, the latest technology is organic PV cells.

3.4 Ideal PV model

The equivalent circuit of the ideal model of a PV cell is shown in Figure 3.4 [5]. The output current I is:

$$I = I_{pv,cell} - I_d$$

(3.1)

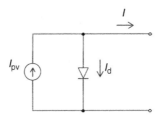

Figure 3.4 Ideal PV model.

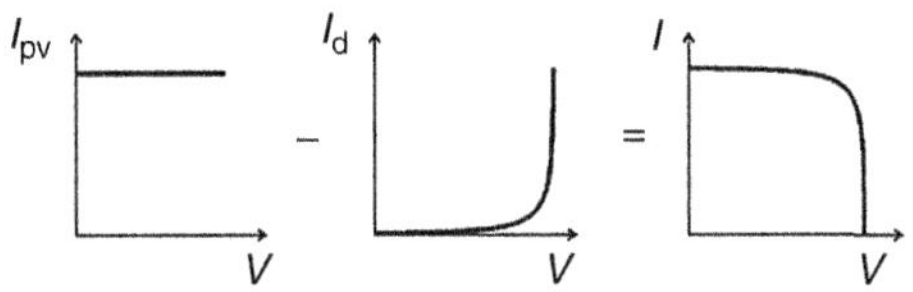

Figure 3.5 Origin of an *I–V* curve.
Reproduced from Ref. [5].

where,

$$I_d = I_o \left[\exp\left(\frac{qV}{akT} \right) - 1 \right] \tag{3.2}$$

$I_{pv,cell}$ is the current generated by the sunlight, I_d is the diode current given by Equation (3.2), I_o is the reverse saturation current of the diode, q is the electron charge ($1.60217646 \times 10^{-19}$ C), k is the Boltzmann constant ($1.3806503 \times 10^{-23}$ J/K), T is the temperature of the diode (K), and a is the diode ideality constant.

Figure 3.5 shows the origin of the *I–V* curve as given by Equation (3.1).

3.5   Practical PV model

In practical applications, PV cells are connected in series and parallel combinations. PV cells are connected in series to increase the output voltage and the output current. This arrangement is known as an array. The basic equation (3.1) is the ideal equation of one PV cell. However, in practical applications PV arrays are used and therefore other parameters are to be considered. The equivalent circuit of a practical PV model is shown in Figure 3.6.

The equation for a practical PV array is:

$$I = I_{pv} - I_o \left[\exp\left(\frac{V + R_s I}{V_t a} \right) - 1 \right] - \left(\frac{V + R_s I}{R_p} \right) \tag{3.3}$$

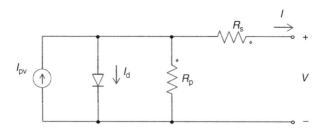

Figure 3.6 Practical PV model.

$$I_o = I_{or} \exp[q E_{GO} / bk((1/T_r) - (1/T))][T/T_r]^3 \tag{3.4}$$

$$I_{pv} = S[I_{sc} + K_1(T - 25)]/100 \tag{3.5}$$

where, I_{pv} and I_o are the PV and leakage current, respectively; a and b are ideality factors; K_I is the short circuit current temperature coefficient at I_{SC}; S is the solar irradiation (W/m^2); I_{SC} is the short circuit current at 25°C and 1000 W/m^2; E_{GO} is the bandgap energy for silicon; T_r is the reference temperature; I_{or} is the saturation current at temperature T_r; $V_t = N_s kT/q$ is the thermal voltage of the array; and N_s is the number of PV cells connected in series.

If the array is composed of PV cells connected in parallel, then $I_{pv} = I_{pv,cell}N_p$ and $I_o = I_{o,cell}N_pR_s$ is the equivalent series resistance of the array and R_p is equivalent parallel resistance of the array.

The plot of Equation (3.3) is shown in Figure 3.7 (for a particular solar PV array model) and is known as the *I–V* characteristic curve of solar PV.

Manufacturers provide electrical parameters of PV rather than equations. These specifications are labeled and highlighted in Figure 3.7.

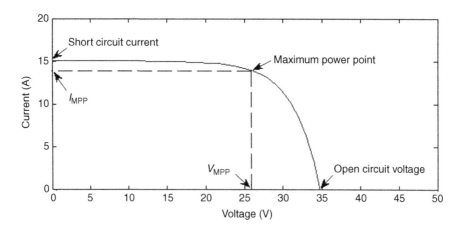

Figure 3.7 *I–V* characteristic curve of a solar PV.

The open circuit voltage is the maximum voltage across the PV array when no external circuit is connected across it. The short circuit current is the current when the PV array terminal is shorted. The other parameters are V_{MPP} and I_{MPP}, that is, the PV array voltage and current at maximum power point.

3.6 Effect of irradiance and temperature on solar cells

The characteristic equations for solar PV are given by Equations (3.3–3.5). The I–V (PV array output current and voltage) and P–V (PV array power and voltage) characteristics are shown in Figures 3.8 and 3.9, respectively. The data for PV for the characteristic curves are taken from Kyocera model no. KC200GT, summarized in Table 3.1.

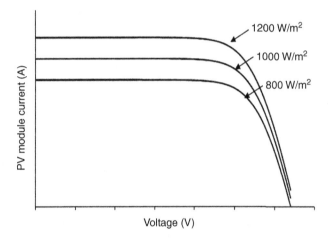

Figure 3.8 *I–V* **characteristic curve in different ambient conditions.**

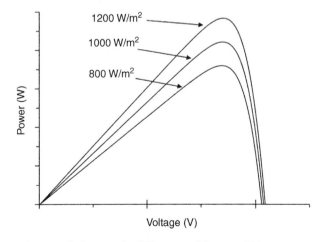

Figure 3.9 *P–V* **characteristic curve in different ambient conditions.**

Table 3.1 **Parameters of solar PV Kyocera model no. KC200GT @ 1000 W/m², 25°C**

Parameters	Value
V_{oc}	32.9 V
I_{sc}	8.21 A
I_{MPP}	7.61 A
V_{MPP}	26.3 V
P_{max}	200 W
N_s	54
K_I	0.0032 A/K

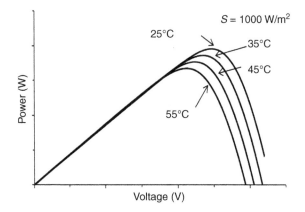

Figure 3.10 *P–V* **characteristic curve in different ambient temperatures.**

As seen in Figure 3.8, the plot between PV array output current and voltage is shown. It can be seen that with the increase in solar irradiation at constant ambient temperature the current for the same voltage increases. The same trend is seen in Figure 3.9 where the maximum power point is increasing with the increase in solar irradiation. The power of solar PV increases with the increase in solar irradiation and it varies nonlinearly.

The other case is when ambient temperature varies and solar irradiation is constant. Figure 3.10 is the curve showing this variation. It is seen that with the increase in ambient temperature both the maximum power and the open circuit voltage decrease.

The best climate condition for PV is high solar irradiation and low ambient temperature. However, the rating of PV should be decided based on the accurate solar irradiation levels.

3.7   PV module

The basic unit of solar PV is a solar cell. Solar cells can produce only a small amount of power. Power generated by a solar cell depends on its efficiency. Depending on the cell efficiency the power generated per unit area varies in the range 10–25 mW/cm²,

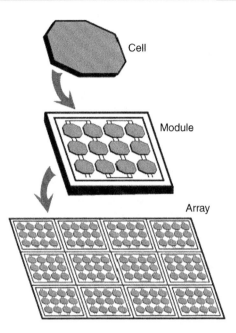

Figure 3.11 Solar PV cell, module, and array.

which corresponds to 10–25% cell efficiency [6]. The typical area of a single solar cell is 225 cm^2. With 10% cell efficiency the maximum power generated would be 2.25 Wp. To meet the requirement of high power, these cells are connected in series/parallel combinations to form modules. Currently solar PV modules are available whose power rating range is from 3 Wp to 200 Wp. These modules can be connected further to form arrays, as shown in Figure 3.11.

A solar PV array can provide power ranging from a few hundred watts to several megawatts.

3.7.1   Series and parallel connections of cells

Series and parallel connections of solar cells are made to generate high power. To increase the output voltage solar cells are connected in series and to increase the output current they are connected in parallel. It is assumed that all the parameters of solar cells are identical for making series and parallel connections.

Figure 3.12 is the I–V characteristic curve of solar PV connected in series. It can be seen that open circuit voltage is increased. In Figure 3.13 solar PV cells are connected in parallel for comparison. It can be seen that short circuit current is increased.

3.7.2   Mismatch in solar cell parameters

In solar PV modules, the solar cells are connected in series or parallel combination with the assumption that all the parameters of connected solar cells have identical

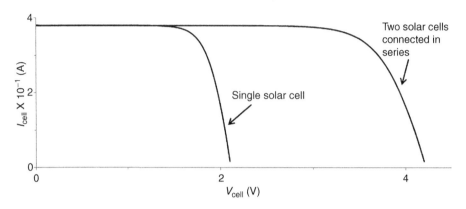

Figure 3.12 *I–V* **characteristic curve PV cells connected in series.**

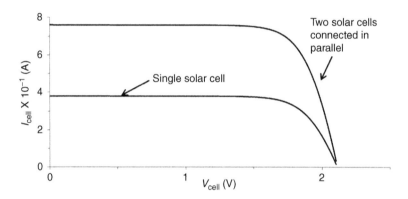

Figure 3.13 *I–V* **characteristic curve PV cells connected in parallel.**

parameters. However, in practice the mismatch may happen due to the following reasons:

- Cells or modules have the same ratings but are manufactured differently.
- Cells are processed differently.
- Outside conditions, that is, partial shading conditions, are different.
- Glass covers, etc., can be broken.

Mismatch in a connection can be caused by electrical parameters but the most common mismatch is seen in two parameters, that is, V_{oc} and I_{sc}. Among these two parameters, the short circuit current mismatch is a matter of concern particularly when solar cells are connected in series, which is common.

The mismatch in V_{oc} is an issue when solar cells/modules are connected in parallel.

Figure 3.14 is the analysis of the effect of I_{sc} mismatch. It shows the case when two solar cells are connected in series and their I_{sc} is not the same. Cell 1 has the higher I_{sc} compared to Cell 2. As per the convention the short circuit current flowing through the external circuit will be equal to the lower value of I_{sc}, that is, Cell 2 in this case. If the combination operates in short circuit condition the sum of the voltage across both the cells is zero, that is, on the Y axis (where the dotted line is crossing). To meet

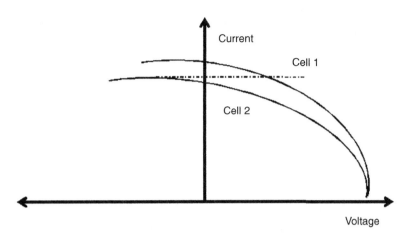

Figure 3.14 *I–V* **characteristic curve cell parameters are mismatched.**

this value of current the solar cell of lower I_{sc} value is forced to operate in reverse bias condition. Due to this effect, there will be significant power loss.

3.7.3 Hot spot due to partial shading

In the solar PV module there are many cells connected in series. It may happen in cloudy weather that one or more cells of the string receive no sunlight as shown in Figure 3.15. Under short circuit conditions the shaded cells of the string will become reverse biased and will be forced to work at V_{bias} to maintain the same current (Figure 3.16). This may lead to heavy power loss in the partially shaded cell and due to excess heat this cell may break down completely. Also, the negative voltage may bring the diode to reverse breakdown voltage and may lead to the same result, that is, breakdown. Due to this effect, the string will become open circuit and the solar power generating system could fail.

Due to the reverse bias operation the cell generates heat, which is treated as hotspot on PV array. These hotspots and failure can be avoided by using bypass diodes. These diodes are connected in parallel to the cells to limit the reverse voltage and hence the power loss in the shaded cells [7]. It is proposed by researchers that in a 36 series

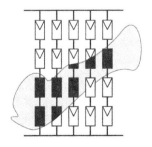

Figure 3.15 Partial shading condition.
Reproduced from Ref. [7].

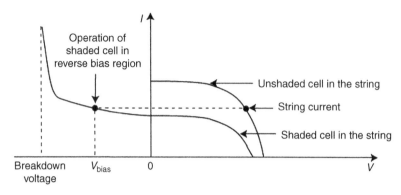

Figure 3.16 *I–V* curve under partial shading condition.
Reproduced from Ref. [7].

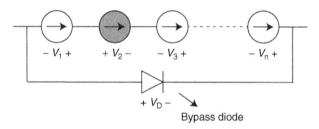

Figure 3.17 Bypass diode.
Reproduced from Ref. [7].

connected cell two bypass diodes can be connected across 18 series cells (Figure 3.17). The bypass diode restricts the voltage of the shaded cell to reach reverse breakdown voltage and also provides an alternate path for the current.

3.8 Daily power profile of PV array

As discussed in the previous section, the power output of the solar PV system depends mainly on solar irradiation. Also, the output power varies nonlinearly with ambient conditions. Based on the climate condition the output power profile of the PV array varies. The total solar irradiation in the plane of the solar array is known as incident solar irradiation. The power profile is measured in terms of performance ratio. It is defined as the ratio of the daily PV system electricity generation effectiveness to the rated array efficiency. It could be done on a daily, monthly, or annual basis. The performance ratio is a PV system metric that is normalized by both PV system capacity and incident solar radiation. An idealized performance ratio of 1.0 would imply that the PV system operated at standard test conditions over the reported period, without any losses [8]. Figure 3.18 shows the PV power production profile for the whole year on an hourly basis. It can be seen that power is delivered when Sun is available. Figure 3.19 shows daily performance ratio over a year.

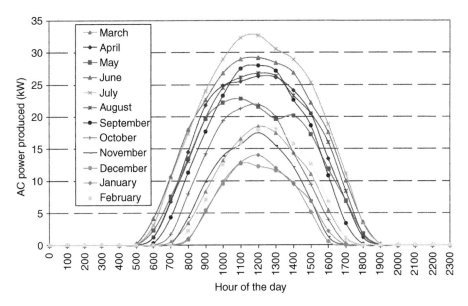

Figure 3.18 PV power production profile.
Reproduced from Ref. [8].

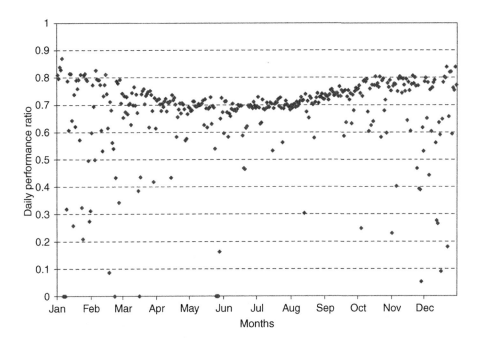

Figure 3.19 PV daily performance ratio.
Reproduced from Ref. [8].

3.9   Photovoltaic system integration

The photovoltaic system is broadly classified into three parts:

1. Standalone (off-grid) PV system
2. Grid connected (on-grid) PV system
3. Network connected solar power plants

3.9.1   Standalone (off-grid) PV system

These systems are used in areas where there is no electricity available from the national grid, that is, rural areas. In this type of system, the energy is stored in a battery generated by solar PV. The DC–DC converter is used to regulate the generated DC voltage and an inverter is connected to convert DC to AC for standard home appliances. The basic structure is shown in Figure 3.20.

3.9.2   Grid connected (on-grid) PV system

In this type of PV system, the solar PV generates the power, which is regulated by a DC–DC converter followed by a DC–AC inverter to supply the AC power to standard home appliances. If the power generated by the solar PV is more than the load

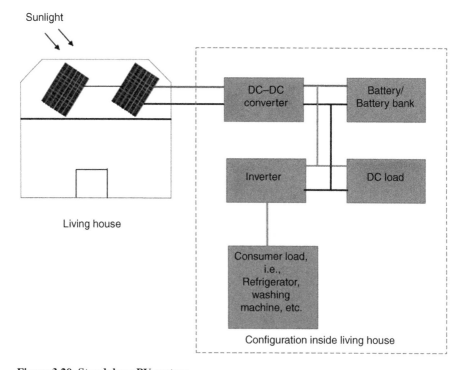

Figure 3.20 Standalone PV system.

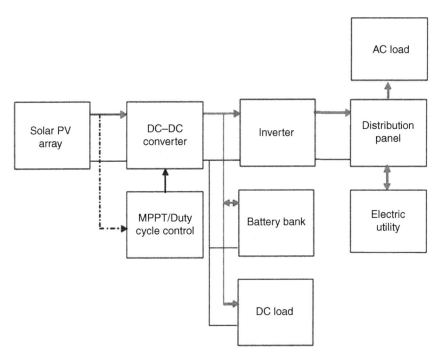

Figure 3.21 Grid connected PV system.

requirement of a house, then it is supplied to the grid measured by the smart meters and with proper protection. The typical structure is shown in Figure 3.21.

3.9.3 Network connected solar power plants

This system is used to generate large amounts of power and the capacity of power generation ranges from several hundred kilowatts to hundreds of megawatts. The photovoltaics are installed on a localized large area and are also connected to the network. These systems are installed on large industrial facilities or terminals. A photograph of one such installation is shown in Figure 3.22.

3.10 Evaluation of PV systems

The following parameters are calculated to evaluate the PV systems.

3.10.1 Net PV system production

The net electricity produced by the PV system and delivered to facility loads or exported to the grid.

Reported as: monthly totals, annual totals, and average daily totals per month.

Units: kWh/month, kWh/year, kWh/day.

Figure 3.22 Large-scale grid connected PV system.
Courtesy NREL website photos: http://www.nrel.gov./esi

3.10.2   Equivalent daily hours of peak rated PV production

The daily PV production normalized by rated PV capacity.
 Reported as: monthly average hours per day per rated PV capacity.
 Units: h/day/kWp of rated capacity.
 Calculated by: Net PV system production (kWh/day)/rated PV capacity (kWp)

3.10.3   Equivalent annual hours of peak rated PV production

The annual PV production normalized by rated PV capacity.
 Reported as: annual thousands of hours per year per rated PV capacity.
 Units: 1000 h/year/Wp of rated capacity.
 Calculated by: Net PV system production (kWh/year)/rated PV capacity (Wp)

3.10.4   Facility electrical load offset by PV production

The net electricity produced by the PV system compared to the total facility electricity use.
 Reported as: percentage of annual and monthly electrical facility load met by the PV system.
 Units: nondimensional value expressed as a percentage $\times$ 100(kWh/month, kWh/year).
 Calculated by: Net PV system production/total facility electricity use

3.10.5   Facility total energy load met by PV production

The net electricity produced by the PV system compared to the total facility energy use. A building that produces more energy than it uses would result in 100% or greater total facility energy load met by the PV system.
 Reported as: percentage of annual and monthly total facility energy load met by the PV system.
 Calculated by: Net PV system production/total facility energy use

3.10.6   Total electricity delivered to utility

When the PV system produces more AC electricity than is used at the facility, the excess production typically will be exported to the utility grid.
Reported as: monthly and annual total electricity that is exported to the utility grid.
Units: kWh/month, kWh/year.

3.10.7   Total incident solar radiation

The solar radiation in the plane of the solar array. It is calculated by summing the time-series solar radiation flux per unit area and multiplying by the PV array area.
Reported as: annual and monthly total incident solar radiation.
Units: kWh/year, kWh/month.
Calculated by: PV array area/total incident solar radiation

3.10.8   PV system AC electricity generation effectiveness

The time-series, monthly, and annual effectiveness of the PV system in converting incident solar resources to AC electricity used in the building or exported to the grid.
Calculated by: Net PV system production/total incident solar radiation

3.10.9   PV system performance ratio

The ratio of the daily, monthly, and annual PV system AC electricity generation effectiveness to the rated PV module efficiency. The performance ratio is a PV system metric that is normalized by both PV system capacity and incident solar radiation. The performance ratio can indicate the overall effect of losses on the rated PV capacity due to system inefficiencies such as cell temperature effects.
Reported as: daily, monthly, and annual average performance ratio.
Calculated by: Rated PV module efficiency (at standard test conditions)/PV system AC electricity generation effectiveness

3.10.10   Reduction of peak demand resulting from the PV system

The monthly peak demand reduction that resulted from the PV system supplying AC electricity to the facility. If a building electrical system *does not* feature demand-responsive controls, it is possible and straightforward to measure the demand reduction afforded by the PV system.
Reported as: monthly demand reduction resulting from PV system.
Units: kW or kVA.
Calculated by monthly values of: Peak demand of total facility electricity use without PV system (kW or kVA) − peak demand of net facility electricity use (kW or kVA)

3.10.11   Energy cost savings resulting from the PV system

The energy cost savings that are a result of the PV system supplying useful electricity to the building. The energy cost saving is the difference between the calculated energy costs without the PV system and the actual utility bills.

Reported as: monthly and annual facility electricity cost savings accruing from the PV system.

Units: $/month, $/year.

Calculated as: Facility electricity costs without PV system − facility electricity costs

3.11   Advantages of solar energy

Solar energy is seen as most reliable source of energy among all renewable energy sources. It has the following advantages.

3.11.1   Price saving and earning

The installation of a solar PV system saves on electricity and other energy bills and can generate money if supplied to the grid.

3.11.2   Energy independence

The consumer receives full energy independence by using rooftop solar PVs.

3.11.3   Jobs and economy

Due to the large integration of solar systems, new industries are setup, which creates new job and strengthening the economy of the country.

3.11.4   Security

The guarantee of energy security is very high, which is a major advantage.

3.12   Disadvantage

The disadvantage is that Sun is not available 24 h. So during times when there is no Sun, alternative arrangements need to be made.

3.13   Summary

This chapter provided the fundamentals of solar energy application in terms of passive and active (PV) solar energy systems. Details about the working principle of the PV system, its dependence on ambient conditions, power profile, and evaluation parameters were given and discussed. PV system integration was also described. The advantages and disadvantages of solar energy were deliberated.

Problems

1. What is a passive solar energy system?
2. What is an active solar energy system?
3. What is the importance of the orientation of house in a passive solar energy system?
4. What is a photovoltaic? Explain its working principle?
5. What is the difference between solar a module and an array?
6. What are the issues regarding series connection of cells if a mismatch in parameters happens?
7. What is a bypass diode and why is it used?
8. What is partial shading?
9. What is the difference between a standalone and grid connected PV system?
10. What are the advantages and disadvantages of solar PV systems?

References

[1] Technology Fact Sheet. Passive solar design. Energy, efficiency and renewable energy, USA, Department of Energy, December 2000.
[2] Solar Information Module. Photovoltaic, principles and methods. Solar Energy Research Institute, USA, Department of Energy, February 1982.
[3] AEO. Annual Energy Report. Department of Energy, USA. Available from: http://www.eia.gov.; 2013.
[4] De Brito MAG, Galotto L, Poltronieri L, Guilherme de Azevedo e Melo M, Canesin Carlos A. Evaluation of the main MPPT techniques for photovoltaic applications. IEEE Trans Ind Electron 2013;60(3) pp. 1156–1167, vol. 3.
[5] Villalva MG, Gazoli JR, Filho ER. Comprehensive approach to modelling and simulation of photovoltaic arrays. IEEE Trans Power Electron 2009;5:1198–208.
[6] Solanki SC. Solar photovoltaics: fundamentals, technologies and applications. New Delhi, India: PHI; 2012.
[7] Bidram A, Davoudi A, Balog RS. Control and circuit techniques to mitigate partial shading effects in photovoltaic arrays. IEEE Trans Photovoltaic 2012;2(4):532–46.
[8] Technical Report. Procedure for measuring and reporting the the performance of photovoltaic systems in buildings. NREL, USA, October 2005.

Wind energy

4

Abdul R. Beig, S.M. Muyeen
Department of Electrical Engineering, The Petroleum Institute, Abu Dhabi, UAE

Chapter Outline

4.1   Introduction

Conventional energy sources such as natural gas, oil, coal, or nuclear are finite but still hold the majority of the energy market. However, renewable energy sources like wind, fuel cells, solar, biogas/biomass, tidal, geothermal, etc. are clean and abundantly available in nature and hence are competing with conventional energy sources. Among the renewable energy sources wind energy has a huge potential of becoming a major source of renewable energy for this modern world. Wind power is a clean, emissions-free power generation technology. As per the Global Wind Energy Council (GWEC) 2013 statistics, cumulative global capacity has reached to a total of 318 GW, which shows an increase of nearly 200 GW in the past 5 years. GWEC predicts that wind power could reach nearly 2000 GW by 2030, supply between 16.7% and 18.8% of global electricity and help save over 3 billion tons of CO_2 emissions annually. From this scenario, it is clear that wind power is going to dominate the renewable as well as the conventional energy market in the not too distant future. Wind energy is the only power generation technology that can deliver the necessary cuts in CO_2 emissions from the power sector in the critical period up to 2020, when greenhouse gas emissions must peak and begin to decline if we are to have any hope of avoiding the worst impacts of climate change. However, grid integration, voltage, and power fluctuation issues should adequately be addressed due to the huge penetration of wind power to the grid.

4.2 Wind turbine

The wind turbine is the most essential part of the wind energy system. It converts the kinetic energy associated with wind (known as wind energy) into mechanical energy and then to electrical energy. Historically windmills are used for lifting water, where wind energy is converted to mechanical energy [1]. The application of windmills for water lifting purpose dates back to as early as 644AD [1]. Several different types of windmills were in use until the twentieth century in different parts of the world for different types of application, such as lifting water, pumping water, lifting heavy materials such as logs, milling grains, etc. In 1891, Poul La Cour of Denmark first produced electricity in direct current (DC) form from a wind turbine [1]. His wind turbine was mainly based on the traditional windmill technology and was capable of producing small amounts of electricity. Since then there have been major improvements in wind turbine technology and currently there are several wind farms successfully installed in different parts of world generating large amounts of power of the order of a few thousand megawatts. The following section will briefly explain different parts of the wind turbine: rotor blades, gearbox, generator, tower, yawing, brakes, cables, anemometer, and pitch angle. These parts for the horizontal axis type turbine are shown in Figure 4.1.

4.2.1 Rotor blades

Rotor blades are the most important parts of a wind turbine in terms of performance and cost of the wind power system. The shape of the rotor blades has a direct impact on performance as this decides the conversion of kinetic energy associated with the wind to mechanical energy (torque). In these types of wind turbines, the blades are designed to have a high lift to drag ratio, based on aerodynamic principles. The number of blades is selected for aerodynamic efficiency, component costs, and system

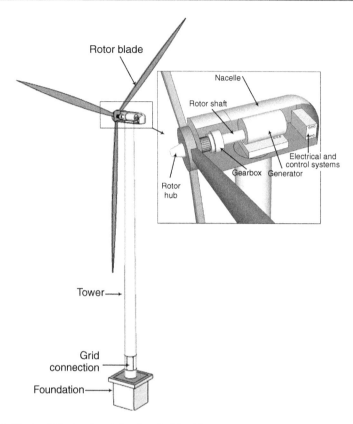

Figure 4.1 Parts of the horizontal axis wind turbine.

reliability. Theoretically, an infinite number of blades of zero width are the most efficient, operating at a high value of tip speed ratio. But other considerations such as
manufacturing, reliability, performance, and cost restrict wind turbines to only a few
blades. The majority of wind turbines are the horizontal axis type with three blades.
The turbine blades must have low inertial and good mechanical strength for durable
and reliable operation. The blades are made up of aluminum or fiberglass reinforced
polyester, carbon fiber reinforced plastics, or wood or epoxy laminates [2,3]. A schematic diagram of a rotor blade is given in Figure 4.2. The exterior shape of the blades
is based on aerodynamics but the interior is determined by attention to strength. In
low power turbines, the blades are directly bolted to the hub and hence are static. In
high power turbines, the blades are bolted to the pitch mechanism, which adjusts
their angle of attack according to the wind speed to control their rotational speed. The
pitch mechanism is bolted to the hub. The blade consists of a spar, which is a continuously tapered longitudinal beam that provides necessary stiffness and strength to withstand the wind load and carry blade weight. The spar is integrated to the hub. Around
the spar two aerodynamically shaped shells are mounted and two edges of the shells
are sealed. The hub is fixed to the rotor shaft, which drives the generator directly or

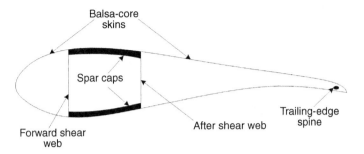

Figure 4.2 Schematic diagram of the blade.

through a gearbox. Dirt will deposit on the rotor blade surfaces, which will affect the performance of the turbine. Therefore, the blade surfaces will need frequent cleaning and, if required, polish to keep the blade performance uniform.

4.2.2 Nacelle

A nacelle is a box type structure that sits at the top of the tower and is attached to the rotor. The nacelle houses all of the generating components in a wind turbine, including the generator, gearbox, drive train, and brake assembly. The nacelle is made up of fiberglass and protects the wind turbine components from the environment. Modern large farms have a helicopter-hoisting platform built on top of the nacelle, capable of supporting service personnel.

4.2.3 Gearboxes

Mechanical power from the rotation of the wind turbine rotor is transferred to the generator rotor through the main shaft, the gearbox, and high-speed shaft. The wind turbine rotates at very slow speed. This requires a large number of poles to the generator. For economic and optimal design it is necessary to have the gearbox between the wind turbine shaft and generator shaft to increase the speed. The gearbox is of a fixed speed ratio and mainly increases speed. The gear ratio is in the range 20–300 [3]. Lubricating oil is used in the gearbox to reduce friction.

4.2.4 Generators

The generators convert the mechanical energy into electrical energy. The generator has widely varying mechanical input. It is usually connected to the grid in high power systems. In low power ratings the generator may be working in isolation supplying power to the local grid. The generator generally produces variable frequency, variable voltage, three-phase alternating current (AC). This voltage is usually converted to DC, then to regulated and fixed frequency AC using an AC-to-DC/DC-to-AC converter.

4.2.5   Tower

Towers are used to mount the wind turbine. Wind energy yield increases with the height. But optimal design limits the height of the tower, as the cost of the tower will be very high if it is too tall. Towers are usually made of tubular steel or concrete. Tubular towers are conical in shape with their diameter decreasing toward the tip. Steel towers are expensive. An alternate solution is concrete towers.

4.2.6   Yaw mechanism

Horizontal axis wind turbines use forced yawing where generators and gearboxes keep the rotor blades perpendicular to the direction of the wind. The upwind machines use brakes on the yaw mechanism. The yaw mechanism is activated by automatic control, which monitors the rotor. Cable carries the current from the wind turbine down through the tower. The yaw mechanism also should be equipped to protect the cable should it become twisted. Besides the role of tracking wind direction, the yaw mechanism also places an important role on connecting the tower with the nacelle.

4.2.7   Brakes

There are three main types of braking mechanism, namely aerodynamic brakes, electro brakes, and mechanical brakes. In the case of aerodynamic brakes, the blades are tuned such that the lift effect disappears. In electro blades, the electrical energy is dumped into a resistor bank. In mechanical type brakes, the disc or drum brakes are used to lock the blades.

4.2.8   Protection of turbines

Wind turbines need to be protected against overheating, overspeed, and overloading. Vibration is one of the main sources of turbine failure. So a vibration monitoring and protection system is to be installed with the turbine. Wind direction and wind speed are also important for the satisfactory operation of the turbine. The cup anemometer is used for this purpose. Since the wind turbine has rotating parts, lubricating systems are required. This is either a forced circulation or pressurized lubricating system. The other measuring or sensors are temperature of the gearbox, temperature of the generator, voltage–frequency measurement, speed measurement, etc.

4.3   Kinetic energy of wind

Kinetic energy in a parcel of air of mass m flowing at speed v_w in the x direction is:

$$E_w = \frac{1}{2}mv_w^2 = \frac{1}{2}(\rho Ax)v_w^2 \tag{4.1}$$

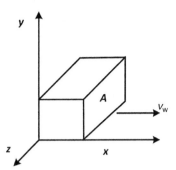

Figure 4.3 Packet of air moving in the x direction with speed v_w.

where, E_w is the kinetic energy in joules, A is the cross-sectional area in m^2, ρ is the air density in kg/m^3, and x is the thickness of the parcel in m.

If we visualize the parcel as in Figure 4.3 with side x moving with speed v_w (m/s), and the opposite side fixed at the origin, we see the kinetic energy increasing uniformly with x, because the mass is increasing uniformly. The power in the wind P_w is the time derivative of the kinetic energy, given by (4.2):

$$P_w = \frac{dE_w}{dt} = \frac{1}{2}\rho A v_w^2 \frac{dx}{dt} = \frac{1}{2}\rho A v_w^3 \qquad (4.2)$$

Thus, the wind power is directly proportional to the cross-sectional area and the cube of the wind velocity.

4.4 Aerodynamic force

4.4.1 Ideal wind turbine output

Ideal wind turbine output can be viewed as the power being supplied at the origin to cause the energy of the parcel to increase according to Equation (4.1). A wind turbine will extract power from side x with Equation (4.2) representing the total power available at this surface for possible extraction.

The physical presence of a wind turbine in a large moving air mass modifies the local air speed and pressure as shown in Figure 4.4. The picture is drawn for a conventional horizontal axis propeller type turbine.

Consider a tube of moving air with initial or undisturbed diameter d_1, speed v_{w1}, and pressure p_1, as it approaches the turbine. The speed of the air decreases as the turbine is approached, causing the tube of air to enlarge to the turbine diameter d_2. The air pressure will rise to the maximum just in front of the turbine and will drop below atmospheric pressure behind the turbine. Part of the kinetic energy in the air is converted

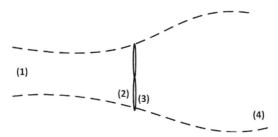

Figure 4.4 Circular tube of air flowing through ideal wind turbine.

to potential energy in order to produce this increase in pressure. Still more kinetic energy will be converted to potential energy after the turbine, in order to raise the air pressure back to atmospheric. This causes the wind speed to continue to decrease until the pressure is in equilibrium. Once the low point of wind speed is reached, the speed of the tube of air will increase back to $v_{w4} = v_{w1}$ as it receives kinetic energy from the surrounding air [2,4,5].

It can be shown [2,4,5] that under optimum conditions, when maximum power is being transferred from the tube of air to the turbine, the following relationships hold:

$$v_{w2} = v_{w3} = \frac{2}{3}v_{w1}; \qquad v_{w4} = \frac{1}{3}v_{w1} \tag{4.3a}$$

$$A_2 = A_3 = \frac{3}{2}A_1; \qquad A_4 = 3A_1 \tag{4.3b}$$

The mechanical power extracted is then the difference between the input and output power in the wind:

$$P_{m, ideal} = P_1 - P_4 = \frac{1}{2}\rho\left(A_1 v_{w1}^3 - A_4 v_{w4}^3\right) = \frac{1}{2}\rho\left(\frac{8}{9}A_1 v_{w1}^3\right) \tag{4.4}$$

This states that eight-ninths of the power in the original tube of air is extracted by an ideal turbine. This tube is smaller than the turbine, however, and this can lead to confusing results. The normal method of expressing this extracted power is in terms of the undisturbed wind speed v_{w1} and the turbine area, A_2. This method yields:

$$P_{m, ideal} = \frac{1}{2}\rho\left[\frac{8}{9}\left(\frac{2}{3}A_2\right)v_1^3\right] = \frac{1}{2}\rho\left(\frac{16}{27}A_2 v_1^3\right) \tag{4.5}$$

The factor $16/27 = 0.593$ is called the Betz coefficient. It shows that an actual turbine cannot extract more than 59.3% of the power in an undisturbed tube of air of the same area. In practice, the fraction of power extracted will always be less because of mechanical imperfections. A good fraction is 35–45% of the power in the wind under optimum conditions. A turbine, which extracts 40% of the power in the wind, is extracting about two-thirds of the amount that would be extracted by an ideal turbine. This is rather good, considering the aerodynamic problems of constantly changing wind speed and direction as well as the frictional loss due to blade surface roughness [5].

4.5 Power output from practical turbines

The fraction of power extracted from the power in the wind by a practical wind turbine is usually given by the symbol C_p, standing for the coefficient of performance or power coefficient. Using this notation and dropping the subscripts of Equation (4.3), the actual mechanical power output can be written as:

$$P_m = C_p \left(\frac{1}{2} \rho A v_w^{\,3} \right) = \frac{1}{2} \rho \pi R^2 v_w^{\,3} C_p (\lambda, \beta) \tag{4.6}$$

where, R is the blade radius of wind turbine (m), v_w is the wind speed (m/s), and ρ is the air density (kg/m^3). The coefficient of performance is not constant, but varies with the wind speed, the rotational speed of the turbine, and turbine blade parameters like angle of attack and pitch angle. Generally it is said that power coefficient C_p is a function of tip speed ratio λ and blade pitch angle β (°).

4.6 Tip speed ratio

Tip speed ratio is the ratio of the circumferential velocity of the rotor at the end of the blade, that is, the maximum velocity v_m and the wind velocity v_w in front of the rotor blade. Originally it was defined as:

$$\lambda = \frac{v_m}{v_w} \tag{4.7}$$

A more popular form of tip speed ratio in the wind industry is as follows:

$$\lambda = \frac{\omega_R R}{v_w} \tag{4.8}$$

where, ω_R is the mechanical angular velocity of the turbine rotor in rad/s and v_w is the wind speed in m/s.

4.7   Coefficient of performance and turbine efficiency

There will be energy loss in the mechanical components of the rotor, gear system, and generator. So the overall efficiency can be obtained as:

$$\eta = C_p \eta_m \eta_g$$

where η_m is the mechanical efficiency and η_g is the generator efficiency.

$$\eta = \frac{P_0}{(1/2)\rho A v_w^3}$$

where P_0 is the electrical output power.

Modeling of a wind turbine rotor is somewhat complicated. According to the blade element theory, modeling of blade and shaft needs complicated and lengthy computations. Moreover, it also needs detailed and accurate information about rotor geometry. For that reason, considering only the electrical behavior of the system, a simplified method of modeling of the wind turbine blade and shaft is normally used.

Typical C_p–λ curves for MOD-2 wind turbine is shown in Figure 4.5 for different values of β [6,7].

4.8   Operating range of wind turbine

Wind turbines are allowed to run only in a well-defined range of wind speed. A minimum wind speed is required for the blades to overcome inertia and friction. This minimum speed is called cut-in wind speed ($v_{\text{cut-in}}$). The typical value of cut-in wind speed is 3–5 m/s. At very high wind speed, say 25 m/s, in order to avoid damage to

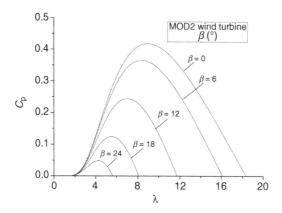

Figure 4.5 C_p–λ curves for different pitch angles.

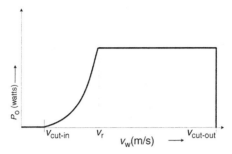

Figure 4.6 Output power versus wind speed.

wind turbines, the wind turbines are stopped from rotating. This is called cut-out speed ($v_{cut-out}$). The operating range of a wind turbine can be best explained by a wind power curve as shown in Figure 4.6.

In the normal wind speed range the turbine will be able to produce rated power. Rated wind speed (v_r) is the wind speed at which the turbine generated rated power. A typical value is about 12–16 m/s. This corresponds to the point at which the conversion efficiency is maximum.

4.9   Classifications of wind turbines

Wind turbines can be classified into two categories, namely (1) horizontal axis and (2) vertical axis wind turbines based on their constructional design.

4.9.1   Horizontal axis wind turbine

Almost all of the commercially established wind energy systems use horizontal type wind turbines. The axis of rotation is horizontal. The major advantage of the horizontal type wind turbine is that by using blade pitch control, the rotor speed and power output can be controlled. Also blade pitch control protects the wind turbine against overspeed when the wind speed becomes dangerously high. The basic principle of a horizontal axis wind turbine is based on propeller-like concepts, so the technological advances of the propeller design are readily incorporated to develop modern highly efficient wind turbines [1]. Figure 4.1 shows the schematic arrangement of a horizontal axis wind turbine. Details of the different parts of the horizontal wind turbine are given in Section 4.1.

4.9.2   Vertical axis wind turbine

The vertical turbine proposed in 1925 by Darrius had some promising features for modern wind energy farms. The blades are curved and the rotor has vertical axis rotation. Figure 4.7 shows the schematic diagram of the Darrius type vertical axis turbine.

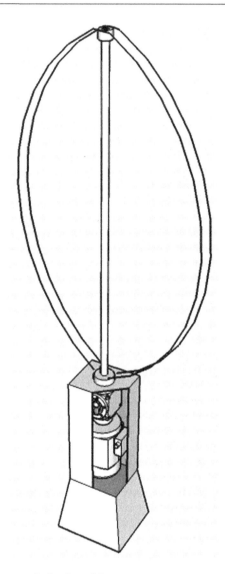

Figure 4.7 Darrius type vertical axis turbine.

Compared to the horizontal axis the shape of vertical axis blades is complicated, making them difficult to manufacture. The H-rotor vertical axis wind turbine uses straight blades instead of curved blades as shown in Figure 4.8. The blades are fixed to a rotor though struts. There are other types of vertical axis wind turbines, namely the Savonius type and V-shaped vertical axis turbines [1,2]. These have very low tip speed ratio and low power coefficient, hence they are used only in very low power wind energy systems.

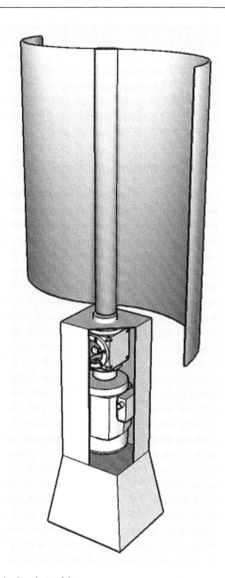

Figure 4.8 H type vertical axis turbine.

The vertical axis type generator has a simple design. The shaft is vertical, so the generator is mounted on the ground and the tower is required only to mount the blades. The disadvantages are the tip speed ratio and power output are very low compared to horizontal axis generators. The turbine needs an initial push to start; it is not self-starting. Also it is not possible to control the power output by pitching the rotor blades. Support wires or guy wires are required in addition to the tower. Due to these reasons not much attention is given to vertical axis wind turbines.

4.10   Types of wind turbine generator systems

A wind turbine generator system (WTGS) transforms the energy present in the wind into electrical energy. As wind is a highly variable resource that cannot be stored, operation of the WTGS must be done according to this feature. Based on the rotational speed of the wind turbine, WTGSs can be broadly classified under two major categories, namely fixed speed and variable speed.

4.10.1   Fixed speed wind turbine generator system

A fixed speed WTGS consists of a conventional, directly grid coupled squirrel cage induction generator, which has some superior characteristics such as brushless and rugged construction, low cost, maintenance free, and operational simplicity. The slip, and hence the rotor speed of a squirrel cage induction generator, varies with the amount of power generated. These rotor speed variations are, however, very small, approximately 1–2% from the rated speed. Therefore, this type of wind energy conversion system is normally referred to as a constant speed or fixed speed WTGS. The advantage of a constant speed system is that it is relatively simple. Therefore, the list price of constant speed turbines tends to be lower than that of variable speed turbines. However, constant speed turbines must be more mechanically robust than variable speed turbines [3]. Since the rotor speed cannot be varied, fluctuations in wind speed translate directly into drive train torque fluctuations, causing higher structural loads than with variable speed operation. This partly cancels the cost reduction achieved by using a relatively cheap generating system (Figure 4.9).

4.10.2   Variable speed wind turbine generator system

The currently available variable speed wind turbine (VSWT) generator system topologies are shown in Figures 4.10 and 4.11. To allow variable speed operation, the mechanical rotor speed and the electrical frequency of the grid must be decoupled. Therefore, a power electronic converter is used in the variable speed wind generator system. In the doubly fed induction generator, a back-to-back voltage source converter feeds the three-phase rotor winding. In this way, the mechanical and electrical

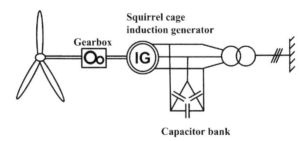

Figure 4.9 Schematic diagram of a fixed speed wind turbine generator system.

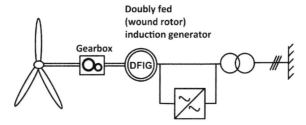

Figure 4.10 Schematic diagram of a variable speed wind turbine generator system using a wound rotor induction motor.

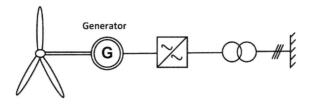

Figure 4.11 Schematic diagram of a variable speed wind turbine generator system using a synchronous generator.

rotor frequencies are decoupled and the electrical stator and rotor frequencies can be matched independently of the mechanical rotor speed. In the direct drive synchronous generator system (PMSG or WFSG), the generator is completely decoupled from the grid by a frequency converter. The grid side of this converter is a voltage source converter, that is, an insulated gate bipolar transistor (IGBT) bridge. The generator side can either be a voltage source converter or a diode rectifier.

The generator is excited using either an excitation winding (in the case of a WFSG) or permanent magnets (in the case of a PMSG). In addition to these three mainstream generating systems, there are some other varieties as explained in Refs [6,7]. One that must be mentioned here is the semivariable speed system. In a semivariable speed turbine, a wound rotor induction generator is used. The output power is regulated by means of rotor resistance control, which is achieved by means of power electronics converters. By changing the rotor resistance, the torque/speed characteristic of the generator is shifted and about a 10% rotor speed decrease from the nominal rotor speed is possible. In this generating system, a limited variable speed capability is achieved at relatively low cost. Other variations are a squirrel cage induction generator or a conventional synchronous generator connected to the wind turbine through a gearbox and to the grid by a power electronics converter of the full rating of the generator.

For each instantaneous wind speed of a VSWT, there is a specific turbine rotational speed, which corresponds to the maximum active power from the wind generator. In this way, the maximum power point tracking (MPPT) for each wind speed increases the energy generation in the VSWT [8,9]. This is illustrated in Figure 4.12.

When the wind speed changes, the rotational speed is controlled to follow the maximum power point trajectory. It should be mentioned here that the measurement of the

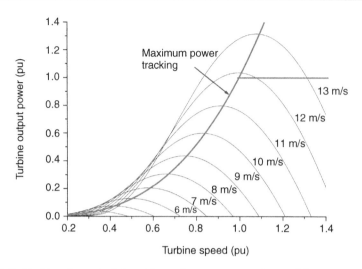

Figure 4.12 Turbine characteristic with maximum power point tracking.

precise wind speed is difficult. Therefore, it is better to calculate the maximum power, P_{max}, without the measurement of wind speed as shown in the succeeding section.

4.11   Wind farm performance

Output of a wind farm mainly depends on wind speed pattern. In the case of a VSWT, maximum power is extracted from the wind, which maximizes the wind generator output power. The voltage fluctuation issue is handled by the power electronics converters equipped with variable speed wind generator [9].

However, in the case of a fixed speed wind generator, the power and terminal voltage of a wind generator or wind farm varies randomly. This is because wind is stochastic and intermittent, which causes power variation at the wind generator or wind farm terminal. Figure 4.13 shows wind speeds that are used to generate electric power in a wind farm where there exist five wind generators. Figure 4.14 shows the output power of a wind farm. A capacitor bank is usually connected to the terminal of a fixed speed wind generator, which is designed in such a way to maintain unity power factor at rated wind speed. Therefore, at lower wind speed we have excess or surplus reactive power at the terminal of the wind generator, which will cause an overvoltage issue in the wind farm. This is depicted in Figure 4.15.

4.12   Advantages and disadvantages

4.12.1   Advantages

- Wind energy is environment friendly as no fossil fuels are burnt to generate electricity from wind energy.

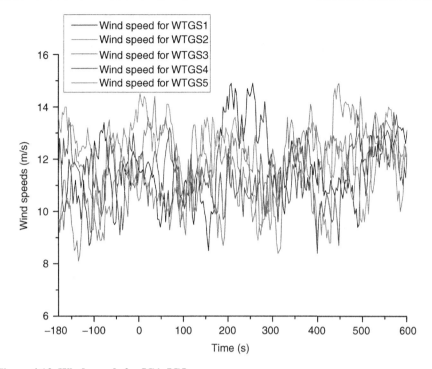

Figure 4.13 Wind speeds for IG1–IG5.

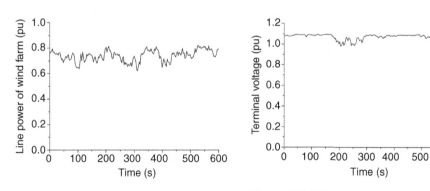

Figure 4.14 Line power of a wind farm.   **Figure 4.15 Wind farm terminal voltage.**

- Wind turbines take up less space than the average power station.
- Modern technologies are making the extraction of wind energy much more efficient. Wind is free, so only installation cost is involved and running costs are low.
- Wind energy is the most convenient resource to generate electrical energy in remote locations, where conventional power lines cannot be extended due to environmental and economic considerations.

4.12.2   Disadvantages

- The main disadvantage of wind energy is varying and unreliable wind speed. When the strength of the wind is too low to support a wind turbine, little electricity is generated.
- Large wind farms are required to generate large amounts of electricity, so this cannot replace the conventional fossil fueled power stations. Wind energy can only substitute low energy demands or isolated low power loads.
- Larger wind turbine installations can be very expensive and costly to surrounding wildlife during the initial commissioning process.
- Noise pollution may be problem if wind turbines are installed in the densely populated areas.

4.13   Summary

This chapter explained the various parts of a wind turbine. Different types wind turbines were presented. Wind dynamics and basic principles of converting wind energy to electric energy were discussed and the different types of wind turbine configurations and generation systems were described.

Problems

1. Find the diameter of a wind turbine rotor that will generate 100 kW of electrical power in a steady wind speed of 8 m/s. Assume that the air density is 1.225 kg/m^3, $C_p = 16/27$, and $\eta = 1$.
2. A 50-m diameter, three-bladed wind turbine produces 800 kW at a wind speed of 15 m/s. The air density is 1.225 kg/m^3. Find:
   a. The rotational speed of the rotor at a tip speed ratio of 5.0.
   b. The tip speed.
   c. The gear ratio if the generator speed is 1600 rpm.
   d. The efficiency of the wind turbine system.
3. A wind turbine is operating in steady winds at a power of 1800 kW and a speed of 30 rpm. Suddenly the connection to the electrical network is lost and the brakes fail to apply. Assuming that there are no changes in the aerodynamic forces, how long does it take the operating speed to double? Take $J = 4 \times 10^2$ kg/m^2.
4. A standalone single-phase wind turbine generator generates 220 V, 50 Hz AC. The output of the generator is connected to a diode bridge full-wave rectifier, which produces a fluctuating DC voltage.
   a. Find the average DC voltage if the output of the generator is rectified using a full-wave diode rectifier.
   b. Find the average DC voltage if the output of the generator is rectified using a full-wave fully controlled SCR rectifier. The firing angle is 45°.
5. A four-pole induction generator is rated at 500 kVA and 400 V. It has the following parameters $X_{LS} = X_{LR} = 0.15$ Ω, $R_S = 0.014$ Ω, $R_R = 0.013$ Ω, $X_M = 5$ Ω.
   a. How much power does it produce at a slip of -0.025 (take synchronous speed as 1500 rpm)?
   b. Find its speed.
   c. Find the torque and power factor.

References

[1] Hau E. Wind turbines, fundamentals, technologies, applications, economics. 3rd ed. Berlin, Heidelberg: Springer; 2013.

[2] Manwell JF, Mcgown JG, Rogers AL. Wind energy explained: Theory, design and application, 2nd edition, Chichester, West Sussex, UK, John Wiley and Sons Ltd; 2009.

[3] Wagner H-J, Mathur J. Introduction to wind energy systems. Berlin, Heidelberg: Springer; 2013.

[4] Johnson GL. Wind energy systems. Loose Leaf, University Reprints; 2006, ASIN: B007U79DJK.

[5] Golding E. The generation of electricity by wind power. New York: Halsted Press; 1976.

[6] Muyeen SM. Wind energy conversion systems. Berlin, Heidelberg: Springer; 2012.

[7] Muyeen SM, Tamura J, Murata T. Stability augmentation of a grid-connected wind farm. London: Springer-Verlag; 2008.

[8] Slootweg JG. Wind power: modelling and impact on power system dynamics. PhD thesis, Delft University of Technology, Netherlands, 2003.

[9] Heier S. Grid integration of a wind energy conversion system. Chichester, UK: John Wiley & Sons Ltd; 1998.

Hydroelectricity

Sreenivas S. Murthy*, Sriram Hegde**
*Department of Electrical Engineering, Indian Institute of Technology, Delhi;
CPRI, Bengaluru, India
**Department of Applied Mechanics, Indian Institute of Technology, Delhi, India

Chapter Outline

5.1   Introduction

This chapter deals with the process of producing electricity from water. While historically this is one of the oldest methods of power generation, the topic has assumed increased importance in the context of renewed interest in renewable energy due to greenhouse gas (GHG) emissions, global warming, and depleting fossil fuels. Hydro energy is a major component of renewable energy-based power generation. Unit sizes of hydrogenerators vary widely from a few kW to a few hundred MW – technologies also differ accordingly. Unit sizes are based on size and power rating. Typical unit sizes are pico – a few kW, micro – hundreds of kW, mini – a few MW, and mega – hundreds of MW. Pico-, micro-, and minihydro plants are listed under small hydro while mega comes under large hydro. Small hydro plays an important role in remote areas, community development, and multipurpose infrastructure – applicable to both developed and developing countries. All large hydro units feed generated power to the grid, while small hydro may be grid fed or off-grid, standalone type. All minihydro

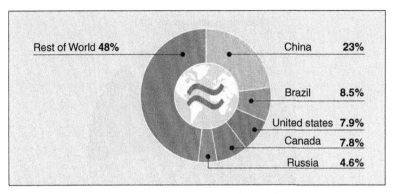

Figure 5.1 Hydropower global capacity – a typical distribution (2008).

plants feed power to the grid, while microhydro plants may be grid fed or off-grid type, and pico is always "standalone." Large hydro involves the building of large dams with considerable civil works. They are not considered environmentally friendly due to large-scale deforestation and displacement of habitats. Further, they too cause GHG emissions. In comparison small hydro is more sustainable as they need minimal civil works and cause little ecodisturbance.

Globally, hydropower is 87% renewables as it provides power in 160 countries. However, It is unevenly distributed. Figure 5.1 shows a typical distribution. Brazil, Canada, China, Russia, USA make up more than 50% of hydro production. A 10-fold increase in Africa, 3-fold in Asia, and doubling in South America are expected in the near future. Some lead countries and their theoretical hydro potential in TWh/year in parentheses are as follows:

- Brazil (3040)
- Canada (2216)
- China (6083)
- Colombia (1000)
- Congo (1397)
- Ethiopia (650)
- Greenland (800)
- India (2638)
- Indonesia (2147)
- Japan (718)
- Nepal (733)
- Norway (563)
- Peru (1577)
- Russia (2295)
- USA (4485)

Figure 5.2 presents typical distribution of renewable energy with predominant hydro energy.

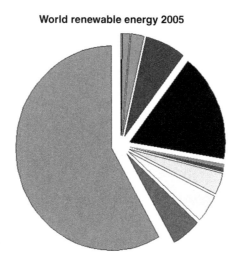

World renewable energy 2005

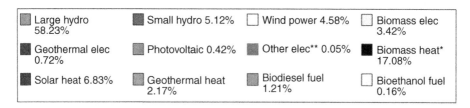

Large hydro 58.23%	Small hydro 5.12%	Wind power 4.58%	Biomass elec 3.42%
Geothermal elec 0.72%	Photovoltaic 0.42%	Other elec** 0.05%	Biomass heat* 17.08%
Solar heat 6.83%	Geothermal heat 2.17%	Biodiesel fuel 1.21%	Bioethanol fuel 0.16%

Figure 5.2 Typical distribution of renewable energy (2008).

5.2   Process of hydroelectricity

The process of producing electricity from water power is accomplished in several stages. Hydropower is available in nature by natural water head in suitable terrains, which can be enhanced through dams and fast flowing rivers. This power is converted to mechanical energy through hydraulic turbines or pumps used as turbines, which is then converted to electrical energy through electric generators. Thus, broadly two stages of energy conversion take place: (1) hydro energy to mechanical energy and (2) mechanical energy to electrical energy. These are presented in detail in the following sections.

5.3   Basics of pumps and turbines

5.3.1   Turbo machine

Any device that extracts energy from or imparts energy to a continuously moving stream of fluid (liquid or gas) can be called a turbo machine. The energy stored by a fluid mass appears in the form of potential, kinetic, and molecular energy. A rotating shaft usually transmits the kinetic energy. A turbo machine is either a power generating (turbine) or head generating (pump) machine, which employs the dynamic action of a rotating element, the rotor. The action of the rotor changes the energy level of the continuously flowing fluid through the machine. Turbines, pumps, compressors, and fans are all members of this family of machines.

5.3.2 Pump

A pump is a device that converts the mechanical energy into hydraulic energy for fluids (liquids or gases), or sometimes slurries, by mechanical action. Pumps can be classified into different groups according to the mode of movement of the fluid.

5.3.2.1 Classification of pumps

Pumps can be classified as:

1. Rotodynamic pump: Dynamic action of the fluid movement is due to the mechanical energy imparted into the system by a rotating element.
2. Reciprocating pump: The liquid is trapped in a cylinder by suction and then pushed against pressure using mechanical energy.
3. Rotary positive displacement pump: The liquid is trapped in a volume and pushed out against pressure in a rotating environment.

Rotodynamic pumps, that is, centrifugal and axial flow pumps, can be operated at high speeds often directly coupled to electric motors. These pumps can handle small as well as very large volumes. They can handle corrosive and viscous fluids and even slurries. The overall efficiency is high for these pumps compared to other types of pumps. Hence, these pumps are the most popular pumps. Rotodynamic pumps can be of radial flow, mixed flow, and axial flow type according to the flow direction. Radial flow or purely centrifugal pumps generally handle lower volumes at higher pressures. Mixed flow pumps handle comparatively larger volumes at medium pressures. Axial flow pumps can handle very large volumes, but the pressure against which these pumps operate is very limited. The overall efficiency of these three types of pumps will depend on the flow and specific speed.

5.3.2.2 Centrifugal pump

In a centrifugal pump, energy is imparted to the fluid by the centrifugal action of moving blades, that is, impeller vanes from the inner radius to the outer radius. The main components of centrifugal pumps are impeller, casing, and rotating shaft with gland and packing. Additionally, a suction pipe with a one-way valve (foot valve) and a delivery pipe with a delivery valve complete the system.

The liquid enters the eye of the impeller axially due to the suction created by the impeller motion. The impeller blades guide the fluid and impart momentum to the fluid, which increases the total head (or pressure) of the fluid, causing the fluid to flow out. The casing is a simple volute type or a diffuser. The volute is a spiral casing of gradually increasing cross-section. A part of the kinetic energy in the fluid is converted to pressure in the casing.

Gland and packing or so-called stuffing box is used to reduce leakage along the driving shaft. By the use of the volute casing or diffuser the kinetic head can be recovered as useful static head of the centrifugal pump unit.

5.3.3 Turbine

A turbine is a rotary mechanical device that extracts energy from a flowing fluid stream and converts it into useful work. It has one moving component called the rotor assembly (with blades), which is mounted on a shaft. Moving fluid acts on the blades so that they move and impart rotational energy to the rotor. The rotor is coupled to an induction motor or a generator, which converts mechanical energy into electrical energy.

Types of turbines include steam turbines, wind turbines, gas turbines, or water turbines. Steam turbines are driven by oil, coal, or by the extraction of nuclear power and are the most common methods of producing electricity. Green electricity applications include wind turbines and water turbines used in applications for wind power and hydel power.

Because of the turbine's many applications in a wide variety of technologies, research is still ongoing to perfect turbine and rotor efficiency, and blade lifespans. In this section, the discussion will concentrate on the types and operational details of hydraulic turbines.

5.3.3.1 Classification of turbines

The main classification depends upon the type of action of the water on the turbine. These are:

1. Impulse turbine: The potential energy is converted to kinetic energy in the nozzles. The impulse provided by the jets is used to turn the turbine wheel. The pressure inside the turbine is atmospheric. This type is found suitable when the available potential energy is high and the flow (discharge) available is comparatively low.
2. Reaction turbine: The available potential energy is progressively converted in the turbine rotors (stages) and the reaction of the accelerating water causes the turning of the wheel. These machines are again divided into radial flow, mixed flow, and axial flow depending upon the head available. Radial flow machines are found suitable for moderate levels of head and medium quantities of flow. The axial machines are suitable for low levels of head and large flow rates.

5.3.3.2 Hydraulic turbine

The hydraulic turbine is very useful when enough flowing water is available. Though initial capital cost is a little high, once it operates it can provide a constant and predictable power supply, whereas other technologies (specifically wind and solar power) provide intermittent or unpredictable energy.

5.3.3.3   Microhydel system and turbine

A complete microhydro system consists of the following major components:

1. Filter mechanism
2. Penstock with valves
3. Turbine and draft tube
4. Power-converting device (generator or direct drive)

A major aspect of system design that often is not considered is the removal of solid bodies from the water before it enters the turbine. If no such system is installed, the turbine could suffer from damage, poor performance, and even stalling. Therefore, it is usual to take a side stream from the main flow to install a microhydel system.

The intake length of the pipeline, that is, penstock, is needed to direct the water to the turbine. Depending on the pressure in the pipeline it may be made strong enough to withstand the water pressure caused by the change in head. The pipeline is sometimes buried in order to protect the pipe from mechanical damage.

The turbine is situated after the penstock where fluid energy is converted into mechanical energy to drive the rotor. The discharge from the turbine is collected in a pressure recovery device called the draft tube. This can be further delivered to the discharge tank for fluid collection at the downstream. The turbine and generator are coupled with a shaft. Thus, when the shaft is rotating due to rotor movement, mechanical energy is converted to the electrical energy by a suitable device and supplied to the grid or microgrid for delivery.

In many experiments different turbines in the small-/mini-/microhydro range, namely cross-flow turbine, Turgo turbine, single and multijet Pelton turbines, and Francis turbine, have been used in the past. There is still no definite recommendation for the use of a specific type of turbine.

The electricity generating device, either a generator or motor operating as a generator, may be used to convert the shaft energy to electricity. This part will be discussed in the electrical section 5.4.

5.3.3.3.1   Specific speed

The concept of specific speed helps us to compare different turbomachines for measuring their efficiency and other operating parameters. The specific speed is used to select a particular type of pump. The performance curves of the pumps supplied by the manufacturers make use of this quantity for preparing such documents. The expression for the dimensionless specific speed is given by:

$$N_s = \frac{N\sqrt{Q}}{(gH)^{3/4}}, \text{ for pump} \tag{5.1}$$

$$N_s = \frac{N\sqrt{(P/Q)}}{(gH)^{5/4}}, \text{ for turbine} \tag{5.2}$$

where N is rotational speed in rpm, Q is discharge in m³/s, P is power in W, H is head in m, and g is gravitational acceleration in m/s².

5.3.3.3.2   Energy equation in hydraulic machines

The basic equation of fluid dynamics relating to energy transfer is the same for all rotodynamic machines and momentum balance applied to a fluid element traversing a rotor. In the following equations (5.3–5.5), fluid enters the rotor at 1, at radius r_1 through the rotor by any path and is discharged at 2, at radius r_2. Fluid is guided by the number of blades and transfer of energy takes place by rotation of blades at the angular velocity. For the sake of simplicity it has been assumed that the flow is steady axisymmetric. The number of blades is infinite in the rotodynamic machine with zero thickness in it. The inlet and outlet velocity triangle of the generalized rotor vane is shown in Figure 5.3.

Variation of the velocity occurs only along the radial direction. It is also observed that the relative velocities are tangential to the blades at entry and exit. The radial component v_f is directed radially through the axis to rotation while the tangential component v_w is directed at right angles to the radial direction and along the tangent to the rotor.

Considering the entire fluid body is within the rotor in a control volume, one can write, from the moment of momentum theorem:

$$T = m\left(v_{w_2} r_2 - v_{w_1} r_1\right) \tag{5.3}$$

where T is the torque exerted by the rotor on the moving fluid and m is the mass flow rate of fluid through the rotor. The rate of energy transfer, that is, power, P to the fluid is then given by:

$$P = T\omega = m\left(v_{w_2} r_2 \omega - v_{w_1} r_1 \omega\right) = m\left(v_{w_2} U_2 - v_{w_1} U_1\right) \tag{5.4}$$

where ω is the angular velocity of the rotor and $U = r\omega$, which represents the tangential velocity of the rotor. Therefore, U_2 and U_1 are the tangential velocities of the rotor at points 2 (outlet) and 1 (inlet), respectively. Equation (5.4) is known as Euler's equation

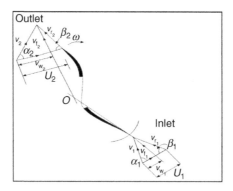

Figure 5.3 Velocity triangle diagram for a generalized rotor vane.

in relation to fluid machines. Equation (5.4) can be written in terms of theoretical head gained H_{th} by the fluid as:

$$H_{th} = \frac{\left(v_{w_2} U_2 - v_{w_1} U_1\right)}{g} \tag{5.5}$$

where shaft power to the rotor $P\ (= \rho g Q H_{th})$ and Q is the discharge through the rotor at the outlet. Equation (5.5) is known as Euler's energy equation for fluid mechanics. The equation relates to the theoretical head developed with the velocities across any type of rotodynamic machine. From the equation, when the right-hand side is positive, that is, $v_{w_2} U_2 > v_{w_1} U_1$, means energy gained by the fluid as in the case of pumps. When the right-hand side is negative, that is, $v_{w_2} U_2 < v_{w_1} U_1$, means H_{th} becomes negative. Thus, energy is being released by the fluid as in the case of turbines. Hence, it can be concluded from Equation (5.5) and Figure 5.1 that for a clockwise rotation, the combination is working as a pump whereas for anticlockwise rotation the same rotodynamic machine is acting as a turbine. This principle has been used in the industry to develop electricity by implementing pump or impeller vanes acting in turbine mode (see Figure 5.2).

5.3.3.3.3 Concept of pump operated as turbine

In many developing countries, the small hydropower stations are in huge demand. Using pump as turbine (PAT) is an attractive and significant alternative for domestic and industrial applications (Figure 5.4). Pumps are mass produced, and as a result have advantages for microhydro compared with custom-made turbines. The main advantages are as follows:

1. Integral with pump and motor combination readily available
2. Wide range of heads and flows
3. Large number of different and standard sizes
4. Low initial cost
5. Short delivery time

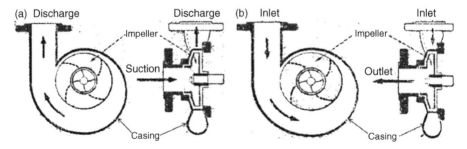

Figure 5.4 The concept of a pump operating as a turbine. (a) A typical pump, (b) a pump working as a turbine.
Adapted from Ref. [1].

6. Spare parts such as seals and bearings are easily available
7. Easy installation

The main limitation is that the range of flow rates over which a particular unit can operate is much less for a conventional turbine. The partial load control in the case of variable load is also not possible as guide vanes are missing for the conventional impeller. Due to inadequate experimental data for pumps working as turbines, field applications of these machines are not yet well defined. Furthermore, the zone of application is very limited as impeller performance is not suitable beyond a certain upper limit.

Still, hydroelectric pumped storage arrangements are some of the most important applications of pumps operating as turbines. This type of arrangement is designed to improve the operation of energy systems.

Not all pumps can be used to work in PAT mode. These reverse running pumps are selected specific to a site based on factors like available range of head, capacity, back pressure at outlet of the turbine, required speed, and the like.

5.3.3.3.4 Selection of appropriate type

This decision on which type of turbine to select depends upon the discharge and head available at the site, and capital and maintenance costs. According to Lueneburg and Nelson, all centrifugal pumps from low to high specific speed, single or multistage, radially or axially split, horizontal or vertical installations can be used in reverse mode. It is possible to use in-line and double suction pumps in turbine mode. However, we need to be aware that they are less efficient in turbine mode.

The general rule of thumb is that multistage radial flow pumps are suitable for high head and low discharge sites, whereas axial flow pumps are appropriate in the low head and high discharge range.

The success of such cost-effective PAT application depends upon the performance prediction before installation. The difficulty to predict performance arises basically because the pump manufacturers generally do not provide the performance curves for reverse mode operation. Though some researchers try to predict PAT performance based on pump performance or specific speeds, these are only approximate and need to be corrected for head and discharge appropriately.

5.4 Electric generators and energy conversion schemes for hydroelectricity

Electric generators are dealt in detail in Chapter 12. As described therein, they may operate in either grid fed or off-grid mode. We might create a microgrid that may involve both. Small hydro systems comprising mini-, micro-, and picohydro units are normally classified under the renewable energy basket, which is our interest here. As already mentioned minihydro units of a few MW operate under grid fed mode, pico units of a few tens of kW operate under off-grid mode and micro units in the mid range may operate in either grid fed or off-grid mode.

The following types of generators are normally employed for grid fed mode:

1. Three-phase synchronous generators (wound field) – both brushed and brushless
2. Three-phase induction generators (squirrel cage)

The following types of generators are normally employed for off-grid mode:

1. Three-phase synchronous generators (wound field) – both brushed and brushless
2. Three -phase capacitor self-excited induction generators (SEIG) with squirrel cage rotor
3. Three -phase SEIG operating in single-phase mode
4. Single-phase SEIG

The following sections describe both grid fed and off-grid small hydro systems.

5.4.1   Grid fed systems

Here, typical unit sizes may vary from a few hundred kW to a few MW feeding converted minihydropower to the local grid. Both wound field synchronous generator (WFSG) and squirrel cage induction generator (SCIG) can be used. In this scheme, the hydro turbine operated by water power drives the generator either directly or through some speed enhancing mechanism. A step-up transformer is interfaced to match the grid voltage with the generator voltage. Power P in W in a hydroturbine is given by:

$$P = \rho g Q H \tag{5.6}$$

where ρ is water density in kg/m^3, g is gravitational acceleration in m/s^2, Q is discharge in L/s, and H is head in m.

Normally, small uncontrolled turbines operate at near constant head and discharge and hence produce near constant power in a given season. Thus, the power fed to the grid from a generator is almost constant. In large hydro schemes turbine blades are controlled to vary power input.

5.4.1.1   Synchronous generator

The WFSG driven by a hydroturbine is shown in Figure 5.5. As explained, power (P) fed to the grid is nearly constant dependent on hydropower input unless there

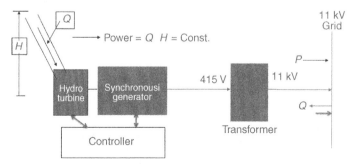

Figure 5.5 Hydroturbine-driven WFSG feeding power to grid.

is turbine control. But the lagging reactive power or VAR(Q) drawn by the grid can be controlled by field current (I_f). With low I_f or an underexcited condition, Q drawn by the grid is positive and PF is lagging at the grid. With high I_f or overexcited condition, Q drawn by the grid is negative and PF is leading at the grid. Thus, there is one value of I_f at which the PF is unity when $Q = 0$. VAR control is an important feature of minihydro generator performance and WFSG provides considerable flexibility. A major drawback is the need for synchronization with the grid every time the synchronous generator is connected to grid through adjusting voltage, frequency, and phase positions. Another lacuna is high short-circuit currents. WFSG needs a slip ring and brush arrangement, which needs regular maintenance unless a brushless arrangement with rotating diodes is used. An automatic voltage regulator (AVR) for stabilizing grid voltage and Q control is an integral part of this system that adds to complexity. The controller shown in Figure 5.5 involves AVR to control generated voltage. In large hydro it may contain turbine control to orient the blades. A controller operates by receiving feedback signals from generator/turbine.

5.4.1.2   Induction generator

SCIG is also suited for grid connected small hydro systems whose schematic is shown in Figure 5.6. To reduce reactive power (VAR) drain from the grid terminal capacitors are used. Since power fed to the grid is constant, slip too is constant. Q control is made through capacitors whose value can be fixed based on desired PF at the grid. SCIG is simpler compared to a synchronous generator due to brushless cage construction with reduced unit cost and ease of maintenance. There is no synchronizing need. Performance under short circuit is better compared to that with a synchronous generator. An induction generator can operate with variable speed controlled by input power. It can start as a motor and run as a generator with input water flow.

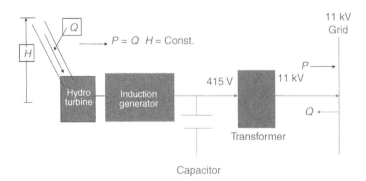

Figure 5.6 SCIG in a minihydro system.

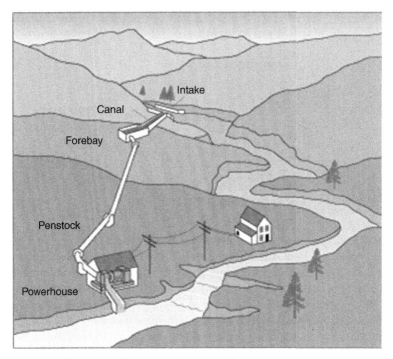

Figure 5.7 An artistic view of a microhydro off-grid scheme.

5.4.2   Off-grid (microgrid) systems

Renewable energy is a preferred option for off-grid systems especially to energize remote and rural areas where grid connection is uneconomical, impracticable, or inaccessible. But the challenge lies in randomly varying power of the source and the load with a robust controller to match the two. The off-grid small hydro systems are often classified as pico- or microhydro as per unit sizes typically varying from 1 kW to 100 kW, the lower end being pico and the rest micro. For such standalone power generation, SEIG is a strong candidate. An artistic view of such a microhydro off-grid scheme is given in Figure 5.7. Water from a natural river is diverted to a fixed level "forebay tank" at a favorable height through a "canal." Water from the forebay is fed to a hydro turbine through a penstock pipe as shown. This drives the generator to distribute generated power to local isolated loads through a microgrid.

A schematic of a standalone micro-/picohydro system is given in Figure 5.8. As explained it is a near constant power (P) scheme as P varies as product of head (H) and discharge (Q) are constant. SEIG is ideally suited to this scheme as shown, which supplies converted power to the variable consumer load. This creates a power unbalance that can be rectified by an electronic load controller (ELC) as shown, which dumps the surplus power (over the demanded load power) to a dummy load. The ELC ensures that the total power P is always constant by adjusting ELC power P_2 as load power P_1

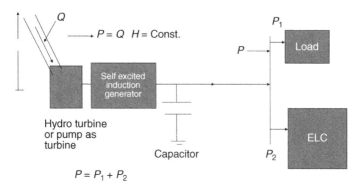

Figure 5.8 Standalone micro-/picohydro system.

varies. A practical scheme with necessary hardware developed by the first author is shown in Figure 5.9. Here, water source drives the microhydroturbine, which in turn drives the SEIG that outputs electric power. The connected capacitor provides self-excitation to generate voltage across the load. The SEIG feeds both consumer loads and the ELC such that the ELC adjusts total power as consumer power changes. We may have both single-phase and three-phase schemes, the latter will use three-phase SEIG and ELC, while loads may be both single phase and three phase. Proper load balancing across phases is desired. Figure 5.10 shows a working model of a turbine/SEIG assembly installed in the field to electrify a remote village.

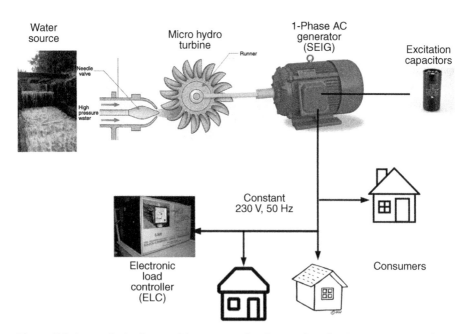

Figure 5.9 A practical scheme with necessary hardware of a microhydro system.

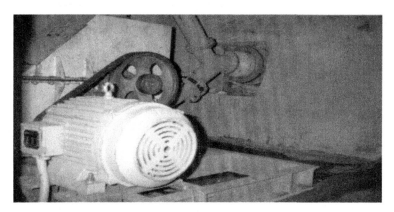

Figure 5.10 Turbine/SEIG assembly installed in the field.

5.5 Summary

Hydroelectricity is an important component of renewable energy systems. Small hydro systems up to a few MW of unit sizes are eco-friendly and installed in large numbers all over the world both under grid fed and off-grid modes. Earth is endowed with enormous water power distributed in almost all regions mainly as mountain based and flowing rivers. Suitable redirecting of water through proper construction is essential. Hydro systems broadly comprise hydroturbines and generators with associated civil works and controls to feed quality power to consumers. This chapter explained different turbines classified under impulse and reaction type. A pump can also be operated as a turbine in reverse power mode. Both synchronous and induction generators can be employed to convert turbine power to electricity. Grid fed systems use both these types while induction generators are preferred for off-grid use. Appropriate electronic controllers must be developed for each application.

Reference

[1] Jain SV, Patel RN. Investigations on pump running in turbine mode: a review of the state-of-the-art. Renew Sustain Energy Rev 2014;30:841–68.

Fuel cells

6

M. Hashem Nehrir*, Caisheng Wang**
*Electrical & Computer Engineering Department, Montana State University, Bozeman, MT, USA
**Electrical & Computer Engineering Department, Wayne State University, Detroit, MI, USA

Chapter Outline

6.1   Introduction

Fuel cells are static energy conversion devices that convert the chemical energy of a fuel directly into direct current (DC) electrical energy. Unlike a battery that stores electrical energy, a fuel cell converts the chemical energy of its input fuel into electrical energy as long as fuel is provided to the cell. Fuel cells can be used, normally via power electronic converters, for a wide variety of applications, for example, electric transportation as main and/or auxiliary power sources, stationary power for buildings, cogeneration applications, and power generation sources for standalone systems (e.g., standalone microgrids) and grid-connected applications.

The main focus of this chapter is stationary power generation applications of fuel cells. Figure 6.1 shows the major processes of a generic fuel cell power generation system, where fuel (e.g., natural gas) containing hydrocarbons is fed to the fuel processor

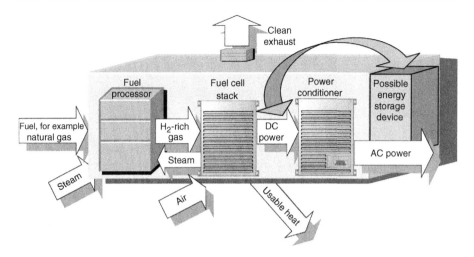

Figure 6.1 Major components and processes in a generic fuel cell energy system [1].

to be cleaned and converted into a hydrogen-rich gas. Through electrochemical energy conversion, the energy of hydrogen is converted into DC electricity. These fuel cells are bundled together in series and parallel combinations (called fuel cell stack) to produce the desired power and voltage for a particular application. The power-conditioning unit converts electric power from DC into regulated DC or AC for consumer use. An energy storage device may also be a part of the fuel cell system for energy management and/or prevention of any type of transient or disturbance, which may adversely affect the performance of the fuel cell system. The energy flow between output of the fuel cell stack and storage system can be bidirectional. The by-products of the fuel cell energy system include heat and clean exhaust, which can be used for water or space heating, or produce additional electricity.

6.2 Fuel cell fundamentals

The main fuel for fuel cells is hydrogen, though some fuel cells can work on natural gas and/or methanol directly (such as a direct methanol fuel cell); it can be pure hydrogen or that obtained from other fuels, such as natural gas and methanol. In such cases, a reformer is required to break up the fuel and generate hydrogen to be used by the fuel cells. The hydrogen fuel combines with oxygen in air inside the fuel cell and produces electricity, water (steam), and heat, as shown in Figure 6.1.

In a simple fuel cell, hydrogen molecules are split at the anode, through electrochemical processes, to generate hydrogen ions and electrons. The hydrogen ions move from the cathode to the anode through the membrane (electrolyte), but the electrons cannot. The electrons travel through an external electrical circuit (load) to recombine with the hydrogen protons and oxygen molecules at the cathode to produce water. The chemical reactions at the anode and cathode are given in (6.1) and (6.2), respectively.

$$2\,H_2 \Rightarrow 4\,H^+ + 4\,e^- \tag{6.1}$$

$$O_2 + 4\,H^+ + 4\,e^- \Rightarrow 2\,H_2\,O + \text{heat} \tag{6.2}$$

The polarity of an ion and its transport direction can differ across different fuel cells, determining the site of water production and removal. If the ions transferred through the electrolyte are positive, as shown in Figure 6.2, then water is produced at the cathode. On the other hand, if the working ion is negative, as in solid-oxide and molten-carbonate fuel cells (discussed later in this chapter), water is formed at the anode. In both the cases, electrons pass through an external circuit and produce electric current.

6.2.1   Fuel cell types

There are six major types of fuel cells, classified by their electrolyte, such as follows.

1. Polymer electrolyte membrane (also called proton exchange membrane) fuel cell (PEMFC)
2. Alkaline fuel cell (AFC)
3. Phosphoric acid fuel cell (PAFC)
4. Direct methanol fuel cell (DMFC)
5. Solid oxide fuel cell (SOFC)
6. Molten carbonate fuel cell (MCFC).

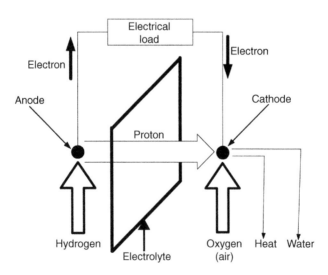

Figure 6.2 Diagram of a simple hydrogen fuel cell.

The first four are low-temperature types and operate at 200°C or lower, whereas the last two operate at temperatures in the 600–1000°C range. The high-temperature fuel cells (SOFC and MCFC) are commonly used in large capacity (greater than 200 kW) distributed power generation applications. The only low-temperature fuel cell, commonly used in smaller-scale (residential level) and dominating in vehicle applications, is PEMFC, which has a solid electrolyte. A brief discussion of the different fuel cell types is given further. The oxidation reactions, range of their operating temperatures, and efficiency are given in Table 6.1.

PEMFC was first developed by General Electric in the 1960s for use by NASA on its first manned spacecraft. It employs a thin (plastic-like wrap) ion-conducting polymer as electrolyte that allows the passage of hydrogen ions and blocks the passage of electrons. The schematic diagram showing a chemical reaction and electron/ion flow of PEMFC is given in Figure 6.3, which is similar to Figure 6.2. PEMFC uses a small amount of platinum as catalyst to speed up the splitting of hydrogen molecules to positive ions and electrons.

AFC is the oldest type of fuel cell, introduced in the early 1900s, but it was in the mid-1900s that it proved to be a viable power source. AFC was used in the Apollo mission that took man to the moon. The electrolyte of AFC is alkaline solution; potassium hydroxide (KOH) solution is the most commonly used. AFC typically operates on pure hydrogen and oxygen to avoid the formation of any carbon dioxide (CO_2), because the reaction of KOH and CO_2 results in potassium carbonate (K_2CO_3), which can poison the fuel cell. This also limits the application of AFC within a controlled environment (e.g., aerospace applications) though it has very high efficiencies.

PAFC was the first type of fuel cell to be produced in commercial quantities. Phosphoric acid, which is used as the electrolyte in PAFC, is the only acid that has enough thermal and chemical stability and low volatility at the operating temperature of PAFC (around 200°C). Unlike alkaline solutions used as electrolytes in AFC, phosphoric acid is tolerant to CO_2, which may be present in the hydrogen fuel. This feature makes PAFC more adaptable in commercial applications. Like PEMFC, hydrogen is split into hydrogen ions (protons) and electrons at the cathode; the hydrogen ions pass through the PAFC's electrolyte and electrons flow through the external load to recombine with hydrogen ions and oxygen at the cathode to produce water.

DMFC is a subcategory of PEMFC in which methanol (CH_3OH) is used as the fuel. Methanol is a high-energy-density alternative fuel, attractive for portable electronic applications, which require low power and high energy (i.e., for a long duration). In addition to methanol fuel, DMFC requires water as an additional reactant at the anode. In DMFC, hydrogen ions are produced at the anode, which transport through the ion-conducting electrolyte to reach the cathode and combine with the electrons, which flow through the external load, to reach the cathode. The by-product (waste) is the unattractive and not environmental friendly CO_2. The efficiency of DMFC is relatively low (up to 40%) compared to other types of fuel cells, discussed here.

SOFC uses a thin ceramic membrane as an electrolyte that allows the passage of oxygen ions (O_2^-) produced at the cathode by the combination of oxygen molecules and the electrons that enter the cathode. It is commonly used in distributed power generation applications. Because of its high operating temperature (600–1000°C), high

Table 6.1 **Comparison of the chemical reaction of major types of fuel cells [1]**

Type	Mobile ion	Cathode reaction	Anode reaction	Overall reaction	Efficiency (%)	Operating temperature range (°C)
AFC	OH^-	$1/2O_2 + H_2O + 2e^- \Rightarrow 2(OH)^-$	$H_2 + 2(OH)^- \Rightarrow 2H_2O + 2e^-$	$H_2 + 1/2O_2 \Rightarrow H_2O$	60–70	50–200
PEMFC	H^+	$1/2O_2 + 2H^+ + 2e^- \Rightarrow H_2O$	$H_2 \Rightarrow 2H^+ + 2e^-$	$H_2 + 1/2O_2 \Rightarrow H_2O$	35–60	30–100
PAFC	H^+	$1/2O_2 + 2H^+ + 2e^- \Rightarrow H_2O$	$H_2 \Rightarrow 2H^+ + 2e^-$	$H_2 + 1/2O_2 \Rightarrow H_2O$	~40	Around 200
DMFC	H^+	$3/2O_2 + 6H^+ + 6e^- \Rightarrow 3H_2O$	$CH_3OH + H_2O \Rightarrow 6H^+ + 6e^- + CO_2$	$CH_3OH + 3/2O_2 \Rightarrow 2H_2O + CO_2$	Up to 40%	50–130
MCFC	CO_3^{2-}	$1/2O_2 + CO_2 + 2e^- \Rightarrow CO_3^{2-}$	$H_2 + CO_3^{2-} \Rightarrow H_2O + CO_2 + 2e^-$	$H_2 + 1/2O_2 + CO_2 \Rightarrow H_2O + CO_2$	50–60*	Around 600
SOFC	O_2^-	$1/2O_2 + 2e^- \Rightarrow O_2^-$	$H_2 + 1/2O_2^- \Rightarrow H_2O + 2e^-$	$H_2 + 1/2O_2 \Rightarrow H_2O$	45–65*	500–1000

AFC, alkaline fuel cell; PEMFC, polymer electrolyte membrane fuel cell; PAFC, phosphoric acid fuel cell; DMFC, direct methanol fuel cell; MCFC, molten carbonate fuel cell; SOFC, solid oxide fuel cell.
*Efficiency of these fuel cells could reach or exceed 80% in CHP operation mode.

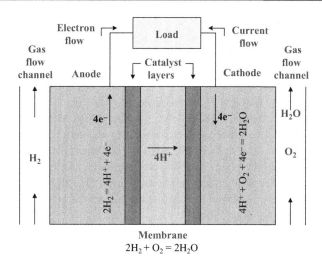

Figure 6.3 Schematic diagram and chemical reactions of a PEMFC [1].

reaction rates can be achieved without expensive catalysts needed for low-temperature fuel cells. Additionally, gases such as natural gas can be used directly as fuel, without the need for a reformer. The high temperature splits the gas and frees hydrogen molecules at the anode to combine with oxygen ions that enter the anode and produce water, as shown in Figure 6.4. Another useful feature of SOFC as a high-temperature fuel cell is the utilization of exhaust heat, which can be used in conjunction with the generated electricity as "combined heat and power" (CHP). The overall efficiency of an SOFC system can be increased from 60–65% (for generated electricity only) to 75–80% in CHP mode.

MCFC is also a high-temperature (typically 600–700°C) fuel cell used for stationary distributed power generation applications. It uses a molten mixture of alkaline

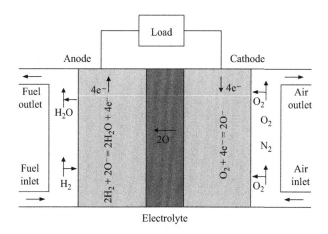

Figure 6.4 Schematic diagram and chemical reactions of a SOFC [1].

metal carbonate as electrolyte and an inexpensive catalyst (nickel). At high temperatures, the alkaline salt mixture is in liquid phase and is an excellent conductor of carbonate ions (CO_3^{2-}). At the cathode, the oxygen and carbon dioxide (CO_2) combine with electrons to produce CO_3^{2-} ions. At the anode, the CO_3^{2-} ions combine with hydrogen fuel and produce carbon dioxide, water, and electrons, which flow through the external load on the fuel cell to produce electricity and reach the anode. The high-temperature operation range of MCFC also gives it fuel flexibility. While MCFCs can run on hydrogen, like other fuel cells, they can also use hydrocarbons (natural gas, methane, or alcohol) as fuel. MCFC operates at a typical electrical efficiency of 50–60%, but its overall efficiency in CHP operation mode can reach as high as 80%. The schematic diagram and chemical reactions of an MCFC are shown in Figure 6.5.

The chemical reaction, mobile ion, and operating temperature range of different types of fuel cells are given in Table 6.1.

The fuel cell is a dynamic research area. Active research is continuously being carried out to improve the durability and reduce cost of fuel cells from various aspects, such as new materials for electrodes, membranes/electrolytes and catalysts, and new manufacturing technologies (e.g., nanotechnology based). The performance parameters and operation characteristics of fuel cells reviewed in this chapter might require updation over time, due to new possible achievements in the near future.

6.3   Modeling of ideal fuel cells

Fuel cell modeling involves the modeling of thermal, chemical, and electric processes. It is difficult to explain these processes in the limited space of this chapter. A brief qualitative description of these processes is given in the following subsections.

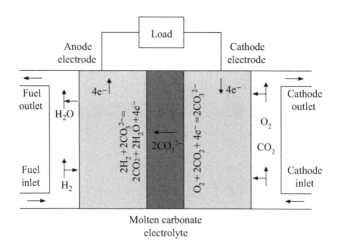

Figure 6.5 Schematic diagram of a MCFC [1].

Interested readers are encouraged to refer textbooks on fuel cell fundamentals, similar to those given in the references [1–3] given at the end of this chapter.

6.3.1 Thermal–electrical processes of fuel cells

Fuel cells are energy conversion devices converting the energy stored in the fuel into electrical energy and heat through electrochemical and thermal processes. The electrochemical processes for different types of fuel cells are briefly covered in the previous section. In this subsection, thermodynamics governing the transformation of energy stored in the fuel to electrical energy, has been briefly described. The thermodynamics of fuel cells can predict the ideal maximum electrical output that can be generated in a reaction.

Rules, called *thermodynamic potentials*, can be written to specify how energy is transferred from one form to another. These potentials are described further [2].

Internal energy (U) is the energy needed to create a system without any change in the temperature and volume of the system. That is, the change in internal energy of the system is equal to the change in heat transferred to the system (dQ), lesser the work done by the system (dW):

$$dU = dQ - dW \qquad (6.3)$$

where dU is the change in internal energy that can be transferred between the system and its surrounding through heat (dQ) or work (dW).

Enthalpy (H) is the energy needed to create a system plus the work needed to make room for it at volume V and pressure P:

$$H = U + PV \qquad (6.4)$$

Helmholtz free energy (F) is the energy needed to create a system, less the energy that can be obtained from the system's environment at a constant temperature T:

$$F = U - TS \qquad (6.5)$$

where S is the entropy of the system. Entropy can be interpreted as how change in heat transfer within a system can take place at constant temperature. For a reversible (ideal) heat transfer at constant pressure, the system entropy can change as:

$$dS = \frac{dQ_{rev}}{T} \qquad (6.6)$$

where dS is the change in entropy of the system for a reversible heat transfer (dQ_{rev}) at a constant temperature T. For a thorough explanation on entropy of a system, interested readers can refer any standard book on thermodynamics.

Gibbs free energy (*G*) is the energy needed to create a system, less the energy that is obtained from its environment due to heat transfer. It represents the work potential of the system:

$$G = U + PV - TS = H - TS \tag{6.7}$$

It can be shown that for an electrochemical reaction (as in fuel cells), the maximum electricity production (W_e) is equal to the change in the Gibbs free energy [2]:

$$W_e = -\Delta G = n_e F E \tag{6.8}$$

where n_e is the number of participating electrons, F is Faraday's constant (96,485.3 C/mol), and E is the potential difference across the electrodes. Therefore, as a result of thermochemical reaction in fuel cells, a potential difference is induced across the fuel cell electrodes:

$$E = \frac{-\Delta G}{n_e F} \tag{6.9}$$

The above potential difference is a function of the fuel cell temperature and pressure of hydrogen and oxygen gases inside the fuel cell [2].

6.3.2   *Fuel cell equivalent circuit*

Like in a battery, the voltage at the terminals of a fuel cell is lower than the internal voltage induced in the cell. This is because of different losses due to electrochemical reactions and the ohmic loss, which is due to a current flow over resistance. There are three different losses (voltage drops) caused due to activation, ohmic, and concentration voltage drops inside a fuel cell, as shown in Figure 6.6. These voltage drops are

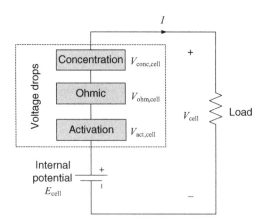

Figure 6.6 Fuel cell voltage drops [1].

functions of load current and fuel cell temperature and/or pressure. The ohmic voltage drop is a linear function of fuel cell load current at a certain operating point, but the ohmic resistance ($R_{ohm,cell}$) is normally a function of fuel cell temperature. The activation and concentration voltage drops are nonlinear functions of load current as well as pressure and/or temperature inside the fuel cell.

From Figure 6.6, the output voltage of a fuel cell can be written as [1]:

$$V_{cell} = E_{cell} - V_{act,cell} - V_{ohm,cell} - V_{conc,cell} \tag{6.10}$$

where V_{cell} and E_{cell} are the fuel cell output voltage and internal voltage, respectively, and $V_{act,cell}$, $V_{ohm,cell}$, and $V_{conc,cell}$ are the voltage drops discussed earlier.

Assuming that the parameters for individual cells can be lumped together to represent a fuel cell stack, we can obtain the output voltage of a fuel cell stack as:

$$V_{out} = N_{cell} V_{cell} = E - V_{act} - V_{ohm} - V_{conc} \tag{6.11}$$

where V_{out} is output voltage of a fuel cell stack (V); N_{cell} is number of cells in the stack; E is fuel cell stack internal potential (V); V_{act} is the overall activation voltage drop (V); V_{ohm} is the overall ohmic voltage drop (V); and V_{conc} is the overall concentration voltage drop (V).

The activation voltage drop is a function of the fuel cell current and temperature, as described empirically by the Tafel equation given here:

$$V_{act} = \frac{RT}{\alpha zF} \ln\left(\frac{I}{I_0}\right) = T\left[a + b\ln(I)\right] \tag{6.12}$$

where α is the electron transfer coefficient; I_0 is the exchange current (A); R is the gas constant, 8.3143 (J/(molK)); T is the temperature in Kelvin; z is the number of participating electrons; and a and b are two empirical constants.

V_{act} can be further described as the sum of V_{act1} and V_{act2} as:

$$V_{act} = \eta_0 + (T - 298)a + Tb\ln(I) = V_{act1} + V_{act2} \tag{6.13}$$

where η_0 is temperature invariant part of V_{act} (V). $V_{act1} = (\eta_0 + (T-298) \times a)$ is the voltage drop affected only by the fuel cell internal temperature (that is not current dependent), and $V_{act2} = (T \times b \times \ln(I))$ is both current and temperature dependent.

The equivalent resistance of activation is defined by the ratio of V_{act2} and the fuel cell current. From Equation (6.14), it is noted that the resistance is both temperature and current dependent:

$$R_{act} = \frac{V_{act2}}{I} = \frac{Tb\ln(I)}{I} \tag{6.14}$$

The overall ohmic voltage drop can be expressed as:

$$V_{ohm} = IR_{ohm} \tag{6.15}$$

where R_{ohm} is a function of current and temperature and can be expressed by [1],

$$R_{ohm} = R_{ohm0} + k_{RI}I - k_{RT}T \tag{6.16}$$

where R_{ohm0} is the constant part of R_{ohm}; k_{RI} is the empirical constant for calculating R_{ohm} (Ω/A); and k_{RT} is the empirical constant for calculating R_{ohm}(Ω/K).

The concentration overpotential in the fuel cell is defined as follows [1]:

$$V_{conc} = -\frac{RT}{zF}\ln(1 - \frac{I}{I_{limit}}) \tag{6.17}$$

where I_{limit} is the fuel cell current limit (A).

The corresponding equivalent resistance for the concentration loss can therefore be defined as:

$$R_{conc} = \frac{V_{conc}}{I} = -\frac{RT}{zFI}\ln(1 - \frac{I}{I_{limit}}). \tag{6.18}$$

Because of the structure of fuel cells, where the electrodes are separated by electrolyte/membrane, which only allows one form of ion (positive or negative) flow through it, there is always charge accumulation on both sides of the electrolyte/membrane, which results in energy stored in the electric field inside cell. This phenomenon is represented by a capacitor, representing the *double-layer charge effect* [1,3]. The equivalent circuit of a fuel cell considering the double-layer charge effect is shown in Figure 6.7, where R_{act}, R_{conc}, and R_{ohm} are resistances corresponding to the activation, concentration, and ohmic voltage drops, C is the capacitance of the double-layer charge effect, and E is the

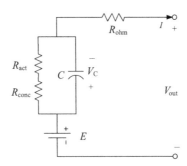

Figure 6.7 Equivalent circuit of the double-layer charging effect inside a fuel cell [1].

cell internal voltage (6.11). From Figure 6.7, the voltage across the capacitor and cell output voltage can be written as follows:

$$V_C = (I - C \frac{dV_C}{dt})(R_{act} + R_{conc}) \tag{6.19}$$

$$V_{out} = E - V_{act1} - V_C - V_{ohm} \tag{6.20}$$

where V_{act1} is the temperature-dependent part of V_{act}.

In practice, the capacitance of double-layer charge effect appears only in the transient response of the fuel cell. The fuel cell voltage is constant DC at steady state, and there is no current through the capacitor. Therefore, the capacitor acts as an open circuit and does not affect the fuel cell steady-state response.

6.3.3 Steady-state electrical characteristics of fuel cells

Assuming that the capacitance in Figure 6.7 is fully charged and acts as an open circuit at steady state, the fuel cell output voltage can then be obtained by (6.11). Since all the voltage drops in (6.11) are a function of load current and temperature, the output terminal voltage of the fuel cell is also a nonlinear function of load current and temperature.

6.3.4 Modeling of actual fuel cells

Modeling of different types of fuel cells depends on their electrochemical characteristics and can be different from one another. Therefore, to be accurate, each fuel cell type should be modeled independently. In Section 6.3.4 the steady-state model response and actual performance of a 500-W Avista-Labs PEM fuel cell, shown in Figure 6.8 and reported in Ref. [1], has been given.

Figure 6.8 Avista Labs (now ReliON) SR-12 500-W PEMFC stack.

The block diagram for building the electrical circuit model for PEMFC, considering the electro-thermo-chemical characteristics of the fuel cell, is shown in Figure 6.9. The output voltage of the fuel cell as a function of load current, which is based on (6.20), and its output power (the product of the output voltage and load current), simulated in Matlab/Simulink©, are shown in Figure 6.10 [1]. The active, ohmic, and concentration zones of the fuel cell are also shown in the Figure 6.10. It is noted that the output voltage of the fuel cell decreases as its load current increases, and the maximum power it can deliver is almost near the rated current of the fuel cell, just before the fuel cell enters the concentration mode. This phenomenon holds for different types of fuel cells.

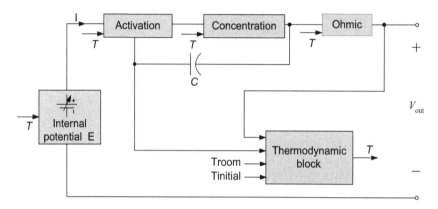

Figure 6.9 Block diagram for building an electrical circuit model for PEMFC.

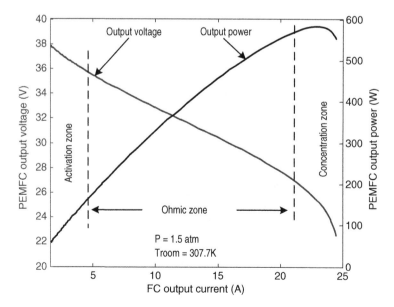

Figure 6.10 Simulated *V–I*- and *P–I*-characteristic curves for the 500-W PEM fuel cell stack [1].

Figures 6.11 and 6.12 show comparisons of the average *V–I* and *P–I* characteristic curves for the SR-12 PEMFC stack, obtained experimentally, and the Simulink and PSpice simulation results obtained for the models developed [1]. The lower and upper curves shown in these figures are the lower and upper ranges of the experimental data, which had ripples. The raw data with ripples have been filtered out and average characteristic curves are shown for easier comparison. For further information about the details of the model development and experimental results interested readers are referred to [1].

6.4   Advantages and disadvantages of fuel cells

Fuel cells have some characteristics and advantages over both internal combustion engines (ICEs) and batteries. Like ICEs, fuel cells convert the energy of a fuel (hydrogen) to electricity, and like batteries, the conversion relies on electrochemistry. However, fuel cells are zero (or near zero) emission-producing and quiet electricity-producing devices, because they directly convert the chemical energy of the fuel into electricity, and do not have any moving components in the core part of the system. They are also generally more efficient than ICEs and can be scaled to provide a desired power capacity as low as 1 W (e.g., in a cell phone) to MW range (e.g., in power applications) by connecting the fuel cell stacks in series and parallel. Unlike batteries, fuel cells do not store electrical energy, but generate electric power. They can generate

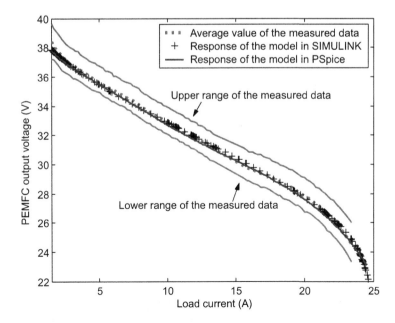

Figure 6.11 *V–I* characteristics of an SR-12 PEMFC and the Simulink and PSpice models.

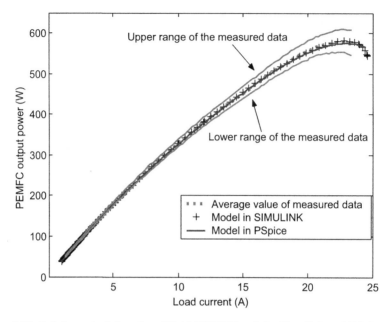

Figure 6.12 *P–I* **characteristics of an SR-12 PEMFC and the Simulink and PSpice models.**

electricity as long as fuels are available while batteries need to be charged after being discharged.

High cost is one of the major disadvantages of fuel cells, which has prevented their widespread use. In addition, fuel cells are not fast-reacting power-producing devices. In general, they have a slow response to fast changing loads and load transients, and because of this limitation, as shown in Figure 6.1, a fast power producing device (e.g., a battery) should be put in parallel with standalone fuel cells, and properly controlled, to respond to fast load demands. Fuel cell durability can also be reduced due to repeated start–stop applications.

6.5   Power applications of fuel cells

The most common type of fuel cells, which have large-scale power applications are PEMFC, SOFC, and MCFC. They can be connected to the power grid or operated independently in standalone mode. In both configurations, they are often referred to as fuel cell distributed generation systems. These two operation modes are discussed below.

6.5.1   Grid-connected fuel cell configuration

The output voltage of a fuel cell is DC. Therefore, a fuel cell power plant is normally interfaced with the power grid through power-electronic interfacing devices.

The interface is very important as it affects the operation of the fuel cell system as well as the power grid. DC/DC converters are necessary to boost and regulate fuel cell output voltage and adapt it to a voltage that can be supplied to a DC/AC converter (inverter) to convert the regulated DC voltage of the fuel cell, to a desired AC voltage. The harmonics of AC voltage are then filtered and the voltage is stepped up, if needed, through a transformer and connected to the grid through a transmission line. The AC voltage must be synchronized with the grid. Figure 6.13 shows the block diagram of a grid-connected fuel cell stack with the power electronic converters, electrical filter, transformer, and transmission line. It should be noted that in a fuel cell power plant, a number of fuel cell stacks are connected in series to provide the required voltage and a number of stacks are connected in parallel to provide the rated current. The regulated output voltage of the DC/DC converter is stabilized with a supercapacitor or battery bank, and converted into AC voltage. The AC voltage is then filtered, stepped up, and connected (synchronized) to the utility grid. The amount of real and reactive power delivered by the fuel cell to the grid can be controlled by the controller of the inverter and by controlling the fuel cell stacks.

6.5.2 Standalone fuel cell configuration

Standalone fuel cell systems have applications in remote areas and islands, as backup power, and in transportation. Block diagram of a standalone fuel cell system is shown in Figure 6.14. The components of these systems are basically the same as those for the grid-connected system, except that the system serves independent loads and is not connected to the utility grid. A battery bank is also used to provide sufficient storage capacity to handle fast power variations due to load transients. An inverter is to invert the DC voltage to AC when fuel cell is to supply AC loads.

6.6 FC and environment: hydrogen production and safety

Hydrogen has the highest energy content or specific energy density (120 MJ/kg) of all fuels. It is also a zero-emission fuel, which can be used in fuel cells to produce DC electricity, or in ICEs to power vehicles, for example. It can also be used in the

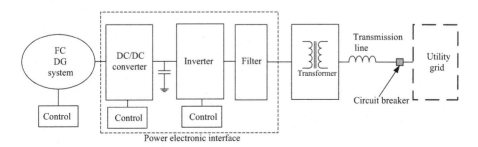

Figure 6.13 Block diagram of a grid-connected fuel cell DG system.

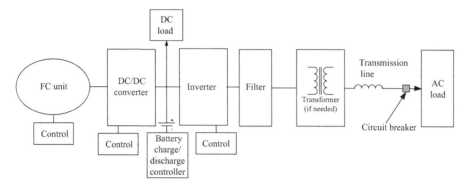

Figure 6.14 Block diagram of a standalone fuel cell DG system.

propulsion of spacecraft and can potentially be mass produced and commercialized for electricity production, passenger vehicles, and aircraft, that is, in a hydrogen economy society.

6.6.1 Hydrogen production

The major existing and potential technologies for producing hydrogen are listed:

- Production by reforming natural gas
- Conversion of coal to hydrogen
- Production by using nuclear energy
- Electrolysis of water through an electrolyzer, using electricity from the grid
- Hydrogen production from renewable energy

In this section the production of hydrogen from electrolysis of water using grid electricity and from renewable energy (the last two bullets) are briefly covered. Hydrogen generation using nuclear energy is also based on water electrolysis. Descriptions of hydrogen production by means other than those listed above are discussed in [1,4].

6.6.1.1 Hydrogen production through water electrolysis

The process of separating water molecules into hydrogen and oxygen is called electrolysis. This process has been in use around the world, primarily in chemical plants, to meet their hydrogen needs. The electrochemical device in which electrolysis takes place is called an *electrolyzer*. The electrochemical process in an electrolyzer is basically the reverse of that in fuel cells. Therefore, an electrolyzer converts DC electrical energy into chemical energy stored in hydrogen. Figure 6.15 shows the schematic diagram of an alkaline electrolyzer, where two water molecules are combined with two electrons arriving at the cathode through the DC source to produce one molecule of hydrogen and two OH$^-$ ions as follows:

$$2\,H_2O + 2\,e^- \rightarrow H_2 \uparrow + 2\,OH^- \tag{6.21}$$

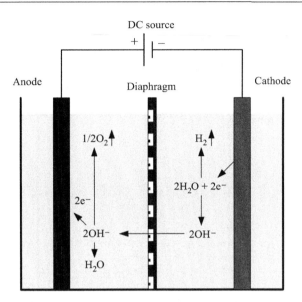

Figure 6.15 Schematic diagram of an alkaline electrolyzer.

The ions are passed through the porous diaphragm (electrolyte) and migrate toward the anode, where the OH⁻ ions are discharged into oxygen gas, water, and two electrons, which migrate to the cathode through the DC source.

In a hydrogen economy, electrolyzers can be used in residential and commercial buildings to produce hydrogen for fuel cells. They can also be placed in the existing service stations to produce hydrogen for fuel cell vehicles.

6.6.1.2 Hydrogen production from renewable energy

Renewable energy resources such as wind, solar, and biomass are promising energy resources to be used for hydrogen production. Wind- and solar photovoltaic (PV)-generated electricity can be used for water electrolysis for the production of a significant amount of hydrogen. Figure 6.16 shows the schematic diagram of a standalone hybrid wind–solar–fuel-cell–electrolyzer energy system with the associated power electronic (AC/DC, DC/AC, and/or DC/DC) converters. The system can be designed such that the wind-/solar-generated power supplies the load, and any excess available power is used first to charge the battery and then supplied to the electrolyzer to generate hydrogen, which is compressed and stored in the reservoir and backup tanks, respectively. During periods when wind- and solar-generated power is not sufficient to supply the load, hydrogen fuel is used to power the fuel cell to generate electricity.

Biomass energy can also be used to produce hydrogen through biomass gasification. There are two types of biomass feedstocks available for conversion to hydrogen: the bioenergy of crops and the organic waste from agricultural farming and wood processing (referred to as biomass residues). The ultimate energy source for hydrogen production from the above sources is actually solar energy.

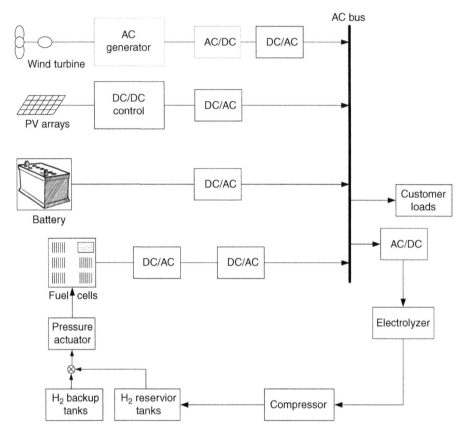

Figure 6.16 Schematic diagram of a standalone hybrid wind-solar-fuel cell energy system.

The challenge facing the above hydrogen production technologies is the low efficiency of the processes and their current costs. Assuming the efficiency of wind and solar energy conversion systems to be around 30% and the efficiency of electrolyzers to be around 50%, the efficiency of wind and solar to hydrogen is around 15%. The efficiency of biomass–hydrogen energy conversion is even lower, due to relatively low hydrogen content and low energy content in biomass.

Hydrogen can also be directly produced through solar-heat energy conversion systems at higher efficiencies. Water (steam) begins decomposing into hydrogen and oxygen around 2000°C without the need for electric current; high-temperature steam can be obtained from concentrated solar thermal plants.

6.6.2 Hydrogen safety

Hydrogen is poised to play an important role in the world's energy future. If handled properly, the production, storage, and distribution of hydrogen is regarded as safe

[4,5]. Hydrogen is 14.4 times lighter than air, rises at about 20 m/s (45 mph), and quickly dilutes and escapes; it diffuses in air about 3.8-times faster and rises about 6-times faster than natural gas. Therefore, hydrogen rapidly escapes upwards if accidentally released, compared to gasoline that is heavy and spreads around. Furthermore, hydrogen combustion is emission free and produces water vapor only. Based on the above features, hydrogen can be safe if handled properly and has the potential to be the fuel of choice for a future emission-free society.

6.7   Hydrogen economy

Hydrogen economy is an economy that relies on hydrogen as the commercial fuel that would deliver a substantial fraction of a nation's energy and services. This vision can become a reality if hydrogen can be produced from domestic energy sources economically and in an environmental-friendly manner. Fuel cell technology should also become mature and economical so that fuel cells and fuel cell vehicles can gain market share in competition with conventional power generation sources and transportation vehicles. In that way, the entire world would benefit from lower dependence on oil and coal as the major sources of energy and improved environmental quality through lower carbon emissions. However, before this vision can become a reality and the transition to such an economy can take place, many technical, social, and policy challenges must be overcome.

Hydrogen is an energy storage medium, that is, an energy carrier, not a primary energy source. It has the potential for use as a fuel in a variety of applications, including fuel cell power generation and fuel cell vehicles. It is combustible and can therefore be used as fuel in conventional ICEs to produce mechanical or electrical power. In this case the overall energy efficiency is higher than ICEs that operate with conventional fuels such as diesel or gasoline. Furthermore, unlike conventional ICEs, which emit pollutant gases as a result of combustion, hydrogen-powered ICEs, fuel cells, and fuel cell vehicles emit only water vapor. For these reasons, it is realistic to have a vision of reaching a hydrogen economy. However, the transition to a hydrogen economy faces multiple challenges that have to be overcome, including large-scale supporting infrastructures similar to those of gasoline and natural gas and the cost of hydrogen production and storage. These challenges can be overcome by will power and persistent research. A good example is that in the early 1900s when conventional cars were developed, there was no infrastructure for gasoline distribution and people could only purchase limited amounts of gasoline at pharmacies. Remembering this could give us hope now that a hydrogen economy society can become a reality in the future. The recent boom of shale gas that started in 2005 makes a hydrogen economy even more promising than before.

As discussed in this chapter, renewable energy resources such as wind and solar, and other types of renewable resources are environment friendly options to produce electricity to be used for hydrogen production. In particular, the potential of solar energy with the approximate total available surface power of 85,000 TW ($85,000 \times 10^{12}$ W)

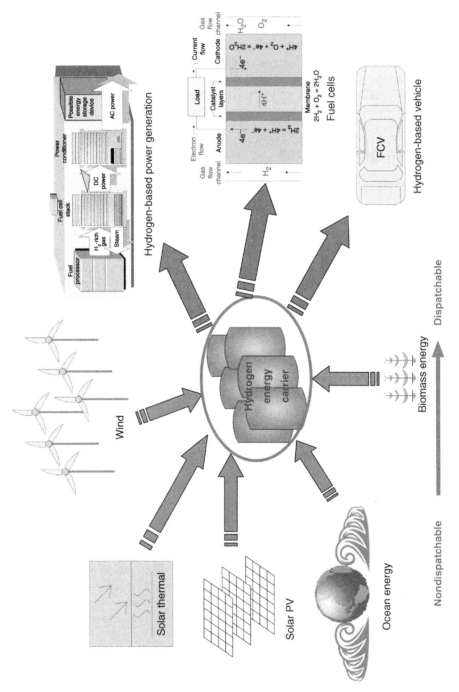

Figure 6.17 Schematic diagram of a vision for hydrogen production from renewable resources in a hydrogen economy society.

[6] is far more than enough to meet human needs, which is currently around 15 TW. Such potential can benefit humanity for producing sustainable electric power (solar PV or solar heat), or by direct use of solar heat to produce hydrogen for fuel cell power generation and as fuel for ICEs. Figure 6.17 shows a schematic diagram of a vision for a hydrogen economy society, where different nondispatchable renewable energy resources are used for hydrogen generation and storage.

References

[1] Nehrir MH, Wang C. Modeling and control of fuel cells: distributed generation applications. Hoboken, New Jersey: IEEE Press-Wiley; 2009.
[2] O'Hayre R, Cha S, Colella W, Prinz F. Fuel cell fundamentals. New York: Wiley; 2006.
[3] Larminie J, Dicks A. Fuel cell systems explained. 2nd ed. Hoboken, New Jersey: John Wiley & Sons, Ltd; 2003.
[4] Abott D. Keeping the energy debate clean: how do we supply the world's energy needs. Proc IEEE 2010;98(1):42–66.
[5] Das LM. Safety aspects of a hydrogen fueled engine system development. Int J Hydrogen Energ 1991;16(9):619–24.
[6] Bull SR. Renewable energy today and tomorrow. Proc IEEE 2001;89(8):1216–21.

Geothermal energy

7

Tubagus Ahmad Fauzi Soelaiman
Mechanical Engineering Department, Faculty of Mechanical and Aerospace Engineering, Thermodynamics Laboratory, Engineering Centre for Industry, Institut Teknologi Bandung, Bandung, Indonesia

Chapter Outline

7.1 Introduction

The term "geothermal energy" is derived from the Greek words "ge," which means *earth*, "therme," which means *heat*, and "energos," which means *active or working* [1]. Therefore, geothermal energy can be considered as the active heat energy from within the earth.

Geothermal energy is considered *renewable* if its demand does not surpass its supply. Among other renewable energies (solar, wind, bio, hydro, and ocean/marine), geothermal is desirable since it has a *high energy density*, unlike wind energy, for example; and it is *available continuously* unlike solar energy, for example. But unlike fossil fuels such as oil, gas, and coal, geothermal energy *cannot be exported or transported* over long distances. It must be used *in situ* to use its heat directly or to produce electricity.

Copyright © 2016 Elsevier Inc. All rights reserved.

Geothermal energy is mostly available in areas close to the *Ring Of Fire*, where Earth's tectonic plates meet. Therefore, the map in Figure 7.1 can be used to determine with relatively high probability the locations of geothermal energy sources. These locations, while benefiting from the availability of geothermal energy, must also endure the risk of volcanic eruptions.

Geologically, geothermal energy may manifest itself in forms of: *volcanoes, lava flows, geysers* (hot springs that spew water into the air intermittently), *fumaroles* (small holes that release dry steam or wet steam), *hot/warm springs* (springs that release hot/ warm water), *hot/warm pools* (pools with a higher temperature than the surroundings, which indicate that a geothermal heat source is present under the surface), *hot lakes* (hot pools with larger surface area), *mud pools* (pools of hot mud, usually with CO_2 bubbling from the ground), *steaming grounds* (grounds that release steam), *warm grounds* (grounds with higher temperature than the surroundings), and *silica sinters* (silvery silica condensation that create silica sinter terraces or sinter platforms) [3,4]. These manifestations reveal that geothermal energy is available and may be tapped from below the ground.

Figure 7.2 shows a typical geothermal field that can be found on Earth. From the center of the Earth, the nearest *magma* to the surface is shown in Figure 7.2. It solidifies into *igneous rock* or *impermeable rock*, also known as *volcanic rock*, if found at the surface. The magma heats the igneous rock by conduction, which in turn heats

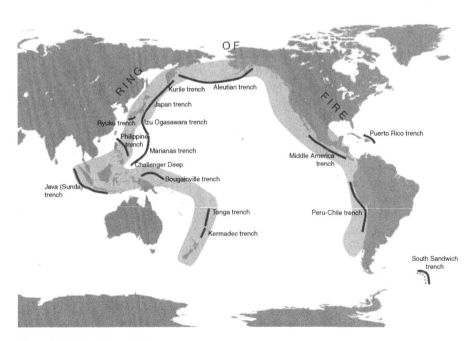

Figure 7.1 The Ring of Fire.
Adapted from Ref. [2].

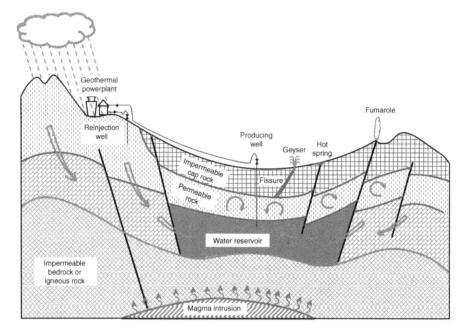

Figure 7.2 A typical geothermal field.

ground water by convection in a *water reservoir* and *permeable rock*. This reservoir is capped by an *impermeable cap rock* on top. The rock may have *fissures* to vent the heated reservoir. The vent may be formed naturally as *geysers*, *fumaroles*, or *hot springs*; or it may be purposely tapped by using a *production well* for a *geothermal power plant*. The cooled fluid can be returned to earth using a *reinjection well*.

Steam originating from the magma itself is called *magmatic steam*, while that from ground water heated by the magma is called *meteoritic steam* [4].

The geothermal source described earlier produces steam. But not all sources produce steam. Some produce warm water, while others produce no water at all, just *hot dry rocks* (HDRs). According to the energy sources, geothermal sources are usually divided into hydrothermal, geopressured, and petrothermal [4].

Hydrothermal systems are those with water heated by the hot rock. If water is heated to create mostly vapor, then it is called *vapor-dominated system*. But if a majority of the water is still liquid, then it is called a *liquid-dominated system* [4].

Geopressured systems are those having water trapped in much deeper underground reservoirs, about 2000–9000 m, with low temperature (about 160°C) and high pressure (about >1000 bar). It has a high salinity of about 4–10% and is highly saturated with natural gas, mostly methane that can be recovered for electric generation. This methane can be combusted to produce electricity, while the heat of the water can also be used to produce electricity [4].

Petrothermal systems are those without naturally occurring water. The heat source is in the form of an HDR. Heat can be extracted by pumping water into the cracked HDR. The steam produced can then be used to produce electricity [4].

7.2   Geothermal energy uses and types

The possible uses of geothermal energy can be found, but are not limited to, as shown in a *Lindal diagram*. The diagram has been modified from the original version in 1973 [5] into several forms such as the one shown in Figure 7.3 [6]. From the diagram, it can be shown that low temperature of a heat source or a reservoir can be used for domestic hot water, greenhouses, copper processing, soil warming, fish farming, etc. In the meantime, high temperature reservoirs can be used to produce electricity by using steam turbines and generators in power plants. This electricity can be easily transmitted using transmission lines to consumers.

When geothermal energy is used as *heat* for bathing, house heating, cultivating plantations, etc., it is known as the *direct use* of geothermal energy. But when geothermal energy is converted into *electricity* through a thermodynamic process/cycle, then it is known as *the indirect use* of geothermal energy.

7.2.1   Direct use of geothermal energy

Direct use of geothermal energy by making *use of the heat directly*, is the oldest and most common use of geothermal energy. People have been bathing for hundreds of years in naturally occurring ponds or lakes that are heated by geothermal sources. By directing the hot water/steam/brine, geothermal energy has also been used directly for space heating, district heating, snow melting, road deicing, agricultural heating/drying, etc., especially in cold countries. Several examples of direct use of geothermal energy are presented in Section 7.2.1.1.

7.2.1.1   Space or object heating

Space or object heating by using geothermal energy is the most obvious use. By directing the steam, water, or brine from a geothermal energy source to a space or an

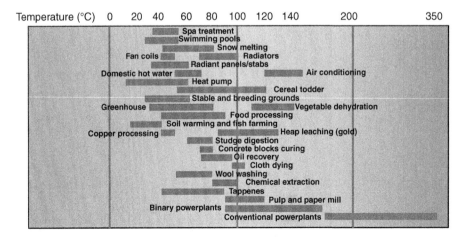

Figure 7.3 Lindal diagram.
Adapted from Ref. [6].

object, the temperature of the space or object can be increased and/or maintained. The heat can be directly obtained from geothermal fluids, or these geothermal fluids can heat another fluid such as water, or air to heat the space or object. When using additional fluids, the system may be more complex but the temperature can be usually controlled more precisely and easily. Furthermore, by using additional fluid, some parts of the system can avoid the undesirable properties of geothermal fluid, such as its corrosiveness.

An example of the use of geothermal energy in a district heating system in Reykjavik, Iceland, is shown in Figure 7.4. Three production wells are used to supply the houses with 80°C fluid. The cooled fluid leaves at 35°C to the drains.

To extract heat from the ground, a piping system can be utilized in several loop configurations as shown in Figure 7.5 [7]. If ground space is available, a horizontal loop configuration can be used by digging low cost trenches in the ground deep enough to obtain a constant temperature. In case ground space is limited, the piping can be in a vertical loop configuration by drilling wells into the ground. If a pond is available nearby, the piping loops can be submerged below the surface of the pond, reducing the cost of digging. Lastly, if underground water is available, the warm water can be pumped up to heat up the system and then the cooled water can be released into a pond.

In *agriculture*, it has been known that certain plants grow rapidly in certain surrounding temperatures. Examples for lettuce, tomato, and cucumber are shown in Figure 7.6 [6]. Of course, other variables are also important such as the type of soil, amount of light, CO_2 concentration, humidity of air and soil, and air movement. By adjusting these variables in a controlled enclosure, one can grow agricultural products optimally in any season or in any weather condition, giving better food security.

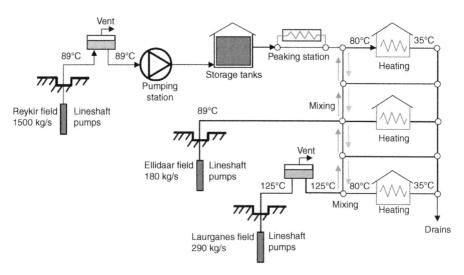

Figure 7.4 An example of direct use of geothermal energy in a district heating system in Reykjavik, Iceland.
Adapted from Ref. [6].

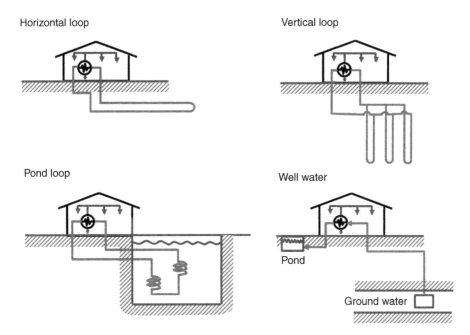

Horizontal loop

Vertical loop

Pond loop

Well water

Pond

Ground water

Figure 7.5 Several configurations of the piping system for geothermal warming.
Adapted from Ref. [7].

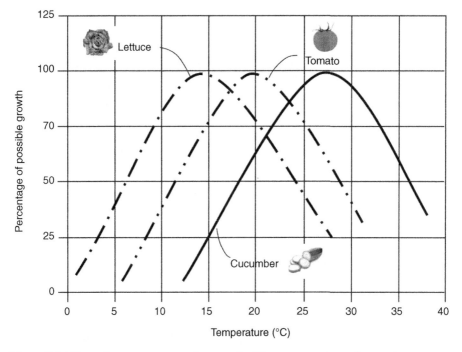

Lettuce

Tomato

Cucumber

Percentage of possible growth

Temperature (°C)

Figure 7.6 Effect of temperature on the growth of lettuce, tomato, and cucumber.
Adapted from Ref. [6].

Geothermal heat can be used to heat these and other agricultural products in green-houses, at the optimum temperature for best growth. Several possible *piping and duct-ing* systems that can be used in greenhouses are shown in Figure 7.7 for natural as well as forced air movements. Additional examples of direct uses of geothermal energy for agricultural products, such as coconut drying, coffee drying, cocoa drying, palm sugar production, and mushroom production, can also be found in Soelaiman [8].

For *farm animals* and *aquatic species*, temperature of the surroundings can also affect the growth rate. Figure 7.8 shows the percentage growth rates of several farm

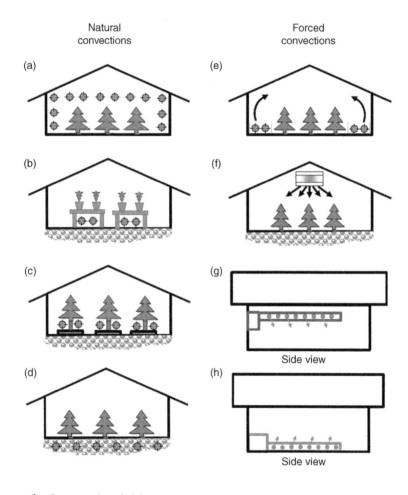

♦ : Cross-section of piping system

Figure 7.7 Examples of using piping/ducting systems for geothermal heating in greenhouses. Heating installations with natural air movement (natural convection): (a) aerial pipe heating; (b) bench heating; (c) low-position heating pipes for aerial heating; and (d) soil heating. Heating installations with forced air movement (forced convection): (e) lateral position; (f) aerial fan; (g) high-position ducts; and (h) low-position ducts. Adapted from Ref. [6].

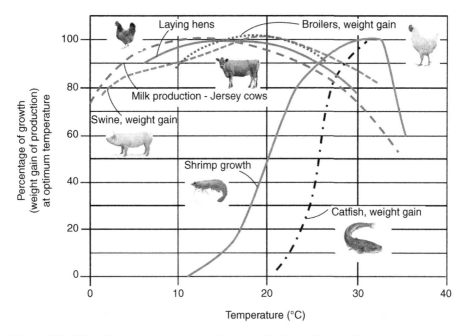

Figure 7.8 Effect of temperature on growth or production of food animals.
Adapted from Ref. [6].

animals and aquatic species. The aquatic species that are typically raised in controlled temperatures are carp, catfish, bass, tilapia, mullet, eel, salmon, sturgeon, shrimp, lobster, crayfish, crab, oyster, clam, scallop, mussel, abalone, etc. [6]. Geothermal heat can of course be used to obtain the optimum growth temperature of the farm or pond with similar installations as shown in Figure 7.5.

7.2.1.2 Geothermal heat pump

A *heat pump* is a device that takes heat from a cold source and releases the heat into a hot sink. The working operation of a heat pump is in the reverse order of refrigeration cycle. In fact, an air conditioning system can be used as a heat pump, just by placing the condenser in the cold source and the evaporator in a hot sink. But rather than physically moving the units, a special valve can be used to reverse the direction of a refrigeration cycle into a heat pump cycle. An example of a heat pump to warm domestic water and a space with the possible use of a geothermal heat source can be seen in Figure 7.9. The heat pump can be used to heat a space with less energy by using reduced electrical energy to run the refrigerant compressor rather than directly heating the space by electricity.

A heat pump can be used to extract heat from the outside of a house into the inside. But during the winter season where the temperature of the outside air is very low, the heat pump has to work harder due to the large temperature difference. In the meantime, the temperature of the soil a few meters below the ground usually remains

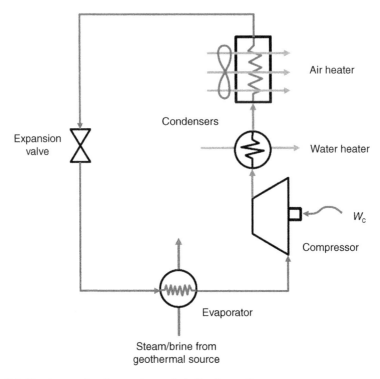

Figure 7.9 Heat pump heating water and air for domestic use.

constant at about 13°C (55°F). Therefore, rather than taking heat from the cold outside air, the heat pump can take the heat from the warmer ground, which will reduce the amount of power needed to heat the house. A schematic diagram of the use of a heat pump for geothermal heating is shown in Figure 7.10 [9].

7.2.1.3 Geothermal cooling

In the summer season where the temperature of air is high (25–40°C) and the temperature of the subsoil is low (about 13°C), water or air can be injected or blown into the ground, cooled, and then resurface into a space or an object to be cooled by a heat exchanger. The loop installation may be similar to the ones in Figure 7.5, but for cooling. This is a type of *ground cooling* that takes advantage of the cooler subsoil compared to the outside air temperature.

Geothermal energy can also be used as *the heat source of an absorption refrigeration cycle*. An example of the absorption cycle is shown in Figure 7.11 where ammonia(NH$_3$) is used as the refrigerant and water (H$_2$O) as the absorber. Alternatively, water (H$_2$O) can be used as the refrigerant and lithium bromide (LiBr) as the absorber. In Figure 7.11, geothermal heat is used to heat the generator while the evaporator is cooling the space or the object to be cooled. Compared to a vapor refrigeration, which uses a compressor, absorption refrigeration is less costly to operate since it uses a

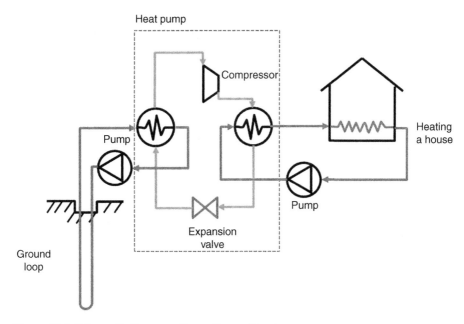

Figure 7.10 Schematic diagram of a geothermal heat pump.
Adapted from Ref. [9].

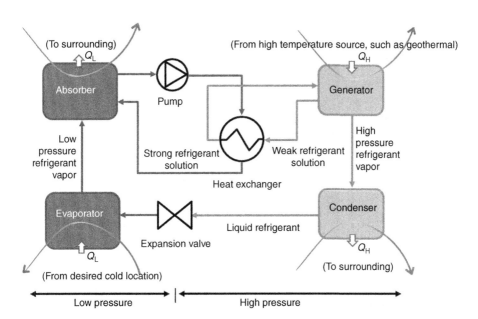

Figure 7.11 Absorption refrigeration cycle that can use geothermal energy as a heat source.

pump. A comprehensive explanation of how an absorption refrigeration system works is beyond the scope of this book. Please refer thermodynamics or refrigeration system books for further information.

7.2.1.4 Calculations for direct use of geothermal energy

The *calculations* for direct use of geothermal energy usually involves *the sizing and material selection of heat exchangers, ducting systems, pipes, pumps, and fans*. This is also beyond the scope of this book. Please refer heat transfer and equipment sizing books for the proper calculation methods. The log mean temperature difference method is usually used to size the equipment, given the states of the working fluids. Alternatively, the number of transfer units-ε method is usually used to calculate whether the size of equipment can transfer the amount of heat as required.

7.2.2 Indirect use of geothermal energy

As stated earlier, *the indirect use of geothermal energy* usually refers to *electricity generation* by using heat from the geothermal source. Basically, this geothermal power plant is similar to steam power plants, but it uses earth as the natural boiler. The first engine used at Larderello, Italy in 1904 to produce electricity from geothermal steam was invented by Prince Piero Ginori Conti [6]. Section 7.2.2.1 analyzes the geothermal power plant from its simplest form to more complex systems.

7.2.2.1 The simplest geothermal power plant

In its *simplest form*, the steam extracted from the geothermal well can be *directly expanded in a turbine*, which turns a generator to produce electricity. The process flow diagram (PFD) and the process diagram on the *T–s* diagram of this geothermal power plant are shown in Figure 7.12. Drawing the process on *h–s* and *P–h* diagrams should also be exercised.

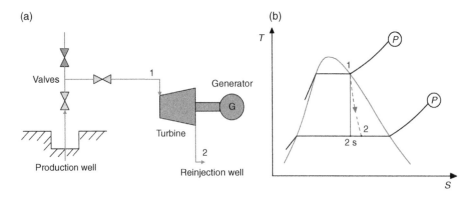

Figure 7.12 The simplest form of a geothermal power plant. (a) Schematic of the PFD. (b) *T–s* process diagram.

To analyze the process that occurs on the working fluid, *mass and energy conservation laws on a control volume* (CV) are needed, which can be found in many thermodynamic books, such as by Moran and Shapiro [10]. They can be written as:

$$\frac{dm_{CV}}{dt} = \sum \dot{m}_i - \sum \dot{m}_e$$

$$\frac{dE_{CV}}{dt} = \dot{Q}_{CV} - \dot{W}_{CV} + \sum \dot{m}_i \left(h_i + \frac{V_i^2}{2} + gz_i \right) - \sum \dot{m}_e \left(h_e + \frac{V_e^2}{2} + gz_e \right)$$

In a steady-state condition, the fluid states do not depend on time, therefore: $dm_{CV}/dt = 0$ and $dE_{CV}/dt = 0$. Then:

$$\sum \dot{m}_i = \sum \dot{m}_e$$

$$0 = \dot{Q}_{CV} - \dot{W}_{CV} + \sum \dot{m}_i \left(h_i + \frac{V_i^2}{2} + gz_i \right) - \sum \dot{m}_e \left(h_e + \frac{V_e^2}{2} + gz_e \right)$$

Applying these equations into a turbine and by assuming that there is no mass leakage on the turbine, the mass conservation law can be written as:

$$\dot{m}_i = \dot{m}_e = \dot{m}$$

Assuming the turbine is adiabatic (no heat loss), and the changes of kinetic and potential energies are negligible, then the energy conservation law on the turbine can be written as:

$$\dot{W}_T = \dot{m}(h_i - h_e) = \dot{m}\left(h_1 - h_2 \right)$$

If the isentropic efficiency (η_i) of the turbine is known from a test, then the real work produced from the turbine can be written as:

$$\dot{W}_T = \eta_i \dot{m}\left(h_1 - h_{2s} \right)$$

The isentropic efficiency of a dry turbine (η_{td}) is usually taken as 85%. But in the case of a geothermal power plant, it is almost certain that the condition of the steam exiting the turbine is at a two-phase region. According to Bauman in DiPippo [11], the Baumann rule states that 1% average moisture causes approximately 1% drop in turbine efficiency. Another way to approximate the efficiency of a wet turbine (η_{tw}) is to use the following equation [11]:

$$\eta_{tw} = \eta_{td} \times \left[\frac{x_{in} + x_{out}}{2} \right]$$

This simplest form of geothermal power plant is seldom utilized since the low quality of steam and presence of impurities within the steam may not be suitable for direct feeding into a turbine. In practice, a *separator* usually has to be used to dry the steam from water droplets and remove the impurities from within the steam near the well or in a header that receives steam from several wells. Assuming that there is no mass leakage in the separator (brine and solids released during blowdown under the separator are negligible), the separator is adiabatic due to the good insulator, and change of kinetic and potential energies can be neglected, then the process in the *separator* can be considered at constant enthalpy (*isenthalpic*). Therefore, the mass and energy conservation laws can be written as:

$$\dot{m}_i = \dot{m}_o = \dot{m}$$
$$h_i = h_o = h$$

If the separator continuously releases liquid or brine, then the separator also separates the liquid from the dry steam at the pressure of the separator. Therefore, the process in the separator will consist of an isenthalpic pressure drop and a constant pressure (*isobaric*) steam and liquid separation. The PFD and process diagram of the simplest geothermal power plant with this type of separator can be seen in Figure 7.13.

Example 7.1

Draw the *h–s* and *P–h* diagrams for the cycle shown in Figure 7.13.

Solution

Following the isenthalpic, isobaric, and isentropic lines, the *h–s* and *P–h* diagrams of the simplest geothermal power plant with a separator can be seen in Figure 7.14.

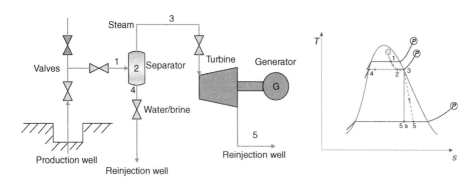

Figure 7.13 Schematic and process diagrams of the simplest geothermal power plant with a separator.

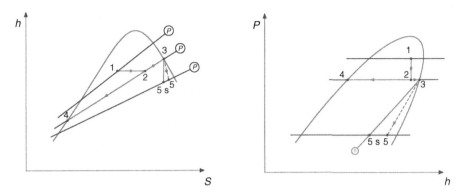

Figure 7.14 The *h–s* and *P–h* diagrams of the simplest geothermal power plant with a separator.

Example 7.2

After the valves of the production well, steam enters a separator at flow rate ($\dot{m}$) = 400 t/h, P = 12 bar (absolute), and x = 70%. If the drop of pressure in the separator is 2 bar, and the isentropic efficiency of the dry turbine is 85%, calculate the power produced by the turbine if the steam is expanded into 1 atm and 0.6 atm. Discuss the results.

Solution

The schematic and *T—s* diagram of the cycle would be similar to those in Figure 7.13.

Mass flow rate ($\dot{m}$) is 400 t/h or 400 t/h × (1000 kg/t) × (1 h/3600 s) = 111.11 kg/s.

At P = 12 bar and x = 70%, h_1 can be calculated as $h_1 = h_f + x\,(h_g - h_f)$. The thermodynamics properties can be found in any steam table or steam calculator. For this example, the steam calculator can be found at http://www.peacesoftware.de/einigewerte/calc_dampf.php5. Entering P = 12 bar (absolute) on the menu "Calculation of thermodynamic properties of saturated steam," and by pressing enter or clicking on the menu "Compute," the properties of thermodynamic properties are calculated.

Calculating $h_1 = h_f + x \times (h_g - h_f)$ = 798.49 + (0.70) × (2787.77 − 798.49)) = 2188.186 kJ/kg. Assuming the drop of pressure occurs at isenthalpic pressure, then $h_2 = h_1$ = 2188.186 kJ/kg. Calculating the quality at 2: $x_2 = (h_2 - h_f)/(h_g - h_f)$ = (21.88.186 − 792.68)/ (2777.12 − 762.68) = 0.7032, or 70.32%. There is a slight increase of quality compared to point 1.

Entering 10 bar at "Calculation of thermodynamic properties of saturated steam" of the online steam calculator, the properties in saturation lines can be found. Calculate s_2 at 10 bar: $s_2 = s_f + x \times (s_g - s_f)$ = 2.1384 + 0.7032 × (6.5849 − 2.1384) = 5.2651788 kJ/kgK.

Point 3s has the same entropy value as point 2, since the ideal process is considered isentropic. Therefore, $s_{3s} = s_2$ = 5.2651788 kJ/kgK. Then steam quality of this point can be calculated as: $x_{3s} = (s_{3s} - s_f)/(s_g - s_f)$ = (5.2651788 − 1.067)/(7.35438 − 1.3067) = 0.6545(65.45%).

Searching the saturated thermodynamics properties at the exit of the turbine at 1 atm (1.01325 bar), then by entering 1.01325 as the pressure, and using the obtained saturated properties, we can calculate $h_{3s} = h_f + x(h_g - h_f)$ = 418.9907 + 0.6545(2675.53 − 418.99) = 1895.896 kJ/kg.

Therefore, assuming the isentropic efficiency of the turbine as 85%, then the power of the turbine is: $\dot{W}_t = \dot{m} \times (\eta_{iT})(h_2 - h_{3s})$ = 111.11 kg/s × (0.85)(2188.186 − 1895.896) = 27604.85 kJ/kg s = 27 MW.

If somehow the outlet of the turbine can be vacuumed into 0.6 atm (0.61 bar), then, using the properties at 0.61 bar:

$x_{3s} = (s_{3s} - s_f) / (s_g - s_f) = (5.2652 - 1.1502) / (7.5255 - 1.1502) = 0.6455(64.55\%).$

$h_{3s} = 361.621 + (0.6455(2,653.5525 - 361.621)) = 1,841.06 kJ / kg.$

$\dot{W}_T = 111.11(0.85(2,188.186 - 1,841.06)) = 32,783.79 kJ / kg \ s = 33 MW.$

Note that by reducing the outlet of the pressure from 1 atm to 0.6 atm, there will be an increase of the power of the turbine by 6 MW or 22%, from 27 MW to 33 MW. Modern geothermal power plants mostly use a condenser to reduce the exit pressure of the turbine in order to increase the power of the turbine and also the cycle's efficiency at the same time. Vacuuming the condenser is usually obtained by using a jet steam ejector or a liquid ring vacuum pump (LRVP).

According to the state of the supplied steam for the turbine, there are several *types of geothermal power plants* that are available, namely:

1. Direct dry steam geothermal power plant
2. Single flash steam geothermal power plant
3. Double and multiflash steam geothermal power plant
4. Binary or *organic Rankine cycle* (OCR) geothermal power plant
5. Kalina cycle
6. Total flow devices
7. Hybrid systems
8. HDR

7.2.2.2   Direct dry steam geothermal power plant

If the *steam* available for the turbine is *dry* (with quality $x = 100\%$), the steam can directly be fed into the turbine, as in the simplest geothermal power plant. In addition, in order to have a larger enthalpy drop within the turbine, which will give higher power from the turbine, a *condenser* can be added to lower the turbine exit's pressure. For this, cooling water is needed to lower the temperature, and therefore the pressure, of the condenser. The condensed water can then be cooled by a *cooling tower*. Overflow from the basin of the cooling tower can then be released into a river. But since the water is usually still warm and releasing warm water into a river can alter the ecosystem of the river, the overflow water is usually reinjected into a *reinjection well*. To decrease the cost, the reinjection well can use a low or no production well that is located outside the area of the predicted reservoir. The reinjection well should be far enough from the reservoir to avoid decreasing the temperature of the reservoir in both the short and long terms. The schematic of this direct dry steam geothermal power plant is shown in Figure 7.15. Note that the process for the separator in this figure assumes there is continuous separation of the liquid and steam. If the pressure drop and the blowdown of the separator are negligible, then point 1 on the process diagram should be at the same location as point 3.

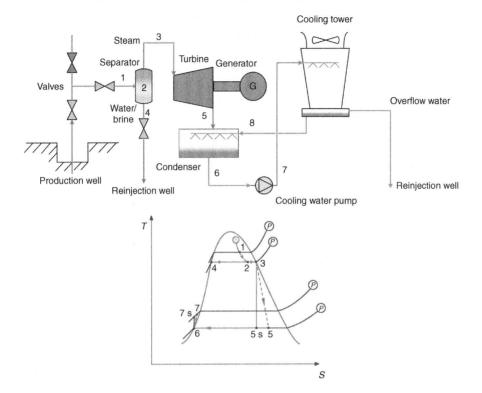

Figure 7.15 Schematic and *T–s* diagrams of a direct dry steam geothermal power plant.

The mass and energy conservation laws of the components of this plant can be written as:

Turbine:

$$\dot{m}_3 = \dot{m}_5$$
$$\dot{W}_T = \dot{m}_5 \left(h_3 - h_5 \right) = \eta_i \dot{m}_5 \left(h_3 - h_{5s} \right)$$

Condenser:

$$\dot{m}_5 + \dot{m}_8 = \dot{m}_6$$
$$\dot{m}_5 h_5 + \dot{m}_8 h_8 = (\dot{m}_5 + \dot{m}_8) h_6$$

where h_8 can be taken as the enthalpy of the saturated liquid at the surrounding pressure and h_6 can be taken as the enthalpy of the saturated liquid at the condenser's pressure.

7.2.2.3 Single flash steam geothermal power plant

For *low quality steam*, which is mostly water, where quality *x* equals or is close to zero, the steam cannot be fed directly into the turbine since the water droplets will shorten the lifetime of the turbine blades, especially in the first few stages. To avoid this, low quality of steam can be *flashed to a lower pressure* and then the produced dry steam can be *separated* and fed into the turbine, while the saturated water can be reinjected into the reinjection well. The process of the steam after exiting the condenser can be the same as the previous type of geothermal power plant. The ideal thermodynamic process during the *flash process* can be considered as *isenthalpic* and the *separation process* as *isobaric*, as shown in Figure 7.16.

Thermodynamic analysis of this type of power plant should be similar to the previous ones. Mass and energy balances of each component should be calculated along with power output and the power plant's efficiency.

7.2.2.4 Double and multiflash steam geothermal power plant

If the low quality steam has sufficiently high pressure, the steam can then be *flashed twice*. Saturated water from the first flasher can then be flashed again. Saturated steam from the first flasher can be fed into a high-pressure turbine, while the saturated steam from the second flasher can be fed into a low-pressure turbine. Both turbines can be

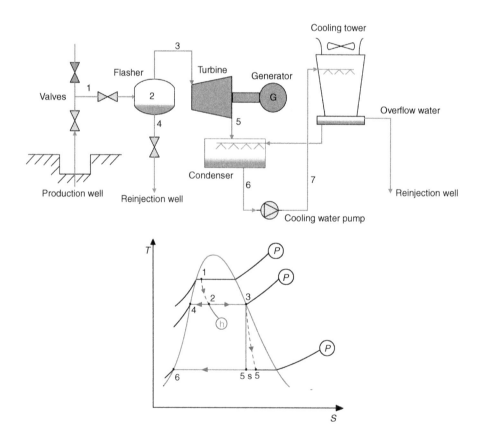

Figure 7.16 Schematic and *T–s* diagrams of a single flash steam geothermal power plant.

constructed in tandem to give the same rotational speed to turn a generator, or be separated. Since there is more steam fed into the turbine, this configuration will give a higher power output at the expense of a more complicated and expensive system. Theoretically, a *triple, quadruple, or multiflash system* can be constructed if considered more economical. A schematic diagram of the double flash steam geothermal power plant can be seen in Figure 7.17.

7.2.2.5   Binary or organic Rankine cycle (ORC) geothermal power plant

If the temperature of a supplied geothermal steam is not high enough, say below 200°C, then the steam is usually not economical enough to be used in a steam turbine. Instead, the hot steam, water, or brine can be used to heat *a secondary fluid with a lower boiling point*, such as *organic fluids* like *propane* or *butane*. The cooled steam or water can then be reinjected into a reinjection well. In the meantime, the organic fluid can be used to create a closed Rankine cycle, usually called ORC to produce work in an organic turbine. Such a turbine is usually physically much smaller than a steam turbine for the same capacity. In the condenser, the working fluid can be cooled by using cooling water or air.

A schematic PFD and process diagram of the OCR can be seen in Figure 7.18. The evaporator can heat the organic fluid until it is a saturated liquid (point 4) or

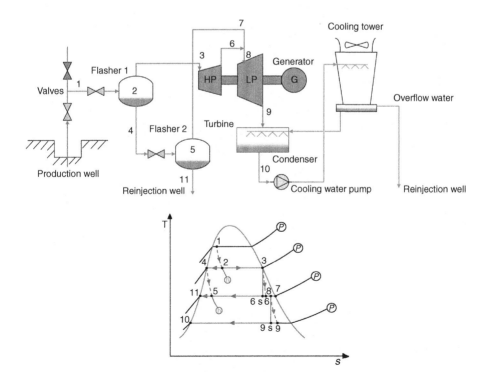

Figure 7.17 Schematic and *T–s* diagrams of a double flash steam geothermal power plant.

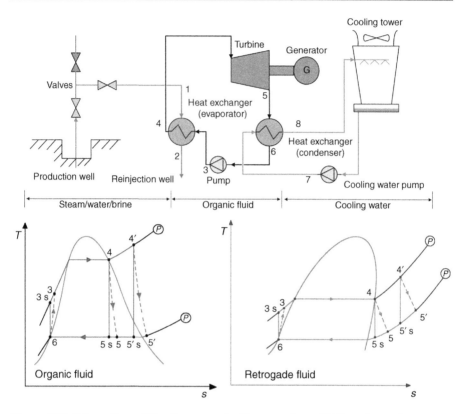

Figure 7.18 Schematic and *T–s* diagrams of a binary cycle or ORC geothermal power plant.

superheated liquid (point 4′). Note that, in *retrograde fluids* such as *n*-butane, *i*-butane, *n*-pentane, and *i*-pentane, the right side of the *T–s* diagram curves, which makes the output of the turbine always in the superheated region.

7.2.2.6 Kalina cycle

Besides the ORC used in the aforementioned geothermal power plants for low temperature geothermal fluid, a *Kalina cycle* can also be used. This cycle was developed by Dr Alexander Kalina in the 1970s and uses *a solution of two fluids with different boiling points*. Although ammonia and water are the two most common working fluids used in the Kalina cycle, other types of fluids can also be used. The closed cycle simply uses low temperature steam, water, or brine from the geothermal well to heat up the working fluid, and cooling water or air to cool it down in closed heat exchangers.

By using two fluids with different boiling points as the working fluid, the solution boils and condenses over a range of temperatures, which gives more heat to be extracted from the geothermal fluid than with a pure working fluid. Depending on the heat input temperature, the range of boiling point of the solution can be adjusted by adjusting the ratio between the components of the solution. As a result, the average temperature of the working fluid in the heating process is higher and the average temperature of the cooling process is lower, thus giving the cycle a higher efficiency. A simple Kalina cycle is shown in Figure 7.19. A more complex Kalina cycle will involve additional separators and recuperators, to gain higher efficiency at the cost of a more complex and expensive system.

Among others, the Kalina cycle is used in the Husavik facility in Iceland, which produces 2 MW of electric power and 20 MW of thermal power, and at the Unter-hatching facility in Germany (near Munich) that produces 3.4 MW of electric power and 38 MW of thermal power [12,13].

The Kalina cycle is not only used in geothermal power plants, but can also used to recover heat from other power plants and industrial process plants in processes, where the heating or cooling process occurs in a liquid or gas phase, where the temperature increases or decreases (not during a phase change, where the temperature remains constant).

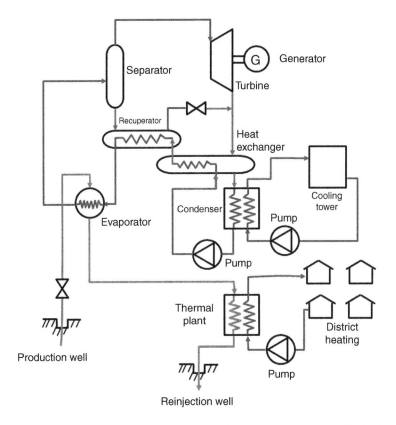

Figure 7.19 Kalina cycle used in a 2 MW geothermal power plant in Husavik, Iceland.
Adapted from Ref. [12].

7.2.2.7 Total flow devices

To gain power from a geothermal fluid, total flow devices can be used to expand the fluid while gaining power to turn a generator. Although the efficiency is usually small, these devices do not usually require additional complex equipment. Examples of these total flow devices are [14]:

1. The Sprankle Hydrothermal Power Company (HPC), Ltd.) prime mover: brine expander
2. The Robertson engine
3. The bladeless turbine (such as the Tesla turbine)
4. The Keller rotor oscillating vane (KROV) machine
5. The Armstead-Hero turbine
6. The gravimetric loop machine
7. Electro-gas-dynamics (EGD)
8. Total flow impulse turbine
9. The biphase turbine

Refer [14] for brief descriptions on these devices.

7.2.2.8 Hybrid systems

In a hybrid system, geothermal heat can be used as a preheat of a fossil fuel power plant. This will reduce the cost of fuel to run the boiler. In another hybrid system, fossil fuel can be used to superheat the geothermal fluid, which will increase the power output and efficiency of the geothermal power plant. Both of these systems require a location where both the geothermal sources and fossil fuels are easily available. The PFD and $T–s$ diagram of these two hybrid systems are shown in Figures 7.20 and 7.21.

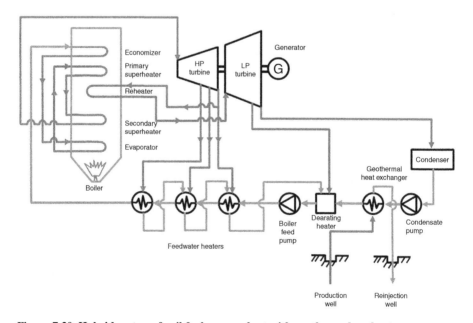

Figure 7.20 Hybrid system: fossil fuel power plant with geothermal preheat.

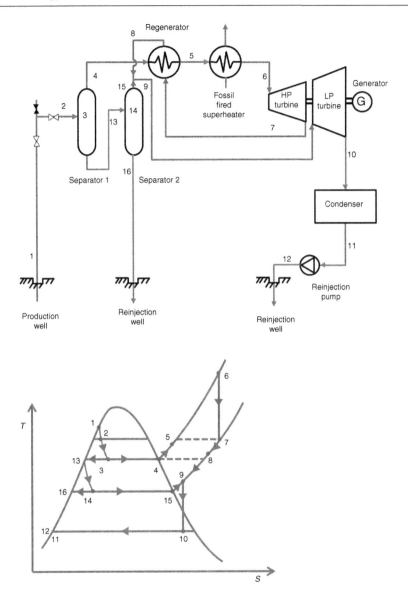

Figure 7.21 Hybrid system: geothermal power plant with fossil fuel superheater.

7.2.2.9   Hot dry rock

HDR, also known as *enhanced geothermal system* (EGS) in a geothermal system is a condition where *water is not naturally present at the site*. The magma only heats dry rock on top of it. In order to tap heat from the dry rock, two wells can be drilled into the rock. One well is used to carry water from the surface down into the HDR. Once the water is heated, steam created is then channeled up through the second well into a turbine above the surface (Figure 7.22). The additional installations will be the same as the previous types of power plant discussed: direct dry steam, flash steam, or OCR.

In order to increase the heat transfer between rock and water, water can be pumped into the rock causing it to *hydraulically fracture*, or the rock can be *control-exploded* first. The explosion should create small rocks that can heat the water into steam more

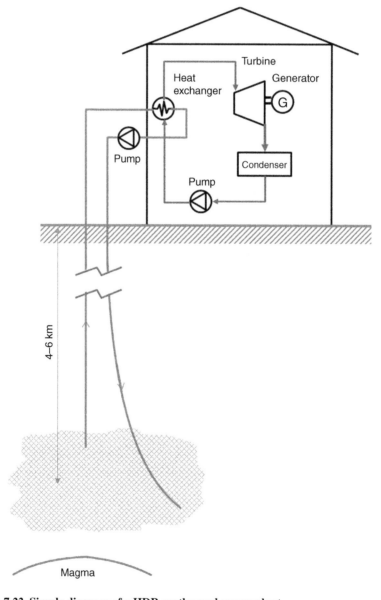

Figure 7.22 Simple diagram of a HDR geothermal power plant.

effectively. Care should be given so that the explosion does not create cracks that can allow the water or steam to leave the reservoir except through the provided well.

HDR system was used for the first time experimentally in Los Alamos, New Mexico, USA in 1970. The experiment was then followed by similar projects in Australia, France, Germany, Japan, and the United Kingdom [4].

7.3   Evaluation of geothermal power plant

When evaluating the geothermal power plant, power output can be considered as the power of the turbine, or $\dot{W}_T$, as in a regular steam power plant without a pump. On the contrary, efficiency of the plant is not so obvious since the boiler is the earth itself. For that, a new definition of efficiency for a geothermal power plant must be used. There are at least two alternatives, namely

$$\eta_{\text{geothermal power plant}} = \frac{\dot{W}_T}{\left(\dot{m}h\right)_{\text{steam exiting the well}} - \left(\dot{m}h\right)_{\text{fluid reinjected}}}$$

or

$$\eta_{\text{geothermal power plant}} = \frac{\dot{W}_T}{\text{Exergy of the inlet steam}}$$
$$= \frac{\dot{W}_T}{\dot{m}\left[\left(h - h_o\right) + T\left(s - s_o\right) + (V^2/2) + gz\right]_{\text{inlet steam}}}$$

where the subscript o is for the dead state condition of the surroundings, usually taken as 1 atm and 25°C.

The last equation is also known as the utilization efficiency [11].

7.4   Summary

The basics of geothermal energy have been covered in this chapter. The definition, possible locations, and types of manifestation of geothermal energy have been described in detail. Direct use of geothermal energy used for heating space, objects, agriculture, farm animals, and aquatic species, have been described. Additional methods such as using a heat pump and absorption cooling were also discussed. The indirect uses of geothermal energy where electricity is generated in geothermal power plants were also covered. Methods of making use of vapor and liquid dominated geothermal sources were discussed. The use of organic fluids in ORC and in the Kalina cycle were also covered. Also use of the total flow devices and how an HDR can be used in a geothermal power plant has been discussed. The chapter concluded with a discussion on how to define the total power and efficiency of a geothermal power plant.

Advance course should include exergy analysis of the plant, analysis of vacuum jet pumps and LRVPs, piping system, cooling tower, handling corrosive fluids, environmental consideration, etc.

Problems

1. Draw the *h–s* and *P–h* diagrams for cycles in Figures 7.15–7.18 (both fluids) and Figure 7.21.
2. Redo Example 7.2 if the steam is dry ($x = 100\%$), therefore no separator is needed. Compare the values with those obtained in Example 7.2.
3. Redo Example 7.2 if the steam is wet with quality of $x = 30\%$. Use one flasher, instead of a separator, and use the exit of the turbine at 0.6 atm only. Vary the flasher's pressure to find the best pressure that gives the turbine's highest power output.

Nomenclature

h	Enthalphy (kJ/kg)
m	Mass (kg)
$\dot{m}$	Mass flow (kg/s)
P	Pressure (Pa)
$\dot{Q}$	Heat transfer rate (kJ/kgs)
s	Entropy (kJ/kgK)
T	Temperature (K)
t	Time (s)
$\dot{W}$	Power (kJ/kgs)

Subscripts

cv	Control volume
e	Exit
f	Saturated liquid
g	Saturated gas
i	Inlet, isentropic
s	Isentropic
t	Turbine
td	Dry turbine
tw	Wet turbine

Acknowledgment

The author wishes to thank Adi Nuryanto, Maesha Gusti Rianta, Putranegara Riauwindu, and Achmad Refi Irsyad for their contributions to construct the figures used in this chapter.

References

[1] Morris N. Geothermal power – facts, issues, the future. Mankato, MN, USA: Franklin Watts; 2008.

[2] "Pacific Ring of Fire" by Gringer (talk) 23:52, (UTC). Licensed under Public Domain via Wikimedia Commons. Available from: http://commons.wikimedia.org/ wiki/File:Pacific_Ring_of_Fire.svg#mediaviewer/File:Pacific_Ring_of_Fire.svg; 2009 [accessed 26.12.2014].

[3] Saptadji NM, Ashat A. Basic geothermal engineering. Jurusan Teknik Perminyakan ITB; 2001.

[4] El-Wakil MM. Powerplant technology. Singapore: McGraw-Hill Publishing Company; 1984. 499–529.

[5] Lindal B. Industrial and other applications of geothermal energy. In: Armstead HCH, editor. Geothermal energy. a review of research and development. Earth science 12. Paris: UNESCO; 1973. p. 135–48.

[6] Dickson MH, Fanelli M. What is geothermal energy? Available from: http://www.unione-geotermica.it/What_is_geothermal_en.html; [accessed 26.12.2014].

[7] Available from: http://www.andrewsauld.com/products-services/geothermal/; [accessed 28.12.2014].

[8] Soelaiman TAF, Geothermal energy development in Indonesia. Institut Teknologi Bandung, Kyoto University, Kyoto University, GCOE, Program of, HSE. The Contribution of Geosciences to Human Security, Logos Verlag Berlin GmbH; 2011. pp. 191–209.

[9] Available from: http://www.nzgeothermal.org.nz/ghanz_heatpumps.html; [accessed 28.12.2014].

[10] Moran MJ, Shapiro HN. Fundamentals of engineering thermodynamics. 6th ed NY, USA: John Wiley & Sons Inc; 2008.

[11] DiPippo R. Geothermal power plants – principles, applications, case studies and environmental impact. 3rd ed Oxford, UK: Butterworth-Heinmann, Elsevier; 2012.

[12] Available from: http://www.mannvit.com/GeothermalEnergy/GeothermalPowerPlants/ Kalinacyclediagram/; [accessed 28.12.2014].

[13] Available from: http://en.gtn-online.de/Projects/Deepgeothermalenergyuse/Projectexam-pleinfo/biggestgeothermalpowerstationinsouthgermany; [accessed 28.12.2014].

[14] Armstead HCH. Geothermal energy, its past, present and future contributions to the energy needs of man. 2nd ed. NY, USA: E. & F.N. Spon; 1983.

Utilization of bioresources as fuels and energy generation

Farid Nasir Ani
Faculty of Mechanical Engineering, Universiti Teknologi Malaysia

Chapter Outline

8.1   Introduction

Bioresources are natural renewable sources like organic wastes and naturally formed or formable raw materials from human and animal activities. In large quantities they are generated by industries or mills in the agriculture, forestry, marine, and municipal sectors. These bioresources feedstocks are taken by processing and manufacturing industries like the oil palm mills. Their bioproducts are made from agricultural plants and may be used as energy carriers, platform chemicals, or specialty products. There

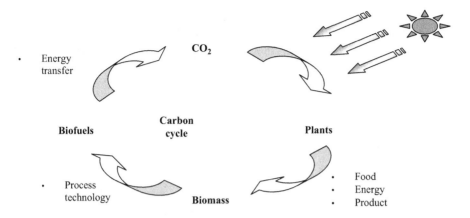

Figure 8.1 Sustainable carbon lifecycle for food, energy, and materials production.

is huge potential for bioproducts in Malaysia and tropical countries from the forestry, agriculture, marine, and municipal sectors. In bioproduct terms, these emerging industries are significantly different from conventional industries in that at the various sector levels, the nature and characteristics of the feedstocks, products, and applications are diverse. The sustainable carbon lifecycle as depicted in Figure 8.1 shows the continuous process of carbon mass transfer into various states of utilization. Energy input from the sun is taken into the carbon cycle, which processes it for food, energy, and materials depending on the requirements of the country.

The bioproducts industry consists of the following sectors: the suppliers of the raw bioresources, manufacturing, and product users. In order to be viable, the bioresources must either have adequate quantities for a long-term basis or production of bioresources in large quantities must be available sustainably. The manufacturing industry must use a conversion process that is based on the best technology, is economically viable, is a sustainable process, and is environmentally friendly. Bioresources in the context in this chapter refer to biomass, organic solid wastes, carbonaceous solid wastes, and agricultural wastes. Most of the research in this chapter deals with solid wastes that are homogeneous in nature rather than heterogeneous wastes due to the handling nature and environmental constraints.

Biomass is gaining increased attention as it is one of the most available renewable energy resources for reducing dependency on fossil fuels. Agricultural wastes are categorized as biomass, and are generated continuously in enormous amounts from agricultural activities. Some of these agricultural wastes are utilized as fuel to generate the heat and electricity required for milling processes. The utilization of biomass for energy conversion is still considered limited due to its poor fuel properties such as high moisture and ash contents, low bulk density, low energy content, and difficulty in storage, handling, and transport. The excess biomass generated not only causes disposal problems but is also considered as a waste of primary resources.

The use of biomass as a renewable energy source is important for countries where there are limited supplies of fossil fuel reserves. The generation of these solid wastes from urban and agricultural sectors is increasing due to the industrialization activities of urban and rural development. The utilizations of biomass or carbonaceous solid wastes from industries for energy and value-added products have contributed to the provision of national energy as well as current and future materials supply. Biomass is the only renewable source of carbon, which is the basic building block for the energy, materials, and chemical industries.

8.2   Biomass characterization

The utilization of biomass for biofuels and materials depends upon chemical and physical properties. Biomass comes from the agricultural, industries, and forest sectors and includes wood cuttings from saw mills and the wood industry. Biomass also comes from the urban sector as heterogeneous wastes, which basically are municipal solid wastes comprised of scrap tires, rubber waste cuttings, refuge-derived solids, unused furniture wastes, and organic and inorganic wastes. Another important characteristic of biomass is their thermal behaviors for usage in value-added products. Knowledge of these parameters will help in properly designing and developing a suitable thermal conversion process, which should be simple, reliable, efficient, and economical for local uses.

Generally, plantation biomass materials contain three major constituents, which are cellulose, hemicellulose, and lignin. Table 8.1 shows the composition of cellulose, hemicellulose, and lignin of various oil palm solid wastes and other biomass materials. The main component of biomass is cellulose with hemicelluloses present in between the cell wall. It is a mixture of polysaccharides, which comprises sugars such as glucose, mannose, xylose, and arabinose, and methlyglucorine and galacturonic acids. Hemicelluloses bind the cellulosic fibers together with lignin to form microfibrils, which enhance the stability of the cell wall [1].

Lignins are highly branched substitutes, mononuclear aromatic polymers in the cell walls of woody types of biomass, and are often adjacent to cellulose fibers to form lignocellulosic complexes. Lignin is regarded as a group of amorphous, high molecular weight, chemically related compounds. The building blocks of lignin are believed to

Table 8.1   Constituents of oil palm solid wastes and other biomass

	Cellulose	Hemicellulose	Lignin
Oil palm shells [2]	31.0	20.0	49.0
Oil palm fibers [3]	40.0	39.0	21.0
Oil palm empty bunches [4]	40.0	36.0	24.0
Softwoods [5,6]	41.0	24.0	27.8
Hardwoods [5,6]	39.0	35.0	19.5
Wheat straws [5,6]	39.9	28.2	16.7
Rice straws [5,6]	30.2	24.5	11.9
Bagasses [5,6]	38.1	38.5	20.2

be three carbon chains attached to rings of six carbon atoms, which are cross-linked with each other with a variety of different chemical bonds giving the cell wall its main mechanical strength [1].

No standard procedure has yet been recommended for the determination of the waste properties but certain ASTM standards recommended for fossil fuel meet the purpose for the characterization of the solid waste. Large quantities of volatiles are available in coconut and oil palm shell, rubber woods, and tire wastes. Moisture content in solid wastes can vary depending upon the type and duration of storage and drying adopted. The energy content varies with the moisture and residual oil content of the solid wastes especially found in oil palm shells. The energy content varies from 13 MJ/kg to 30 MJ/kg and is found highest in coconut shell charcoal and wood charcoal. This is due to higher carbon content and less volatile matter. Ash and volatile matter influence the energy content of the wastes. Coconut shell, oil palm shell, and rubber wood give the next best heating value after charcoal.

Agricultural wastes are usually of high moisture content and low in bulk density, and consequently have relatively low calorific values. The energy content of the wastes varies according to their moisture and residual oil contents. The chemical and physical characteristics of some solid wastes are presented in Table 8.2. The range for the oxygen content of biomass is 38–45% on a moisture and ash-free basis. The result of high oxygen content leads to relatively low calorific values of 14–20 MJ/kg as compared to hydrocarbon fuels, which have calorific values of 40–45 MJ/kg. They have negligible amounts of sulfur content and most of them contain low ash composition compared to coal [7]. Biomass has a significant amount of potassium in it, which leads to ash deposition during combustion. This alkali ash can lead to corrosion or erosion of boiler tubes, heat exchangers, and turbine blades. Biomass also contains a small amount of inorganic minerals such as potassium, sodium, phosphorus, calcium, and magnesium.

Biomass is basically from biological and organic material from living or recently living organisms in equal amounts. As a renewable energy source, biomass can either be used directly or converted into other energy products such as biofuels.

8.3 Pretreatment of biomass

The purpose of biomass pretreatment processes is to reduce the crystallinity of cellulose, which increases the porosity of the biomass and achieves the desired fractions. The various pretreatments are physical, physicochemical, chemical, and biological processes. The pretreatment of lignocellulosic materials have been reported extensively by Keshwani and Cheng [8].

8.3.1 Physical pretreatment

Physical pretreatment of lignocelluloses typically involves size comminution by grinding, milling, or chipping. The aim is to reduce the crystallinity of the cellulose fibers in the biomass. Size reduction is also necessary to eliminate mass and heat

Table 8.2 Analysis of typical biomass wastes

	Elemental composition wt% (dry ash free)				Proximate analysis wt% (air dry)			Gross CV (MJ/kg)	Average bulk density (kg/m^3)
	C	H	N	O	Ash	VM	FC		
Shells	55.35	6.27	0.37	38.01	2.5	77.2	20.3	19.56	440 (size < 18 mm)
Fibers	52.89	6.43	1.08	39.6	7.1	73.3	19.6	19.15	–
Bunches	47.89	6.05	0.65	45.41	6.0	72.3	21.7	17.83	–
Rice husks	55.8	0.31	1.7	42.07	21.0	9.5	19.4	14.1	100
Rubber wood	–	–	–	–	1.0	81.0	18.0	18.6	–
Scrap tires	78.28	6.78	0.17	8.71	5.1 (sulfur: 0.96)	63.2	31.3	36.2	–

transfer limitation during the required reaction. The size of the resulting materials is typically 10–30 mm after chipping and 0.2–2 mm after milling or grinding. Ball milling is used for particle sizes smaller than 90 μm, with cellulose content lower than for larger particle sizes. For hammer mill, the energy requirements for size reduction increase linearly as the particle size is reduced and the moisture content increases, where the energy requirements tend to level off for particle sizes less than 2 mm. Generally, the higher the moisture content of the biomass, the more energy is required for size reduction.

8.3.2   Physicochemical pretreatment

There are three types of physicochemical pretreatment, which are steam explosion, ammonia fiber explosion (AFEX), and CO_2 explosion. In steam explosion, the reduced size of biomass is subjected to high-pressure saturated steam for a short time before a sudden drop in pressure causes an explosive decompression of biomass. The process causes transformation of lignin and degradation of hemicelluloses. Steam explosion is known to be a cost-effective pretreatment for hardwood and agricultural residues. AFEX and CO_2 explosion are similar to steam explosion, where the biomass is exposed to liquid ammonia or CO_2 at high temperatures and pressures for a short period of time, followed by a sudden drop in pressure. AFEX does not solubilize hemicelluloses but does require recovery of ammonia for cost and environmental reasons.

8.3.3   Chemical pretreatment

Chemical pretreatment of biomass includes the use of ozone, acids, alkali, organic solvents, and peroxides. Ozonolysis is carried out at room temperature and is effective for lignin removal without the formation of toxic by-products. Mild acid pretreatment with sulfuric acid is efficient for the removal of hemicelluloses but fails to effectively remove lignin. It also helps in the removal of ash in biomass. Dilute alkali pretreatment using sodium hydroxide devastates the intermolecular bonds between lignin and hemicelluloses and improves the porosity of the biomass. Other studies on dilute alkali pretreatments have examined the use of ammonia water and hydrated lime. The use of methanol, ethanol, acetone, and ethylene glycol along with inorganic and organic acids as catalyst has also been studied but the pretreatment cost is relatively high as compared with physicochemical pretreatments.

8.3.4   Biological pretreatment

This treatment involves the use of microorganisms that selectively degrade lignin and hemicellulose. Biological pretreatments are less energy intensive compared to chemical and physicochemical processes and only require mild reaction conditions. However, the process is very slow, making it unattractive for commercial use.

8.4 Thermal conversion processes

8.4.1 Conversion processes

Awareness of the potential of recovering, utilization of energy, and upgraded products from agricultural wastes must be made known to researchers. There is a huge amount of energy locked in these solid wastes. By processing them in mills, tremendous savings can be realized through the potential of converting them into energy and value-added products. The dependence of industries on conventional energy and fuels can be reduced to a great extent. The potential offered by biomass energy for reducing a country's energy problem seems viable to a certain degree. Figure 8.2 shows the main processes of thermal energy conversion. There are several technologies for conversion of biomass into energy and higher value products. These are mainly classified as biochemical, thermochemical, physical, and liquefaction.

The biological process or the wet process may lead to anaerobic methane generation and ethanol fermentation. In the anaerobic digestion process, biogas and sludge are produced, where both could be processed into fuel and fertilizer, respectively. Selective microbial bacteria produce methane or hydrogen gas, operating at the required temperature and pH level. In the alcohol fermentation route, ethanol, carbon dioxide, and solid residues are produced. Biomass containing sugars and starch is converted into simple glucose using microorganisms to produce ethanol. In the liquefaction process, solvents are used at high pressure and moderate temperature to obtain the liquid product. The process employs a reactive carrier gas to produce hydrogenated liquid fuels.

Densification of biomass involves the physical transformation of the loose biomass into a more compact form, such as briquettes, pelletized fuel, and fuel logs, for ease of handling and storage. These involve the extrusion process of loose biomass particles with or without binder at high pressure and a carbonization process to obtain

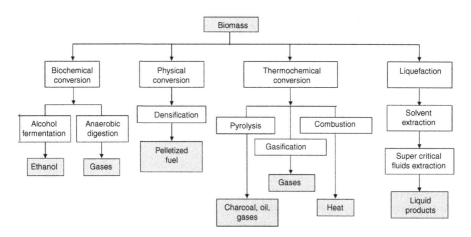

Figure 8.2 The main processes of thermal energy conversion.

the charcoal form. Pelletizing or briquetting basically increases energy density of the initial biomass form.

The thermochemical or dry process is the main process of the thermal conversion process. There are three main thermal processes for converting solid wastes into energy and by-products, that is, pyrolysis, gasification, and combustion. Each process gives a different range of products, that is, gas, liquid, or solid, depending on how the process is controlled. Figure 8.3 shows the various thermal conversion processes with different oxygen requirement for energy and upgraded products [9]. The system involves a different reactor design and configuration for a particular application. With the constraint of harvesting, collecting, and transportation cost of biomass, the best utilization and conversion to upgraded products are at the mill sites where the crops are processed and produce an abundance of biomass.

Biomass could contribute to value-added products after the excess availability of energy from the mills. Therefore, the utilization of this biomass seems appropriate in decentralized technological applications in the rural area scenario. This is an advantage for the rural population, on account of the social development and technological education for rural regions. Priority to their utilization as a fuel, agroresidues can also compete with other conventional sources for use as food, animal feed, fiber applications, fertilizers, chemical applications, etc. The basis of an agroresidue utilization strategy should therefore be processed on site and converted to value-added products, whichcan subsequently be stored and later dispatched.

8.4.2 Combustion process

Presently, direct combustion of biomass is used in medium and large industries to produce electrical power and process heat, which is the simplest route for energy recovery from these wastes. There are various combustion technologies available in the market but their suitability depends on the characteristics of the biomass itself. Generally,

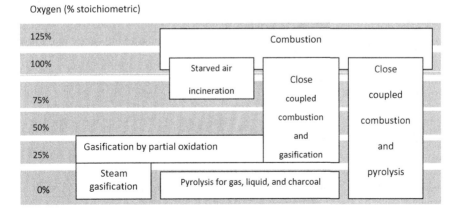

Figure 8.3 Thermal conversion processes and oxygen requirement [9].

biomass burns efficiently in an inclined moving bed combustion unit. Staged combustion techniques are used to improve the emissions standards. The combustion of biomass generally consists of volatile and char combustion. Therefore, the heat transfer and residence time for the two kinds of combustion will be different for different types of biomass. Thus, the design for the combustion process will be different for different types of biomass. The heat from the combustion process can then be used for drying, and steam can be used for thermal heating and steam electrical power generation.

8.4.3 Pyrolysis process

Biomass pyrolysis is a thermochemical decomposition of organic material at medium temperatures in the absence of oxygen. It is one of the oldest processes of making charcoal or biochar, which is often called carbonization that produces solid residues high in carbon content. The aim is to remove moisture and in the process convert the volatiles in the biomass material into higher carbon content. Long residence time and medium heating with an inert environment are the key requirements for the process. There are various techniques for the conversion of biomass to charcoal ranging from burying heated biomass, to beehive stoves, to modern carbonization plants. Table 8.3 shows the various pyrolysis processes with their respective residence time and terminal temperatures.

8.4.3.1 Fast pyrolysis process

Fast pyrolysis has been an advanced and relatively new emerging technology over the last 25 years compared with combustion and gasification. It converts solid bioresources into liquid products that can be processed into liquid fuel and value-added chemicals. It converts high yields of liquid products that can be stored and transported. The production of liquid pyrolysis oil from fast pyrolysis is known as bio-oil. It also utilizes moderate temperature and short vapor residence time, which are optimum for liquid production. The heart of the technology is the reactor itself, which current research and development is focused on. Bio-oil, which yields more than 75 wt% on a dry feed basis, together with the by-product of biochar and gas, could be reused within the process as energy recovery. Depending on the type of process used, the three main products are biochar (solid), pyrolytic oil or bio-oil (liquid), and gaseous fuel of low

Table 8.3 **Classification of pyrolysis process based on process variables [10–13]**

Pyrolysis process	Residence time	Heating rate	T (°C)	Pressure (bar)
Fast pyrolysis	0.1–2 s	High	400–650	~1.01
Flash pyrolysis	<0.5 s	Very high	>1000	~1.01
Slow (carbonization)	Hours–days	Very slow	300–500	~1.01
Slow pyrolysis	Hours	Low	400–600	~1.01
Vacuum pyrolysis	2–30 s	Medium	350–450	~0.15
Liquefaction	<10 s	High	250–325	250–300

heating value. The bio-oil is a high energy density fuel and therefore is easily transported and stored. The treated bio-oil can be burnt directly in steam generation plants, or converted to high-grade fuel in a biorefinery plant. The upgraded bio-oil fuel of higher calorific value and quality could be used in internal combustion engines and gas turbines.

8.4.4 Gasification process

Biomass gasification is used to provide clean gaseous fuel for combustion in furnaces, boilers, and internal combustion engines for power generation and process heating. Biomass in the form of char is usually used rather than in its dried form because the producer gases are relatively free of tar, water, and corroding components. Downdraft gasifiers are of popular design, which specifically eliminates the tars and oils from the gas for gas engine application. In the fixed bed gasifier, the moisture is usually driven off at the top drying zone before entering the pyrolysis zone. The tars and oils pass through the bed of hot char where they are synthesized into simpler gases. The gas velocity is low in the downdraft gasifier and the ash settles through the bottom grate so that very little ash is carried over with the gases. Prior to the utilization of the gas in engines, the gas passes through the dry cleaning system, which usually consists of cyclone, filter bags, and gas coolers. Presently, small-scale biomass gasification is not popular for power generation due to the messy maintenance issue of the system, but commercial scale would be the biomass integrated plasma gasification combined cycle where it was initially designed and operated for coal gasification.

8.4.5 Biochemical process

8.4.5.1 Biogas

Biogas is produced from the anaerobic bacteria or fermentation decomposition of organic matter in the absence of oxygen. It is a renewable energy source using anaerobic digestion of biodegradable materials such as manure, sewage, municipal waste, green waste, plant material, and organic effluents. It primarily contains methane (CH_4), carbon dioxide (CO_2), and trace amounts of hydrogen sulfide (H_2S). Methane, hydrogen, and carbon monoxide (CO) can be combusted or oxidized with oxygen. This energy release allows the clean biogas to be used as a fuel; it can be used for any heating purpose, such as cooking. It can also be used in a gas engine to convert the energy in the gas into electricity and heat.

In Malaysia, the potential source comes from the palm oil mills, which generate about 3.5 tons of liquid effluents per ton of palm oil produced. Anaerobic processes are used for the production of biogas, producing about 28 m^3/ton of POME (palm oil mill effluent). Utilizing the gas in a gas engine can generate 1.8 kWh of electricity per m^3 of biogas. Biogas production from POME is used for power generation and process heating. Biogas with a methane content of 60–70%, 30–40% CO_2, and small traces of hydrogen sulfide is used as fuel in steam boilers and thermal heaters in the palm oil refinery. In a conventional palm oil mill, about 2.5 m^3 of POME are generated for every ton of palm oil produced. The gases are usually distributed near the palm oil mill to other industries that use gases, such as the ceramic industry or palm oil refinery. Biogas can be compressed, the same way natural gas is compressed to compressed natural gas, and used to power motor vehicles.

8.4.6 Physical conversion process

The torrefaction process can be described as a mild form of pyrolysis at temperatures ranging from 200°C to 320°C. During the process, the water contained in the biomass as well as superfluous volatiles are removed. The biomass loses about 20% of its mass and about 10% of its heating value, without any change in its volume, thus decreasing its energy density. The purpose is to obtain a much better fuel quality for combustion and gasification applications as well as allow the material to be easily pelletized or briquetted.

8.4.7 Biomass liquefaction process

Biomass like coal can be converted into higher value hydrocarbons: liquid fuels, methane, and petrochemicals. Biomass to liquid fuels or "BTL" mimics "coal to liquid fuels" or "CTL." Biomass liquefaction is the production of liquid fuels from biomass using high pressure and temperature with the presence of solvents or catalysts. Specific liquefaction technologies generally fall into two categories: direct liquefaction and indirect liquefaction processes. Indirect liquefaction processes generally involve gasification of coal or biomass to a mixture of carbon monoxide and hydrogen (syngas) and then use a process such as the Fischer–Tropsch process to convert the syngas mixture into liquid hydrocarbon. The direct liquefaction processes converts coal or biomass into liquids directly, without the intermediate step of gasification, by breaking down its organic structure with the application of solvents or catalysts in a high pressure and temperature environment. Since liquid hydrocarbons generally have a higher hydrogen–carbon molar ratio than coals or biomass, either hydrogenation or carbon-rejection processes must be used in both direct and indirect liquefaction technologies. Biomass or coal liquefaction generally is a high-temperature/high-pressure process requires significant energy consumption at an industrial scale (thousands of barrels/day), and needs multibillion dollar capital investments. Thus, biomass/coal liquefaction is only economically viable at historically high oil prices, and therefore presents a high investment risk.

8.5 Densification of biomass

The low bulk density of biomass makes it expensive and inefficient to be developed and used. In order to address this issue, biomass can be densified; this is usually achieved with some form of extrusion and increasing the bulk density of the biomass significantly. Briquetting is a densification technique that essentially improves the handling characteristics of the materials for transport and storage. Consequently, biomass is most economically feasible when used close to the source. It produces

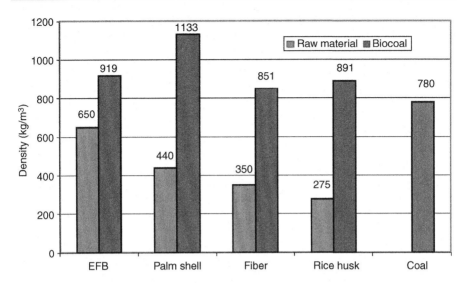

Figure 8.4 Densities of various biomass with respect to their biocoal and coal.

a homogeneous product with a higher energy density than that of the original raw material.

A previous study worked on the process of converting biomass material, selected from a group consisting of oil palm shells, oil palm fibers, empty fruit bunches, dried leaves, rice husks, and wood wastes, to biocoal. The process consists of pyrolysis that involves the application of pressure to the loose biomass particles in a mold that is simultaneously heated, in the absence or with a very limited quantity of oxygen to convert them into a compact and agglomerate form of the desired range of bio-coal products. Various biomass materials can be used and improvements made to the physical and chemical characteristics of the biocoal produced by the application of pressure, temperature, and interval timings. The biomass could be blended with other types of biomass materials or additives used for improved quality. The biocoals could be produced with certain fixed carbon content, ash content, and volatile matter content. The variation of densities is shown in Figure 8.4. It shows an improved fuel characteristic from its raw state. The process provided for each type of biomass had its individual specific carbonization temperature and pyrolysis pressure with timing to produce the required physical characteristic of biocoal [14].

8.6 Biomass gasification

Biomass gasification is one of the chemical processes that convert biomass solid residues into usable gaseous fuel called producer gas. It offers the cleanest, most efficient method available to produce synthesis gas from low value carbon-based feedstocks such as low rank coals, petroleum coke, biochar, and biomass materials. Producer gas is generated when the carbon materials are burned in substoichiometric air conditions. Producer gas contains mainly CO, hydrogen (H_2), CH_4, water vapor, and some inert gases. When mixed with air, producer gas can be used in internal combustion engines with a little modification. It can also be used as a cocombustion fuel with other liquid fuels, such as diesel or biodiesel, to minimize the liquid fuel consumption. There are different types of gasifiers available for the gasification process such as updraft, downdraft, cross-draft bed, and fluidized bed designs.

In previous studies, there have been some efforts to use producer gas as fuel using the cocombustion technique for the drying process. This concept has been applied in a dual fuel burner. In another study, pressurized induced flow was utilized in a downdraft gasifier using a low emission swirl burner. It was coupled with an air ejector, orifice cylinder, and a gas burner. Cogasification of biomass and coal was used in the process with primary compressed air supply to the gasifier. Producer gas flowed out from the gasifier into an orifice cylinder and gas burner. Secondary compressed air was supplied to the gas burner through an air ejector to assist the mixing and burning process. Low emission levels were obtained at various secondary air pressure supplies.

8.7 Biodiesel fuels

Animal and plant fats and oils are composed of triglycerides, which are esters containing three free fatty acids (FFAs) and the trihydric alcohol, glycerol. Methanol is the most common alcohol used because of its low cost and high reactivity as compared to other long-chain alcohols. The reaction requires triglycerides, alcohol, and heat, and a catalyst (acid and/or base) is used to speed up the reaction. It is important to note that the catalysts are not consumed by the reaction process, thus they are not reactants, but catalysts. Biodiesel produced from high FFA requires acid catalysis, which is much slower. Esterification is an acid-catalyzed chemical reaction involving high FFA and alcohol, which yields fatty alkyl ester and water. Sulfuric acid is the common acid catalyst due to its low cost.

Almost all biodiesel is produced from neat vegetable oils using the base-catalyzed technique as it is the most economical process for treating virgin vegetable oils, requiring only low temperatures and pressures and producing more than 98% conversion yield (provided the starting oil is low in moisture and FFAs). Common base catalysts for transesterification include sodium hydroxide, potassium hydroxide, and sodium methoxide.

Biodiesel is produced from vegetable oils and/or animal fat through transesterification with alcohol to convert triglycerides into alkyl esters of the fatty acids (biodiesel) and glycerol using a basic homogeneous catalyst, such as sodium hydroxide, potassium hydroxide, and sodium acetate. Base-catalyzed reactions are very sensitive to the presence of FFA and are unsuitable for crude oils with FFA contents higher than about 3%. To prevent saponification during the transesterification reaction, neat oils with FFA of less than 0.5 and water contents of 0.05 wt% must be used. High FFA levels in the oil feed also deplete the base catalyst through acid–base neutralization reactions. Base-catalyzed reactions also require that the NaOH catalyst be neutralized with acid and removed from the reactor effluent with a water wash. The resulting salt by-product from the acid–base neutralization must then be separated from the biodiesel product.

In Malaysia, palm oil has successfully been processed into palm diesel (palm oil methyl ester) and run in unmodified engines in buses, trucks, taxis, and cars. Crude palm oil is used directly in Germany's Esbett engines fixed in several Mercedes cars, which proved to be successful during the trial demonstration period. Nonedible seed oil such as *Jatropha curcas* has also successfully been used as biodiesel fuel in India and many tropical countries are cultivating and using these plants as a biodiesel substitute.

Homogeneous acid catalyst reactions are generally slow and less suitable for biodiesel processes. Although the performance of the acid catalyst is not affected by the presence of FFAs in the oil or fat feedstock, nevertheless, the process requires a high alcohol-to-oil mole ratio and long reaction times due to the low activity of the acid catalyst. When using a base or an acid catalyst, the transesterification process, which occurs in a corrosive environment, requires costly neutralization, water wash, filtration, and solid waste disposal steps to remove the spent catalyst from the biodiesel and glycerol product streams.

Heterogeneous acid and base catalysts can also be classified as Brönsted or Lewis catalysts. Some solid metal oxides such as those of tin, magnesium, aluminum, and zinc are examples of heterogeneous catalysts. The reaction is performed at a higher temperature than homogeneous catalysis processes, with an excess of methanol. Heterogeneous catalysts can be recycled and used several times with better separation of the biodiesel. They are environmentally friendly and can be used in a continuous process without the need for further purification steps. They are also potentially cheap and available abundantly. The catalysts can easily be tuned to include desired catalyst properties so that the presence of FFAs or water does not adversely affect the reaction steps during the transesterification process.

8.8   Bioethanol from biomass

Sugarcane, corn, and cassava are the first generation bioethanol (food-based) sources and are the best plants for ethanol production. Oil palm empty fruit bunches, bagasse (second generation – nonfood source based), contains cellulose material, which can be converted into simple sugar. The fermentation of sugar produces liquid, which is then distilled to obtained fuel-grade ethanol. Ethanol is a high-octane fuel that can improve engines' performance as well as reduce air emissions. Ethanol can be used in a neat form or blended with petrol, which is called gasohol (22% ethanol). ASEAN countries like Thailand and the Philippines have implemented the use of ethanol in commercial petrol (E5) and will increase its percentage to E11.

Bioethanol production consists of hydrolysis on pretreated lignocellulosic materials, using enzymes to break complex cellulose into simple sugars such as glucose, followed by fermentation and distillation. The stages to produce ethanol using a biological approach are, first, the pretreatment phase, to make the lignocellulosic material such as wood or straw amenable to hydrolysis. The second stage is cellulose hydrolysis, which breaks down the molecules into sugars. The next stage is the separation process of the sugar solution from the residual materials and the reaction of microbial fermentation of the sugar solution. In the last stage distillation produces roughly 95% pure alcohol with dehydration by molecular sieves to bring the ethanol concentration to over 99.5%.

8.9   Present and future utilization scenario of biomass

Presently, biomass from plantation is gathered in the mills such as palm oil mills, sugarcane mills, rice mills, etc., which are easier to manage. Urban waste such as municipal solid waste and industrial waste are heterogeneous types, which need to be segregated and further processed. Small biomass generators need to be able to store, sell, or utilize products for themselves for individual or community needs. The best application to utilize biomass is to know the value of the biomass for a particular application so that maximum benefits can be generated. Ideally, the first approach is the physical recycling process, followed by the thermochemical recycling approach. The last option for surplus biomass is to utilize it as fuel for combustion, which is the thermal recycling process where the biomass is burned in boilers for electrical power generation and for process heating for cogeneration plants.

8.10   Conclusions

Depending on the amount of biomass available, type and the distance to application, indigenous technology should be developed rather than importing foreign technology. Local technologies may be crude, unattractive, and lack of esthetic values but performance can be improved from experience. Funding may be limited but governments and indigenous populations should be proud of having their own technologies. Therefore, it is important for developing countries to nurture their own technologies, standards, and policies. Governments should act swiftly to promote local investors and private companies to develop and improve such technologies since the latest technologies from overseas are not easy to obtain. These activities will help local people to be creative and innovative and improve the living standards for future generations. Research from universities can provide new information to the various agencies regarding future activities and a comprehensive economic transformation plan can be formulated to propel the country's economy for future needs.

References

[1]   Mohan D, Pittman CU Jr, Steele PH. Pyrolysis of wood/biomass for bio-oil: a critical review. Energy Fuels 2006;20:848–89.
[2]   Islam, MN. Pyrolysis of biomass solid wastes and its catalytic upgrading with techno-economics analysis. PhD thesis, Universiti Teknologi Malaysia, 1998.

[3]   Huffman DR, Vogiatzis AJ, Bridgwater AV. The characterisation of fast pyrolysis bio-
      oils. In: Bridgwater AV, editor. Advances in thermochemical biomass conversion. Lon-
      don: Blackie Academic and Professional; 1992.
[4]   Saka S, Munusamy MV, Shibata M, Tono Y, Miyafuji H. 2008. Chemical constituents of
      the different anatomical parts of the oil palm for their sustainable utilization. JSPS-VCC
      Group Seminar, Natural Resources and Energy Environment. Kyoto, Japan; November
      24–25, 2008. p. 19–34.
[5]   Bridgwater AV. Production of high grade fuels and chemicals from catalytic pyrolysis of
      biomass. Catal Today 1996;29:285–95.
[6]   Czernik S, Bridgwater AV. Overview of applications of biomass fast pyrolysis oil. En-
      ergy Fuels 2004;18:590–8.
[7]   Ani FN. Fast pyrolysis of bioresources into energy and other applications. Proceeding of
      Energy from Biomass 2006 Seminar, Conversion of Bioresources into Energy and Other
      Applications. Kuala Lumpur; 2006. p. 1–12.
[8]   Keshwani DR, Cheng JJ. Switchgrass for bioethanol and other value-added application:
      a review. Bioresour Technol 2009;100:1515–23.
[9]   Bridgwater AV, Peacock GVC. Fast pyrolysis process for biomass. Renew Sustain En-
      ergy Rev 2000;4:1–73.
[10] Bridgwater T. Biomass for energy. J Sci Food Agric 2006;86(12):1755–68.
[11] Bulushev DA, Ross JRH. Catalysis for conversion of biomass to fuels via pyrolysis and
      gasification: a review. Catal Today 2011;171(1):1–13.
[12] Huber GW, Iborra S, Corma A. Synthesis of transportation fuels from biomass: chemis-
      try, catalysts, and engineering. Chem Rev 2006;106(9):4044–98.
[13] Vamvuka D. Bio-oil, solid and gaseous biofuels from biomass pyrolysis processes – an
      overview. Int J Energy Res 2011;35:835–62.
[14] Ani and Sarif. Automation and continuous production of biocoal. Malaysian Patent Fil-
      ing, PI 2008 4024; 2008.

Single-phase AC supply

Sameer Hanna Khader, Abdel Karim Khaled Daud
Department of Electrical Engineering, College of Engineering,
Palestine Polytechnic University, Hebron–West Bank, Palestine

9

Chapter Outline

9.1 Introduction

Electric circuits are divided into direct current (DC) and alternating current (AC) circuits, where the voltage and current has constant or alternative characters. These two currents produce electrical energy, whereas much of electrical energy used worldwide is AC energy. Such widespread use of this kind of energy requires a good understanding of the physics of AC systems.

AC has several advantages. The most important is allowing larger-scale power generation, transmission, and distribution than that of DC power transmission [10,12]. Large AC generators are multiphase (mainly three phase) having very high power ratings plus an efficient transformer to step up or step down the alternating voltage. This

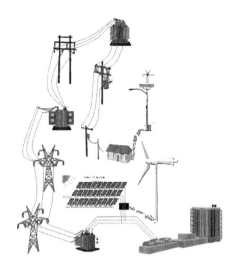

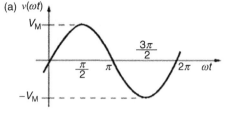

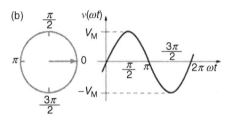

Figure 9.1 Electrical power systems.

Figure 9.2 Plot of waveform. (a) Sinusoidal voltage wave as a function of ωt and (b) sine wave with phase angle θ of 0°.

makes it possible to conduct AC energy economically over long distances from the generating plant (station) to various load centers by way of a high-voltage transmission system as shown in Figure 9.1. Furthermore, since power transformers present the key element in power transmission they are energized only by AC source.

9.2 Alternating current waveform

Equation (9.1) shows a simple sinusoidal voltage:

$$v(t) = V_M \sin \omega t \tag{9.1}$$

where $v(t)$ is the instantaneous value of the voltage sine wave at a given instant of time t; V_M is the amplitude *or* peak value; ω is the angular frequency, expressed in radians per second (rad/s); and ωt is the argument of the sine function. A plot of the function in Equation (9.1) as a function of its argument is shown in Figure 9.2a. Obviously, the function repeats itself every 2π radians. This condition is described mathematically as $v(\omega t + 2\pi) = v(\omega t)$.

Note that this function goes through one period every T seconds. In other words, in 1 s, it goes through $1/T$ periods or cycles. The number of cycles per second, called hertz (Hz), is the frequency f, where $f = 1/T$. Now since $\omega t = 2\pi$, as shown in Figure 9.2a, we find that

$$\omega = \frac{2\pi}{T} = 2\pi f \tag{9.2}$$

This is, of course, the general relationship among periods in seconds, frequency in Hz, and radian frequency. Now that we have discussed some of the basic properties of a sine wave, let us consider the following general expression for a sinusoidal voltage function:

$$v(t) = V_M \sin(\omega t + \theta) \tag{9.3}$$

In this case, $(\omega t + \theta)$ is the argument of the sine function, and θ is called the phase angle.

From Equation (9.3), it is easy to observe that the initial value (i.e., the value at $t = 0$) of the sine wave depends entirely on the phase angle θ because the term ωt equals 0 at $t = 0$. In other words, the phase angle θ determines by how much the value of a sine wave differs from 0 at time $t = 0$, and thus the position in time of the sine wave. Figure 9.2b shows a sine wave with a phase angle θ of $0°$. The initial value of this sine wave is 0 because $V_M \sin(\omega \times 0 + 0) = 0$.

Figure 9.3a shows a sine wave with a phase angle θ of $45°$. As you can see from Figure 9.3, a positive phase angle $(0-180°)$ results in the sine wave having a positive instantaneous value when $\omega t = 0°$. In other words, a positive phase angle shifts the sine wave toward the left, that is, advances the sine wave in time. Figure 9.3b shows a sine wave with a phase angle θ of $-60°$. A negative phase angle $(0°$ to $-180°)$ results in the sine wave having a negative instantaneous value when $\omega t = 0°$.

In other words, a negative phase angle shifts the sine wave toward the right, that is, delays the sine wave in time. Figures 9.2 and 9.3 also show the phasor representations of the sine waves at time $\omega t = 0°$. Notice that, in Figures 9.2 and 9.3, the vertical distance between the tip of the rotating phasor representing the sine wave matches the instantaneous value of the sine wave at $\omega t = 0°$.

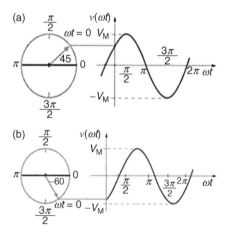

Figure 9.3 Sine waves at different phase angles θ. (a) $45°$ and (b) $-60°$.

Although our discussion has centered on the sine function, we could just as easily have used the cosine function, since the two waveforms differ only by a phase angle; that is,

$$\cos \omega t = \sin(\omega t + \pi/2); \sin \omega t = \cos(\omega t - \pi/2) \tag{9.4}$$

Example 9.1

Given the voltage $v(t) = 120\cos(314t + \pi/4)$, determine the frequency of the voltage in Hz and the phase angle in degrees.

Solution
The frequency in Hz is given by the expression $f = \omega/2\pi = 314/2\pi = 50$ Hz; using Equation (9.4), $v(t)$ can be written as $v(t) = 120\cos(314t + 45°) = 120\sin(314t + 135°)$ where the phase angle of $v(t)$ is 135°. A plot of this function is shown in Figure 9.4.

9.3 Root mean square

The term "RMS" stands for "root-mean-squared," and is defined as the "amount of AC power that produces the same heating effect as an equivalent DC power." It is worth mentioning that the RMS value is the square root of the mean (average) value of the squared function of the instantaneous values. It also called the effective value because it yields a single value that can be used in power calculations. The symbols used for defining RMS value are V_{RMS} or I_{RMS}.

In DC circuits, voltage, current, and corresponding power can be determined easily, while in an AC circuit RMS value can be defined according to Figure 9.5 as follows:

$$V_{RMS} = \sqrt{\frac{1}{T}\int_0^T v^2(\omega t)\, d\omega t} = \sqrt{\frac{1}{T}\int_0^T (V_M \sin(\omega t))^2\, d\omega t} = \frac{V_M}{\sqrt{2}} = 0.707\, V_M \tag{9.5}$$

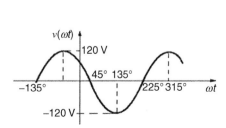

Figure 9.4 Waveform of Example 9.1.

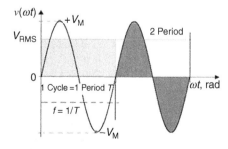

Figure 9.5 The sinusoidal form of induced voltage.

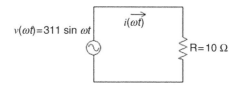

Figure 9.6 Simple single-phase circuit.

Example 9.2

The single-phase circuit shown in Figure 9.6 has a supply source of 311 V peak value at 50 Hz frequency, and energizes a heater with 10 Ω resistance.
  Calculate the RMS value of voltage and current.

Solution
According to the Ohm's law, the load current can be expressed as follows:

$$i(\omega t) = \frac{v(\omega t)}{R} = \frac{V_M}{R}\sin(\omega t) = I_M \sin(\omega t) \qquad (9.6)$$

where I_M is the peak value of the current. Notice that for pure resistive loads, the load current has the same phase angle as the applied voltage. Therefore, the load current can be described using Equation (9.6) as follows:

$$i(\omega t) = \frac{311}{10}\sin(\omega t) = 31.1\sin(\omega t)\ \text{A}$$

Thus, the RMS values of both supply voltage and drawn current are: $V_{RMS} = 311/\sqrt{2} = 220\ \text{V}$; and $I_{RMS} = V_{RMS}/R = 220/10 = 22\ \text{A}$.

9.4 Phase shift

Generally, AC circuits contain resistors, inductors, and/or capacitors. Impedance, represented by the letter "Z" and measured in Ohms, is the total opposition that a circuit offers to the flow of AC current, and is a combination of resistance R and reactance X, where $Z = R + jX$ [2]. In AC circuits, because of the existence of reactive components, the voltage and current may not reach the same amplitude peaks at the same time; they generally have a difference in timing. This timing difference is called phase shift, φ, where $0° \leq \varphi \leq 90°$, and is measured in angular degrees. Figure 9.7 illustrates the current for various loads with different phase shifts, where j is the complex operator [11,13].

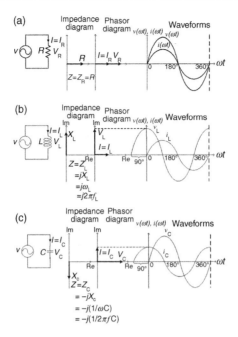

Figure 9.7 (a) V–I relationship of the pure R circuit, (b) V–I relationship of the pure L circuit, and (c) V–I relationship of the pure C circuit.

9.5 Concept of phasors

A phasor represents a time-varying sinusoidal waveform by a fixed complex number [3,13]. In other words a phasor is a complex number representing the amplitude and phase angle of a sinusoidal voltage and current. Phasors and complex impedances are only relevant to sinusoidal sources that have sine or cosine form. For two sine waves, the leading one reaches its peak first; the lagging one reaches its peak second.

Example 9.3

Refer to the single-phase circuit shown in Figure 9.6 with 5 Ω resistance, 20 mH inductance, and 100 μF capacitor. The source voltage is 220 V at 50 Hz frequency. Calculate the impedance, phase angle, and the RMS value of the drawn current for the following circuit configurations: (1) R circuit, (2) R–L circuit, (3) R–C circuit, and (4) R–L–C circuit. (5) Write the MATLAB code and display the voltage and current waveforms.

Solution

1. The circuit has an R character. The drawn current in this circuit has the following expression:

$$i(\omega t) = \frac{v(\omega t)}{R} = \frac{V_M}{R}\sin(\omega t - \theta_R) \tag{9.7}$$

where $V_M = \sqrt{2} \times 220 = 311.126$ V and the impedance angle, $\theta_R = \tan^{-1}(X_L/R) = 0°$. Referring to Equation (9.7), the current has the following value:

$$i(\omega t) = \frac{311.126 \text{ V}}{5\,\Omega} \sin(\omega t - 0°) = 62.225 \sin(\omega t) \text{ A}$$

Thus, the RMS value of the current is: $I_{RMS} = I_M/\sqrt{2} = 62.225/\sqrt{2} = 44$ A.

2. The circuit has an R–L character. The drawn current in this circuit can be expressed as follows:

$$i(\omega t) = \frac{V(\omega t)}{Z} = \frac{V_m}{\sqrt{R^2 + X_L^2}} \sin(\omega t - \theta_L) \tag{9.8}$$

where $V_m = \sqrt{2} \times 220 = 311.126$ V; $X_L = 2\pi f L = 2\pi \times 50 \times 20$ mH $= 6.283\ \Omega$; the circuit impedance is: $Z = \sqrt{R^2 + X_L^2} = 8.029\angle\theta_L°\ \Omega$; $\theta_L = \tan^{-1}(X_L/R) = \tan^{-1}(6.283/5) = 51.487°$.

Referring to Equation (9.8), the current has the following value: $i(\omega t) = 38.750 \sin(\omega t - 51.487°)$ A.

Thus, the RMS value of the current is: $I_{RMS} = I_m/\sqrt{2} = 38.750$ A$/\sqrt{2} = 27.4$ A.

3. The circuit has an R–C character: the drawn current in this circuit has the following expression:

$$i(\omega t) = \frac{V(\omega t)}{Z} = \frac{V_m}{\sqrt{R^2 + X_c^2}} \sin(\omega t - \theta_c) \tag{9.9}$$

where X_C is the capacitor reactance with the following expression:
XC $= 1/2\pi f C = 1/2\pi \times 50 \times 200$ µF $= 15.293\ \Omega$;
$\theta_c = \tan^{-1}(X_c/R) = \tan^{-1}(-15.293/5) = -71.895°$.

Referring to Equation (9.9), the current has the following value: $i(\omega t) = 18.641 \sin(\omega t + 71.895°)$ A.

Thus, the RMS value of the current is: $I_{RMS} = I_m/\sqrt{2} = 18.641/\sqrt{2} = 13.181$ A.

4. The circuit has an R–L–C character. The drawn current in this circuit has the following expression:

$$i(\omega t) = \frac{V(\omega t)}{Z} = \frac{V_m}{\sqrt{R^2 + (X_L - X_c)^2}} \sin(\omega t - (\theta_L - \theta_C)) \tag{9.10}$$

where $Z = \sqrt{R^2 + (X_L - X_c)^2} = 10.304\angle\theta°\ \Omega$ is the total impedance; the impedance angle is $\theta = \tan^{-1}((X_L - X_c)/R) = \tan^{-1}(-9.01/5) = -60.972°$. Referring to Equation (9.10), the current has the following value: $i(\omega t) = \sin(\omega t + 60.972°) = 30.194 \sin(\omega t + 60.972°)$ A.

Thus, the RMS value of the current is: $I_{RMS} = I_m/\sqrt{2} = 30.194/\sqrt{2} = 21.350$ A.

5. The MATLAB code of the described example is stated as follows [9]:

```
clc;
% Example 10.3: Calculating the circuit current at various load
characters and displaying the current waveform.
% The Input Data
% The circuit resistance, Ohm
R=input(' The circuit resistance:     R, Ohm=' );
% The circuit inductance, L
L=input(' The circuit inductance:     L, mH=' );
% The circuit capacitance
C=input(' The circuit capacitance:    C, uF=' );
% The RMS value of supply voltage
V=input(' The RMS value of supply voltage:  V, V =' );
% The supply voltage frequency
f=input(' The frequency of supply voltage:  f, Hz=' );

% Solution
% the circuit reactance XL
XL=2*pi*f.*L/1000;
% the circuit reactance XC
XC=1e+6/(2*pi*f.*C);
% the magnitude of the impedance Zm
Zm_RL=sqrt(R.^2+XL.^2);
Zm_RLC=sqrt(R.^2+(XL-XC)^2);
% the magnitude of the supply voltage Vm
Vm=sqrt(2)*V;
% the magnitude of the circuit current ImR, ImL, ImLC
Im_R=Vm./R;
Im_RL=Vm./Zm_RL;
Im_RLC=Vm./Zm_RLC;
% The RMS values of the obtained current I
I_R=Im_R/sqrt(2);
I_RL=Im_RL/sqrt(2);
I_RLC=Im_RLC/sqrt(2);
%the phase angle of the current Thita, dg
Thita_L_rd=atan(XL/R);
Thita_L=Thita_L_rd*180/pi;
Thita_LC_rd=atan((XL-XC)/R);
Thita_LC=Thita_LC_rd*180/pi;
% generating the instantaneous sinusoidal current waveforms i(wt)
W=2*pi*f;
T=1/f;
dt=T/1000;
ts=0:dt:T;
Vt=Vm.*sin(W*ts);
It_R=Im_R.*sin(W*ts-0);
It_RL=Im_RL.*sin(W*ts-Thita_L_rd);
It_RLC=Im_RLC.*sin(W*ts-Thita_LC_rd);

%Plot the sinusoidal current waveforms i(wt)
plot(ts*1000,It_R , ts*1000,It_RL, ts*1000,It_RLC);
title ('Plot of the drawn currents versus time at various characters');
label('Time, ms');
ylabel( 'i(wt), A ');
axis([0 T*1000 +Im_R -Im_R]);
Hold on;
```

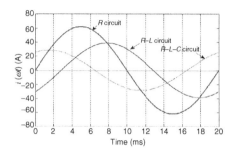

Figure 9.8 The current waveforms at various circuit characters.

The simulation results are displayed in Figure 9.8 where it can be concluded that the magnitude of the current varies at different circuit characters, in addition to the observed change in the current's phase angle with a $0°$ phase shift for a pure R circuit, a lagging phase angle of $51.48°$ for an $R{-}L$ circuit, and a leading phase angle of $60.97°$ for $R{-}L{-}C$ circuit.

9.6 Complex number analysis

Definition: What are complex numbers? These are a combination of a real and an imaginary number in the form $A + jB$, where A and B are real numbers, and j is the "unit imaginary number." The values A and B can be zero [14]. Examples: $1 + j$, $2 - j6$, $-j5$, 4.

What is an imaginary number? The imaginary number is denoted by $j = \sqrt{-1}; \Rightarrow j^2 = -1$; which means a number that, when squared, gives a negative result. If we squared a real number we always get a positive or zero result. In electrical and electronics applications where sinusoidal signals present the core of these applications, the complex numbers approach presents unique solutions for such circuits.

Complex numbers forms: Complex numbers can be written in three different forms (rectangular, polar, and exponential) as illustrated in Table 9.1, with the relationship between these forms.

Table 9.1 Complex numbers written in different forms

Form	General form	Example	Conversion between forms
Rectangular	$X + jY$	$3 + j4$	$R = \sqrt{X^2 + Y^2}$
Polar	$R(\cos\theta + j\sin\theta)$	$5(\cos53° + j\sin53°)$	
Exponential	$Re^{j\theta}$	$5e^{j(53/180)\pi}$	$\theta = \tan^{-1}\dfrac{Y}{X}$

9.7 Complex impedance

The impedance of a circuit "Z" is the total effective resistance to the flow of current by a combination of the elements of the circuit in Ohms. The total voltage across all three elements (resistors, capacitors, and inductors) is written as V_{RLC}. To find this total voltage, we cannot just add the voltages V_R, V_L, and V_C. Because V_L and V_C are considered to be imaginary quantities, we have the voltage of $V_{RLC} = IZ$, so $Z = R + j(X_L - X_C)$ where the magnitude of Z is given by:

$$Z = \sqrt{R^2 + (X_L - X_C)^2} \tag{9.11}$$

Example 9.4

A particular AC circuit given in Figure 9.9a has a resistor of 4 Ω, an inductive reactance of 8 Ω, and a capacitive reactance of 11 Ω.

Express the impedance of the circuit as a complex number in polar form and show that in complex plane.

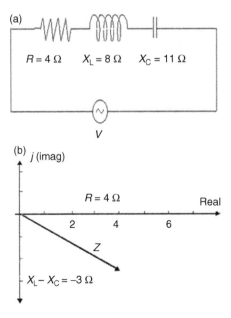

Figure 9.9 Three principle elements of an AC circuit. (a) Electrical circuit of Example 9.4. (b) Complex plane of circuit impedance.

Solution

$R = 4\ \Omega;\ X_L = 8\ \Omega;\ X_C = 11\ \Omega$. The circuit impedance Z and the magnitude of that impedance can be expressed as follows:

$$Z = R + j(X_L - X_c) = 4 + j(8-11) = (4 - j3)\ \Omega;\ \Rightarrow |Z| = 5\ \Omega \tag{9.12}$$

The phase angle of this impedance is $\theta = \tan^{-1}\left((X_L - X_c)/R\right) = \tan^{-1}\left((8-11)/3\right) = -36.87°$; the polar form of the circuit impedance is $Z = |Z|\ \angle\theta = 5\angle - 36.87°\ \Omega$. Based on obtained results for the circuit impedance a complex plane of this circuit is illustrated in Figure 9.9b, where the capacitive character of the circuit is clearly shown, which leads in turn to draw a current with a capacitive character.

9.7.1   Series impedance

Recall Ohm's law for pure resistance $V = IR$, but when a circuit contains an impedance with either inductive or capacitive element, Ohm's law becomes: $V = IZ$. For more than one impedance connected in series the total impedance can be expressed as:

$$Z_T = Z_1 + Z_2 + Z_3...Z_n = \sum_{k=1}^{n} Z_n = R_T + jX_T = \sum_{k=1}^{n} R_k + jX_k;$$
$$R_T = R_1 + R_2 + R_3...R_n;\ \ X_T = X_1 + X_2 + X_3...X_n \tag{9.13}$$

where Z_T is the total impedance with two components of total resistance R_T and total reactance X_T, respectively.

Example 9.5

For the series connected impedance circuit shown in Figure 9.10, assume the source voltage is 100 V at 50 Hz. Determine:

1. The total circuit impedance and corresponding total resistive, reactive components and phase angle.
2. The occurred phase angle and total current drawn by this circuit.
3. The voltage drop across each element.

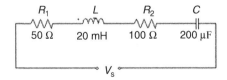

Figure 9.10 Series connected impedances.

Solution

1. Recalling Equation (9.13) the total impedance will be:

$$Z_\mathrm{T} = Z_1 + Z_2 = R_\mathrm{T} + jX_\mathrm{T}; \quad R_\mathrm{T} = R_1 + R_2 = 5 + 10 = 15\,\Omega;$$

$$X_\mathrm{T} = \omega L_1 - \frac{1}{\omega C} = 2\pi f L_1 - \left(\frac{1}{2\pi f C}\right) = 6.28 - 15.923 = -9.643\,\Omega \qquad (9.14)$$

The total impedance and phase angle are:

$$Z_\mathrm{T} = 15 - j9.643 = |Z_\mathrm{T}| e^{-j\theta}\,\Omega \qquad (9.15)$$

where $|Z_\mathrm{T}| = \sqrt{R_\mathrm{T}^2 + X_\mathrm{T}^2} = 17.832\,\Omega; \theta = \tan^{-1}\left(\dfrac{X_\mathrm{T}}{R_\mathrm{T}}\right) = -32.735°.$

Thus, $Z_\mathrm{T} = 17.832\angle - 32.735°\,\Omega = 17.832 e^{+j32.735}\,\Omega.$

2. The phase angle of this circuit is $\theta = -32.735\,\Omega$ with a capacitive character and the drawn current can be determined as follows:

$$i_\mathrm{T} = \frac{V_s\angle 0°}{Z_\mathrm{T}} = \frac{100\angle 0°}{17.832\angle - 32.753°} = 5.607\angle + 32.735°\,\mathrm{A} \qquad (9.16)$$

The magnitude of circuit current is $I_\mathrm{m} = \sqrt{2}\,I_\mathrm{T} = 7.930\,\mathrm{A}$ with leading angle of the current.

3. The voltage drop across each impedance is:

$$V_{z_1} = I_\mathrm{T} Z_1 = (5.607\angle 32.735°)(8.027\angle 51.474°) = 45\angle 84.209°\,\mathrm{V}$$
$$V_{z_2} = I_\mathrm{T} Z_2 = (5.607\angle 32.735°)(18.803\angle - 57.870°) = 105.428\angle - 25.065°\,\mathrm{V} \qquad (9.17)$$
$$V_\mathrm{T} = \sqrt{V_{z_1}^2 + V_{z_2}^2} = \sqrt{45^2 + 105.428^2} = 114.630\,\mathrm{V}$$

Notice that according to Equation (9.17) the total effective voltage of the circuit is ~115 V, which is greater than the source voltage. This is due to the dominant value of the capacitor compared with the inductive value. This fact is used in reducing the voltage drop in transmission lines.

9.7.2 Parallel impedance

Recall also, if we have several resistances $(R_1, R_2, R_3, \ldots, R_n)$ connected in parallel, then the total resistance R_T is given by:

$$\frac{1}{R_\mathrm{T}} = \frac{1}{R_1} + \frac{1}{R_2} + \frac{1}{R_3} + \ldots + \frac{1}{R_n} \qquad (9.18)$$

In the case of AC circuits, this becomes:

$$\frac{1}{Z_\mathrm{T}} = \frac{1}{Z_1} + \frac{1}{Z_2} + \frac{1}{Z_3} + \ldots + \frac{1}{Z_n} \qquad (9.19)$$

Suppose that two impedances Z_1 and Z_2 are to be connected in parallel, then the total resistance Z_T is given by:

$$\frac{1}{Z_T} = \frac{1}{Z_1} + \frac{1}{Z_2} = \frac{Z_1 + Z_2}{Z_1 Z_2} \tag{9.20}$$

Finding the reciprocal of both sides gives us:

$$Z_T = \frac{Z_1 Z_2}{Z_1 + Z_2} \tag{9.21}$$

Example 9.6

For the parallel connected circuit shown in Figure 9.11. Determine:

1. The total combined impedance in polar and rectangular forms.
2. The current flow in circuit branches if the source voltage is 220 V with frequency of 50 Hz.

Solution

1. Call the impedance given by the top part of the circuit Z_1 and the impedance given by the bottom part Z_2, therefore $Z_1 = 70 + j60$ Ω and $Z_2 = 40 - j25$ Ω. The total impedance can be calculated according to Equation (9.21) as follows:

$$Z_T = \frac{(70 + j60)(40 - j25)}{(70 + j60) + (40 - j25)} = \frac{(92.20\angle 40.60°)(47.17\angle - 32.01°)}{(115.4\angle 17.65°)} = 37.69\angle - 9.06° \ \Omega$$

2. To calculate the circuit currents and overall power factor (PF), first, we have to determine the total current drawn by the circuit, then the current of each branch assuming that the source voltage has zero phase shifts taken as reference vector. The currents are as follows:

$$V = I_T Z_T \Rightarrow I_T = \frac{V}{Z_T} = \frac{220\angle 0°}{37.69\angle - 9.06°} = 5.837\angle 9.06° \ A \tag{9.22}$$

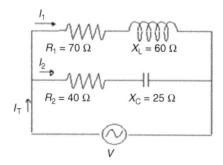

Figure 9.11 Parallel connected impedances.

The overall PF is cos(9.06°) = 0.987 with a lagging character. Since the voltage across each branch of the circuit is the same, therefore the branch current is:

$$I_1 Z_1 = I_2 Z_2 = I_T Z_T \Rightarrow I_1 = \frac{I_T Z_T}{Z_1} = \frac{I_T (Z_1 Z_2)}{Z_1 (Z_1 + Z_2)} = \frac{I_T Z_2}{(Z_1 + Z_2)}$$

$$\therefore I_1 = \frac{(5.837 \angle 9.06°)(47.12 \angle -32.01°)}{(115.4 \angle 17.65°)} = 2.383 \angle -40.6° \text{ A}$$

(9.23)

Similarly for the second branch current:

$$I_2 = \frac{I_T Z_T}{Z_2} = \frac{I_T (Z_1 Z_2)}{Z_2 (Z_1 + Z_2)} = \frac{I_T Z_1}{(Z_1 + Z_2)} = 4.663 \angle 32° \text{ A}$$

(9.24)

The second branch current can also be determined using Kirchhoff's law:

$$I_2 = I_T - I_1 = 5.837 \angle 9.06° - 2.383 \angle -40.6° = 4.663 \angle 32° \text{ A}$$

(9.25)

The obtained results from this example show that one of the branch currents has a leading character, while the second branch has a lagging character, and the overall character of the circuits has a lagging effect.

9.8 Electric power

Electric power is the rate at which electrical energy is transferred by an electric circuit and presents the work done during a certain time [4,8]. Electrical power presents the rate of doing work and is measured in Watt, which is joule per second. Electric power is usually produced by electric generators, but can also be supplied by sources such as electric batteries, wind turbines, photovoltaic generators, etc [5,6]. These terminologies will be discussed in following sections.

9.8.1 Real power

Real power (active) is the power that is dissipated in the resistance of the load. It uses the same formula applied to DC circuits, $P = VI$, while in AC circuits, voltage and current are functions of time. Power at a particular instant in time is given by:

$$P(\omega t) = v(\omega t)i(\omega t) = V_M \sin(\omega t) I_M \sin(\omega t) = \frac{V_M I_M}{2}(1 - \cos 2\omega t)$$

(9.26)

Equation (9.26) has two components, a constant component that presents the average component, which can be expressed as follows:

$$P = \frac{V_M I_M}{2} = \frac{V_M}{\sqrt{2}}\frac{I_M}{\sqrt{2}} = V_{rms} I_{rms} = P_{rms} \text{ W}$$

(9.27)

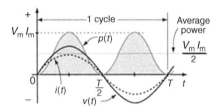

Figure 9.12 The instantaneous values of $v(t)$, $i(t)$, and $p(t)$ for resistive load.

where V_{rms}, I_{rms}, and P_{rms} are the RMS values of current, voltage, and power, respectively. Figure 9.12 illustrates the instantaneous values of voltage, current, and produced power, where it is shown that the power has a positive character with average value called real or effective value.

9.8.2 Reactive power

Reactive power (Q) is the power that is exchanged between reactive components, inductors, and capacitors that can be expressed as follows:

$$Q = I^2 X = \frac{V_s^2}{X} \text{ VAR} \tag{9.28}$$

unit of reactive power is volts-amps-reactive (VAR). By convention, Q is negative for capacitors and positive for inductors. Depending on the reactive power character, the reactive power for each element can be expressed as follows:

$$
\begin{aligned}
&V_L(\omega t) = V_M \sin(\omega t + 90°); \quad I_L(\omega t) = I_M \sin(\omega t) \\
&\therefore \Rightarrow P_L(\omega t) = V_L(\omega t) I_L(\omega t) = \frac{V_M}{\sqrt{2}} \frac{I_M}{\sqrt{2}} \sin(2\omega t) = V_{rms} I_{rms} \sin(2\omega t) \\
&V_C(\omega t) = V_M \sin(\omega t - 90°); \quad I_C(\omega t) = I_M \sin(\omega t); \\
&\therefore \Rightarrow P_C(\omega t) = V_C(\omega t) I_C(\omega t) = -\frac{V_M}{\sqrt{2}} \frac{I_M}{\sqrt{2}} \sin(2\omega t) = -V_{rms} I_{rms} \sin(2\omega t)
\end{aligned}
\tag{9.29}
$$

where V_L, V_C, I_L, and I_C are the voltage across and the current flows in the inductor and capacitor, respectively. The obtained power is displayed in Figure 9.13 for both elements where they have equally positive and negative portions among one period, which leads to zero average power. The reactive components Q_L and Q_C are the portions of reactive power that flow into the load and then back out. They contribute nothing to average power.

9.8.3 Complex power

Apparent power is the power that "appears" to flow to the load. The magnitude of apparent power can be calculated using similar formulas to those for active or reactive power:

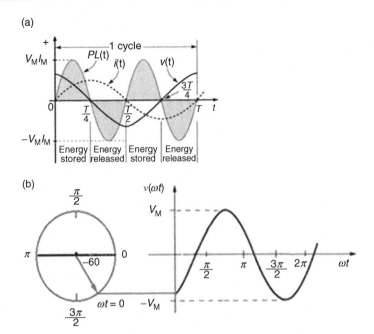

Figure 9.13 The reactive power for various loads. (a) Inductive load and (b) capacitive load.

$$S = P + jQ$$
$$|S| = VI = I^2 Z \text{ VA} \tag{9.30}$$

Complex power can also be determined using the current conjugate as follows:

$$S = VI^* = (V \angle \theta_v)(V \angle -\theta_I) = \underbrace{VI \cos(\theta_v - \theta_I)}_{P} + j\underbrace{VI \sin(\theta_v - \theta_I)}_{Q} \tag{9.31}$$

$$P = \text{Re}[\vec{S}] = S \cos\theta \text{ W}; \quad Q = \text{Im}[\vec{S}] = S \sin\theta \text{ VAR}$$

where I^* is complex current conjugate, θ_v is the phase angle of supply voltage taken as a reference with zero degree, and $\theta_I = \theta$ is the current phase angle.

9.8.4 Total power in AC circuits

The total power real (P_T) and reactive power (Q_T) are simply the sum of the real and reactive powers for all of the individual circuit elements. Figure 9.14 shows such a case where a few loads are connected together in parallel and series configuration. How much will the total active and reactive power be? It is worth mentioning that how

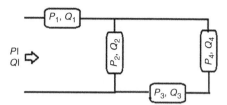

Figure 9.14 Parallel and series connected loads.

elements are connected does not matter for computation of total power. Thus, the total active and reactive power can be expressed as follows:

$$P_T = \sum_{n=1}^{\infty} P_n = P_1 + P_2 + P_3 + \dots; \quad Q_T = \sum_{n=1}^{\infty} Q_n = Q_1 + Q_2 + Q_3 + \dots$$

$$S_T = \sqrt{P_T^2 + Q_T^2} \quad \text{and} \quad \theta_T = \tan^{-1}\left(\frac{Q_T}{P_T}\right) \tag{9.32}$$

where P_T, Q_T, S_T, and θ_T are the total active, reactive, apparent power, and phase angle of the circuit, respectively.

Example 9.7

Recalling Example 9.6, calculate the consumed active and reactive power by the circuit elements. Calculate the total power and power angle.

Solution

From the previous example the apparent power of each current branch can be calculated taking into consideration the same branch voltage and calculated branch currents. Two approaches are used.

First approach: The branch current is used to determine the total power components:

$$I_1 = 2.383\angle -40.6° \text{ A}$$
$$S_1 = V_s I_1^* = (220\angle 0°)(2.383\angle 40.6°) = 524.26\angle 40.6°$$
$$= 398.055 + j341.175 \text{ VA}$$
$$I_2 = 4.663\angle 32° \text{ A}$$
$$S_2 = V_s I_2^* = (220\angle 0°)(4.663\angle -32°) = 1025.86\angle -32°$$
$$= 869.978 - j543.623 \text{ VA} \tag{9.33}$$
$$I_T = 5.837\angle 9.06°$$
$$S_T = V_s I_T^* = (220\angle 0°)(5.837\angle -9.06°) = 1284.14\angle -9.06°$$
$$= 1268.119 - j202.211 \text{ VA}$$
$$\therefore P_T = 1268.119 \text{ W}; \quad Q_T = -202.211 \text{ VAR}; \text{ and } \theta_T = -9.06°$$

Second approach: If the system's active and reactive powers are already known, the total values of these quantities can be directly obtained by summing the power quantities according to proposed methodology stated in Equation (9.32). Having the results of the first approach, the total power can be calculated as follows:

$$P_T = P_1 + P_2 = 398.055 + 869.978 = 1268.033 \text{ W}$$
$$Q_T = Q_1 + Q_2 = +341.175 - 543.623 = -202.448 \text{ VAR} \qquad (9.34)$$
$$S_T = S_1 + S_2 = P_T + jQ_T = 1268.11 - j202.21 \text{ VA}$$
$$= 1284.134 \angle -9.065 \text{ VA}$$

As expected the same results are obtained by using either the first or second approach. The total character of the circuit is resistive–capacitive as shown in Figure 9.15 where the total current leads the source voltage and the circuit consumes capacitive–reactive energy and returns inductive–reactive energy to the source. This is the main concept of PF correction systems that will be described in succeeding section.

9.8.5 Power factor

Power factor (PF) presents a major concern of any electrical company when delivering electrical energy to the industrial and domestic sectors. It is one of the measures of the

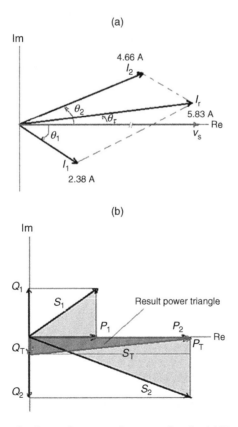

Figure 9.15 Current and voltage phasors and power triangle. (a) Phasor diagram and (b) power triangles.

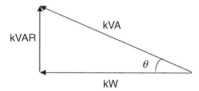

Figure 9.16 Power triangle.

quality of delivered electrical energy and has a direct effect on the cost of that energy [7]. The PF presents the ratio between the actual working power and the total apparent power:

$$PF = \frac{\text{Working power}}{\text{Apparent power}} = \frac{kW}{kVA} = \frac{kW}{kW + kVAR} = \frac{P}{S} = \cos\theta \qquad (9.35)$$

Referring to the previous paragraph, the relationships between the PF and power components are illustrated in Figure 9.16.

Depending on the load character, the value of the obtained PF and the formulated current phase angle can be classified into three categories:

$$PF \begin{cases} > 0 \text{ with } \theta > 0° \text{ for inductive loads} \Rightarrow Q = Q_L - Q_C > 0; \\ = 1 \text{ with } \theta = 0° \text{ for resistive loads} \Rightarrow Q = 0; \\ > 0 \text{ with } \theta < 0° \text{ for capacitive loads} \Rightarrow Q = Q_L - Q_C < 0. \end{cases} \qquad (9.36)$$

Example 9.8

Recalling Example 9.7, calculate the PF for each current branch and the total PF of the circuit.

Solution
The PFs of the illustrated circuit are calculated as follows:

$$PF_1 = \frac{P_1}{S_1} = \frac{398.055}{524.60} = 0.758 \Rightarrow \theta_1 = \cos^{-1}(0.758) = 40.64°$$

$$PF_2 = \frac{P_2}{S_2} = \frac{869.978}{1025.86} = 0.848 \Rightarrow \theta_2 = ? \qquad (9.37)$$

$$\sin\theta_2 = \frac{Q_2}{S_2} = \frac{-543}{1025.86} = -0.529 \Rightarrow \theta_2 = -32°$$

The total circuit PF is:

$$PF_T = \frac{P_T}{S_T} = \frac{1268.119}{1284.14} = 0.987 \Rightarrow \theta_T = ?$$

$$\sin\theta_T = \frac{Q_T}{S_T} = \frac{-202.11}{1284.14} = -0.157 \Rightarrow \theta_T = -9.056° \qquad (9.38)$$

For both equations $\sin\theta$ is determined in the case of a negative component of reactive power in order to determine the character of the circuit inductive of capacitive.

9.8.6 Power factor correction

Since PF is defined as the ratio of kW to kVA, therefore low PF results when kW is small in relation to kVA. So, what causes a large kVAR in a system? The answer is inductive loads. These loads include: transformers, induction motors, induction generators, welding machines, high intensity discharge lighting, power electronics converters, etc. The benefits that should be realized by improving the system PF are:

- Avoidance of utility penalties when the PF falls below the minimum permissible
- Reduced transformer, cabling, and motor losses
- Reduced total plant kVA for the same kW working power
- Lowered utility electrical billing
- Improved voltage regulation
- Minimized electrical distribution system investment costs

How is the PF corrected? The PF can be corrected by installing loads that consume capacitive reactive energy and returns to the utility inductive reactive energy, which in turn reduces the total apparent power resulting in increasing the PF. Such loads are capacitor banks and synchronous motors. So, several methods are used to improve the PF. The most popular method is installing a capacitor bank (set of capacitors) where the connected capacitor bank generates inductive reactive energy that compensates part of required by most of utility loads energy. Figure 9.17 illustrates the effect of adding a capacitor bank to the overall consumed apparent power. The value of a capacitor bank required to improve the PF can be calculated based on the consumed active power with the actual PF and the desired PF as follows [1]:

$$Q_C = P\left(\tan\theta_1 - \tan\theta_2\right) \text{kVAR} \tag{9.39}$$

where P is the active power of the load in kW; θ_1 is the PF's angle before correction (actual); and θ_2 is the PF's angle after correction (desired).

Example 9.9

It is required to install a capacitor bank energized from a single-phase utility with 220 V/50 Hz. Calculate the amount of capacitance required to improve the PF from 0.82 to 0.95 with a maximum power consumption of 200 kW. Calculate the freed-up kVA due to PF correction.

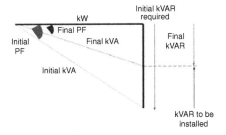

Figure 9.17 The effect of installing a capacitor bank.

Solution

Recalling Equation (9.39), the required capacitor bank can be calculated as follows:

$$Q_C = P(\tan\theta_1 - \tan\theta_2) = 200\left[\tan(\cos^{-1}0.82) - \tan(\cos^{-1}0.95)\right]$$
$$= 200\left[\tan 34.915° - \tan 18.198°\right] = 73.867\,\text{kVAR}$$

(9.40)

$$C = \frac{1,000\,\text{kVAR}}{2\pi f (kV)^2} = \frac{1,000 \times 73,867}{2\pi 50(0.22)^2} = 4,860.438\,\mu F$$

$$X_C = \frac{10^6}{2\pi f C} = 6.552\,\Omega \text{ and } I_C = \frac{V}{X_C} = 33.577\,A$$

(9.41)

The freed-up kVA can be calculated as follows:

$$kVA = kW\left(\frac{1}{\text{actualPF}} - \frac{1}{\text{desiredPF}}\right)$$

$$= 200\,kW\left(\frac{1}{0.82} - \frac{1}{0.95}\right) = 33.34\,kVA$$

(9.42)

which means that 33.34 kVA apparent power is saved by keeping the same active power, resulting in increased utility capacity and a reduced consumption bill.

9.9 Electrical energy

Electrical energy is the work done resulting from the flow of charged particles, such as electrons or ions, from one electric potential to another causing heat, movement, light, sound, etc [11,13]. The conducted work is measured in joules, which defines how much electric power is consumed by being converted into mechanical force to push these electrons to travel between electric potential ends for a unit of time, thus:

$$\text{Energy} = \text{Power} \times \text{Time} = \text{Volt} \times \text{Current} \times \text{Second}$$
$$= \frac{\text{Joule}}{\text{Coulomb}} \times \frac{\text{Coulomb}}{\text{Second}} \times \text{Second} = \text{Joule}$$

Electrical energy is measured in watt-hours (Wh) or kilowatt-hours (kWh), which means the amount of energy spent for doing certain work. The main equation of electric energy in an AC circuit is:

$$\text{Energy} = \frac{VI\cos\theta}{1000}\,h = ...\,kWh$$

(9.43)

Example 9.10

For a 100-Ω incandescent light bulb with efficiency of 75%, 0.85 lagging PF, 5 h daily operation, and 220 V/50 Hz source voltage, determine:

1. The bulb power consumed from the source power and how much luminance this bulb should produce per m².
2. The produced annual energy and annual energy losses.
3. The annual bill of consumed energy, at a price of 1 kWh, is $0.06.

Solution

The light bulb has a resistance of 100 Ω, but also has some kind of reactance, which is indirectly represented throughout the existing PF. The consumed energy is calculated based on the source power P_{inp}, which depends on the output power P_{out} and the bulb's efficiency η, thus:

1. The input power and output power are calculated as follows:

$$P_{out} = VIPF = \frac{V^2 PF}{R} = \frac{(220V)^2(0.85)}{100\,\Omega} = 411.4\,W$$

$$P_{inp} = \frac{P_{out}}{Efficieny\,\%} = \frac{411.4\,W}{0.75} = 548.534\,W = 0.5485\,kW \tag{9.44}$$

The amount of illumination produced by this lamp depends on the luminance intensity. For incandescent bulbs this intensity is around 17 lm/W [15,16].

$$Luminance = P_{out} \times \frac{lm}{W} = 411.4\,W \times \frac{17\,lm}{W} = 6993.8\,lux = 6993.8\,lm/m^2 \tag{9.45}$$

2. The produced annual energy E_{an} and energy losses E_{ls} are:

$$\begin{aligned} E_{an} &= P_{inp}(\text{operation hours/year}) \\ &= 0.5485\,kW \times 5\,h \times 365\,days = 1001.012\,kWh/year \\ E_{ls} &= (1-\eta)E_{an} \\ &= (1-0.75)1001.012\,kWh/year = 250.253\,kWh/year \end{aligned} \tag{9.46}$$

3. The annual bill of consumed energy is:

$$\begin{aligned} \text{Energy bill} &= E_{an} \times price/kWh \\ &= (1001.012\,kWh/year) \times (0.06\$/kWh) \cong 60\$/year \end{aligned} \tag{9.47}$$

9.10 Advantages and disadvantages of a single-phase supply

Single-phase AC supply sources present the reference for comparison of various alternating power supplies and lay in the basement of these supplies. However, single-phase sources have major advantages and disadvantages when compared with three-phase supply sources as follows.

9.10.1   Advantages

- Presents simple and cheap power supplies that are capable of covering the industrial and domestic consumption needs to a certain limit.
- Requires the use of one transformer for power conversion and transfer.
- Single-phase current peaks twice during one cycle, whereas three-phase current peaks six times during one cycle resulting in stronger mechanical force and motor torque.
- Single-phase supply service can be less expensive than a three-phase current in certain situations.
- With respect to mathematical modeling and processing the physical process single-phase supply is simple compared with three-phase supply.

9.10.2   Disadvantages

- Single-phase power transmission requires a live and a return wire, so power is lost due to current flowing through two wires for each (there is one) driving electromotive force, while in three-phase power transmission the sum of all currents at either end is zero, and there is no return of current in the case of symmetrical loading, which means a return conductor is available "for free."
- A balanced three-phase, three-wire circuit with equal voltages uses 75% of the copper required for conductors compared with single-phase supply circuits.
- As the industrial loads increase in size and require more power single-phase supply cannot meet this demand.
- Due to the same current return in the second wire, single-phase voltages drop much more quickly than do three-phase voltages.
- Single-phase supply produces a breathing magnetic field in single-phase motors, which means these motors (split motor and capacitor motor) cannot self-start without an additional coil, called a starting coil, capacitor, and centrifugal switch. A three-phase supply produces a revolving magnetic field, which in turn causes self-start torque in three-phase motors.
- Three-phase motors are less expensive and usually lighter and smaller than single-phase motors of the same horsepower rating. There is also a wider choice of enclosures available than for single-phase motors.
- Some of single-phase loads generate a spectrum of current harmonics causing excess conductor heat, power loss, and low efficiency, while the effect of current harmonics when a three-phase supply is used is light due to three-phase transformer connections (delta connection) resulting in less loss and high efficiency.
- Interrupting the single-phase source causes power loss for the loads, while in three-phase power supply with a special transformer connection (open delta), the loss of one phase can be recovered by the other two phases, which means loads continue to have supply power with reduced rate.
- With respect to PF correction only the capacitor bank method is available for single-phase sources, while in three-phase sources three methods for power correction are available. Furthermore keeping the same kVAR single-phase capacitor banks incurs much greater cost compared to three-phase capacitor banks.

Finally, despite there being quite a number of advantages for using single-phase supply, the disadvantages of using this source are much greater resulting in huge investments in three-phase sources, applying state-of-the-art technologies in order to enhance the source quality, and providing unlimited electrical energy to consumers with excellent service quality.

9.11 Summary

Single-phase power supply has found widespread use mainly in domestic and industrial systems on a small scale. These sources are simple to design, implement, and regulate. However, they have major advantages related to power capacity, signal quality, and capability compared with three-phase sources. The applied methodology in this chapter was based on accumulative knowledge for understanding how these supply sources functioned and how the connected loads behave with respect to loading character, power consumption losses, PF correction, and energy cost.

This chapter covered the following subjects:

- A review of electrical power and power generation.
- The application of AC fundamentals for reactance, impedance, current, current phase shift, and power.
- The application of complex numbers for solving AC tasks and problems.
- An understanding of the principles for series and parallel connections of loads.
- An appreciation of the concept of consumed power, yielded losses, and produced energy.
- An understanding of the principles of PF correction and cost analysis.
- Finally, a discussion of the main advantages and disadvantages of single-phase supply.

Problems

1. A 2 kWh heater is energized by a single-phase source of 220 V RMS value at 50 Hz.
   a. What is the heater resistance and effective consumed current?
   b. What is the consumed power?
2. A single-phase source of 220 V RMS at 50 Hz energizes an R–L–C load with various characters of 10 Ω, 15 mH, and 100 μF.
   a. Determine the drawn current of the load at rated frequency.
   b. Suppose that frequency is reduced by 20% of its rated value, determine the drawn current.
   c. What is the value of the source frequency that causes the current to be in phase with voltage?
   d. Write the MATLAB code and display the results for the previously mentioned tasks.
3. Series connected four elements $R_1 = 6\ \Omega$; $R_2 = 12\ \Omega$; the circuit inductance is 15 mH; and the capacitor is 150 μF. The effective value of supply voltage is 100 V at 50 Hz. Determine:
   a. The total circuit impedance, corresponding total resistive, reactive component, and phase angle.
   b. The occurred phase angle and total current drawn by this circuit.
   c. The voltage drop across each element.
4. Parallel connected four elements combining two branches, first branch with $R_1 = 6\ \Omega$ and $L = 15$ mH; while the second branch is with $R_2 = 12\ \Omega$ and $C = 150\ \mu$F. The effective value of the supply voltage is 100 V at 50 Hz. Determine:
   a. The total circuit impedance, corresponding total resistive, reactive component, and phase angle.
   b. The occurred phase angle and total current drawn by this circuit.
   c. The voltage drop across each element.
   d. The results obtained in this problem with previous one, what do you conclude?

5. Write the MATLAB code and answer the following:
  a. Vary the value of connected inductance and observe how the current and voltage behaves across the circuit elements.
  b. Eliminate the value of connected capacitor and observe the circuit behaviors.
  c. At what value of supply voltage frequency will the current be maximum?
  d. Plot the obtained result for all tasks.
6. Series connected impedance circuit with $Z_1 = 5 + j6\,\Omega$; $Z_2 = 3 - j4\,\Omega$; and $Z_3 = 5\,\Omega$ energized by a source voltage of 100 V at 50 Hz. Determine:
  a. The total circuit impedance, corresponding total resistive, reactive component, and phase angle.
  b. The total current drawn by this circuit and the corresponding PF.
  c. The voltage drop across each element.
7. Parallel connected impedance circuit with three elements $Z_1 = 5 + j6\,\Omega$; $Z_2 = 3 - j4\,\Omega$; and $Z_3 = 3 - j8\,\Omega$ energized by a source voltage of 100 V at 50 Hz. Determine:
  a. The total circuit impedance, corresponding total resistive, reactive component, and phase angle.
  b. The total current drawn by this circuit and the corresponding PF.
  c. The voltage drop across each element.
8. A combination circuit of series and parallel connected impedances is illustrated in Figure 9.18. Determine:
  a. The total circuit impedance, corresponding total resistive, reactive component, and phase angle.
  b. The total current drawn by this circuit and the corresponding PF.
  c. The voltage drop across each element.
9. Based on obtained results of problem 8, determine:
  a. The power components of the circuit elements and the total circuit power.
  b. The overall character of the illustrated circuit. Draw the current phasors and power triangles.
  c. The MATLAB code and display the current, voltage, and power waveforms.
10. For the circuit illustrated in Figure 9.19, determine:
  a. The value of the resistance R and total active and reactive power P_T and Q_T, respectively.
  b. S and draw the power triangle.
  c. The total PF and the character of the obtained PF.

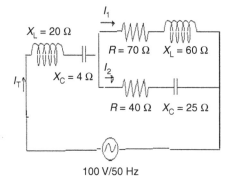
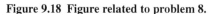

100 V/50 Hz

Figure 9.18 Figure related to problem 8.

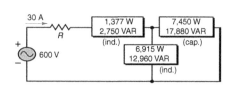

Figure 9.19 Figure related to problem 10.

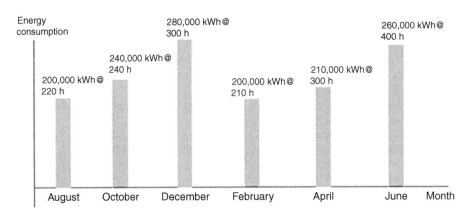

Figure 9.20 Figure related to problem 11.

11. It is required to install a capacitor bank energized from a single-phase utility with 220 V/50 Hz. Calculate the capacitor bank and capacitor current:

   a. To improve the PF using the monthly bill approach with the annual consumption diagram shown in Figure 9.20 with constant PF of 0.88.

   b. To improve the PF from 0.85 to 0.96 with a maximum power consumption of 300 kW.

   c. Repeat the previous task by replacing the source voltage with another three-phase source with 380 V energized delta connected capacitor bank, what do you conclude?

References

[1] Alpes Technologies. How to calculate the power of capacitors. Available from: http://www.alpestechnologies.com/sites/all/images/pdf_EN/Calculate.pdf; 2011.

[2] Bakshi UA, Bakshi VU. Electrical circuits & machines. 3rd ed. Pune: Technical Publication; 2007.

[3] Boylestad LR. Introductory circuit analysis. 4th ed. Ohio: Charles E. Merrill Publishing Company; 1982.

[4] Chapman JS. Electric machinery fundamental. 4th ed. Australia: McGraw Hill; 2005.

[5] Weedy BM, Cory BJ, Jenkins N, Ekanayake JB, Strbac G. Electric power systems. Sussex, England: John Wiley & Sons; 2012.

[6] Chowdhury H. Load flow analysis in power systems. University of Missouri-Rolla, USA: McGraw-Hill; 2000. p. 11.1–11.16.

[7] Christopher AH, Sen PK, Morroni A. Power factor correction: a fresh look into today's power systems. 2012 IEEE-IAS PCA San Antonio, TX. 54th Cement Industry Technical Conference, May 13–17, 2012.

[8] Hirofumi A, Watanabe EH, Mauricio A. Instantaneous power theory and applications to power conditions. Hoboken, NJ: John Wiley & Sons Inc.; 2007.

[9] Matlab User's Guide, version 8, R2008a.

[10] Power systems solutions. Available from: http://www.pssamerica.com/resources/PF Correction, Application Guide.pdf

[11] Rajput RK. Alternating current machines. New Delhi, India: Firewall Media; 2002. p. 260.

[12] Herman SL. Alternating current fundamentals. 8th ed. NY: Delmar Cengage Learning; 2011. Available from: CENGAGE brain.com.

[13] Kuphalt TR. Alternating current electric circuit'. 6th ed. 2007. Available from: www. gnu.org/licenses/dsl.html.

[14] Yagel EA. Complex numbers and phasors. Michigan: Department of EECS, University of Michigan; 2005.

[15] Smith WJ. Modern optical engineering – the design of optical systems. 3rd ed. Kaiser Electro-Optics Inc., Carisbad, California, USA: McGraw-Hill; 2000 [chapter 8].

[16] Candela, Lumen, Lux: the equations – CompuPhase. http://www.compuphase.com/ electronics/candela_lumen.htm

Three-phase AC supply

Abdul R. Beig
Department of Electrical Engineering, The Petroleum Institute,
Abu Dhabi, UAE

10

Chapter Outline

Electric Renewable Energy Systems

10.1 Introduction

The three-phase system is an economical way of bulk power transmission over long distances and for distribution. The three-phase system consists of a three-phase voltage source connected to a three-phase load by means of transformers and transmission lines. Two types of connections are possible, namely delta (Δ) connection and star or wye (Y) connection. The load and the source can be either in delta or star. The transmission line will be in delta connection. Normally the power system is operated in balanced three-phase condition. Most of the bulk loads such as industrial loads are three-phase balanced loads in nature. However, when there is a mix of single-phase and three-phase loads such as residential loads, the load will be unbalanced. Nevertheless, an unbalanced load can be resolved into a set of more than one balanced system. Hence, the analysis of a balanced system at steady state will be enough to understand three-phase systems.

10.2 Generation of three-phase voltages

The bulk of electricity is generated using three-phase alternating current (AC) generators also known as alternators. The AC generators are driven by hydroturbines, gas turbines, steam turbines, or internal combustion engines. AC generators have three separate windings distributed around the inner periphery of the stator [1]. The three windings corresponding to three phases of the system are electrically separated by 120°. The rotor consists of field windings or electromagnets and is rotated at synchronous speed by the prime mover. The rotatory magnetic field cuts the windings in the stator and according to Faraday's law [1,2] voltages are induced in these windings.

The three-phase windings are designed such that the induced three-phase voltages are sinusoidal, have same frequency, same peak voltage, and are phase displaced by 120°. The three-phase voltages under balanced conditions are defined as in (10.1) and the waveforms are shown in Figure 10.1.

$$v_a = V_m \sin \omega t \quad v_b = V_m \sin\left(\omega t - \frac{2\pi}{3}\right) \quad v_c = V_m \sin\left(\omega t - \frac{4\pi}{3}\right) \text{ or}$$

$$v_c = V_m \sin\left(\omega t + \frac{2\pi}{3}\right)$$

(10.1)

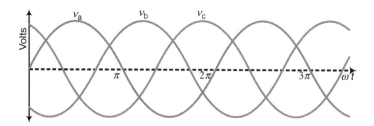

Figure 10.1 Waveforms of induced voltage in three-phase windings.

Let us denote these three phases as a, b, and c. Phase a voltage is taken as reference, then phase b voltage lags phase "a" by 120° (or $2\pi/3$ rad) and phase "c" voltage lags phase "a" voltage by 240° (or $4\pi/3$ rad). The phase sequence is a–b–c. Instead of a–b–c notation, the phases can also be donated by R (red)–Y (yellow)–B (blue) color code. The earlier mentioned phase sequence is referred to as positive phase sequence. If any of the phases are interchanged (e.g., phase "b" and phase "c" are interchanged) that is phase "c" lags phase "a" by 120° and phase "b" lags phase "a" by 240° then the resulting a-c-b phase sequence is referred as negative phase sequence.

For the past the few decades, major emphasis has been given on renewable energy sources such as wind energy, photovoltaic systems, fuel cells, etc. The generation of electricity from those sources is explained in Chapters 3, 4, and 6, respectively. In these types of electric sources three-phase electricity is generated with the help of direct current (DC) to AC converters [3]. The generation of balanced three-phase sinusoidal power from DC sources is explained in Chapter 15.

10.3   Connections of three-phase circuits

The three-phase sources can be connected either in delta or wye (or star-Y). Similarly the three-phase load can be connected either in delta or wye [4,5].

10.3.1   Wye-connected balanced source

A wye-connected balanced source is shown in Figure 10.2. In Y connection, the common terminal "n" is known as the neutral terminal of the source. Z_{sa}, Z_{sb}, and Z_{sc} are the source impedance of the sources in phases a, b, and c. For balanced sources, $Z_{sa} = Z_{sb} = Z_{sc} = Z_s$. The voltages v_a, v_b, and v_c are the source internal voltages (or the induced emf [electromotive force]) of the three phases a, b, and c. For balanced source, all three voltages will have the same frequency and amplitude, and are phase displaced by 120° ($2\pi/3$ rad) as defined in (10.1).

10.3.1.1   Phase voltages

Phase voltages are the voltages at the terminals "a," "b," and "c" measured with reference to neutral point "n."

In Figure 10.2, under open circuit conditions:

$$v_{an} = V_m \sin \omega t$$
$$v_{bn} = V_m \sin\left(\omega t - \frac{2\pi}{3}\right) v_{cn} = V_m \sin\left(\omega t - \frac{4\pi}{3}\right) \tag{10.2}$$

The phase voltages in phasor form are:

$$V_{an} = V_\varnothing \; \angle 0°, \quad V_{bn} = V_\varnothing \; \angle -120° \quad \text{and} \quad V_{cn} = V_\varnothing \; \angle -240°$$

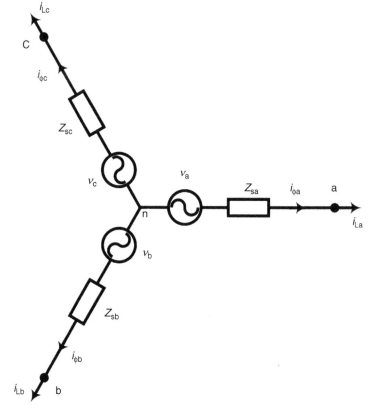

Figure 10.2 Wye-connected balanced source.

where $V_\varnothing$ is the RMS value of the phase voltage and $V_\varnothing = \dfrac{V_m}{\sqrt{2}}$.

Under balanced conditions, $v_{an} + v_{bn} + v_{cn} = 0$ and $V_{an} + V_{bn} + V_{cn} = 0$.

10.3.1.2 Line voltages

The voltage difference between two line terminals or two phases is the line voltage. Following the Kirchhoff's voltage law, line voltages v_{ab}, v_{bc}, and v_{ca} can be expressed in terms of phase voltages as in (10.3):

$$v_{ab} = v_{an} - v_{bn} = V_m \sin(\omega t) - V_m \sin\left(\omega t - \frac{2\pi}{3}\right) = \sqrt{3}V_m \sin\left(\omega t + \frac{\pi}{6}\right)$$

$$v_{bc} = v_{bn} - v_{cn} = V_m \sin\left(\omega t - \frac{2\pi}{3}\right) - V_m \sin\left(\omega t - \frac{4\pi}{3}\right) = \sqrt{3}V_m \sin\left(\omega t - \frac{\pi}{2}\right) \quad (10.3)$$

$$v_{ca} = v_{cn} - v_{an} = V_m \sin\left(\omega t - \frac{4\pi}{3}\right) - V_m \sin(\omega t) = \sqrt{3}V_m \sin\left(\omega t - \frac{7\pi}{6}\right)$$

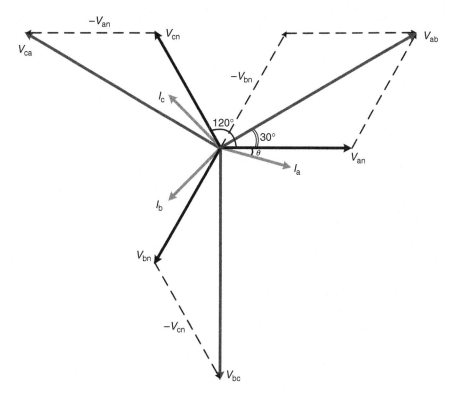

Figure 10.3 Phasor diagram of line and phase voltage vectors of wye-connected sources.

The line voltages in phasor form are:

$$V_{ab} = V_L \ \angle 30°, \quad V_{bc} = V_L \ \angle -90° \quad \text{and} \quad V_{ca} = V_L \ \angle -210°,$$

where V_L is the RMS value of the line voltage and $|V_L| = \left(\dfrac{\sqrt{3} V_m}{\sqrt{2}} \right) = \sqrt{3} \ | V_\varnothing |$.

Note: Under balanced conditions, $v_{ab} + v_{bc} + v_{ca} = 0$ and $V_{ab} + V_{bc} + V_{ca} = 0$.

The phasor diagram of the phase voltages and line voltages in a wye-connected source is shown in Figure 10.3.

10.3.1.3 Phase currents

The current flowing in the source winding or the source is the phase current. Let us assume that the phase current in the source is given by (10.4)

$$i_{\varnothing a} = I_m \sin(\omega t - \theta)$$

$$i_{\varnothing b} = I_m \sin\left(\omega t - \frac{2\pi}{3} - \theta \right) \tag{10.4}$$

$$i_{\varnothing c} = I_m \sin\left(\omega t - \frac{4\pi}{3} - \theta \right)$$

where θ is the phase angle between the phase voltage and corresponding phase current. In phasor form the above can be written as:

$$I_{\varnothing a} = I_{\varnothing} \ \angle -\theta°, \quad I_{\varnothing b} = I_{\varnothing} \ \angle -120° -\theta°, \quad \text{and} \quad I_{\varnothing c} = I_{\varnothing} \ \angle -240° -\theta°,$$

where $|I_{\varnothing}| = \dfrac{I_m}{\sqrt{2}}$. Note: $i_{\varnothing a} + i_{\varnothing b} + i_{\varnothing c} = 0$

10.3.1.4 Line currents

The current flowing out of the terminals is known as the line current. For a wye-connected source the line current is equal to the phase current:

$$I_{La} = I_{\varnothing a}, I_{Lb} = I_{\varnothing b}, I_{Lc} = I_{\varnothing c}, \quad \text{and} \quad I_L = I_{\varnothing} \tag{10.5}$$

Note: In a balanced wye-connected system:

- The line voltage leads the corresponding phase voltage by 30° and the magnitude of the line voltage is $\sqrt{3}$ times the magnitude of the phase voltage.
- The phase current is equal to the line current.

10.3.2 Delta-connected balanced source

A delta-connected balanced source is shown in Figure 10.4. Z_{sa}, Z_{sb}, and Z_{sc} are the source impedance of phases a, b, and c and for balanced sources, $Z_{sa} = Z_{sb} = Z_{sc} = Z_s$. The voltages v_a, v_b, and v_c are internal voltages of the source (or the induced emf) of the three phases a, b, and c. For balanced source, all three voltages will have the same frequency, equal amplitude, and phase displaced by 120° ($2\pi/3$ rad) as defined in (10.1).

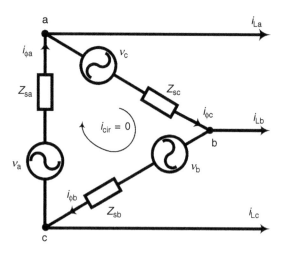

Figure 10.4 Delta-connected balanced source.

10.3.2.1 Line voltages

The windings of each phase in delta connections now appear between two lines as shown in Figure 10.4. Under open-circuit conditions, the phase voltage for a delta-connected source is given by:

$$v_{ab} = v_a = V_m \sin \omega t$$
$$v_{bc} = v_b = V_m \sin\left(\omega t - \frac{2\pi}{3}\right) \tag{10.6}$$
$$v_{ca} = v_c = V_m \sin\left(\omega t - \frac{4\pi}{3}\right)$$

The line voltages in phasor form are:

$$V_{ab} = V_L \angle 0°, \quad V_{bc} = V_L \angle -120°, \quad \text{and} \quad V_{ca} = V_L \angle -240°$$

where V_L is the RMS value of the line voltage and $|V_L| = \dfrac{V_m}{\sqrt{2}}$.

Note: $v_{ab} + v_{bc} + v_{ca} = 0$ so $i_{cir} = 0$.

10.3.2.2 Phase voltages

For a delta-connected source, the phase voltages are equal to line voltages:

$$V_\emptyset = V_L \tag{10.7}$$

10.3.2.3 Phase currents

The current flowing in the source winding or the source is the phase current. Let us assume that the phase current in the source is given by (10.8):

$$i_{\emptyset a} = I_m \sin(\omega t - \theta)$$
$$i_{\emptyset b} = I_m \sin\left(\omega t - \frac{2\pi}{3} - \theta\right) \tag{10.8}$$
$$i_{\emptyset c} = I_m \sin\left(\omega t - \frac{4\pi}{3} - \theta\right)$$

where θ is the phase angle between the phase voltage and corresponding phase current. The phase currents in phasor form are:

$$I_{\emptyset a} = I_\emptyset \angle -\theta°, \quad I_{\emptyset b} = I_\emptyset \angle (-120° - \theta)°$$

$$\text{and } I_{\emptyset c} = I_\emptyset \angle (-240° - \theta°), \quad \text{where } |I_\emptyset| = \frac{I_m}{\sqrt{2}}.$$

10.3.2.4   Line currents

The current flowing out of the terminals is the line current. For a delta- connected source the line current is given by (10.9).

Applying Kirchhoff's current law at nodes a, b, and c, respectively, we get:

$$i_{\text{La}} = i_{\varnothing a} - i_{\varnothing c} = I_m \sin(\omega t - \theta) - I_m \sin\left(\omega t - \frac{2\pi}{3} \pm \theta\right) = \sqrt{3}I_m \sin\left(\omega t - \frac{\pi}{6} \pm \theta\right)$$

$$i_{\text{Lb}} = i_{\varnothing b} - i_{\varnothing a} = I_m \sin\left(\omega t - \frac{2\pi}{3} \pm \theta\right) - I_m \sin(\omega t \pm \theta) = \sqrt{3}I_m \sin\left(\omega t - \frac{5\pi}{6} \pm \theta\right)$$

$$i_{\text{Lc}} = i_{\varnothing c} - i_{\varnothing b} = I_m \sin\left(\omega t - \frac{4\pi}{3} \pm \theta\right) - I_m \sin(\omega t \pm \theta) = \sqrt{3}I_m \sin\left(t - \frac{3\pi}{2} \pm \theta\right)$$

$$(10.9)$$

The line currents of a delta-connected source in phasor form are written as:

$$I_{\text{La}} = I_L \ \angle \pm \theta^\circ, \quad I_{\text{Lb}} = I_L \ \angle -120^\circ \pm \theta^\circ \quad \text{and} \quad I_{\text{Lc}} = I_L \ \angle -240^\circ \pm \theta^\circ,$$

where $|I_L| = \sqrt{3}\,|I_\varnothing| = \dfrac{I_m}{\sqrt{2}}$

Note: In a balanced delta-connected system:

• The line current lags the corresponding phase current by 30° and the magnitude of the line current is $\sqrt{3}$ times the magnitude of the phase current.
• The phase voltage is equal to the line voltage.

The phasor diagram of voltages and currents in a delta-connected source is given in Figure 10.5.

10.3.3   Wye-connected balanced load

A wye-connected load is shown in Figure 10.6. A, B, and C are the terminals of phase a, phase b, and phase c, respectively The three loads are connected at common point "N" called the neutral point of the load. Z_A, Z_B, and Z_C are the source impedance of phases a, b, and c. For a balanced load, $Z_A = Z_B = Z_C = Z$. If we take the phase voltage V_{AN} as reference, then

$$V_{\text{AN}} = V_\varnothing \ \angle 0^\circ, \quad V_{\text{BN}} = V_\varnothing \ \angle -120^\circ \quad \text{and} \quad V_{\text{CN}} = V_\varnothing \ \angle -240^\circ.$$

The line voltages are:

$$V_{\text{AB}} = V_{\text{AN}} - V_{\text{BN}} = \sqrt{3}\,|V_\varnothing| \ \angle 30^\circ = |V_L| \ \angle 30^\circ,$$

$$V_{\text{BC}} = V_{\text{BN}} - V_{\text{CN}} = \sqrt{3}\,|V_\varnothing| \ \angle -90^\circ = |V_L| \ \angle -90^\circ,$$

$$V_{\text{CA}} = V_{\text{CN}} - V_{\text{AN}} = \sqrt{3}\,|V_\varnothing| \ \angle -210^\circ = |V_L| \ \angle -210^\circ$$

and

$$|V_L| = \sqrt{3}\,|V_\varnothing|$$

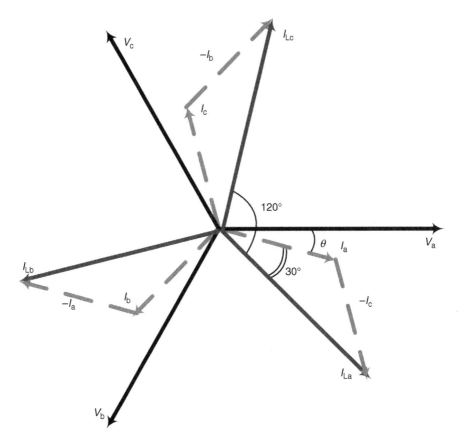

Figure 10.5 Phasor diagram of line currents and phase currents in a delta-connected source.

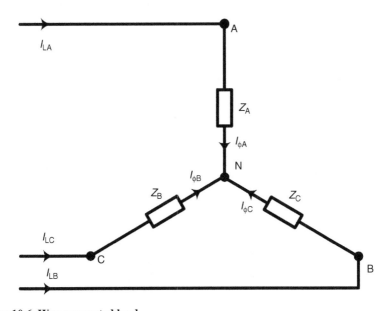

Figure 10.6 Wye-connected load.

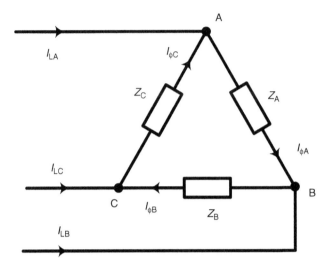

Figure 10.7 Delta-connected load .

When a wye-connected balanced load is supplied by the balanced source then the phase currents in the load will also be balanced.

$$I_{\varnothing A} = I_{\varnothing} \ \angle \pm \theta°, \quad I_{\varnothing B} = I_{\varnothing} \ \angle -120° \pm \theta°, \quad \text{and} \quad I_{\varnothing C} = I_{\varnothing} \ \angle -240° \pm \theta°,$$

where θ is the impedance angle, same as the phase angle between the phase voltage and the load current.

For a wye-connected load, the line currents are given by:

$$i_{LA} = i_{\varnothing A}, i_{LB} = i_{\varnothing B}, \ i_{LC} = i_{\varnothing C}, \ \text{and} \ I_{L} = I_{\varnothing}.$$

10.3.4 Delta-connected balanced load

A delta-connected load is shown in Figure 10.7. A, B, and C are the terminals of phase a, phase b, and phase c, respectively. Z_A, Z_B, and Z_C are the respective load impedances. For balanced sources, $Z_A = Z_B = Z_C = Z$. When the balanced load is supplied by a balanced source, the line currents as well as the phase currents in the load are balanced.

For a delta-connected load, the line voltages are given by:

$$V_{AB} = V_{L} \ \angle 0°, \quad V_{BC} = V_{L} \ \angle -120°, \quad \text{and} \quad V_{CA} = V_{L} \ \angle -240°,$$

For a delta-connected load, $V_{L} = V_{\varnothing}$.

When a delta-connected balanced load is supplied by a balanced source then the phase currents in the load will also be balanced:

$$I_{\varnothing A} = I_{\varnothing} \ \angle \pm \theta°, \quad I_{\varnothing B} = I_{\varnothing} \ \angle -120° \pm \theta°, \quad \text{and} \quad I_{\varnothing C} = I_{\varnothing} \ \angle -240° \pm \theta°,$$

where θ is the impedance angle.

For a delta-connected load, the line currents are given by:

$$I_{LA} = I_{\oslash A} - I_{\oslash B} = \sqrt{3}|I_{\oslash}|\angle - 30 \pm \theta°$$
$$I_{LB} = I_{\oslash B} - I_{\oslash C} = \sqrt{3}|I_{\oslash}|\angle - 150 \pm \theta°$$
$$I_{LC} = I_{\oslash C} - I_{\oslash A} = \sqrt{3}|I_{\oslash}|\angle - 270 \pm \theta°$$
$$\text{and } |I_L| = \sqrt{3}|I_{\oslash}|$$

10.4 Circuits with mixed connections

There are mainly four types of networks in three-phase systems and these are listed in Table 10.1.

The delta-connected source or load can be transformed into an equivalent wye-connected source or load, respectively [4,5]. Hence, all the other three connections can be analyzed by converting them into an equivalent wye–wye network. In the following subsection analysis of a wye–wye network is presented.

10.4.1 Wye–wye network

Line currents (I_L) are the currents flowing in each line and phase current ($I_{\oslash}$) is the current flowing in an individual source or load. For simplicity we limit our analysis to passive loads, that is, loads consisting of R, L, and C components only.

The circuit in Figure 10.8 shows a typical wye–wye network of the power system. $Z_{sa} = Z_{sb} = Z_{sc} = Z_s$ and $Z_{La} = Z_{Lb} = Z_{Lc} = Z_L$ are the impedances of the line connecting the source to load; these are equal and referred to as line impedances.

$Z_A = Z_B = Z_C = Z$ are the load impedances connected in wye. N is the neutral point of the load.

Applying Kirchhoff's current law at source neutral "n," the node equation is:

$$\frac{V_{a'n} - V_{Nn}}{Z_{sa} + Z_{La} + Z_A} - \frac{V_{b'n} - V_{Nn}}{Z_{sb} + Z_{Lb} + Z_B} + \frac{V_{c'n} - V_{Nn}}{Z_{sc} + Z_{Lc} + Z_C} = \frac{V_{Nn}}{Z_{Ln}}$$
$$\frac{V_{a'n} + V_{b'n} + V_{c'n}}{Z_s + Z_L + Z} = \frac{V_{Nn}}{Z_{Ln}} + \frac{V_{Nn}}{Z_s + Z_L + Z}$$

Table 10.1 Types of three-phase networks

Type of connection	Source	Load
Wye–wye connection	Wye	Wye
Wye–delta connection	Wye	Delta
Delta–wye connection	Delta	Wye
Delta–delta connection	Delta	Delta

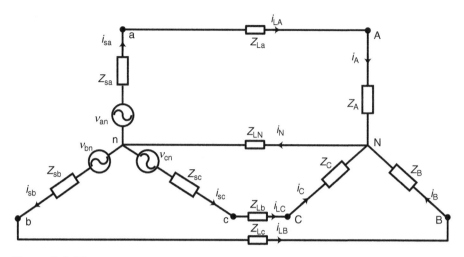

Figure 10.8 Wye–wye network.

For a balanced source $V_{a'n} + V_{b'n} + V_{c'n} = 0$:

$$\frac{V_{Nn}}{Z_{Ln}} + \frac{V_{Nn}}{Z_s + Z_L + Z} = 0$$

So, for a balanced three-phase circuit, $V_{Nn} = 0$.

This means that there is no voltage difference between the source neutral "n," and the load neutral "N." Consequently, $I_N = 0$, Hence, we can remove the neutral line from the balanced wye–wye configuration resulting in a three-phase –three-wire wye–wye connection or we can connect n to N resulting in a three-phase–four-wire wye–wye connection [1,4,5].

10.4.1.1 Line currents and phase currents

The line currents (I_{LA}, I_{LB}, and I_{LC}) are the currents flowing out of source terminals and into the load terminals through the line impedance.

In Figure 10.8, the voltages at the source in each phase are defined as:

$$V_{a'n} = |V_\varnothing| \angle 0°, \quad V_{b'n} = |V_\varnothing| \angle -120°, \quad \text{and} \quad V_{c'n} = |V_\varnothing| \angle -240°$$

The voltage $V_{a'n}$ is taken as the reference vector.

Let us define:

$$Z_\varnothing = Z_s + Z_L + Z = |Z_\varnothing| \angle \theta$$

$$I_{sA} = I_{LA} = I_A = \frac{V_{a'n}}{Z_{sa} + Z_{La} + Z_A} = \frac{V_\varnothing}{Z_{sa} + Z_{La} + Z_A} = \frac{|V_\varnothing| \angle (0 - \varnothing)°}{|Z_\varnothing|} = I_\varnothing \angle -\varnothing°,$$

where $|I_\varnothing| = \dfrac{|V_\varnothing|}{|Z_\varnothing|}$

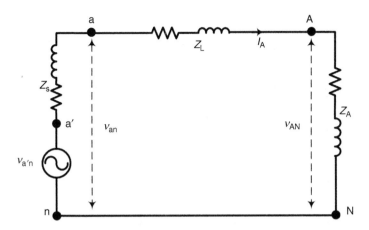

Figure 10.9 Per-phase equivalent circuit.

$$I_{sB} = I_{LB} = I_B = \frac{V_{b'n}}{Z_{sb} + Z_{Lb} + Z_B} = \frac{V_\varnothing}{Z_{sb} + Z_{Lb} + Z_B} = \frac{|V_\varnothing| \angle 0° - \varnothing°}{|Z_\varnothing|} = I_\varnothing \angle - \varnothing°$$

$$I_{sC} = I_{LC} = I_C = \frac{V_{c'n}}{Z_{sc} + Z_{Lc} + Z_C} = \frac{V_\varnothing}{Z_{sc} + Z_{Lc} + Z_C} = \frac{|V_\varnothing| \angle 0° - \varnothing°}{|Z_\varnothing|} = I_\varnothing \angle - \varnothing°$$

$$I_A + I_B + I_C = 0$$

For a wye-connected load or source, the line current = phase currents.
So $I_L = I_\varnothing$.
Note: For a balanced three-phase wye–wye network:

- The load currents are balanced, that is, they are equal in magnitude and phase displaced by 120°.
- There is no current in the neutral wire.

We can represent the previously mentioned system by an equivalent per phase network using phase quantities. The per phase equivalent circuit for any one phase (say, for phase a) is given in Figure 10.9.

10.4.1.2 Line voltages at the load terminals

In Figure 10.8, the phase voltages at load terminal are given by:

$$V_{AN} = I_A Z$$
$$V_{BN} = I_B Z$$
$$V_{CN} = I_C Z$$

It can be seen that all the three-phase voltages are balanced.

$$V_{AN} + V_{BN} + V_{CN} = 0$$

The line voltages are given by:

$$V_{AB} = V_{AN} - V_{BN}$$
$$V_{BC} = V_{BN} - V_{CN}$$
$$V_{CA} = V_{CN} - V_{AN}$$

It can be seen that all the three-line voltages are balanced. Hence, $V_{AB} + V_{BC} + V_{BC} = 0$ and as seen in Section 10.3.1, the line voltages will be leading the corresponding phase voltage by 30° and $|V_L| = \sqrt{3}|V_\varnothing|$.

10.4.1.3 Line voltages at the source terminals

In Figure 10.8, the phase voltages at the phase terminal are given by:

$$V_{an} = V_{a'n} - I_{sa}Z_{sa} = V_{a'n} - I_A Z_s$$
$$V_{bn} = V_{b'n} - I_{sb}Z_{sb} = V_{b'n} - I_B Z_s$$
$$V_{cn} = V_{c'n} - I_{sc}Z_{sc} = V_{c'n} - I_C Z_s$$

It can be seen that all the three-phase voltages are balanced:

$$V_{an} + V_{bn} + V_{cn} = 0$$

The line voltages at the source terminals are given by:

$$V_{ab} = V_{an} - V_{bn}$$
$$V_{bc} = V_{bn} - V_{cn}$$
$$V_{ca} = V_{cn} - V_{an}$$

It can be seen that all the three-line voltages are balanced.

As seen in Section 10.3.1, the line voltage is leading the corresponding phase voltage by 30° and $|V_L| = \sqrt{3}|V_\varnothing|$.

Example 10.1

A balanced three-phase wye-connected generator has an internal impedance of $0.2 + j0.5\ \Omega$ per phase and the phase to neutral voltage is 220 V. The generator feeds a balanced three-phase wye-connected load having a load impedance of $10 + j8\ \Omega$ in each phase. The impedance of the line connecting the generator to the load is $0.2 + j0.8\ \Omega$ in each line. The a-phase internal voltage of the generator is specified as the reference phasor.

1. Construct a phase equivalent circuit of the system.
2. Calculate the line currents I_{LA}, I_{LB}, and I_{LC}.
3. Calculate the phase voltages at the load, V_{AN}, V_{BN}, and V_{CN}.
4. Calculate the line voltages V_{AC}, V_{BC}, and V_{CA} at the load terminals.

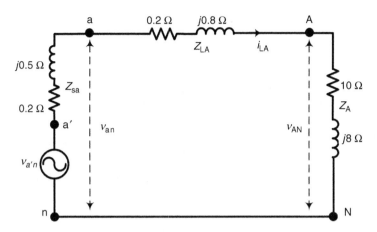

Figure 10.10 Per-phase circuit for Example 10.1.

5. Calculate the phase voltages at the generator terminals, V_{an}, V_{bn}, and V_{cn}.
6. Calculate the line voltages at the generator terminals, V_{ab}, V_{bc}, and V_{ca}.

Solution

1. a-Phase equivalent circuit (Figure 10.10):
2. Line currents I_{LA}, I_{LB}, and I_{LC}:

$$I_{LA} = \frac{V_{an}}{Z_\varnothing} = \frac{220\angle 0°}{(0.2+0.2+10)+j(0.5+0.8+8)} = \frac{220\angle 0°}{13.95\angle 41.8°} = 15.77\angle -41.8° \text{ A}$$

$$I_{LB} = 15.77\angle(-41.8°-120°) = 15.77\angle -161.8° \text{ A}$$

$$I_{LC} = 15.77\angle(-41.8°+120°) = 15.77\angle 78.2° \text{ A}$$

3. Phase voltages at the load, V_{AN}, V_{BN}, and V_{CN}:

$$V_{AN} = I_{LA}Z_A = (15.77\angle -41.8°)(10+j8) = 202\angle -3.14° \text{ V}$$

$$V_{BN} = 202\angle(-3.14°-120°) = 202\angle -123.14° \text{ V}$$

$$V_{CN} = 202\angle(-3.14°+120°) = 202\angle 116.86° \text{ V}$$

4. Line voltages V_{AB}, V_{BC}, and V_{CA} at the load terminals:

$$V_{AB} = \sqrt{3}V_{AN}\angle 30° = \sqrt{3}\times 202\angle(-3.14°+30°) = 349.9\angle 26.86° \text{ V}$$

$$V_{BC} = \sqrt{3}V_{BN}\angle 30° = \sqrt{3}\times 202\angle(-123.14°+30°) = 349.9\angle -93.14° \text{ V}$$

$$V_{CA} = \sqrt{3}V_{CN}\angle 30° = \sqrt{3}\times 202\angle(116.86°+30°) = 349.9\angle 146.86° \text{ V}$$

5. Phase voltages at the generator terminals, V_{an}, V_{bn}, and V_{cn}.
   Phase current in the source $= I_{LA}$:

$$V_{an} = 220\angle 0° - I_{sa}Z_s = 220\angle 0° - (15.77\angle -41.8°)(0.2+j0.5) = 212.43\angle -1.02° \text{ V}$$

$$V_{bn} = 212.43\angle(-1.02°-120°) = 212.43\angle -121.02° \text{ V}$$

$$V_{cn} = 212.43\angle(-1.02°+120°) = 212.43\angle 118.98° \text{ V}$$

6. Line voltages at the generator terminals, V_{ab}, V_{bc}, and V_{ca}:

$$V_{ab} = \sqrt{3}V_{an}\angle 30° = \sqrt{3} \times 212.43\angle(-1.02° + 30°) = 367.94\angle 28.98° \text{ V}$$
$$V_{bc} = \sqrt{3}V_{bn}\angle 30° = \sqrt{3} \times 212.43\angle(-121.02° + 30°) = 367.94\angle -91.02° \text{ V}$$
$$V_{bc} = \sqrt{3}V_{bn}\angle 30° = \sqrt{3} \times 212.43\angle(118.98° + 30°) = 367.94\angle 148.98° \text{ V}$$

10.4.2 Delta to wye conversion

A delta-connected load (or source) can be converted to a wye-connected load (or source) and vice versa. This transformation is particularly useful to convert all the other types of configurations to get equivalent wye–wye configurations.

Assume that Z_D is the load connected in delta as shown in Figure 10.11. The supply voltage is $V_{AB} = V\angle 0°$, $V_{BC} = V\angle -120°$, and $V_{CA} = V\angle -240°$.

Let us use the principle of superposition by considering only one source at a time. First consider source V_{AB}:

$$V_{AB} = V\angle 0°; \quad V_{BC} = 0 \text{ (open circuit)}; \quad V_{CA} = 0 \text{ (open circuit)}$$

The current due to the delta-connected load is:

$$I_{LA} = \frac{V_{AB}}{Z_D \parallel 2Z_D}$$

Let Z_Y be the equivalent wye-connected load in each phase supplied from the same source.

The current due to the wye-connected load is:

$$I_{LA} = \frac{V_{AB}}{Z_Y + Z_Y}$$

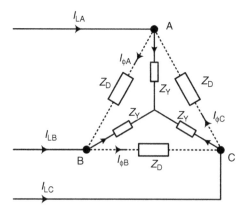

Figure 10.11 Delta-connected load and its wye equivalent.

Both the currents should be equal, hence:

$$\frac{V_{AB}}{(Z_D \times 2Z_D)/3Z_D} = \frac{V_{AB}}{2Z_Y}$$
$$\frac{3}{2Z_D} = \frac{1}{2Z_Y} \qquad\qquad (10.10)$$
$$3Z_Y = Z_D$$
$$Z_Y = \frac{Z_D}{3}$$

10.4.3 Wye-connected source/delta-connected load

The wye-connected source and delta-connected load are shown in Figure 10.12. The delta-connected load can be replaced by an equivalent wye-connected load using delta to wye conversion. In Figure 10.12, for a balanced load the load in each phase, $Z_A = Z_B = Z_C = Z_D$.

Using equation (10.9), the previously mentioned load can be replaced by an equivalent wye-connected load of $Z_Y = \dfrac{Z_D}{3}$.

Now the resulting wye–wye network can be solved using the procedure given in Section 10.4.1.

The magnitude load current in each phase in a delta-connected load is given by $I_\varnothing = \dfrac{I_L}{\sqrt{3}}$ and the phase current will be leading the line current by 30°.

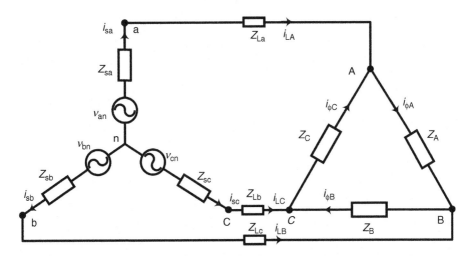

Figure 10.12 Wye-connected source and delta-connected load.

Example 10.2

A balanced three-phase wye-connected generator has an internal impedance of $0.2 + j0.5 \; \Omega/\varnothing$ and an internal voltage 220 V per phase. The generator feeds a delta-connected load having an impedance of $30 + j24 \; \Omega$ in each phase. The impedance of the line connecting the generator to the load is $0.1 + j0.9 \; \Omega$ in each line. The a-phase internal voltage of the generator is specified as the reference phasor.

1. Construct the a-phase equivalent circuit of the system.
2. Calculate the line currents I_{LA}, I_{LB}, and I_{LC}.
3. Calculate the phase voltages at the load, V_{AN}, V_{BN}, and V_{CN}.
4. Calculate the phase currents of the load.
5. Calculate the line voltages at the generator terminals, V_{ab}, V_{bc}, and V_{ca}.

Solution

1. Construct the a-phase equivalent circuit of the system (Figure 10.13):

$$Z_Y = \frac{Z_D}{3} = \frac{30 + j24}{3} = 10 + j8$$

2. Line currents I_{LA}, I_{LB}, and I_{LC}:

$$I_{LA} = \frac{V_{an}}{Z} = \frac{220\angle 0°}{(0.2 + 0.1 + 10) + j(0.5 + 0.9 + 8)} = \frac{220\angle 0°}{14\angle 42.38°} = 15.78\angle -42.38° \; A$$

$$I_{LB} = 15.78\angle(-42.38° - 120°) = 15.78\angle -162.38° \; A$$

$$I_{LC} = 15.78\angle(-42.38° + 120°) = 15.78\angle 77.62° \; A$$

3. Phase voltages at the load, V_{AN}, V_{BN}, and V_{CN}:

$$V_{AN} = I_{aA}Z_Y = (15.78\angle -42.38°)(10 + j8) = 202\angle -3.72° \; V$$

$$V_{BN} = 202\angle(-3.72° - 120°) = 202\angle -123.72° \; V$$

$$V_{CN} = 202\angle(-3.72° + 120°) = 202\angle 116.38° \; V$$

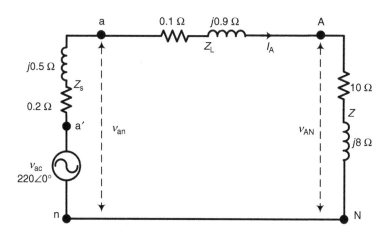

Figure 10.13 Single-phase equivalent circuit for Example 10.2.

4. Phase currents of the load I_{AB}, I_{BC}, and I_{CA}:

$$I_{\varnothing A} = \frac{I_{aA}\angle 30°}{\sqrt{3}} = \frac{15.78\angle(-42.38°+30°)}{\sqrt{3}} = 9.11\angle -12.38° \, A$$
$$I_{\varnothing B} = 9.11\angle(-12.38°-120°) = 9.11\angle -132.38° \, A$$
$$I_{\varnothing C} = 9.11\angle(-12.38°+120°) = 9.11\angle 107.62° \, A$$

5. Line voltages at the generator terminals V_{ab}, V_{bc}, and V_{ca}.
We start by calculating the phase voltages at the generator terminals, V_{an}, V_{bn}, and V_{cn}:

$$V_{an} = 220\angle 0° - I_{LA}Z_{an} = 220\angle 0° - (15.77\angle -42.38°)(0.2+j0.5) = 212.38\angle -0.998° \, V$$
$$V_{ab} = \sqrt{3}V_{an}\angle 30° = \sqrt{3}\times 212.38\angle(-0.998°+30°) = 367.86\angle 29.002° \, V$$
$$V_{bc} = 367.86\angle(29.002°-120°) = 367.86\angle -90.998° \, V$$
$$V_{ca} = 367.86\angle(29.002°+120°) = 367.86\angle 149.002° \, V$$

10.4.4 Delta–wye network

The delta–wye network is shown in Figure 10.14. The delta-connected source can be represented by the equivalent wye-connected source. The resulting equivalent circuit can be solved similar to the wye–wye network.

10.4.5 Delta–delta network

The delta–delta network is shown in Figure 10.15. The delta-connected source and delta-connected load can be represented by the equivalent wye-connected source and wye-connected load, respectively. The resulting equivalent circuit can be solved similar to the wye –wye network.

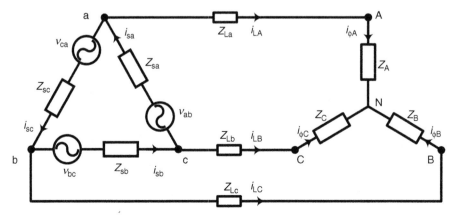

Figure 10.14 The delta–wye network.

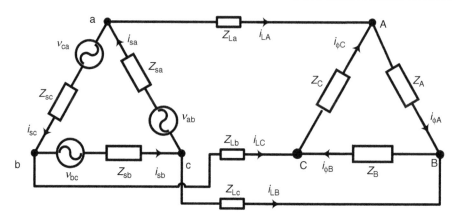

Figure 10.15 The delta–delta network.

10.5 Power calculation of balanced three-phase circuit

10.5.1 Balanced wye loads

Consider a wye-connected balanced load (refer to Figure 10.6) with load impedance in each phase, $Z = |Z| \angle \pm \theta$.

Phase voltage across the load is $V_{\varnothing} = V_{\varnothing} \angle 0°$.

Current in the load in each phase is $I_{\varnothing} = \dfrac{V_{\varnothing}}{Z}$.

Active power per phase: $P_{\varnothing} = V_{\varnothing} I_{\varnothing} \cos\theta$ (10.11)

Reactive power per phase: $Q_{\varnothing} = V_{\varnothing} I_{\varnothing} \sin\theta$ (10.12)

Apparent power per phase: $S_{\varnothing} = P_{\varnothing} + jQ_{\varnothing}$, $\; |S_{\varnothing}| = V_{\varnothing} I_{\varnothing}$ (10.13)

In a wye-connected system:

$V_{\mathrm{L}} = \sqrt{3} V_{\varnothing}$ and $I_{\mathrm{L}} = I_{\varnothing}$

Total active power in a three-phase circuit:

$P = 3P_{\varnothing} = 3V_{\varnothing} I_{\varnothing} \cos\theta = \sqrt{3} V_{\mathrm{L}} I_{\mathrm{L}} \cos\theta$ (10.14)

Total reactive power in a three-phase circuit:

$$Q = 3Q_\varnothing = 3V_\varnothing I_\varnothing \sin\theta = \sqrt{3}V_L I_L \sin\theta \tag{10.15}$$

Total apparent power in three phases:

$$S = P + jQ, \quad |S| = 3V_\varnothing I_\varnothing = \sqrt{3}V_L I_L \tag{10.16}$$

10.5.2 Balanced delta loads

Consider a delta-connected balanced load (refer to Figure 10.7) with load impedance in each phase, $Z = |Z| \angle \pm \theta$.

Active and reactive power consumed by the load across nodes A and B is:

$$P_{AB} = V_{AB} I_{\varnothing A} \cos\theta \qquad Q_{AB} = V_{AB} I_{\varnothing A} \sin\theta$$

Active and reactive power consumed by the load across nodes B and C is:

$$P_{BC} = V_{BC} I_{\varnothing B} \cos\theta \qquad Q_{BC} = V_{BC} I_{\varnothing B} \sin\theta$$

Active and reactive power consumed by the load across nodes C and A is:

$$P_{CA} = V_{CA} I_{\varnothing C} \cos\theta \qquad Q_{CA} = V_{CA} I_{\varnothing C} \sin\theta$$

$$|V_{AB}| = |V_{BC}| = |V_{CA}| = |V_\varnothing|$$
$$|I_{\varnothing A}| = |I_{\varnothing B}| = |I_{\varnothing C}| = |I_\varnothing|$$

The total active and reactive power consumed by the three-phase delta-connected load is:

$$P = P_{AB} + P_{BC} + P_{CA} = 3V_\varnothing I_\varnothing \cos\theta$$
$$Q = S_{AB} + S_{BC} + S_{CA} = 3V_\varnothing I_\varnothing \cos\theta$$

In delta-connected system, $V_L = V_\varnothing$ and $I_L = \sqrt{3}I_\varnothing$.

$$P = 3V_\varnothing I_\varnothing \cos\theta = \sqrt{3}V_L I_L \cos\theta \tag{10.17}$$

$$Q = 3V_\varnothing I_\varnothing \sin\theta = \sqrt{3}V_L I_L \sin\theta \tag{10.18}$$

$$S = P + jQ \qquad |S| = 3V_\varnothing I_\varnothing = \sqrt{3}V_L I_L \tag{10.19}$$

Example 10.3

Consider the wye–wye network in Example 10.1:

1. Calculate the active power, reactive power, and apparent power in each phase of the load.
2. Calculate the total active power, reactive power, and apparent power consumed by the load.
3. Calculate the total active power, reactive power, and apparent power supplied by the source.

Solution
From Example 10.1:
  Load current $I_A = 15.77\angle - 41.8°$ A.
  Phase voltage at load terminals is $V_{AN} = 202\angle - 3.14°$ V.

1. Calculate the active power, reactive power, and apparent power in each phase of the load:

$P_\varnothing = V_\varnothing I_\varnothing \cos\theta = 202\times15.77\times\cos(41.8° - 3.14°) = 202\times15.77\times\cos(38.66°) = 2.487\,\text{kW}$
$Q_\varnothing = V_\varnothing I_\varnothing \sin\phi = 202\times15.77\times\sin(41.8° - 3.14°) = 202\times15.77\times\sin(38.66°) = 1.989\,\text{kVAR}$
$S_\varnothing = V_\varnothing I_\varnothing = 202\times15.77 = 3.185\,\text{kVA}$

2. Calculate the total active power, reactive power, and apparent power consumed by the load:

$P = 3P_\varnothing = 3\times2.487 = 7.461\,\text{kW}$
$Q = 3S_\varnothing = 3\times1.989 = 5.967\,\text{kVAR}$
$S = 3S_\varnothing = 3\times3.185 = 9.555\,\text{kVA}$

3. Calculate the total active power, reactive power, and apparent power supplied by the source.
   Source current $= I_{LA} = 15.77\angle - 41.8°$ A.
   Phase voltage at source $v_{an} = 220\angle 0°$ V.

$P = 3V_\varnothing I_\varnothing \cos\theta = 3\times220\times15.77\times\cos(41.8°) = 7.759\,\text{kW}$
$Q = 3V_\varnothing I_\varnothing \sin\theta = 3\times220\times15.77\times\sin(41.8°) = 6.937\,\text{kVAR}$
$S = 3V_\varnothing I_\varnothing = 3\times220\times15.77 = 10.408\,\text{kVA}$

Example 10.4

Consider the wye–delta network in Example 10.2:

1. Calculate the active power, reactive power, and apparent power in each phase of the load.
2. Calculate the total active power, reactive power, and apparent power consumed by the load.
3. Calculate the total active power, reactive power, and apparent power supplied by the source.

Solution
From Example 10.2,

1. Calculate the active power, reactive power, and apparent power in each phase of the equivalent wye connected load

$I_\varnothing = 15.78\angle - 42.38°$ A
$V_\varnothing = 202\angle - 3.72°$ V

For delta-connected load.
Phase voltage = voltage across load,

$$V_{AB} = \sqrt{3}\, V_{AN}\angle 30° = \sqrt{3} \times 202 \times \angle(-3.72+30)° = 349.87\angle 26.28°$$

Load current per phase

$$I_{AB} = \frac{I_{\varnothing}}{\sqrt{3}}\angle(\theta+30)\,\text{A}$$

$$I_{AB} = 9.11\angle -12.38°\,\text{A}$$

$$P_{\varnothing} = V_{\varnothing}I_{\varnothing}\cos\theta$$
$$= 349.87 \times 9.11 \times \cos(26.28°+12.38°) = 349.87 \times 9.11 \times \cos(38.66°) = 2.488\,\text{kW}$$

$$Q_{\varnothing} = V_{\varnothing}I_{\varnothing}\sin\theta$$
$$= 349.87 \times 9.11 \times \sin(26.28°+12.38°) = 1.990\,\text{kVAR.}$$

$$S_{\varnothing} = V_{\varnothing}I_{\varnothing} = 348.87 \times 9.11 = 3.186\,\text{kVA}$$

2. Calculate the total active power, reactive power, and apparent power consumed by the load.

$$P = 3P_{\varnothing} = 3 \times 2.487 = 7.461\,\text{kW}$$
$$Q = 3S_{\varnothing} = 3 \times 1.901 = 5.703\,\text{kVAR}$$
$$S = 3S_{\varnothing} = 3 \times 3.186 = 9.558\,\text{kVA}$$

3. Calculate the total active power, reactive power, and apparent power consumed by the source.
Voltage per phase at source $V_{\varnothing} = 220\angle 0°\,\text{V}$
Phase current in source $I_{\varnothing} = I_A = 15.78\angle -42.38°\,\text{A}$

$$P = 3V_{\varnothing}I_{\varnothing}\cos = 3 \times 220 \times 15.78 \times \cos(42.38°) = 7.693\,\text{kW}$$
$$Q = 3V_{\varnothing}I_{\varnothing}\sin\theta = 3 \times 220 \times 15.78 \times \sin(42.38°) = 7.020\,\text{kVAR}$$
$$S = 3V_{\varnothing}I_{\varnothing} = 3 \times 220 \times 15.78 = 10.414\,\text{kVA}$$

10.6 Advantages and disadvantages of three-phase supply

10.6.1 Advantages

- Power transmission will be three times that of a single-phase supply.
- Higher reliability: if one or two of the phases fail, the healthy phase can supply power.
- Two voltage levels: possible to connect single-phase and three-phase with two voltage levels (line and phase).

10.6.2 Disadvantages

- Three lines are required; expensive compared to single phase.
- Difficult to maintain exact balanced load when single-phase and three-phase loads are used.

10.7   Summary

- In a balanced three-phase system, all the three phases or line voltages will have the same frequency, equal voltage magnitude, and are phase displaced by 120°.
- In a balanced three-phase system, all the three phases or line currents will have the same frequency, equal voltage magnitude, and are phase displaced by 120°.
- In a wye-connected system the magnitude of the line voltage will be $\sqrt{3}$ times the magnitude of the phase voltage and the line voltage will lead the corresponding phase voltage by 30°. The line current is equal to the phase current.
- In a wye-connected system the magnitude of the line current will be $\sqrt{3}$ times the magnitude of the phase current and the line current will lag the corresponding phase current by 30°. The line voltage is equal to the phase voltage.
- In a balanced three-phase system the sum of all three-phase (line) voltages is equal to zero and the sum of all three phase (line) currents is equal to zero.
- A balanced wye–wye circuit can be reduced to a single-phase equivalent single-phase circuit and can be analyzed similar to a single-phase circuit using per phase quantities.
- A delta-connected source or load can be converted to an equivalent wye-connected source or load. So the wye–delta, delta–wye, and delta–delta network can be converted to an equivalent wye–wye network.

Problems

1. For each set of voltages, state whether or not the voltages form a balanced three-phase set. If the set is not balanced, explain why.

    a. $v_a = 156\cos 314t$ V,
    $v_b = 156\cos(314t - 120°)$ V,
    $v_c = 156\cos(314t + 120°)$ V.

    b. $v_a = 600\sqrt{2}\sin 377t$ V,
    $v_b = 600\sqrt{2}\sin(377 - 240°)$ V,
    $v_c = 600\sqrt{2}\sin(377 + 240°)$ V.

    c. $v_a = 1796\sin(\omega t + 10°)$ V,
    $v_b = 1796\cos(\omega t - 110°)$ V,
    $v_c = 2796\cos(\omega t + 130°)$ V.

2. The time domain expressions for the three line-to-neutral voltages at the terminals of a wye-connected load are:

    $v_{AN} = 1796\cos\omega t$ V,
    $v_{BN} = 1796\cos(\omega t + 120°)$ V,
    $v_{CN} = 1796\cos(\omega t - 120°)$ V.

    Write the time domain expressions for the three line-to-line voltages v_{AB}, v_{BC}, and v_{CA}.

3. The magnitude of the phase voltage of an ideal balanced three-phase wye-connected source is 2200 V. The source is connected to a balanced wye-connected load by a distribution line that has an impedance of 2 + j16 per line. The load impedance is 20 + j40 Ω in each phase. The

phase sequence of the source is a–b–c. Use the a-phase voltage of the source as the reference. Specify the magnitude and phase angle of the following quantities:

   a. The three line currents.

   b. The three line voltages at the source.

   c. The three phase voltages at the load.

   d. The three line voltages at the load.

4. A balanced delta-connected load is powered from a wye-connected source. The load imped-ance is $36 + j10 \ \Omega$ in each phase. The line impedance of each line connecting the source to the load is $0.1 + j1 \ \Omega$. The line voltage at the terminals of the load is 33 kV.

   a. Calculate the three phase currents of the load.

   b. Calculate the three line currents.

   c. Calculate the three line voltages at the source.

5. In a balanced three-phase system, the source is balanced wye with line voltage = 190 V. The load is a balanced wye in parallel with a balanced delta. The phase impedance in each phase of the wye-connected load is $4 + j3 \ \Omega$ and the impedance of the delta is $3 − j9 \ \Omega$ per phase. The line impedance is $1.4 + j0.8 \ \Omega$ per line.

   a. Calculate the current in the wye-connected load.

   b. Calculate the current in the delta-connected load.

   c. Calculate the current in the source.

6. The delta-connected source is connected to a wye-connected load. The load impedance per phase is $100 + j200 \ \Omega$, and the line impedance is $1.2 + j12 \ \Omega$. The line voltage of the source is 6600 V. Ignore the internal impedance of the source.

   a. Construct a single-phase equivalent circuit of the system.

   b. Determine the magnitude of the line voltage at the terminals of the load.

   c. Determine the magnitude of the phase current in the delta-connected source.

7. A three-phase delta-connected generator has an internal impedance of $0.6 + j4.8 \ \Omega$ per phase with an open-circuit terminal voltage of 33,000 V. The generator feeds a delta-connected load through a transmission line with an impedance of $0.8 + j6.4 \ \Omega$. The impedance of the load is $2877 − j864 \ \Omega$ per phase.

   a. Calculate the magnitude of the line current.

   b. Calculate the magnitude of the line voltage at the terminals of the load.

   c. Calculate the magnitude of the line voltage at the terminals of the source.

   d. Calculate the magnitude of phase current in the load.

   e. Calculate the magnitude of phase current in the source.

8. The output of the balanced positive-sequence three-phase source is 100 kVA and power factor is 0.8 lead. The line voltage at the source is 400 V. The impedance of each line connect-ing the source to the load is $0.05 + j0.2 \ \Omega$.

   a. Find the magnitude of the line voltage at the load.

   b. Find the total complex power at the terminals of the load.

9. A balanced wye-connected load is supplied from a balanced three-phase voltage source. The three-phase voltage source is 220 V. Load impedance is $8 + j6 \ \Omega$ at 50 Hz. Neglect the source internal impedance and impedance of the line connecting the source and load. Find the cur-rent in the load and source.

10. The current in a balanced wye-connected load is 8 A. The source is wye connected with a phase voltage of 230 V. The load power factor is 0.8 lagging. Neglect the source internal im-pedance and impedance of the line connecting the source and load. Find the resistance and reactance of the load.

11. The p and Q in a three-phase wye-connected system are 1000 W and 1000 VAR, respectively. The source voltage is 220 V. Neglect the source internal impedance and impedance of the line connecting the source and load. Find the impedance per phase and current.

12. Show that the total instantaneous power in a balanced three-phase circuit is constant and equal to $1.5V_mI_m \cos \theta_\emptyset$, where V_m and I_m represent the maximum amplitudes of the phase voltage and phase current, respectively.

13. A balanced three-phase source is supplying 90 kVA at 0.8 lagging to two balanced wye-connected parallel loads. The distribution line connecting the source to the load has negligible impedance. Load 1 is purely resistive and absorbs 60 kW. Find the per-phase impedance of load 2 if the line voltage is 415.69 V and the impedance components are in series.

14. The total apparent power supplied in a balanced three-phase wye–delta system is 10 kVA. The line voltage is 190 V. if the line impedance is negligible and the power factor angle of the load is 25°, determine the impedance of the load.

15. In a balanced three-phase system, the source has an *a–b–c* sequence, is wye connected, and $V_{an} = 110 \angle 20°$ V. The source feeds two loads, both of which are wye connected. The impedance of load 1 is $8 + j6$ Ω per phase. The complex power for the a-phase of load 2 is $600 \angle 36°$ VA. Find the total complex power supplied by the source.

References

[1] Hayt W, Buck J. Engineering electromagnetics. 8th ed. New York: McGraw-Hill International Edition; 2011.

[2] Chapman SJ. Electric machinery fundamental. New York: McGraw-Hill International Edition; 2005.

[3] Nilsson J, Riedel SA. Electric circuits. 9th ed. New Jersey: Pearson International Edition; 2011.

[4] Dorf RC, Sroboda JA. Introduction to electric circuits. New Jersey: John Wiley and Sons, Inc.; 2006.

[5] Rashid MH. Power electronics – circuits, devices and applications. 4th ed. New Jersey: Pearson International Edition; 2013.

Magnetic circuits and power transformers

Easwaran Chandira Sekaran

Associate Professor, Department of Electrical and Electronics Engineering, Coimbatore Institute of Technology, Coimbatore, INDIA

Chapter Outline

11.1   Introduction

Magnetic circuits and magnetic components such as inductors and transformers are an important, integral, and indispensable part of most power electronics and renewable energy systems. Magnetic components fall into two categories:

1. Energy storage devices
2. Energy transfer devices

Energy storage devices store kinetic energy of a desirable quality of current flowing through it. Devices in this category are called inductors.

Energy transfer devices transfer the power from one energy port to another energy port without storing or losing energy in the process. These devices are called transformers.

11.2   Magnetic circuits

Magnetic circuits [1] have a magnetic material of standard geometrical shape called a core and a coil with conducting material having a number of turns (N) wound over the core. The coil is also called the exciting coil. When the current flow through the coil is zero there is no magnetic field of lines or lines of forces present inside the core. The incidence of the current in the coil produces magnetic lines of force and the path of these magnetic lines of force can be thought of as a magnetic circuit. The various terminologies used in the magnetic circuit are briefly discussed in the subsequent sections.

11.2.1   Magnetic field and magnetic flux

Magnetic fields are produced due to the movement of electrical charge and they are present around permanent magnets and current carrying conductors (magnetic circuit) as shown in Figure 11.1. In permanent magnets, revolving electrons produce an external field. If a current carrying conductor is wounded to form a coil of multiturns, the magnetic field is stronger than that of a single conductor. The magnetic field of the electromagnet is intensified when the coil is wound on an iron core. The strength of magnetic fields is varied in many applications through electromagnets. Magnetic fields form the basis for the operation of transformers, generators, and motors.

A magnetic field that can be visualized as the lines of force is referred to as magnetic flux, which is the "current" of the magnetic circuit. Magnetic flux is a measure

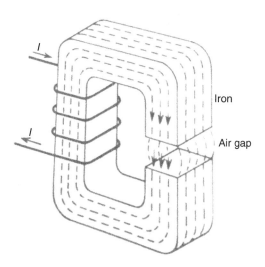

Figure 11.1 Magnetic circuit.

of the amount of magnetic field passing through a given surface (such as a conducting coil). The unit of flux is the weber (abbreviated as Wb) and the mathematical symbol for the number of webers of flux is Φ.

11.2.2 Magnetomotive force

Magnetomotive force (MMF) is the flux-producing ability of an electric current in a magnetic circuit. MMF is analogous to the electromotive force (EMF) of the electric circuit. The MMF developed is proportional to the current and to the number of turns in the coil in which current is flowing.

The expression for the MMF is:

$$\mathcal{F} = NI \qquad (11.1)$$

where N is the number of turns and I is the coil current in amperes and has the unit of ampere-turn.

11.2.3 Magnetic flux density, magnetic field strength, and flux linkage

The flux density is defined as the concentration of uniformly distributed flux per unit area of the cross-section through which it acts. Flux density is denoted as B and it is given by:

$$B = \frac{\Phi}{A} \qquad (11.2)$$

where Φ is the flux in webers and A is the cross-sectional area in square meters. The unit of the flux density is Wb/m^2. This is also called as tesla (T).

The magnetic field strength is defined as the force that creates the flow of lines of flux and is denoted as H, expressed in units of oersteds. H is a magnetic field strength gradient along a magnetic path roughly equivalent to the voltage drop around an electric circuit loop. It is measured in ampere-turns per meter:

$$H = \frac{4\pi NI}{l} \tag{11.3}$$

where N is the number of turns in the winding, I is the instantaneous current flowing through the wire, and l is the mean magnetic length of the core. Magnetic field strength is directly proportional to the winding current and the number of turns in the winding, and is inversely proportional to the magnetic length over which the flux must travel. This also indicates that the value of H is independent of the material used within the core.

The relation between B and H is given by:

$$B = \mu H \tag{11.4}$$

The flux passing through the surface bounded by a coil is said to link the coil, and for a coil of conductors, the flux passing through the coil is the product of the number of turns N and the flux passing through a single turn, Φ. This product is called the magnetic flux linkage of the coil, λ.

Flux linkage $(\lambda) = \Phi N$ Weber-turns $\tag{11.5}$

11.2.4 Reluctance and permeance

Reluctance is defined as the measure of how difficult it is to develop flux from the MMF in a given magnetic circuit. This is magnetic analogy to the electrical resistance. The symbol for reluctance is given by:

$$\Re = \frac{\mathcal{F}}{\Phi} \tag{11.6}$$

where $\Re$ is reluctance in ampere-turns per weber (A-t/Wb), $\mathcal{F}$ is MMF in ampere-turns, and Φ is flux in webers.

Materials referred to as "magnetic" have relatively low reluctance when used in a magnetic circuit. In other words, with these materials in the magnetic circuit, a smaller coil with less current is needed to provide a given number of webers of flux than would be the case if the path were through air or some magnetic material like copper, glass, or plastic.

Permeance is the reciprocal of reluctance, which is a measure of the quantity of flux for a given number of ampere-turns in a magnetic circuit. Thus, the permeance is given by:

$$P = \frac{1}{\Re} \tag{11.7}$$

where P is permeance and $\Re$ is reluctance. The unit for the permeance is webers per ampere-turn. From Equations (11.6) and (11.7):

$$P = \frac{1}{\Re} = \frac{1}{(\mathcal{F}/\Phi)} \tag{11.8}$$

$$P = \frac{\Phi}{\mathcal{F}} \tag{11.9}$$

The magnetic analogy of permeance is electrical conductance. Permeance, which increases as reluctance decreases, is an expression of ease with which the flux is developed in a given magnetic circuit for a given MMF. Thus, a magnetic circuit in which the path of the lines of force is almost through iron, the permeance of the circuit is relatively high compared with the permeance when air, plastic, or other nonmagnetic material is substituted for iron.

The space or material inside a coil is called the core. A coil that is wound around only a thin hollow tube of nonmagnetic material as shown in Figure 11.2 is simply known as an "air core" and such a coil generates only a relatively small amount of magnetic flux. To develop as much flux as possible it is preferred to use a core of magnetic material.

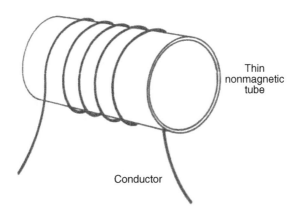

Thin
nonmagnetic
tube

Conductor

Figure 11.2 Typical air core coil.

For any specified magnetic material, shape, and dimension reluctance and permeance are the important characters that indicate the magnetic properties of a magnetic circuit. These characters correspond to the resistance and conductance of an electric circuit.

11.3   Equivalent circuit of a core excited by an AC MMF

When a current passes through a wire, a magnetic field is set up around the wire. If the wire is wound on a rod, its magnetic field is greatly intensified. The magnetic circuit [2] is the space in which the flux travels around the coil. The magnitude of the flux is determined by the product of the current, I, and the number of turns, N, in the coil. The force, NI, required to create the flux is MMF. The relationship between flux density, B, and magnetizing force, H, for an air core coil is linear ($B = \mu H$ and μ is unity for air core coil). If the coil is excited with an AC source, as shown in Figure 11.3, the relationship between B and H would have the characteristics shown in Figure 11.4. The linearity of the relationship between B and H represents the main advantage of air core coils. Since the relationship is linear, increasing H increases B, and therefore the flux in the coil, and, in this way, very large fields can be produced with large currents. There is obviously a practical limit to this, which depends on the maximum allowable current in the conductor and the resulting rise. To achieve an improvement over the air coil, as shown in Figure 11.4, the coil can be wound over a magnetic core. In addition to its high permeability, the advantages of the magnetic core over the air core are that the magnetic path length is well defined, and the flux is essentially confined to the core, except in the immediate vicinity of the winding. There is a limit as to how much magnetic flux can be generated in a magnetic material before the magnetic core

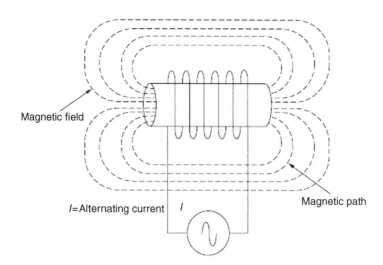

Figure 11.3 Air core coil driven from an AC source.

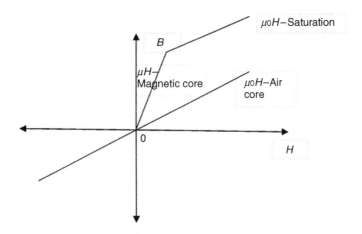

Figure 11.4 Relationship between *B* and *H* with AC excitation.

goes into saturation, and the coil reverts back to an air core, as shown in Figure 11.4. In transformer design, it is useful to use flux density.

11.4 Principle of operation of a transformer

Transformers [3] are devices that transfer energy from one circuit to another by means of a common magnetic field. An ideal transformer in its simplest form is shown in Figure 11.5. When an AC voltage is applied to the primary winding, time-varying current flows in the primary winding and causes an AC magnetic flux to appear in the transformer core. This flux links with the secondary winding due to the mutual

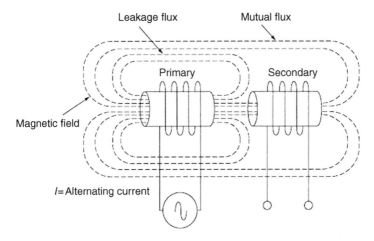

Figure 11.5 Simple transformer.

magnetic coupling, and induces a voltage in the secondary winding (Faraday's law). Depending on the ratio of turns in the primary and secondary winding, the root mean square (RMS) secondary voltage can be greater or less than the RMS primary voltage.

This transformer has two air coils that share a common flux. The flux diverges from the ends of the primary coil in all directions. It is not concentrated or confined. The primary is connected to the source and carries the current that establishes a magnetic field. The other coil is open circuited. It is inferred that the flux lines are not common to both coils. The difference between the two is the leakage flux; that is, leakage flux is the portion of the flux that does not link both coils.

11.4.1   Air core transformer

Some small transformers for low-power applications are constructed with air between the two coils. Such transformers are inefficient because the percentage of the flux from the first coil that links the second coil is small. The voltage induced in the second coil is determined as follows:

$$E = N \frac{d\Phi}{dt}$$ (11.10)

where N is the number of turns in the coil, $\frac{d\Phi}{dt}$ is the time rate of change of flux linking the coil, and Φ is the flux in lines.

At a time when the applied voltage to the coil is E and the flux linking the coils is Φ, the instantaneous voltage of the supply are:

$$e = \sqrt{2}E \cos \omega t = N \frac{d\Phi}{dt}$$ (11.11)

$$\Phi = \frac{\sqrt{2}E}{2\pi fN}$$ (11.12)

Since the amount of flux Φ linking the second coil is a small percentage of the flux from the first coil, the voltage induced into the second coil is small. The number of turns can be increased to increase the voltage output, but this will increase costs. The need then is to increase the amount of flux from the first coil that links the second coil.

11.4.2   Iron or steel core transformer

The ability of iron or steel to carry magnetic flux is much greater than air. This ability to carry flux is called permeability. Modern electrical steels have permeability in the order of 1500 compared with 1.0 for air. This means that the ability of a steel core to carry magnetic flux is 1500 times that of air. Then, the equation for the flux in the steel core is:

$$\Phi = \frac{\mu_0 \mu_r NAI}{d} \tag{11.13}$$

where μ_r is the relative permeability of steel ≈ 1500.

Since the permeability of the steel is very high compared with air, all of the flux can be considered as flowing in the steel and is essentially of equal magnitude in all parts of the core. The equation for the flux in the core can be written as follows:

$$\Phi = \frac{0.225E}{fN} \tag{11.14}$$

where E is the applied alternating voltage, f is the frequency in hertz, N is the number of turns in the winding.

For analyzing an ideal transformer, the following assumptions are made:

- The resistances of the windings can be neglected.
- All the magnetic flux is linked by all the turns of the coil and there is no leakage of flux.
- The reluctance of the core is negligible.

The equations for sinusoidal voltage for the ideal transformer are as follows.
The primary winding of turns N_p is supplied by a sinusoidal voltage v_p:

$$v_p = V_{pm} \cos(\omega t) \tag{11.15}$$

From Faraday's law, the voltage across the primary winding terminals can be written as:

$$v_p = N_p \frac{d\Phi}{dt} \tag{11.16}$$

Therefore:

$$v_p = V_{pm} \cos(\omega t) = N_p \frac{d\Phi}{dt} \tag{11.17}$$

Rearranging and integrating, the equation for common flux can be written as:

$$\Phi = \frac{V_{pm}}{N_p \omega} \sin(\omega t) \tag{11.18}$$

This common flux passes through both the windings.

11.5   Voltage, current, and impedance transformations

11.5.1   Voltage relationship

This common flux flows through the transformer core, links with the secondary winding, and induces voltage across the secondary winding according to Faraday's law. The primary and secondary voltage relationships are specified by:

$$v_p = N_p \frac{d\varphi}{dt} \text{ and } v_p = N_s \frac{d\varphi}{dt} \tag{11.19}$$

The polarities are defined by Lenz's law. From the previous relationship:

$$\frac{v_p}{v_s} = \frac{N_p}{N_s} = k \text{ (transformation ratio)} \tag{11.20}$$

The turns ratio or transformation ratio determines the amount of voltage transformed. If $k = 1$, it is an isolation transformer, if $k > 1$, it is a step-up transformer, and if $k < 1$, it is a step-down transformer.

11.5.2   Current relationship

If a load is connected to the secondary side, current passes through the secondary as the circuit is complete. The MMF corresponding to the current flowing in the secondary side is given by $N_s i_s$. The input coil is forced to generate an MMF to oppose this MMF and therefore the resultant MMF is $\mathcal{F} = N_p i_p - N_s i_s$ and is related to flux and reluctance. As the reluctance is zero for an ideal transformer, $\mathcal{F} = N_p i_p - N_s i_s = 0$.
Therefore:

$$\frac{i_s}{i_p} = \frac{N_p}{N_s} = k \text{ (transformation ratio)} \tag{11.21}$$

11.5.3   Power in an ideal transformer

The power delivered to the load by the secondary winding is $p_s = v_s i_s$ and using the voltage and current relationship with the primary winding, power in the secondary is:

$$1/k \, v_p \times k i_p = v_p i_p. \tag{11.22}$$

Therefore, power between primary and secondary is equal.

11.5.4 Impedance in an ideal transformer

Considering the load impedance Z_L connected across the secondary winding, the impedance across the secondary circuit is derived from the voltage and current flowing through the secondary circuit, $Z_L = V_s/I_s$.

Substituting for V_s and I_s:

$$Z_L = \frac{(N_s/N_p)V_p}{(N_p/N_s)I_p} = \left(\frac{N_s}{N_p}\right)^2 \frac{V_p}{I_p} = k^2 Z_L \tag{11.23}$$

11.6 Nonideal transformer and its equivalent circuits

An actual transformer differs from an ideal transformer in the following contexts:

- Copper losses are present in both the primary and secondary windings.
- Not all the flux produced by the primary winding links the secondary winding, and vice versa. This gives rise to some leakage of flux.
- The core requires a finite amount of MMF for its magnetization.
- Hysteresis and eddy current losses cause power loss in the transformer core.

The equivalent circuit of an ideal transformer can be modified to include these effects:

- Resistances R_p and R_s can be added on both the primary and secondary side to represent the actual winding resistances.
- The effect of leakage flux can be included by adding two inductances, L_p and L_s, respectively, in the primary and secondary winding circuits.
- Nonzero reluctance value is included by adding a magnetizing inductance, L_m. The corresponding reactance of the iron core is X_m $(=2\pi f L_m)$.
- To account for the hysteresis and eddy currents, which cause iron losses in the core, a resistance R_c is added in the transformer equivalent circuit.

A simple two-winding transformer is shown in the schematic diagram of Figure 11.6. A primary winding of N_p turns is on one side of a ferromagnetic core loop, and a similar coil having N_s turns is on the other. Both coils are wound in the same direction with the starts of the coils at H_1 and X_1, respectively. When an alternating voltage V_p is applied from H_2 to H_1, an alternating magnetizing flux ϕ_m flows around the closed core loop. A secondary voltage $V_s = V_p \times N_s/N_p$ is induced in the secondary winding and appears from X_2 to X_1 and very nearly in phase with V_p. With no load connected to X_1–X_2, I_p consists of only a small current called the magnetizing current. When load is applied, current I_s flows out of terminal X_1 and results in a current $I_p = I_s \times N_s/N_p$ flowing into H_1 in addition to magnetizing current. The ampere-turns of flux due to current $I_p \times N_p$ cancel the ampere-turns of flux due to current $I_s \times N_s$, so only the magnetizing flux exists in the core for all the time the transformer is operating normally.

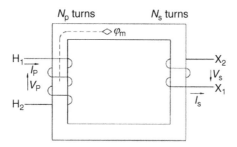

Figure 11.6 Schematic of a two-winding transformer.

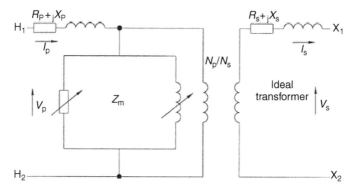

Figure 11.7 Complete transformer equivalent circuit.

Figure 11.7 shows a complete equivalent circuit of the transformer. An ideal transformer is inserted to represent the current- and voltage-transformation ratios. A parallel resistance and inductance representing the magnetizing impedance are placed across the primary of the ideal transformer. Resistance and inductance of the two windings are placed in the H_1 and X_1 legs, respectively.

11.7   Tests on transformers

Transformers are tested (IS:2026 Part I: 1977) before they reach consumers. The tests are classified as type tests, routine tests, and special tests. The tests are conducted to measure:

1. Resistance of windings
2. Voltage ratio
3. Voltage vector relationship
4. Short-circuit impedance
5. Load loss
6. Insulation resistance

7. Dielectric test
8. Temperature rise
9. Zero sequence impedance
10. Acoustic noise level
11. Harmonics

11.7.1 Design tests

Tests performed by manufacturers on prototypes or production samples are referred to as "design tests [4]." These tests may include sound-level tests, temperature-rise tests, and short-circuit current withstand tests. The purpose of a design test is to establish a design limit that can be applied by calculation to every transformer built. In particular, short-circuit tests are destructive and may result in some invisible damage to the sample, even if the test is passed successfully. The IEEE standard calls for a transformer to sustain six tests, four with symmetrical fault currents and two with asymmetrical currents. One of the symmetrical shots is to be of long duration, up to 2 s, depending on the impedance for lower ratings. The remaining five shots are to be 0.25 s in duration. The long-shot duration for distribution transformers 750 kVA and above is 1 s. The design passes the short-circuit test if the transformer sustains no internal or external damage (as determined by visual inspection) and minimal impedance changes. The tested transformer also has to pass production dielectric tests and experience no more than a 25% change in exciting current.

11.7.2 Production tests

Production tests are given to and passed by each transformer made. Tests to determine ratio, polarity or phase displacement, iron loss, load loss, and impedance are done to verify that the nameplate information is correct. Dielectric tests specified by industry standards are intended to prove that the transformer is capable of sustaining unusual but anticipated electrical stresses that may be encountered in service. Production dielectric tests may include applied voltage, induced voltage, and impulse tests.

11.7.2.1 Applied-voltage test

Standards require application of a voltage of (very roughly) twice the normal line-to-line voltage to each entire winding for 1 min. This checks the ability of one phase to withstand voltage it may encounter when another phase is faulted to ground and transients are reflected and doubled.

11.7.2.2 Induced-voltage test

The original applied-voltage test is now supplemented with an induced-voltage test. Voltage at higher frequency (usually 400 Hz) is applied at twice the rated value of the winding. This induces the higher voltage in each winding simultaneously without saturating the core. If a winding is permanently grounded on one end, the applied-voltage test cannot be performed. In this case, many IEEE product standards specify that the induced primary test voltage be raised to 1000 plus 3.46 times the rated winding voltage.

11.7.2.3   Impulse test

Distribution lines are routinely disturbed by voltage surges caused by lightning strikes and switching transients. A standard 1.2×50 μs impulse wave with a peak equal to the basic impulse insulation level of the primary system (60–150 kV) is applied to verify that each transformer will withstand these surges when in service.

11.7.3   Performance test

In order to determine the losses, and to calculate the efficiency and voltage regulation at different loads, open circuit, short circuit, load tests are conducted.

11.7.3.1   Open-circuit test

To carry out an open-circuit test, the low voltage (LV) side of the transformer, where rated voltage at rated frequency is applied, and the high voltage (HV) side are left opened as shown in Figure 11.8. The voltmeter, ammeter, and wattmeter readings are taken as V_0, I_0, and W_0, respectively. During this test, rated flux is produced in the core and the current drawn is the no-load current, which is quite small, about 2–5% of the rated current. Therefore, a low range ammeter and wattmeter current coil should be selected. Strictly speaking, the wattmeter will record the core loss as well as the LV winding copper loss. But the winding copper loss is very small compared to the core loss as the flux in the core is rated. In fact this approximation is built-in in the approximate equivalent circuit of the transformer, referred to as the primary side, which is the LV side in this case. The approximate equivalent circuit and the corresponding phasor diagrams are shown in Figures 11.9 and 11.10 under no-load condition.

The resistance of the primary winding is R_0. Therefore, the copper loss in the primary winding at no-load is $I_0^2 R_0$.

Hence, the iron losses of the transformer $= W - I_0^2 R_0$ (11.24)

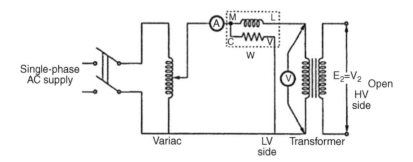

Figure 11.8 Circuit diagram for an open-circuit test.

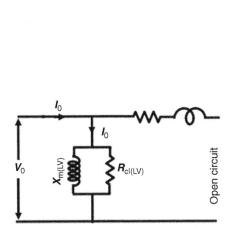

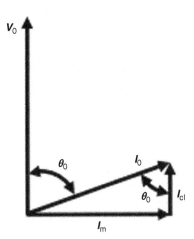

Figure 11.9 Equivalent circuit under no-load condition.

Figure 11.10 Phasor diagram under no-load condition.

The power factor of the transformer at no-load is:

$$\cos \theta_0 = \text{Resistance/impedance} = (R_0 I_0)/V_1$$

where V_1 is the supply voltage, indicated by the voltmeter. Alternatively, power factor:

$$\cos \theta_0 = \frac{\text{Wattmeter reading}}{\text{Voltmeter reading} \times \text{Ammeter reading}} = \frac{W}{V_1 I_0} \qquad (11.25)$$

From the values of the power factor $\cos \theta_0$, the magnetizing component (I_μ) and wattless component (I_w) of the no-load current (I_0) can be calculated as follows:

$$I_\mu = I_0 \sin \theta_0 \qquad (11.26)$$

and

$$I_w = I_0 \cos \theta_0 \qquad (11.27)$$

11.7.3.2 Short-circuit test

The connection diagram for a short-circuit test on a transformer is shown in Figure 11.11. A voltmeter, wattmeter, and ammeter are connected to the HV side of the transformer as shown. The voltage at rated frequency is applied to the HV side with the help of a variac of variable ratio autotransformer. Usually the LV side of the transformer is short circuited. Now with the help of the variac applied voltage is

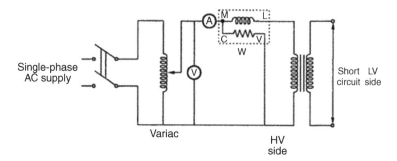

Figure 11.11 Circuit diagram for a short-circuit test.

slowly increased (usually 5–10% of the normal primary voltage) until the ammeter gives a reading equal to the rated current of the HV side. After reaching the rated current of the HV side, the readings of all three instruments (voltmeter, ammeter, and wattmeter) are recorded. The ammeter reading gives the primary equivalent of a full load current, I_L.

Since the impressed voltage (equal to only a few percent of the rated value) is merely that required to overcome the total impedance of the windings, the mutual flux produced in the core is only a small percentage of its normal value (because the flux is proportional to the voltage). Consequently the iron core losses are very small. The wattmeter reading W equals the total full load copper losses in both the primary and secondary windings of the whole transformer. If V_{sc} is the voltage required for circulating the rated load current in the short-circuited transformer, then equivalent impedance:

$$Z_{01}(\text{or } Z_{02}) = V_{sc}/I_1 \tag{11.28}$$

$$W = I_1^2 R_{01} \tag{11.29}$$

Resistance of the transformer:

$$R_{01} = W/I_1^2 = (R_1 + R_2/k^2) \tag{11.30}$$

And leakage reactance:

$$X_{01}(\text{or } X_{02}) = X_1 + X_2/k^2 = \sqrt{Z_{01}^2 - R_{01}^2} \tag{11.31}$$

From the knowledge gained from Z_{01} (or Z_{02}), the total voltage drop in the transformer referred to as primary (or secondary) can be computed and hence the voltage regulation of the transformer can be calculated.

11.7.3.3   Load test

To determine the total losses in a transformer, a load test is performed. The load test gives information about the rated load of the transformer and temperature rise. Efficiency and regulation can also be determined from the load test. Nominal voltage is applied across the primary and rated current is drawn from the secondary. Load is applied continuously observing the steady-state temperature rise. Based on the different insulation and cooling methods incorporated in the transformer, different loading levels are permitted for the same transformer.

Efficiency of a transformer:

1. Commercial efficiency of a transformer at a particular load and power factor is defined as the ratio of output power to input power. Thus, efficiency:

$$\eta = \frac{\text{Output power}}{\text{Input power}} = \frac{\text{Output power}}{\text{Output power} + \text{Losses}}$$

$$= \frac{\text{Output power}}{\text{Output power} + (\text{Iron} + \text{Copper})\text{Losses}} = \frac{\text{Input power} - \text{Losses}}{\text{Input power}}$$

$$= 1 - \frac{\text{Losses}}{\text{Input power}}$$

$$\text{Primary input} = V_1 I_1 \cos\theta_0 \tag{11.32}$$

$$\text{Primary copper loss} = I_1^2 R_1 \tag{11.33}$$

$$\text{Iron losses} = (\text{Hysteresis} + \text{Eddy current})\text{Losses} = W_h + W_e = W_i \tag{11.34}$$

$$\therefore \text{Efficiency } \eta = \frac{V_1 I_1 \cos\theta_0 - I_1^2 R_1 - W_i}{V_1 I_1 \cos\theta_0} = 1 - \frac{I_1^2 R_1 - W_i}{V_1 \cos\theta_0} - \frac{W_i}{V_1 I_1 \cos\theta_0} \tag{11.35}$$

Differentiating with respect to I_1:

$$\frac{d\eta}{dI_1} = 0 = \frac{R_1}{V_1 \cos\theta_0} + \frac{W_i}{V_1 I_1^2 \cos\theta_0} \tag{11.36}$$

For maximum η, $\dfrac{d\eta}{dI_1} = 0$ or copper loss = iron loss.

Efficiency of a transformer depends on both load and power factor. Input and output power depends on power factor of load, so transformers are usually specified kVA rating only.

Load at maximum efficiency, let W_i and W_c be the full load iron loss and copper loss, respectively:

$$\therefore W_c \propto \left(\text{Full load kVA}\right)^2 \tag{11.37}$$

If x is the load, when the efficiency is maximum, then:

$$W_i \propto x^2 W_c \qquad (11.38)$$

$$\therefore \frac{W_c}{W_i} = \frac{\left(\text{Full load kVA}\right)^2}{x^2} \qquad (11.39)$$

Load at maximum efficiency (x) is:

$$x = \left(\text{Full load kVA}\right)\left(\frac{\text{Iron lossses } W_i}{\text{Full load copper losses } W_c}\right)^{1/2} \qquad (11.40)$$

11.7.3.3.1   All day efficiency

Transformers are employed for energy distribution 24 h a day. In a day the secondary is loaded throughout 24 h contributing to iron losses. But only during peak loaded conditions are copper losses significant. The performance of a transformer is judged by its operational efficiency, called all day efficiency, based on the load cycle in 24 h.

$$\text{All day efficiency } \eta_{\text{all day}} = \frac{\text{Output in kWh}}{\text{Input in kWh}} \qquad (11.41)$$

11.7.3.3.2   Regulation

Whenever a transformer is loaded, terminal voltage across the secondary changes with the load variations, with the primary voltage supply held constant. The change in secondary terminal voltage from no load to full load, expressed as a percentage of no-load voltage, is known as voltage regulation of a transformer:

$$\% \text{ Regulation} = \frac{\text{Secondary voltage at no load} - \text{Secondary voltage at full load}}{\text{Secondary voltage at no load}} \times 100 \qquad (11.42)$$

11.8   Transformer polarity

The phase relationship of single-phase transformer voltages is described as "polarity." The polarity of a transformer can be either additive or subtractive. These terms describe the voltage that may appear on adjacent terminals if the remaining terminals

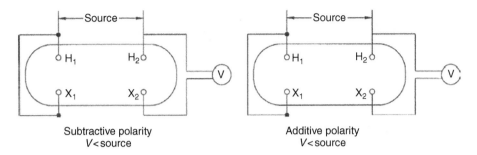

Figure 11.12 Single-phase polarity.

are jumpered together. Although the technical definition of polarity involves the relative position of primary and secondary bushings, the position of primary bushings is always the same according to standards. Therefore, when facing the secondary bushings of an additive transformer, the X_1 bushing is located to the right (of X_3), while for a subtractive transformer, X_1 is farthest to the left. To complicate this definition, a single-phase pad-mounted transformer built to IEEE standard Type 2 will always have the X_2 mid-tap bushing on the lowest right-hand side of the low-voltage slant pattern. Polarity has nothing to do with the internal construction of the transformer windings, only with the routing of leads to the bushings. Polarity only becomes important when transformers are being paralleled or banked. Single-phase polarity is illustrated in Figure 11.12.

11.9 Transformers in parallel

For supplying a load in excess of the rating of an existing transformer, two or more transformers may be connected in parallel with the existing transformer. It is usually economical to install another transformer in parallel instead of replacing the existing transformer by a single larger unit. The cost of a spare unit in the case of two parallel transformers (of equal rating) is also lower than that of a single large transformer. In addition, it is preferable to have a parallel transformer because of reliability. With this, at least half the load can be supplied with one transformer out of service. For parallel connection of transformers, primary windings of the transformers are connected to source bus-bars and secondary windings are connected to the load bus-bars. There are various conditions that must be fulfilled for the successful parallel operation of transformers. These are as follows:

1. The line voltage ratios of the transformers must be equal (on each tap): If the transformers connected in parallel have slightly different voltage ratios, then due to the inequality of induced EMFs in the secondary windings, a circulating current will flow in the loop formed by the secondary windings under the no-load condition, which may be much greater than the normal no-load current. The current will be quite high as the leakage impedance is low. When the secondary windings are loaded, this circulating current will

tend to produce unequal loading on the two transformers, and it may not be possible to take the full load from this group of two parallel transformers (one of the transformers may become overloaded).

2. The transformers should have equal per-unit leakage impedances and the same ratio of equivalent leakage reactance to the equivalent resistance (X/R): If the ratings of both the transformers are equal, their per-unit leakage impedances should be equal in order to have equal loading of both the transformers. If the ratings are unequal, their per-unit leakage impedances based on their own ratings should be equal so that the currents carried by them will be proportional to their ratings. In other words, for unequal ratings, the numerical (ohmic) values of their impedances should be in inverse proportion to their ratings to have current in them in line with their ratings. A difference in the ratio of the reactance value to resistance value of the per-unit impedance results in a different phase angle of the currents carried by the two paralleled transformers; one transformer will be working with a higher power factor and the other with a lower power factor than that of the combined output. Hence, the real power will not be proportionally shared by the transformers.

3. The transformers should have the same polarity: The transformers should be properly connected with regard to their polarity. If they are connected within correct polarities then the two EMFs, induced in the secondary windings that are in parallel, will act together in the local secondary circuit and produce a short circuit.

The previous three conditions are applicable to both single-phase as well as three-phase transformers. In addition to these three conditions, two more conditions are essential for the parallel operation of three-phase transformers:

4. The transformers should have the same phase sequence: The phase sequence of line voltages of both the transformers must be identical for parallel operation of three-phase transformers. If the phase sequence is incorrect, in every cycle each pair of phases will be short circuited.

5. The transformers should have the zero relative phase displacement between the secondary line voltages: The transformer windings can be connected in a variety of ways, which produce different magnitudes and phase displacements of the secondary voltage. All the transformer connections can be classified into distinct vector groups. Each vector group notation consists of an uppercase letter denoting HV connection, a second lowercase letter denoting LV connection, followed by a *clock number* representing LV winding's phase displacement with respect to HV winding (at 12 o'clock). There are four groups into which all possible three-phase connections can be classified:
   a. Group 1: Zero phase displacement (Yy0, Dd0, Dz0)
   b. Group 2: 180° phase displacement (Yy6, Dd6, Dz6)
   c. Group 3: −30° phase displacement (Yd1, Dy1, Yz1)
   d. Group 4: +30° phase displacement (Yd11, Dy11, Yz11)

In the previously mentioned notations, letters y (or Y), d (or D), and z represent star, delta, and zigzag connections, respectively. In order to have zero relative phase displacement of secondary side line voltages, the transformers belonging to the same group can be paralleled. For example, two transformers with Yd1 and Dy1 connections can be paralleled. The transformers of groups 1 and 2 can only be paralleled with transformers of their own group. However, the transformers of groups 3 and 4 can be paralleled by reversing the phase sequence of one of them. For example, a transformer with Yd11 connection (group 4) can be paralleled with that having Dy1 connection (group 3) by reversing the phase sequence of both primary and secondary terminals of the Dy1 transformer.

Table 11.1 **Various three-phase transformer connections**

S. no.	Primary configuration	Secondary configuration	Symbolic representation	Primary or secondary	
				Line voltage	**Line current**
1.	Delta (mesh)	Delta (mesh)	Δ–Δ	$V_1 = n V_1$	$I_1 = \dfrac{I_1}{n}$
2.	Delta (mesh)	Star (wye)	Δ–Y	$V_1 = \sqrt{3}\, n V_1$	$I_1 = \dfrac{I_1}{\sqrt{3}\, n}$
3.	Star (wye)	Delta (mesh)	Y–Δ	$V_1 = \dfrac{n V_1}{\sqrt{3}}$	$I_1 = \sqrt{3}\,\dfrac{I_1}{n}$
4.	Star (wye)	Star (wye)	Y–Y	$V_1 = n V_1$	$I_1 = \dfrac{I_1}{n}$
5.	Interconnected star	Delta (mesh)	⅄–Δ		
6.	Interconnected star	Star (wye)	⅄–Y		

11.10   Three-phase transformer connections

Transformer power levels range from low-power applications, such as consumer electronics power supplies, to very high power applications, such as power distribution systems. For higher power applications, three-phase transforms are commonly used. A three-phase transformer is constructed as a single unit with a bank of transformers, that is, three numbers of identical single-phase transformers connected in required form. In a single-phase transformer, only two coils, namely primary and secondary, are available whereas in a three-phase transformer there will be three numbers of primary coils and three numbers of secondary coils. The coils are connected in various methods as listed in Table 11.1 to obtain different voltage levels.

- An advantage of Δ–Δ connection is that if one of the transformers fails or is removed from the circuit, the remaining two can operate in the open Δ or V connection. This way, the bank still delivers three-phase currents and voltages in their correct phase relationship. However, the capacity of the bank is reduced to 57.7% of its original value.
- In the Y–Y connection, only 57.7% of the line voltage is applied to each winding but full line current flows in each winding. The Y–Y connection is rarely used.
- The Δ–Y connection is used for stepping up voltages since the voltage is increased by the transformer ratio multiplied by 3.

11.11   Special transformer connection

An *air core transformer* is a special transformer, used in radio frequency circuits. As the name implies, its windings are wrapped around a nonmagnetic material in the form of a hollow tube. Though the degree of coupling (mutual inductance) is much less, ferromagnetic cores (eddy current loss, hysteresis, saturation, etc.) are completely eliminated. In high-frequency applications, the effects of iron losses are more problematic. An example of air core transformers is the tesla coil, which is a resonant, high-frequency, step-up transformer used to produce extremely high voltages.

A *Scott-connected transformer* is a type of circuit device used to convert a three-phase supply (3-φ, 120-degree phase rotation) into a two-phase (2-φ, 90-degree phase rotation) supply, or vice versa. The Scott connection evenly distributes a balanced load between the phases of the source.

11.12   Parallel operation of three-phase transformers

Ideal parallel operation between transformers occurs when (1) there are no circulating currents on open circuit, and (2) the load division between the transformers is proportional to their kVA ratings. These requirements necessitate that any two or more three-phase transformers, which are desired to be operated in parallel, should possess:

1. The same no-load ratio of transformation
2. The same percentage impedance
3. The same resistance to reactance ratio
4. The same polarity
5. The same phase rotation
6. The same inherent phase-angle displacement between primary and secondary terminals
   The previously mentioned conditions are characteristic of all three-phase transformers whether two winding or three winding. With three-winding transformers, however, the following additional requirement must also be satisfied before the transformers can be designed suitable for parallel operation:
7. The same power ratio between the corresponding windings

Table 11.2 gives the possible combinations of transformers that can be operated in parallel.

The methods to check for synchronization of transformers are done using a synchroscope or synchronizing relay. The advantages of parallel operation of transformers are:

Table 11.2 Parallel operation of three-phase transformers

S. no.	Vector group of transformers that will operate in parallel		Vector group of transformers that will NOT operate in parallel	
	Transformer 1	Transformer 2	Transformer 1	Transformer 2
1.				
2.	ΔΔ	ΔΔ or Yy	ΔΔ	Δy
3.	Yy	Yy or ΔΔ	Δy	ΔΔ
4.	ΔY	Δy or YΔ	YΔ	Yy
5.	YΔ	YΔ or Δy	Yy	YΔ

1. Maximize electrical system efficiency
2. Maximize electrical system availability
3. Maximize power system reliability

But during the parallel operation of three-phase transformers, the magnitude of short-circuit currents, risk of circulating currents, bus rating, and reduction in transformer impedance make the circuit complex in providing protective mechanisms.

11.13   Autotransformers

In an autotransformer, the primary and secondary windings are linked together both electrically and magnetically. Therefore it is economical for the same VA rating as windings are reduced, but the disadvantage is that it does not have isolation between primary and secondary windings. The winding can be designed with multiple tapping points, to provide different voltage points along its secondary winding. The winding diagram and the number of windings in primary and secondary (N_p and N_s, respectively), current, and voltage across primary and secondary are shown in Figure 11.13.

11.14   Three-winding transformers

In certain high rating transformers, one winding in addition to its primary and secondary winding is used, called a tertiary winding transformer. Because of this third winding, the transformer is called a three-winding transformer. The advantage of using a tertiary winding in a transformer is to meet one or more of the following requirements:

1. It reduces the unbalancing in the primary due to unbalancing in three-phase load.
2. It redistributes the flow of fault current.

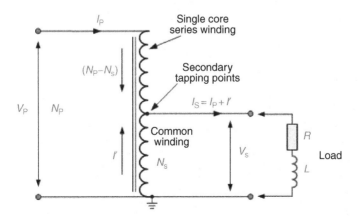

Figure 11.13 Winding diagram of an autotransformer.

3. Sometimes it is required to supply an auxiliary load at different voltage levels in addition to its main secondary load. This secondary load can be taken from the tertiary winding of the three-winding transformer.
4. As the tertiary winding is connected in delta formation in a three-winding transformer, it assists in limiting fault current in the event of a short circuit from line to neutral.
5. A star–star transformer comprising three single units or a single unit with five limb core offers high impedance to the flow of unbalanced load between the line and neutral.

11.15   Instrument transformers

Instrument transformers are used for transforming the magnitude of current (typically 1 A or 5 A at secondary) and voltage (typically 120 V at secondary) from one level to another. Also instrument transformers can be used as an isolation transformer for safety purposes. Various types of instrument transformers, namely current transformers, inductive voltage transformers, capacitive voltage transformers, combined current/voltage transformers, and station service voltage transformers, are designed to transform high current and high voltage levels down to low current and low voltage outputs in a known and accurate proportion for a specific application.

Potential transformers consist of two separate windings on a common magnetic steel core. One winding consists of fewer turns of heavier wire on the steel core and is called the secondary winding. The other winding consists of a relatively large number of turns of fine wire, wound on top of the secondary, and is called the primary winding.

Current transformers are constructed in various ways. One method is quite similar to that of the potential transformer in that there are two separate windings on a magnetic steel core. But it differs in that the primary winding consists of a few turns of heavy wire capable of carrying the full load current while the secondary winding consists of many turns of smaller wire with a current carrying capacity of between 5 A and 20 A, dependent on the design. This is called the wound type due to its wound primary coil. Another very common type of construction is the so-called "window," "through," or donut type current transformer in which the core has an opening through which the conductor carrying the primary load current is passed. This primary conductor constitutes the primary winding of the current transformer (one pass through the "window" represents a one turn primary), and must be large enough in cross-section to carry the maximum current of the load.

The operation of instrument transformers differs from power transformers. The secondary winding of an instrument transformer has a very small impedance called burden, such that the instrument transformer operates under short-circuit conditions. The burden (expressed in ohms) across the secondary of an instrument transformer is also defined as the ratio of secondary voltage to secondary current. This helps to determine the volt-ampere loading of the instrument transformer.

The major applications of instrument transformers include:

1. Revenue metering for electric utilities, independent power producers, or industrial users
2. Protective relaying for use with switchgear to monitor system current and voltage levels
3. High accuracy wide current range use for independent power facilities
4. Station service power needs within substations or for power needs at remote sites

11.16 Third harmonics in transformers

In addition to the operation of transformers on the sinusoidal supplies, harmonic behavior is important as the size and rating of the transformer increases. In recent times, transformers have been designed to operate at closed levels of saturation in order to reduce the weight and cost of the core used. Because of this and hysteresis, the transformer core behaves as a nonlinear component and generates harmonic currents. If a sinusoidal voltage is applied to the primary of a transformer, the flux wave will vary as a sinusoidal function of time, but the no-load current wave will be distorted because the hysteresis loop contains a pronounced third harmonic. Figure 11.14 represents the hysteresis loop taken to the maximum flux density and the manner in which the shape of the magnetizing current can be obtained and plotted. In Figure 11.14, at any instant of the flux density wave the ampere-turns required to establish are read out and plotted by traversing the hysteresis loop.

To produce a flux density of NP requires 0N ampere-turns per meter:

$$N'P' = NP$$

The ampere-turns per meter are plotted as $N'A$ in Figure 11.14b. To produce a flux density MR requires 0M ampere-turns per meter:

$$0M = M'B$$
$$MR = M'R'$$

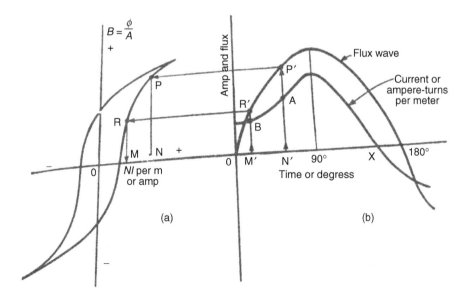

Figure 11.14 (a) Hysteresis loop and (b) magnetizing current.

In brief, the various abscissas of Figure 11.14a are plotted as ordinates to determine the shape of the current wave on Figure 11.14b. This is continued until a sufficient number of points are obtained. The use of a suitable constant changes the wave BAX from ampere-turns per meter to amperes. Such a wave represents the magnetizing component and the hysteresis component of the no-load current. It reaches the maximum at the same time as the flux wave, but the two waves do not go through zero simultaneously. The sinusoidal flux density represents the sinusoidal voltage, whereas the plot of magnetizing current rises sharply, saturates quickly, and is thereby distorted. This magnetizing current can be analyzed with Fourier series. The harmonic components are obtained from this Fourier analysis. The harmonic spectrum of the magnetizing current waveform reveals a very high percentage of third harmonics. These harmonic currents produce harmonic fields in the core and harmonic voltages in the windings. A relatively small value of harmonic fields produces substantial magnitudes of harmonic voltages. For example, a 10% magnitude of third harmonic flux produces a 30% magnitude of third harmonic voltage. These effects become even more pronounced for higher order harmonics. The no-load current can be considered to be made up of approximately two sine components (loss component and magnetizing component) and the nonsinusoidal component with dominant third harmonics. The sum of sine and nonsine waves forms a distorted waveform.

In the case of single-phase transformers connected to form a three-phase star-connected bank, the fundamental voltages supplied will produce voltages in the individual transformers that contain third harmonics. These third harmonics for all the three transformers will be in time phase. Each voltage between the neutral of the primaries and the lines will contain both a fundamental component and a third harmonic component, and as a result the secondary voltage of each transformer will contain both fundamental and third harmonic components.

The effects of the harmonic currents are:

1. Additional copper losses
2. Increased core losses
3. Increased neutral current and overheating of neutral conductor
4. Increased EMI with communication networks

On the other hand, the harmonic voltages of the transformer cause:

1. Increased dielectric stress on insulation
2. Resonance between winding reactance and feeder capacitance

11.17   Transformers in a microgrid

With technological improvements, design, materials, etc. of power transformers, autotransformers, and instrument transformers, performance has increased. In renewable energy integrated microgrids, AC and direct current (DC) supply are available and hence solid-state transformers and smart transformers are being developed and marketed to cater for the features of microgrids and smart grids.

11.17.1   Solid-state transformers

The advantages and limitations of conventional transformers when power quality is a major concern are as follows:

Advantages:

1. Relatively economical
2. Highly reliable
3. Quite efficient

Limitations:

1. Sensitive to harmonics
2. Voltage drop under load
3. No protection from system disruptions and overloads
4. Environmental concerns regarding mineral oil
5. Poor performance under DC-offset load unbalances
6. No power factor improvement

The solid-state transformers (SSTs) are designed with different topologies based on its application:

1. AC to AC buck converter: the salient points are that,
   a. Transformation of the voltage level directly without any isolation transformer
   b. Switches must be capable of blocking full primary voltage during OFF state and conducting full secondary current during ON state
   c. Difficult to control series-connected devices
   d. Lack of magnetic isolation
   e. Inability to correct load power factor
2. SST without a DC link:
   a. Transformer weight and size reduced
   b. Provides isolation
   c. No power factor improvement is possible
3. SST with a DC link:
   a. Reduced size due to a high-frequency transformer
   b. Power factor improvement is possible
   c. Multilevel converter topologies can be applied to achieve high voltage levels (e.g., 11 kV, 22 kV)
   d. High cost and low efficiency
   e. It is a three-stage topology: most popular now

SSTs that can be widely used in microgrid applications are relatively advantageous with respect to power quality.

Advantages:

1. An excellent utilization of distributed renewable energy resources and distributed energy storage devices
2. Power factor control
3. Fast isolation under fault conditions due to a controlled SST
4. Control of both AC and DC loads can be done using the SST scheme
5. Improved power quality
6. DC and alternative frequency AC service options

7. Integration with system monitoring and advanced distribution
8. Reduced weight and size
9. Elimination of hazardous liquid dielectrics

Limitations:

1. Multiple power conversion stages can lower the overall efficiency
2. DC-link capacitors are required
3. The transformer lifetime can be shorter due to storage devices

11.17.2 The smart transformers

The smart transformer is a smart device for integrating with the distribution grid, solar-wind renewable energy storage, and electric vehicles. The components of a smart transformer include power conversion system and built-in STATCOM functions. The smart transformer can be fully controlled through Internet or wireless communication systems. It uses high-voltage semiconductor switches based on an AC/DC rectifier, DC/DC converter, high-voltage and high-frequency transformer, DC/AC inverter, and their switching control circuitry. This is expected to be a critical component in the development of smart grids.

11.18   Summary

In this chapter, the fundamentals of magnetic circuits and basic concepts of power transformers, types, and testing methods were discussed in detail. The development of equivalent circuit efficiency and regulation calculation of the transformer was elaborated.

References

[1] Johnson JR. Electric circuits – Part-I direct current. San Francisco, CA: Rinehart Press; 1970.
[2] Puchstein AF, Lloyd TC, Conrad AG. Alternating current machines. Mumbai, India: Asia Publishing House; 1950.
[3] Richardson DV. Rotating electric machinery and transformer technology. Richmond, VA: Reston Publishing Company, Inc; 1982.
[4] Deshpande MV. Design and testing of electrical machines. New Delhi, India: Prentice Hall of India; 2010.

Renewable energy generators and control

12

Sreenivas S. Murthy

Department of Electrical Engineering, Indian Institute of Technology, Delhi;
CPRI, Bengaluru, India

Chapter Outline

12.1 Introduction – general

Electricity generation using renewable energy (RE) requires suitable energy converters to deliver power of required quality to consumer loads. Wind, bio, and hydro
energy use prime movers, which in turn rotate electric generators that generate electricity to feed different types of loads. In contrast, solar energy converters using photovoltaic panels do not require rotating prime movers as they employ direct energy
conversion to produce direct current (DC). But if solar thermal systems are used to
produce steam, electricity can be produced through rotating steam turbines. Wind
energy drives suitable wind turbines, which in turn drive generators with or without a
speed changing gearbox. Bio energy can produce gas to run an engine that drives the
generator. An alternative is to use heat in bio energy to produce steam driving a steam
turbine coupled to a generator. Hydro energy uses a suitable hydroturbine or operates
a pump as a turbine, which in turn rotates a generator. Typical mechanical prime movers for different RE sources are listed as follows:

- Oil/gas engine: bio energy
- Steam turbine: bio, solar-thermal, geothermal
- Wind turbine: wind energy
- Hydro turbine: small hydro energy
- Pump as turbine: small hydro energy

They are either of fixed speed or variable speed needing suitable energy converters
to produce electricity. Generally most of the loads consuming electricity need alternating currents (AC) of fixed voltage and frequency. We need single-phase or three-phase
output from the generators at required voltage and frequency to feed either a load or
a grid. Thus, we need suitable generators tailored to the prime mover to produce the
above quality power.

There are varieties of such generators in engineering practice. However, only a few
of them are of interest to RE systems and are explained in this chapter. They are classified under electromechanical energy conversion that converts mechanical energy
of the prime mover to electrical energy. These electric machines (EM) effect energy
conversion through electromagnetic fields produced by current carrying conductors
embedded in ferromagnetic cores. There are both DC and AC generators producing
corresponding currents. However, only AC generators are of interest in RE to tailor to
loads or grids. Further, the DC power if required can easily be obtained through power
electronic converters or rectifiers. Therefore, we discuss only AC generators suited to
RE here. They may operate at fixed or variable speed. There are mainly two types of
AC generators: synchronous generators (SGs) operating at fixed speed and frequency
and induction generators (IGs) operating at near fixed or variable speed. Large conventional thermal, large hydro, nuclear, and gas power plants of several megawatts
(MW) range feeding utility grids mainly use fixed speed SG at fixed grid frequency.
Among RE too, they find applications in wind, small hydro, bio, solar thermal, and
geothermal energy. However, IGs are finding extensive use in RE both for grid connected and off-grid standalone applications, typically for wind, small hydro, and bio
energy.

12.2   General features of electric machines

The previously mentioned generators for RE come under the family of electric machines (EM) having the following common features and phenomena. They are all rotating electromechanical energy converters changing the form of energy from electrical to mechanical and vice versa. There are varieties of rotating machines at power levels of a few watts to a few hundred MW.

1. "Motoring" and "generating" action: All electrical rotating machines normally operate either as a motor or a generator. In a motor electrical energy is fed as input and mechanical energy is taken as output; the reverse process takes place in a generator. The direction of flow of electrical and mechanical energy is reversed between a motor and a generator, while the direction of losses remains the same.
2. Supply system and load: Each EM is interfaced between a supply and load with energy flowing from supply to load. In a motor, the supply is electrical and the load is mechanical and vice versa in a generator. Types of supply and load vary with application. In RE systems, the type of mechanical supply is decided by the source and prime mover that outputs mechanical energy. The load is of an electric nature to feed either to isolated loads or grids.
3. Conductor and core: All machines will have current carrying conductors made of copper or aluminum and a flux carrying core made of ferromagnetic materials.
4. Electromagnetic field: Magnetic flux lines produced by current carrying conductors or permanent magnets are present in all machines traversing through iron and air and represent electromagnetic field and magnetic field energy.
5. Electrostatic field: A small electrostatic field is also present in all machines represented by electric flux and charges representing electric field energy.
6. Losses, heating, noise, and vibrations: There are losses of electrical, mechanical, and field origin in all machines. Losses cause heating, which needs special cooling to keep temperature rise in check. Noise and vibration caused by mechanical and electrical forces are observed in all machines and need special measures to keep them within limits.

12.3   Basic construction

Figure 12.1 depicts the basic construction of an electric machine, which is a cylindrical structure. Here the stator is the stationary member and the rotor the rotating member. Air gap is the narrow spacing between the stator and rotor surfaces essential for the rotor to move with respect to the stator. Normally the rotor is inside the stator separated by the air gap; in special cases the rotor may be the external part. The frame is the outer casing in which the machine is housed. Any machine consists of a mass of copper and iron, respectively forming the electric and magnetic circuits. Core is the high permeability part of the machine consisting of ferromagnetic material and meant to carry flux with least reluctance. Windings are conducting parts meant for carrying currents. Normally both stator and rotor will have cores in which windings are suitably embedded to form electromagnetic structures. The rotor is mounted on a mechanical shaft supported by bearings for smooth rotation. Frame and shaft are mechanical parts supporting stator and rotor, respectively. Windings are electrical ports

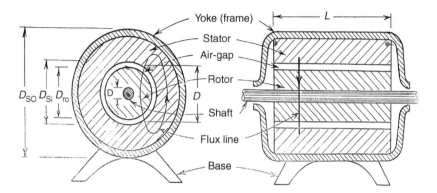

Figure 12.1 Basic construction of an electric machine.

to feed or extract electrical energy while the shaft is the mechanical port to feed or extract mechanical energy. Windings are brought to a terminal box fixed to the frame to facilitate access to electrical energy. The shaft handles mechanical energy fed by the external prime mover through a suitable mechanical power transmission system such as direct coupling, belts, and gears. Windings comprise a suitable configuration of coils placed in slots formed in the core of both stator and rotor. The nature of currents in these windings may be AC or DC. Machines are wound for different even numbers of magnetic poles. Connecting the stator winding to the external network is easy, but the same is not true for the rotating rotor winding, which needs a special method called "slip ring and brush arrangement" as explained later. In some machines, the rotor may be inherently short circuited through conducting end rings fixed on both ends of the rotor. This is a called "squirrel-cage" winding.

Slip rings are circular brass or copper rings fixed to and insulated from the shaft. Each slip ring is electrically connected to a terminal of the rotor winding such that current can be injected or extracted through the stationary brushes resting on the rotating slip rings. Due to this direct contact, the nature of voltage and current in the rotor winding would be the same as those at the brushes connected to the external circuit. Figure 12.2 shows a slip-ring arrangement in which the terminals a–b of the rotor winding are electrically connected to slip rings S_1, S_2 touching a pair of stationary

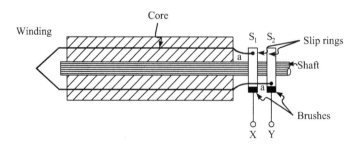

Figure 12.2 Slip ring and brush arrangement.

brushes forming the external terminals XY. Rotating winding terminals ab are connected to stationary terminals XY, such that the current in the external circuit is the same as that in the winding.

Different types of electric machines can be created by different types of stator and rotor, winding configurations, and nature of excitation of windings.

12.4 Type of electric supply and load

Any RE system is invariably connected to an electric supply. Mostly they are an AC supply of fixed frequency, though a DC supply may be relevant with solar energy. There are two types of AC supply, namely single phase and three phase. A single-phase supply will have a fixed voltage across two terminals brought out through two supply lines: 230 V, 50 Hz, and 120 V, 60 Hz are common with alternating sinusoidally time varying voltages of corresponding frequency available across the lines. Three-phase supply is universally used in power systems. It represents three AC voltages (called phase voltages) of equal peak magnitude displaced by a time phase angle of 120°. These phase voltages can be connected in "star" or "delta" configuration with a neutral point in the former. There are "line" voltages between two lines and "phase" voltages across each phase. Line voltage is $\sqrt{3}$ times phase voltage in "star" and equal to phase voltage in "delta": 415 V, 50 Hz, 230/460 V, 60 Hz are common three-phase supply systems at the consumer end. Higher three-phase voltages (11 V, 33 V, 220 kV, etc.) are employed at transmission and distribution stages. With increased application of solar photovoltaic, the DC grid is assuming importance; a 380-V DC bus is considered suitable based on which connected load gadgets must be tailored.

Similar to the supply, there are single-phase, three-phase, and DC loads. Broadly they are classified as linear, nonlinear, and dynamic loads. In three-phase systems there could be unbalanced loads with unequal current in phases. In linear (static) loads, the voltage–current relation is linear, for example, lighting, heating. Here the waveform of voltage and current would be identical. Thus, such static loads typically can be represented by a fixed impedance, which is mostly resistive or inductive (rarely capacitive). Depending on this impedance, the phase angle between its voltage and current waveforms will vary from 0° to 90°, lag or lead. Dynamic loads are typically motor loads comprising either single-phase or three-phase motors driving mechanical loads applying fixed or varying torque on the motor. Thus, effective load on the generator may vary with time dynamically posing dynamic or time varying impedance. Mechanical loads are in the form of fans, blowers, compressors, pumps, crushers, impact loads, mixers, food processors, drills, juicers, robots, servo drives, conveyers, hoists, cranes, etc. Here current drawn by the load is dynamic in nature, varying with time dependent on connected load driven by the motor. Such drive motors are subjected to starting, stopping, speed, and torque control. Dynamic loads pose abnormal loading on the generator demanding varying current (I_L), active power (P), and reactive power (Q). Normally a generator cannot start a motor of the same power rating and a rated motor power must be lower (typically one-third). Nonlinear loads draw nonsinusoidal current from the generator; typical ones are computers, rectifiers, power electronic converters (AC–DC, AC–AC), and variable frequency drives. In nonlinear loads, the voltage–current relation is nonlinear with distorted current waveform even with sinusoidal voltage. Nonlinearity is introduced by electronic devices used in connected loads. Here the generator must be designed to supply harmonic currents, which introduce losses and reduce efficiency. In unbalanced loads, impedance across the three phases will be unequal due to unequal connected load across each phase, which results in uneven currents leading to current in the neutral wire. These result in so-called negative sequence currents and induced rotor currents with associated losses and vibrations.

12.5 Basic energy conversion principles

The energy converter of interest or electric generator converts mechanical energy of the RE-driven prime mover to electrical energy to be fed to a grid in a grid-fed system or isolated loads in an off-grid system. In a steady-state condition, the prime mover feeds torque to the generator through the shaft and rotates at a suitable speed. Based on Newtonian laws of mechanics, there is a mechanical equilibrium in which there is a torque balance wherein input torque is opposed by internally developed generator torque, inertial torque, and friction torque. Electrical energy output of the generator will be in the form of generated voltage, based on Faraday's laws, feeding a current to the grid or the load. There is an electrical equilibrium wherein the internally generated voltage is opposed by the grid or load voltage and drops across resistances and reactances based on circuit laws such as Kirchhoff's voltage/current laws and Ohm's law. Thus, basic factors of energy conversion are generation of voltage and torque caused by electromagnetic interaction phenomena as explained later.

When stator and rotor windings are excited their surfaces are magnetized to a magnetic potential, called magnetomotive force (mmf), resulting in a magnetic potential difference across the air gap. This causes a radial magnetic flux in the air gap that closes through the stator and rotor core. Due to high permeability and low reluctance of the core there is negligible potential drop. This air gap magnetic flux links both stator and rotor windings. Based on Faraday's law, voltage is induced in each winding due to time rate of change of flux linkage given by $N(d\Phi/dt)$, where N is the number of turns and Φ is the time varying flux linking it. Thus, the machine must have a certain number of turns N made to link a time varying flux Φ. In some machines one of the members, stator or rotor, carries the coils while the other member produces Φ, either by having permanent magnets or electromagnets with DC excited coils. In a simplistic form, we can say that the part carrying N turns is called the "armature" and that producing the flux Φ is called the "field." We must note that in all machines, by Faraday's law, voltage is induced in various coils so long as there is rate of change of flux linkage in the coil. Often the term "armature" is used to designate the set of coils that handle power in the machine by carrying current through them coupled with induced or impressed voltage.

The phenomenon of torque production acting on the rotor may be explained in different ways based on laws of physics. Bil law defining force on a current carrying conductor placed in a magnetic field of flux density B may be extended to develop torque expressions. A more general form is to derive the same from field energy. The presence of a magnetic field results in a field energy of density ($\frac{1}{2}BH$) in the machine. Since H in the core is negligible, the bulk of the energy is in the air gap. By the concept of virtual work it can be proven, assuming linearity, that the torque is the space derivative of magnetic field energy W_{fld} given by $T = [\delta W_{fld}/\delta\theta]$, where θ is the space angle. For example, if the machine has a single coil of inductance L carrying a current i the energy is ($\frac{1}{2}Li^2$) and the torque is $T = (\frac{1}{2}i^2)[\delta L/\delta\theta]$.

As mentioned earlier, two types of AC generators, SGs and IGs, are predominantly employed in RE. In the following sections, we study the basic working of these generators and associated energy conversion principles.

12.6   Synchronous generators

Here the stator is normally the armature carrying a three-phase balanced winding, each displaced by 120°, and the rotor is the field. These are built from a few kilowatts (kW) to hundreds of MW. However, low powered SGs in the kW range are also made with the stator as the field and the rotor as the armature having a three-phase winding, the power being extracted through slip rings. There are also low power (kW range) single-phase generators with special construction.

The stator core is normally made up of silicon steel (ferromagnetic) circular laminations, the inside portion being cut to form slots and teeth. The three-phase distributed windings are housed in the slots in a symmetric pattern as per design. Figure 12.3 shows the schematic of a three-phase synchronous machine symbolizing stator (armature)

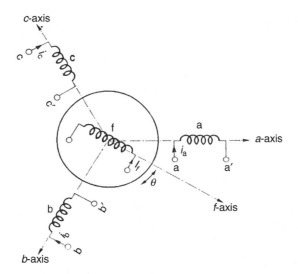

Figure 12.3 Schematic of a synchronous machine.

windings a–b–c housed inside a suitable frame and rotor (field) winding f in a core mounted on a shaft.

There are two types of rotors, namely (1) salient pole rotor and (2) cylindrical rotor. In the salient pole rotor with protruding poles, field windings are placed on the poles. With this type of configuration, there is scope for installing a large number of poles. However, the air gap is highly nonuniform, minimum at the pole axis and maximum along the interpolar axis. In the cylindrical rotor, field windings are distributed in slots such that the air gap would be uniform ignoring slotting. These can be used for a small number of poles; typically two or four poles. There is nonuniform slotting of the rotor surface, as the slotting is done only in interpolar space to accommodate the winding conductors carrying unidirectional currents. In both types of rotors, the field winding is excited by DC through a pair of slip rings. There are also brushless exciters where DC is fed through rotating diodes fixed to the rotor, by converting AC voltages fed to its input terminals through AC exciters.

A synchronous generator is also called an alternator. As the rotor is driven at a fixed speed, voltages are induced in the three-phase armature winding in the stator. Due to the alternating flux pattern linking the stator windings, alternating voltages would be induced. Since three-phase windings are space displaced by 120°, the voltages induced in them would be phase displaced in time by 120°, thus generating three-phase balanced voltages commonly used in power supply. Alternatively, by injecting three-phase balanced currents in the stator windings with the rotor excited by DC, a motoring action is achieved in the process of the magnetized rotor trying to align with the field produced by the stator windings.

The major application of the synchronous machine is a "generating" operation (i.e., an alternator), as the bulk of the electrical power generated in the world is through SGs driven mainly by hydro, steam, or gas turbines. However, limited application of synchronous motors can be found in constant speed drives, and in power systems to improve system power factor (PF) by using it as a leading PF load.

SGs find wide applications in RE – wind, bio, small hydro, geothermal – both for grid-fed and off-grid operation.

12.6.1   Voltage generation in synchronous machines

Referring to Figure 12.3, the three-phase balanced stator windings abc are represented by aa′, bb′, cc′ with their magnetic axes as shown. The rotating field winding is represented by f (housed in salient pole or cylindrical rotor) whose axis is at an angle θ ahead of the a axis at any instant. Since the rotor is rotating at a constant speed of ω rad/s, we may write:

$$\theta = \omega t \tag{12.1}$$

assuming that at $t = 0$, the f axis is aligned with the a axis (for any arbitrary position δ of the f axis at $t = 0$, $\theta = \omega t + \delta$).

When the field winding is excited by DC, it produces a flux density **B** distributed in the air gap. For simplicity, this **B** variation can be assumed to be sinusoidal in space such that:

$$B = B_{max} \cos \alpha \qquad (12.2)$$

where: B_{max} is the maximum flux density along the f axis, α is the angle from the f axis along the air gap.

Maximum flux produced by the field would link the stator windings a, b, or c when the f axis is aligned with the respective stator winding axis. This maximum flux Φ_{max} would be nothing but the total flux/pole given by:

$$\Phi_{max} = \int_{-\pi/2}^{\pi/2} B_{max}(D/2)L \cos \alpha . d\alpha$$

$$\text{or} \quad \Phi_{max} = B_{max}DL \qquad (12.3)$$

This expression can be directly obtained by multiplying the pole area with the average flux density B_{av} over a pole. For a sinusoidal spatial distribution of B:

$$B_{av} = \left(\frac{2}{\pi} \right) B_{max}$$

and pole area is:

$$A_p = \frac{\pi DL}{2} \text{ for 2-pole machine,}$$

$$A_p = \frac{\pi DL}{P} \text{ for } P\text{-pole machine}$$

Thus, flux/pole is:

$$\Phi_{max} = B_{av}A_p = B_{max}DL$$

For a P-pole machine:

$$\Phi_{max} = \left(\frac{2B_{max}DL}{P} \right) \qquad (12.4)$$

As the field rotates, the flux linking each stator winding will change with time. For one revolution of rotor, flux linkage goes through one cycle. For sinusoidal

B distribution, the cyclical variation of flux linkage would also be sinusoidal varying between λ_{max} and $-\lambda_{max}$ given by:

$$\lambda_{max} = N_{se}\Phi_{max} \tag{12.5}$$

where N_{se} is the number of series turns of the coil.
Flux linkage of a winding λ_a is:

$$\lambda_a(\theta) = \lambda_{max}\cos\theta$$

Since θ changes with time, flux linkage as a function of time is:

$$\lambda_a(t) = \lambda_{max}\cos\omega t \tag{12.6}$$

Induced voltage e_a in this winding, by Faraday's law, is $[d\lambda_a(t)/dt]$:

$$e_a = \omega\lambda_{max}\sin\omega t \tag{12.7}$$

Thus, we obtain an AC voltage induced in the stator winding, whose maximum value is:

$$E_{max} = \omega\lambda_{max}$$
$$= \omega N_{se}\Phi_{max}$$

The rms value of the induced voltage E is: $E = \dfrac{\omega N_{se}\Phi_{max}}{\sqrt{2}}$

ω is the frequency of the induced voltage in rad/s, which can be expressed in Hz by relating $\omega = 2\pi f$, therefore:

$$E = 4.44\, f N_{se}\Phi_{max} \tag{12.8}$$

For a 4-pole configuration, waveforms of B, λ_a, and e_a are modified giving two cycles for each rotation of the rotor. We must now discriminate between the mechanical angle θ_m and electrical angle θ_e and corresponding speeds ω_m and ω_e related by:

$$\theta_m = [2/P]\theta_e \text{ or } \omega_m = [2/P]\omega_e \tag{12.9}$$

To generate a frequency f, the speed of rotation corresponds to ω elec rad/s, that is, $P/2\ \omega_m$ mech rad/s for a P-pole machine.

$$f = \frac{\omega}{2\pi} = \frac{P\omega_m}{4P\pi} \tag{12.10}$$

If N_s is the speed of rotation in rpm ($\omega_m = (2\pi N_s/60)$), we get:

$$f = \frac{N_s P}{120}$$

$$\text{or} \quad N_s = \frac{120 f}{P}$$

(12.11)

This is an important relation, which decides the speed of the machine to generate voltage at required frequency f. Since P for the machine is fixed, there is only one speed N_s for the specified generator frequency f. This speed is called the synchronous speed of the machine, and the synchronous machine has to run only at this speed.

By a similar analogy flux linkages and generated voltages of b and c windings can be written down by noting that these are identical windings but displaced by 120 electrical degrees. Thus:

$$\lambda_b = \lambda_{max} \cos\left(\omega t - \frac{2\pi}{3} \right)$$

(12.12a)

$$\lambda_c = \lambda_{max} \cos\left(\omega t + \frac{2\pi}{3} \right)$$

(12.12b)

$$e_b = -\omega\lambda_{max} \sin\left(\omega t - \frac{2\pi}{3} \right) = -\sqrt{2}E \sin\left(\omega t - \frac{2\pi}{3} \right)$$

(12.12c)

$$e_b = -\omega\lambda_{max} \sin\left(\omega t + \frac{2\pi}{3} \right) = -\sqrt{2}E \sin\left(\omega t + \frac{2\pi}{3} \right)$$

(12.12d)

Note that the rms value of the induced voltage in all the phases a, b, and c would be the same given by Equation (12.8), only their instantaneous values change and the voltages e_a, e_b, and e_c form a balanced three-phase set of voltages displaced in time phase by 120 electrical degrees in the sequence abc.

12.6.2 Equivalent circuit

To analyze the performance of a synchronous generator and to obtain relevant characteristics, it is useful to have suitable modeling based on the physical phenomena involved. Under normal operating conditions the SG will supply three-phase currents to an external network.

The following air-gap magnetic fields in synchronism can be identified under such a condition:

1. Stator mmf wave of peak value F_s produced by three-phase balanced windings carrying three-phase balanced currents rotating at ω elec rad/s where ω is the radian frequency of winding currents. F_s causes a peak flux density B_s.

2. Rotor mmf wave of peak value F_r due to the DC field winding current, which is stationary with respect to the rotor. Since the rotor is driven at ω elec rad/s the rotor field also rotates at the same speed as the stator field with a phase angle δ_{sr} between them.

3. Resultant mmf wave of peak value F_{sr} that is the resultant of the previous two fields, also rotating at ω elec rad/s.

F_s, F_r, and F_{sr} will cause peak flux densities B_s, B_r, and B_{sr}, respectively. Assuming uniform air gap length g, B is related to F by:

$$B = \mu_o F / g \tag{12.13}$$

At any given instant all the previously mentioned fields are assumed sinusoidally distributed in space with a phase angle between them having the peak values as denoted earlier. These three fields will be in equilibrium vectorially in space with one being the resultant of the other two. We may denote δ_s, δ_r respectively as the phase position of F_s, F_r, with respect to F_{sr}. These three synchronously rotating phase shifted fields (F_s, F_r, and F_{sr}) in the air gap will produce three voltages (V_s, E_f, and V_{ta}) in each phase, shifted in time phase as per the space position of fields that cause them. In each phase, these three voltages will be in equilibrium – the phasor sum of the voltages will be zero and hence can be represented by the per-phase equivalent circuit of Figure 12.4 where:

$$V_s + E_f + V_{ta} = 0$$

Here V_s is the voltage in each phase due to F_s, which is proportional to per-phase current I_s. Hence, per-phase rms voltage V_s is proportional to per-phase rms current I_s. When a-phase has peak value, the peak of B_s is in line with its axis and hence peak flux links this phase; consequently induced voltage leads this flux by 90°and hence the current. Thus, we may write:

$$V_s = jX_{as} I_s$$

where X_{as} is a constant reactance. It is called the synchronous reactance associated with the air-gap flux per unit stator current. In the equivalent circuit of Figure 12.4 V_s is replaced by jX_{as} with a current I_s flowing through it. To account for leakage flux and

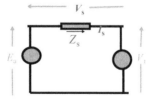

Figure 12.4 Equivalent circuit of a synchronous machine.

winding resistance, we add leakage reactance X_{ls} and resistance R_s to the previously mentioned reactance X_{as} to form a composite synchronous impedance $Z_s = R_s + jX_s$ to represent voltage due to stator currents or load. We observe that the synchronous machine can be effectively modeled (under steady-state balanced supply) by a per-phase equivalent circuit, which has a voltage source E_f ($=E_s$) in series with an impedance, say Z_s, which equals $(R_s + jX_s)$ and is termed "synchronous impedance." Here E_f is only due to field winding current (often called open circuit (OC) or no-load voltage), V_s is the drop due to synchronous impedance (proportional to load), and V_t is the terminal voltage of the generator that appears across the external network.

12.6.3 Performance equations and phasor diagram

From the equivalent circuit of Figure 12.4, the following phasor (complex) equations relating the phasor values of per-phase voltage and current can be written. Here both space and time phasors interconnected through proper phasor diagrams are relevant. Winding positions, air-gap mmf, flux density, etc., are represented by space phasors as they are displaced in space. In fact, space displaced air-gap mmfs rotating at synchronous speed cause time displaced flux linkages and voltages in the windings at fixed frequency. In a generating mode:

$$E_f = V_t + I_s(R_s + jX_s) \tag{12.14}$$

Taking the terminal voltage V_t as reference, a phasor diagram can be drawn as in Figure 12.5 using Equation (12.14) for a given current I_s at any arbitrary phasor position θ with respect to V_t called the power factor angle.

From this phasor diagram, we can identify the output power P of the generator as:

$$P = 3V_t I_s \cos\theta \tag{12.15}$$

where voltage and current are rms values.

An alternative expression of power neglecting resistance can be proven as:

$$P = 3\frac{V_t E_f \sin\delta}{X_s} \tag{12.16}$$

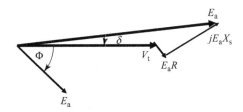

Figure 12.5 Phasor diagram of synchronous generator.

12.6.4 Operation of synchronous generators in different modes

SGs operate mainly in the following two modes:

- Standalone or off-grid mode
- Grid-connected mode

12.6.4.1 Standalone or off-grid mode

Off-grid standalone units are found in large numbers to energize local loads independent of grid supply using different energy sources. Diesel generating (DG) sets are the most common type of standalone units driven by diesel engines. Commercial and industrial establishments maintain them as standby units in the event of grid failure. They are also used in emergency services such as in hospitals. For remote or island communities or defense establishments they form the main source of electricity. Recently, a large number of such DG sets have been deployed to energize telecom towers to serve mobile telephone networks. Apart from fossil fuels such as diesel, petrol, kerosene, and natural gas, renewable sources such as bio and small hydro energy systems too are increasingly being used for such off-grid generation. Typical prime movers are engines run by diesel, petrol, kerosene, biofuel or turbines run by gas or hydro energy. Characteristics of such prime movers will have a bearing on system performance. Typical applications include lighting, heating, and motor loads used in domestic, commercial, industrial, and service sectors. Cogeneration is another application using industrial by-products such as steam or other fuels; for example, in sugar mills bagasse is a useful by-product to produce heat or electricity. While most of these generators are the rotating field type (up to a few MW) smaller units may also be of the rotating armature type. Interestingly, the bulk of the units use brushless excitation with rotating diodes.

A schematic of a standalone unit is shown in Figure 12.6. It will have an energy source such as fossil fuel, biofuel, or small hydro potential driving a prime mover (an engine or a turbine) that rotates the SG at the desired (fixed) synchronous speed to generate the required frequency. On excitation of the field winding by an appropriate field current I_f, the voltage V_t will come across the load delivering a load current I_s. Invariably the generator is placed close to the load and seldom needs long transmission lines or transformers, as the generated voltage is normally the same as the one

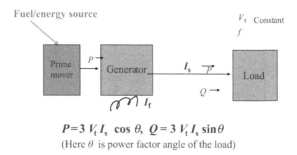

$$P=3\ V_t I_s\ \cos\ \theta,\ \ Q=3\ V_t I_s\ \sin\theta$$
(Here θ is power factor angle of the load)

Figure 12.6 Schematic of a standalone unit.

needed by the load. The generator must provide the desired power quality to the load in terms of constant, balanced harmonic-free voltage and fixed frequency. While the prime mover must ensure constant speed for the required frequency at all loads, field current must be adjustable for the required terminal voltage at all loads. All such units are fitted with automatic voltage regulators (AVRs) to adjust I_f as load changes, which employ power electronic-based controlled rectifiers to feed the desired DC value of I_f. Normally terminal voltage is sensed and compared with a reference voltage so that the error signal with due control processing adjusts the so-called firing angle of thyristors of the control rectifiers to cause the desired field current at the output to make the previously mentioned error signal zero. Thus, voltage across the load is regulated at all loads. In brushless generators the error voltage adjusts the DC current in the stator winding of the exciter such that the rotor voltage and in turn the DC output current of the rotating diode rectifier is adjusted to make the error voltage zero. As per standards, permissible variation in the terminal voltage is within ±5%.

The loads connected to a standalone generator as explained earlier are:

- Static
- Dynamic
- Nonlinear
- Unbalanced

We notice that a standalone SG will be called upon to supply fixed or variable current, which may even be nonsinusoidal. The total system must be correctly modeled to assess the performance and effect proper design. For simplicity the effective load on the generator can be modeled as impedance demanding both active and reactive power, P and Q in Figure 12.6. P will be provided by the prime mover while Q has to be supplied by generator excitation. Often these could change in time while supplying dynamic loads.

12.6.4.2   Grid-connected mode

Grid-connected synchronous generators (GCSGs) are large in size (up to a few hundred MW) feeding generated power to large power (utility) grids to energize connected loads – domestic, commercial, industrial, and agricultural – through national or regional grid networks. They are driven by hydro, steam, gas, or wind turbines with corresponding energy sources such as hydro potential, nuclear fuel, fossil fuels (coal, gas, oil), and wind. Among RE sources wind, bio, geothermal, and solar-thermal energy systems employ GCSGs driven by suitable prime movers – turbines or engines.

A schematic of a typical GCSG unit is shown in Figure 12.7. Energy source (fossil or nuclear fuel, hydro or wind power) is fed to the prime mover (steam/hydro/wind/gas turbine or engine) to run the generator at synchronous speed (based on poles and frequency). All grid-connected generators are of the rotating field type; low speed hydro generators use salient poles to have large poles and high speed thermal generators use cylindrical rotors with two or four poles. An excitation system causes the field current I_f in the rotor to generate the voltage (decided by the grid). A "generator

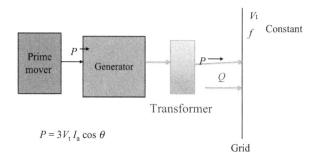

Figure 12.7 Schematic of a grid-connected unit.

transformer" is normally interfaced between the generator and the grid to step up the voltage to the grid voltage.

Grid-connected generators operate at constant voltage and frequency decided by the grid; such systems are often termed the machine connected to infinite bus. The generator feeds active and reactive power P and Q to the grid. P is decided by the prime mover input power and Q by excitation. There may be many generators and loads connected to the common grid, which form a multimachine power system. Here P and Q fed to the grid by different generators may differ. The total of P and Q may balance with the total of P and Q demanded by the loads.

12.6.5 Performance characteristics under load

Performance of SGs under load depends on the mode – stand alone or grid connected. As the generator is loaded, electrical and mechanical responses would be altered from the no-load status that leads to many performance characteristics.

12.6.5.1 Load characteristics

It is often defined as the variation of terminal voltage with load current. This variation is dependent on machine impedance and load PF. Load characteristics can be determined by the equivalent circuit and phasor diagram explained earlier. Simplified equivalent circuit neglecting armature resistance in Figure 12.6, results in corresponding phasor diagrams for lagging, unity, and leading load PF as in Figure 12.10a, b, and c, respectively, based on the following equation:

$$E_f = V_t + jX_s I_s \qquad (12.17a)$$

We notice that for the same terminal voltage (V_t) and load current (I_s) the internally generated (no-load) voltage E_f is maximum at lagging PF (E_{f3}) and decreasing for unity (E_{f2}) and leading PF (E_{f1}). Based on these phasor diagrams, typical load characteristics can be derived as shown in Figure 12.8. In Figure 12.9a load voltage (V_t) is kept the same at rated value at rated load current at all PF by adjusting I_f at different PF.

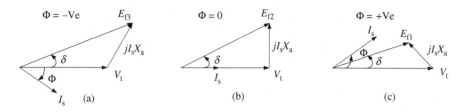

Figure 12.8 Phasor diagram at different PF. (a) Lagging PF, (b) unity PF, and
(c) leading PF.

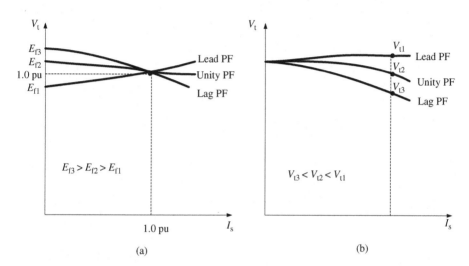

Figure 12.9 (a) **Rated voltage at rated current at all PFs and** (b) **Constant field current
at all PF load characteristics.**

Thus, no-load voltage (E_f) would be different with high value at lagging PF and low
at leading PF with a middle value at unity PF. In Figure 12.9b I_f is kept constant at all
PF, which makes no-load voltage the same and load voltage lower at lagging PF (V_{t3})
compared to unity PF (V_{t2}) or lead PF (V_{t1}). Load characteristics exhibit increased
drooping at lagging PF due to the demagnetizing effect of lagging load currents. On
the other hand, leading (capacitive) load currents have a magnetizing effect so that the
load voltage may even exceed no-load voltage.

12.6.5.2 Excitation characteristics

Variation of field current to keep terminal voltage the same at all loads at a given
load PF can be defined as excitation characteristics, often called compounding curves.
These can be derived from the phasor diagrams and voltage equations presented earlier and shown in Figure 12.10.

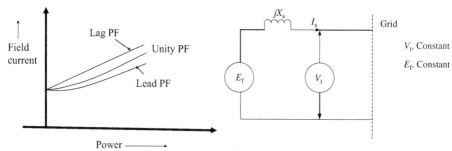

Figure 12.10 Excitation characteristics.

Figure 12.11 Simplified equivalent circuit of grid-connected generator.

Since I_f nearly varies with E_f at any load current it is high at lagging PF, which decreases at unity and leading PF. At no load I_f is same at all PFs. These curves dictate the way the field current must be varied with load to keep terminal voltage constant, a task cut out for an AVR. Increased I_f causes increased supply of reactive power demanded by lagging loads.

12.6.5.3   Performance characteristics of a grid-connected generator

Here terminal voltage (V_t) and frequency are always constant, dictated by the grid. A simplified equivalent circuit is shown in Figure 12.11, satisfying Equation (12.17) at constant V_t. Here X_s may include reactance of line connecting the machine with the grid.

Normally two ports of control exist, one is the power fed to the shaft and the other is the current fed to the field winding. Generator characteristics with variations of these two quantities are of interest.

1. Varying input power at constant field current: Here V_t, E_f, I_f, and frequency are constant and only the input power (P) through the shaft is varied by the prime mover, which causes the variation of current I_s fed to the grid. Such a variation is affected by varying input to the driving turbine or engine such as oil (engine), gas (gas turbine), steam (steam turbine), water (hydroturbine), or wind speed (wind turbine). Neglecting all losses, a phasor diagram for different input power is shown in Figure 12.12a, where V_t phasor is kept fixed. Due to the constant magnitude of E_f, the locus of the tip of E_f will be an arc (part of a circle) with center at O. From Equation (12.16) P is proportional to sin δ since all other quantities are constant and the P–δ curve is a sine curve as in Figure 12.12b with δ varying from 0° to 90° in a stable region. At zero power V_t and E_{f0} are in phase ($\delta = 0$) and I_{s0} lags voltages by 90° representing zero PF condition. As power is increased to P_1 causing I_{s1} (lagging), E_f phasor moves to E_{f1} (B) with power angle δ_1. A further increase in P moves E_f to C causing I_{s2} at unity PF and power angle at δ_2. Maximum power P_{max} is reflected by E_f at D with $\delta = 90°$ causing current phasor I_{s3} (leading).

2. Varying field current at constant input power: Extending the previous argument, a phasor diagram at different field currents can be drawn. Since power is constant, E_f sin δ is constant for all field current from Equation (12.16). Thus, the locus of the tip of E_f phasor will be a horizontal line. Further I_s cos Φ is constant and the tip of I_s phasor will move in a vertical line. We can show that increased field current causes more lagging current at the same

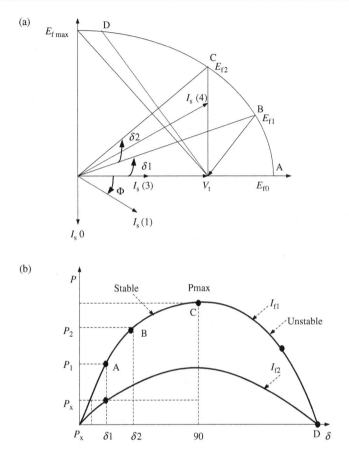

Figure 12.12 (a) Phasor diagram at different input power and (b) Power-angle characteristics of SG.

power and can supply more lagging VAR (volt-ampere reactive) (Q) to the grid. On the other hand, decreased field current causes more leading current at the same power and can supply more leading VAR to the grid as the current fed to the grid becomes more leading. We can plot the variation of armature current with field current as in Figure 12.13. As explained, at low I_f, I_s is leading and as I_f is increased it becomes lagging. The magnitude of I_s is

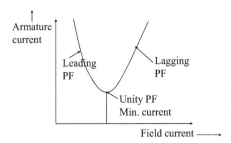

Figure 12.13 Variation of armature (load) current with field current at constant power.

minimum at unity PF for one field current beyond which the current increases with lagging PF as the phase angle increases with I_f. Thus, by controlling the field current, the reactive power fed to the grid can be controlled. If the grid is supplying more lagging PF loads, I_f has to be increased.

From the earlier discussions, we notice that both P and Q can be controlled as required thereby varying the magnitude and phase angle of current at the point of feeding to the grid. Since these controls are automatic, sophisticated mechanisms (instrumentation, hardware, and software) have to be put in place.

12.6.6   Permanent magnet synchronous generator

Permanent magnet synchronous generators (PMSGs) are generators of special interest to RE, as they are widely employed for wind energy-based grid-fed systems directly coupled to wind turbines without a gearbox. They are generally of low speed generating low frequency, which is converted to grid frequency via DC using power electronic converters to interface with the grid. Here the stator or armature is similar to a conventional SG. But the field structure (rotor) is completely different involving permanent magnets mostly fabricated from rare earth materials of high energy density and fixed to the ferromagnetic core. There are embedded and surface mounted magnets based on design. Unlike in wound field SGs, here field flux cannot be varied. Generated or no-load voltage E_f remains constant making Q control difficult. Operating temperature and load affect magnet property and may demagnetize. Construction of PMSGs is based on design. But its "brushless" configuration without slip rings is a merit. Sometimes, mainly for wind systems, they are built "inside-out" with a rotor (permanent magnetic field) as the outer member and the polyphase stator inside. Analysis and modeling may follow similar procedures as SGs, but magnetic circuit calculations are involved often needing finite element methods using software for flux computations.

12.7   Induction machines

The main difference between the synchronous and induction machines is rotor construction, as the stators would be identical having three-phase distributed windings placed in slots formed in a laminated core. Whereas currents are supplied to the field (rotor) winding of a synchronous machine through slip rings, rotor currents in an induction machine are induced by the stator mmf. The rotor is normally short circuited, except in what is known as a doubly fed machine. When balanced three-phase voltages are applied to the stator windings a rotating magnetic field in the air gap is created. Currents are induced in the rotor except at one speed when the rotor conductors appear stationary with respect to the air-gap field produced by the stator excitation. Obviously at this speed no voltage or current is induced in the rotor. This speed is termed the "synchronous speed" and hence the induction machine will not operate at this speed.

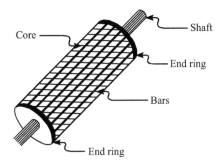

Figure 12.14 Squirrel-cage rotor.

There are two types of rotors: (1) squirrel-cage rotor and (2) slip ring or wound rotor. Figure 12.14 shows a typical squirrel-cage rotor. The core is formed by stacking laminations of suitable shape, which are mounted and keyed on to the shaft. Copper or aluminum bars are placed inside the slots, and these bars are short circuited at both ends by what are known as end rings; this entire winding is called a squirrel-cage winding.

There are two types of squirrel-cage rotors: "die cast" and "brazed." In die cast rotors, used in low power range of motors, bars and end rings are formed by pouring liquid aluminum in the core placed in a proper mold. The rotor is then machined to obtain a smooth surface and to provide an air gap of required dimensions. In brazed rotors, copper bars are manually inserted in the slots and then brazed on either side to end rings made up of copper or an alloy of copper. Squirrel-cage rotors are rugged, compact, and need no maintenance due to the absence of sliding parts and brushes. Due to the large number of pieces produced and the ease of manufacturing, squirrel-cage induction motors are the cheapest among equally powered machines. Squirrel-cage rotors effectively form a multiphase short-circuited winding.

One disadvantage of such rotors is the absence of controllability in the rotor to obtain desired performance. Slip ring or "wound" rotors, on the other hand, have three-phase balanced distributed windings brought out to external circuits through slip rings, explained earlier. Rotors are externally short circuited. Control of the motor through a rotor is possible by connecting suitable electrical controllers at the slip-ring terminals. Windings are internally connected in star or delta arrangement and only three or four terminals are brought out requiring an equal number of slip rings.

The basic working of an induction motor (IM) can be understood from transformer principles. If voltages are applied to a stator when the rotor is at stand still, currents are induced at line frequency in the short-circuited rotor by transformer action. By the interaction of impressed stator currents and induced rotor currents a torque is developed on the rotor shaft as a reaction to the resultant air-gap magnetic field and stored energy. Thus, the motor accelerates. Due to the changed relative speed of the rotor, its frequency and current magnitudes change resulting in different torques at different speeds. This process continues till the motor accelerates to such a speed when the developed electromagnetic torque equals the connected load torque. As mentioned earlier, there is one speed called synchronous speed when the rotor current is zero, causing zero developed torque. The motor can run at this speed if the load torque is zero. Due to inherent friction, there is some opposing torque even when no external torque is connected. Since zero load torque is not practicable, the motor can never run in equilibrium at synchronous speed, therefore, the IM is also termed an "asynchronous motor."

Asynchronous machines can also be operated as a generator by driving the rotor above the synchronous speed so that the direction of induced current reverses with respect to the voltage, causing generator action. Such generators are finding increased application these days in small hydro, bio, and wind energy conversion systems. Due to the rugged brushless rotor, they offer certain advantages over alternators for operation in remote unattended regions.

12.7.1 Modeling and analysis

Induction machines are modeled through an equivalent circuit shown in Figure 12.15, comprising resistances, reactances, and a term called per unit slip (s) defined as the difference between synchronous speed and actual speed as a fraction of synchronous speed.

The parameters are:

- R_s = Stator resistance/phase
- x_{ls} = Stator leakage reactance/phase
- x_m = Magnetizing reactance
- R_c = Core loss resistance
- R_r = Rotor resistance/phase (ref. to stator)
- x_{lr} = Rotor leakage reactance/phase (ref. to stator)
- s = pu slip
- v = pu speed
- $S = 1 - v$

Here voltages and currents are:

- V_s = Supply stator voltage/phase
- I_s = Stator current/phase
- E_s or V_g = Induced stator voltage or air-gap voltage
- I_0 = No-load current

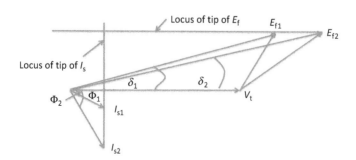

Figure 12.15 Phasor diagram at different field current.

- I_m = Magnetizing current
- I_c = Core loss current
- I_r = Rotor current/phase (ref. to stator)

The equivalent circuit may be simplified as follows:
Impedances:

$$\left.\begin{array}{l} \text{Stator impedance, } Z_s = (R_s + jX_{ls}) \\ \text{Rotor impedance, } Z_r = (R_r/s + jX_{lr}) \\ \text{Magnetizing impedance, } Z_m = (jX_m)(R_c)/(R_c + jX_m) \end{array}\right\} \qquad (12.17b)$$

Equivalent per-phase impedance looking from the terminals is:

$$Z_{eq} = Z_s + [Z_m Z_r/(Z_m + Z_r)] \qquad (12.18)$$

The per-phase stator current is given by:

$$I_s = V_s/Z_{eq} \qquad (12.19)$$

From the previous equivalent circuit, the air-gap voltage can be obtained by deducting the stator impedance drop as:

$$V_g = E_s = V_s - I_s Z_s \qquad (12.20)$$

Currents in the shunt magnetizing branch are given by:

$$\left.\begin{array}{l} I_o = V_g/Z_m \quad \} \\ I_m = V_g/x_m \quad \} \\ I_c = V_g/R_c \quad \} \end{array}\right. \qquad (12.21)$$

The rotor current is given by:

$$I_r = V_g/Z_r = I_s - I_o \qquad (12.22)$$

The performance equations can be derived as follows. The impedance angle Φ_s of Z_{eq} is:

$$\Phi_s = \tan^{-1}(X_{eq}/R_{eq}) \qquad (12.23)$$

such that the input PF:

$$PF_s = \cos\Phi_s$$

Input power is:

$$P_{in} = 3V_s I_s \cos \Phi_s \qquad (12.24)$$

Stator copper loss is:

$$P_{cus} = 3(I_s^2 R_s) \qquad (12.25)$$

Core loss attributed to loss in the core due to time varying flux is:

$$P_c = 3\left(V_g^2\right)/R_c \qquad (12.26)$$

In motoring mode, the power transferred from stator to rotor through the air gap, called the air-gap power, is given by:

$$P_g = P_{in} - P_{cus} - P_c \qquad (12.27)$$

From the equivalent circuit, we can infer that this power is also given by:

$$P_g = 3\left(I_r^2 R_r/s\right) \qquad (12.28)$$

Now the rotor copper loss can be written as:

$$P_{cur} = 3\left(I_r^2 R_r\right) = sP_g \qquad (12.29)$$

Now the developed power available for external work is:

$$\begin{aligned}
P_d &= P_g - P_{cur} \\
&= (1-s)P_g \\
&= (1-s)3\left(I_r^2 R_r/s\right)
\end{aligned} \qquad (12.30)$$

This power will not be fully available as output power due to mechanical losses, often called friction and windage (FW) losses, such that output power is:

$$P_{out} = P_d - FW \qquad (12.31)$$

Developed torque is given by:

$$T = (\text{Developed power})/(\text{Speed})$$

where speed is $(1 - s)$ ω_s; ω_s is the mechanical synchronous speed in rad/s. Thus:

$$T = [(1-s)3(I_r^2 R_r/s)]/[(1-s)\omega_s]$$
$$= 3/\omega_s (I_r^2 R_r/s) \qquad (12.32)$$

Expressions for efficiency can be written as:

Electrical efficiency, $\eta_e = $ (Power dev.)/Power input $= P_d/P_{in}$ $\qquad$ (12.33)

Total efficiency is:

$\eta = $ Power output/Power input $= P_{out}/P_{in}$

Input reactive power from the grid mainly to magnetize the machine is:

$$Q_{in} = 3 V_s I_s \sin \Phi_s \qquad (12.34)$$

Input volt-ampere is:

$$S = 3 V_s I_s \qquad (12.35)$$

so that:

$$P_{in} = S \cos \Phi_s$$
$$Q_{in} = S \sin \Phi_s$$

12.7.2 Characteristics with speed under different modes

Using the previous performance equations, it is possible to obtain relevant character-istics by computing the performance at different speeds. The speed range of interest is from -1 pu to $+2$ pu with synchronous speed as 1.0 pu. Variation of torque with speed in the previously mentioned range provides interesting information on how the machine behaves in different speed zones. The typical torque/speed characteristic over the above speed range is given in Figure 12.16 indicated as ABCDE.

In the speed range of 0–1 pu, the characteristic is denoted by BCD. Here both torque and speed are positive indicating "motoring" mode as the machine outputs positive torque through the shaft. Output mechanical power being the product of torque and speed is positive as it is motoring. B is the torque at zero speed or "starting" torque. D is the maximum torque T_{max} at the slip $s_{maxT}(m)$. D is the torque at synchronous speed, which is zero. It can be proven that BC is the unstable zone and CD is the stable zone

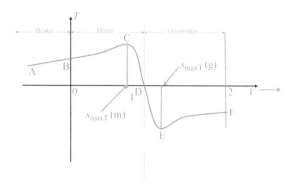

Figure 12.16 Typical torque–speed characteristics of IM.

where the motor operates in a torque range from 0 to T_{max} and the slip range from 0 to s_{maxT}(m). In this mode input electric power to the terminals from Equation (12.24) is positive since voltage, current, and PF are all positive. Normally current lags voltage by Φ_s (which is less than 90°) such that cos Φ_s is always positive.

In the speed range of 1–2 pu, the characteristic is denoted by DEF. Here speed is positive and torque is negative indicating "generating" mode as the machine outputs negative torque through the shaft or receives positive torque from the shaft. Output mechanical power being the product of torque and speed is negative as it is generating. D is the torque at synchronous speed, which is zero. E is the maximum torque T_{max}(g) at the slip s_{maxT}(g), which is negative as the machine operates at supersynchronous speed. F denotes torque when $s = -1$. It can be proven that EF is the unstable zone and DE is the stable generating zone where the generator operates in a torque range from 0 to $-T_{max}$(g) and the slip range from 0 to s_{maxT}(g). In this mode, input electric power to the terminals from Equation (12.24) is negative since voltage and current are positive but the PF is negative. Normally current lags voltage by Φ_s (which is more than 90°) such that cos Φ_s is always negative. Thus, electric power is fed to the grid from the terminals and hence is generating.

In the negative speed range denoted by AB, speed is negative and torque is positive. Torque effectively acts to bring down the speed, which in effect is braking. This is the braking mode.

The previous discussions are summarized as follows:

Motoring: $0 < v < 1$
            $1 > s > 0$
Generating: $1 < v < 2$
            $0 > s > -1$
Braking: $-1 < v < 0$
            $2 > s > 1$

Motoring:

- Stable: $s_{max} > s > 0$
- Unstable: $1 > s > s_{maxT}$

Generating:

- Stable: $0 > s > -s_{maxT}(g)$
- Unstable: $-S_{maxT}(g) > s > -1$

Braking:

- Stable: nil
- Unstable: $2 > s > 1$

12.7.3 Induction generators

Two recent factors have brought IGs to center stage. One is the need to have suitable energy converters for RE applications, typically wind, hydro, and bio energy. The second is the advances in power electronics technology that facilitates control and modification of electrical energy as per need. Unlike conventional energy based on fossil or nuclear fuels, RE tends to be sporadic, localized, time varying, random, or seasonal needing proper and involved control. The system developed must be simple and user friendly and usable even in remote places. IGs especially with squirrel-cage rotors tend to be simple, rugged, and economical compared to SGs. They are also found attractive with variable speed prime movers.

Broadly IGs can be divided into grid-fed and off-grid systems. Grid-fed induction generators (GFIGs) will have two types based on the type of rotor used: squirrel-cage induction generators (SCIGs) and wound rotor induction generators (WRIGs), or doubly fed induction generators (DGIGs). SCIGs will have control and operation options only from the stator while DFIGs have access to both stator and rotor for control that makes them more suitable for variable speed prime movers through change of slip power. The stator of SCIGs can have a single or dual winding to handle different power levels especially for wind applications. SCIGs can also be made doubly fed by having two sets of stator windings of different poles with the flexibility of additional control. The major requirement of IGs is to have reactive power Q (VAR) to magnetize the machine as there is no other source of flux. In GFIGs VAR is supplied from the grid. In off-grid or standalone systems VAR is supplied by additional terminal capacitors to make them work as self-excited induction generators (SEIGs). The following sections present the basic working and analysis of the previously mentioned types of IG.

12.7.3.1 Grid-connected squirrel-cage induction generators

Here the prime mover drives the IG connected to the grid. Normally, they are started as motors and the prime mover will then overspeed beyond synchronous speed to operate at a suitable negative slip dependent on input power. The machine can be analyzed using the well-known motor equivalent circuit (Figure 12.15) with the associated equations of Section 7.1 for motoring conventions by making the slip negative

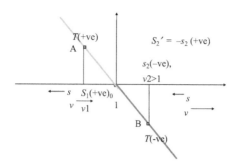

Figure 12.17 Stable part of *T–N* characteristics: A – motoring, B – generating.

and employing a computer program. The torque/speed characteristics of Figure 12.16 present motoring, generating, and braking regions. The stable generating region at a small negative slip is now of interest. Figure 12.17 shows only the stable part of the previously mentioned characteristics with points A and B, respectively, in motoring and generating zones at slips s_1 (positive) and s_2 (negative).

For a qualitative understanding of generator performance we may simplify the equivalent circuit by neglecting core loss current I_c and rotor leakage reactance x_{lr}. If we closely examine the *T–N* characteristics in the stable region in motoring and generating modes our points of interest will be (Figure 12.17) motor stable point A and generator stable point B at the following slips (s) and pu speeds (v).

Motoring (A): $s = s_1$ (positive), $v = v1 < 1$
Generating (B): $s = s_2$ (negative), $v = v2 > 1$

The motor phasor diagram at slip s_1 (A) is in Figure 12.18 (neglecting core loss) drawn starting from V_g and I_r phasors in line. By adding leakage impedance drop to V_g we get stator voltage phasor V_s leading to the input power factor angle Φ_s as shown.

12.7.3.1.1 Understanding induction generators using motoring conventions

To extend the operation to generation operation at negative slip s_2 at (B) we may use the generator equivalent circuit drawn with motoring current conventions (neglecting x_{lr} and I_c). Here $V_s = V_g + I_s (R_s + jx_{ls})$.

Relevant equations with negative slip s_2 lead to the phasor diagram of Figure 12.19.

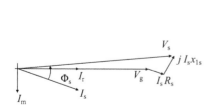

Figure 12.18 Motor phasor diagram at s_1.

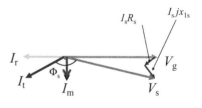

Figure 12.19 Phasor diagram of generator at s_2 with motoring conventions.

Here we observe that the power factor angle Φ_s is such that:

$$90° < \Phi_s < 180°$$

And I_s lags V_s by more than 90°.
Thus, active power input at the terminal of the GCIG is:

$$P_{in} = P_{elec} = 3V_s I_s \cos\Phi_s$$

which is negative (since Φ_s is more than 90°). Thus, P_{in} will be opposite to that of motor and power will be fed to the grid from the machine.
Reactive power input at the terminal of the GCIG is:

$$Q_{in} = 3V_s I_s \sin\Phi_s$$

which is positive and Q has to be fed to the machine from the grid as in a motor.
The power flow will have the following sequence:

Output power, $P_{elec} = 3V_sI_s \cos\Phi_s$
Stator copper loss, $P_{cus} = 3(I_s^2R_s)$
Core loss, $P_c = 3(V_g^2)/R_c$
Air-gap power, $P_g = P_{elec} + P_{cus} + P_c = P_{mech} - \text{FW} - P_{cur} = 3(I_r^2R_r/s)$
Rotor copper loss, $P_{cur} = 3(I_r^2R_r) = sP_g$
As defined for developed power in the motor, $P_d = P_g - P_{cur} = (1 - s_2) P_g = (1 - s_2) 3(I_r^2R_r/s_2)$
Since slip is negative, P_d is positive and is greater than P_g.

P_g is negative, P_{cur} is positive, and power through the shaft is negative compared to the motor. Power is fed through the shaft equal to P_d as mentioned earlier.

Mechanical loss = Friction windage = FW

In the motor, output power, $P_{out} = P_d - \text{FW}$
Since P_d is negative, output power is negative. This is the input power from prime mover, P_{in}.
Torque in IG:
Developed torque, $T = (\text{Developed power})/(\text{Speed})$

$$\text{Speed} = (1-s)\omega_s$$
$$T = (1-s_2)3(I_r^2R_r/s_2)/[(1-s_2)\omega_s]$$
$$= 3/\omega_s(I_r^2R_r/s_2)$$

which is negative since slip is negative. Here ω_s is the mechanical synchronous speed in rad/s.

Efficiency in IG:

Electrical efficiency, η_e = (Power developed)/Power input
$$= P_d/P_{in}$$
Total efficiency, η_t = Power output/Power input = P_{out}/P_{in}
Input reactive power, $Q_{in} = 3V_sI_s \sin \Phi_s$
Input volt-ampere, $S = 3V_sI_s$
$P_{elec} = S \cos \Phi_s \ldots$negative
$Q_{in} = S \sin \Phi_s \ldots$positive

12.7.3.2 Doubly fed induction generators

Unlike SCIGs doubly fed induction generators (DFIGs) with wound rotor and slip rings provide greater flexibility and variable speed operation facilitated by an additional rotor port through which energy can be fed or extracted. Here the machine can operate in motoring and generating modes in both subsynchronous and supersynchronous speeds by adjusting the direction of power flow through the rotor.

A schematic of doubly fed induction machine (DFIM) is shown in Figure 12.20. Here the stator is directly connected to the grid at nominal voltage and frequency f, drawing power P_e from the grid. From rotor slip rings, power P_r is extracted at rotor frequency f_r equal to sf where s is the slip. A bidirectional power electronic converter transforms power P_r at rotor frequency to mains frequency f via a DC link, which in turn is fed to the grid through a transformer. Often a three-winding transformer is used with two primary windings connected respectively to stator and rotor (through a converter) and one secondary winding connected to the grid. This facilitates suitable voltage transformation to interface the machine with the grid. P_g is the air-gap power transferred from stator to rotor equal to P_e minus stator copper loss (P_{cus}) and core loss. P_m is the mechanical power in the shaft transferred to an external mechanical port – load or prime mover to extract or feed power – which is equal to P_g minus rotor copper loss (P_{cur}).

Steady-state analysis of the previously mentioned DFIM can be carried out using an equivalent circuit shown in Figure 12.21, which is an extension of the equivalent

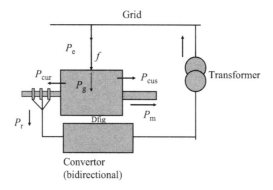

Figure 12.20 Schematic of DFIM.

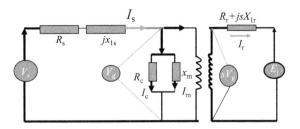

Figure 12.21 Equivalent circuit of DFIM.

circuit of a squirrel-cage induction motor by including a voltage source E_j at slip frequency in the rotor circuit. Here the stator loop is at mains frequency f and the rotor loop is at slip frequency s_f as reflected in the rotor impedance. The air-gap voltage phase V_g in the stator changes to sV_g in the rotor at the same turns.

From the previous equivalent circuit, the following power relations hold true:

$$sP_g = P_r + P_{cur}$$
$$P_g = P_r + (1-s)P_g + P_{cur}$$

In squirrel-cage machines P_r is zero and the air-gap power is used only for rotor copper loss and mechanical power to the shaft. But here P_r is attributable to injected voltage E_j. From the equivalent circuit, the rotor equations are:

$$sV_g - I_r(R_r + jsx_{lr}) - E_j = 0$$
$$\text{or } sV_g - E_j = I_r(R_r + jsx_{lr})$$

based on which the phasor diagram with injected emf E_j (referred to as stator turns) at slip frequency can be drawn as in Figure 12.22. Here β and Φ_r are the phase position of E_j and I_r with respect to sV_g as shown in Figure 12.22. Figures 12.23–12.26 show power distribution of DFIM in sub-synchronous and super-synchronous motoring and generating modes.

Resolving voltages and drops along the I_r axis, we can write:

$$sV_g \cos\Phi_r = I_r R_r + E_j \cos(\Phi_r + \beta)$$

Multiplying both sides by I_r:

$$sV_g I_r \cos\Phi_r = I_r^2 R_r + I_r E_j \cos(\Phi_r + \beta)$$

or:

$$V_g I_r \cos\Phi_r = (1-s)V_g I_r \cos\Phi_r + I_r^2 R_r + I_r E_j \cos(\Phi_r + \beta)$$

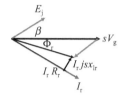

Figure 12.22 Rotor phasor diagram of DFIM with E_j.

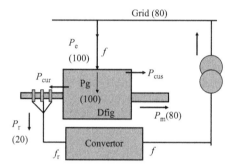

Figure 12.23 Power distribution under subsynchronous motoring.

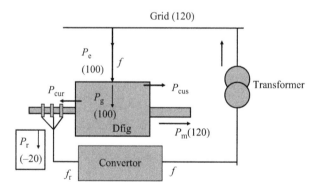

Figure 12.24 Power distribution under supersynchronous motoring.

On a three-phase basis:

$$P_g = 3V_g I_r \cos\Phi_r$$
$$P_{cur} = 3I_r^2 R_r$$

Power fed to the converter through rotor slip rings:

$$P_r = 3I_r E_j \cos(\Phi_r + \beta)$$

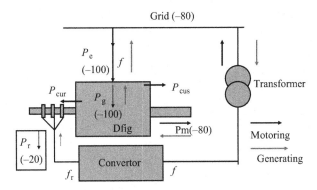

Figure 12.25 Power distribution under subsynchronous generating.

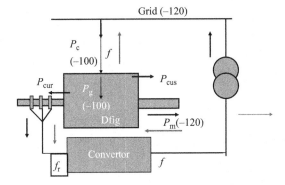

Figure 12.26 Power distribution under supersynchronous generating.

The previous equation yields the power balance relation as:

Air-gap power = Mechanical power + Rotor copper loss + Power fed to converter

It is important to note that mechanical power is $(1 - s)$ times the air-gap power P_g.

Let us now consider the following four cases, neglecting all losses P_{cus}, P_{cur}.

1. Mode I: Subsynchronous motoring. Here:

$0 < s < 1$, s is positive.

P_g, P_m, P_e, P_r are positive.

$P_m = (1 - s)P_g$ is positive.

For example, consider $s = 0.2$. For 100 units of air-gap power, the distribution of power will be $P_g = +100$, $P_m = +80$, $P_r = +20$, $P_e, = +100$ and power drawn from grid = $P_{grid} = 80$. Here slip power P_r is extracted from the rotor and fed to the grid through the converter. This distribution is shown in Figure 12.27.

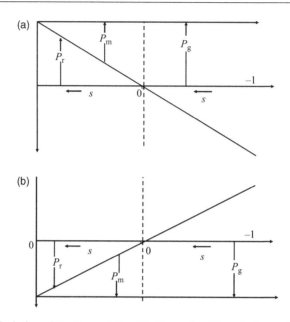

Figure 12.27 Variation of P_g, P_m, and P_r with slip under (a) motoring and (b) generating.

2. Mode II: Supersynchronous motoring. Here:

$-1 < s < 0$, s is negative.
P_g, P_m, P_e are positive.
$P_m = (1-s)P_g$ is positive and is more than P_g.

For example, consider $s = -0.2$. For 100 units of air-gap power, distribution of power will be $P_g = +100$, $P_m = 120$, $P_r = -20$, $P_e = +100$ and power drawn from grid = $P_{grid} = 120$. Here slip power P_r is extracted from the grid and fed to the rotor through the converter. This distribution is shown in Figure 12.28.

3. Mode III: Subsynchronous generating. Here:

$0 < s < 1$, s is positive.
P_g, P_m, P_e, P_r are negative.
$P_m = (1-s)P_g$ is negative and is less than P_g.

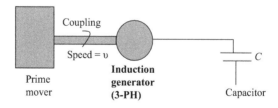

Figure 12.28 Basic SEIG scheme.

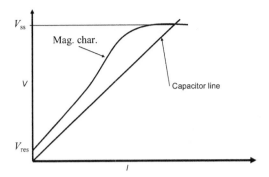

Figure 12.29 Voltage buildup in SEIG.

For example, consider $s = +0.2$. For 100 units of air-gap power, distribution of power will be $P_g = -100$, $P_m = -80$, $P_r = -20$, $P_e = -100$ and power fed to the grid = $P_{grid} = 80$. Here slip power P_r is extracted from the grid and fed to the rotor through the converter. This distribution is shown in Figure 12.29.

4. Mode IV: Supersynchronous generating. Here:

$-1 < s < 0$, s is negative.

P_g, P_m, P_e are negative.

$P_m = (1-s)P_g$ is negative and is more than P_g.

For example, consider $s = -0.2$. For 100 units of air-gap power, distribution of power will be $P_g = -100$, $P_m = -120$, $P_r = +20$, $P_e = -100$ and power fed to the grid = $P_{grid} = 120$. Here slip power P_r is extracted from the rotor and fed to the grid through the bidirectional converter. This distribution is shown in Figure 12.30.

Based on the earlier logic, variation of different power with slip under motoring and generation is shown in Figure 12.27a,b, respectively.

In subsynchronous motoring, power is extracted from the rotor and fed to the grid and thus slip power is recovered. In supersynchronous motoring power is extracted from the grid and fed to the rotor and thus slip power augments air-gap power to transfer to shaft.

In subsynchronous generating, power is extracted from the grid and fed to the rotor slip rings. Since DFIGs are used predominantly for variable speed grid-fed wind

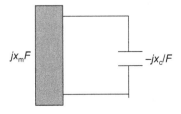

Figure 12.30 A simplified model (equivalent circuit) of SEIG.

turbines the previously mentioned operation is affected in lower wind speeds when P_m is low and is augmented by slip power P_r to get the required P_g. On the other hand, DFIGs operate at supersynchronous speeds at higher wind speeds when P_m is high and surplus power is diverted to the grid via the slip rings keeping P_g at the required value, which is transferred to the grid via the stator.

12.7.3.3 Self-excited induction generators

Both SCIGs and DFIGs discussed earlier need a grid to operate and hence are unsuitable for off-grid applications. We noticed that reactive power (VAR) must be supplied by the grid to magnetize the machine and terminal capacitors are often employed to reduce the VAR drain. As a corollary, if all the needed VARs are supplied by the capacitors, the grid need not supply any VAR and the IG can work in off-grid self-excited mode. In other words, if sufficient terminal capacitors are connected as needed at the rotational speed of the machine, an induction machine can operate as a self-excited induction generator (SEIG). However, it is necessary to know the phenomenon of self-excitation that produces a voltage across the machine decided by the capacitor and speed. A basic SEIG scheme is shown in Figure 12.28. The prime mover drives the three-phase induction machine at a pu speed v and a capacitor C per phase is connected at its terminals.

Voltages and currents are induced in the stator due to self-excitation and exchange of energy between the electromagnetic machine and electrostatic capacitor similar to resonance. A rise in voltage and current is arrested by magnetic saturation in the machine. For voltage induction we need a source of flux and its time variation. An unexcited IM cannot produce a voltage. An important condition to initiate self-excitation is remnant flux in the machine or charge in the capacitor. A remnant flux in the rotor induces a small balanced voltage in three-phase stator windings; the connected capacitor has a magnetizing effect as in any AC generator. As flux rises, voltage increases leading to an avalanche effect with the perpetual increase limited by saturation. This process of voltage buildup can be observed in Figure 12.29 wherein magnetization characteristics of the machine and voltage/current variation across the capacitor (called capacitor line) are drawn. Initially the residual flux causes a residual voltage (V_{res}) dependent on the speed, which in turn results in a current in the capacitor. This current causes a voltage decided by the magnetizing characteristics. This increased voltage causes an increased capacitor current decided by the capacitor line. This process continues until the capacitor line intersects with the magnetizing curve when no further rise in voltage is possible and a steady-state voltage (V_{ss}) is reached.

Thus, at a constant speed voltage gradually builds from a low value to the steady-state value on switching the capacitor and can be easily observed experimentally by recording the voltage waveform. Under steady state the magnetic circuit is saturated. The magnetizing reactance x_m, which is the ratio of voltage and current of the no-load magnetizing characteristics, will assume different values from unsaturated to saturated gradually decreasing as voltage builds up.

A simple and approximate analysis can be made to understand the effect of speed and capacitance on voltage and to have an idea of the minimum capacitance required for self-excitation at any speed in a given machine. Let us neglect all resistances and leakage reactances so that the machine on no load is represented by x_m only, which will be in parallel with the capacitor C with reactance x_c at base frequency as shown in Figure 12.30. If the self-excited induced pu frequency is F (rated IM frequency is taken as base frequency and corresponding synchronous speed as base speed) effective reactances will be jFx_m and $-x_c/F$ as in Figure 12.30.

12.7.3.3.1 Effect of capacitor at constant speed – minimum capacitor for self-excitation

We have learnt that at constant speed the steady-state voltage is the intersecting point of the magnetization curve and the capacitor line. If capacitance is increased the capacitor line will shift to the right with a new intersecting point with higher steady voltage with the magnetizing curve unaltered. Figure 12.31 shows a family of capacitor curves for different capacitors (C1, C2, C3, C4) intersecting the magnetizing curve resulting in different voltages – higher for higher capacitance. At C4 the capacitor line is tangential to magnetizing curve below which no self-excitation is possible due to no intersection. Thus, C4 is the minimum capacitance (C_{min}) for self-excitation at this speed when the machine has unsaturated magnetizing reactance corresponding to so-called air-gap line. As an approximation, we may assume F to be equal to pu speed v, taking synchronous speed at base frequency as 1.0 pu under minimum capacitance conditions from the equivalent circuit so that:

$$v\, x_m = x_c/v$$
$$= 1/(\omega\, C_{min})v$$
$$C_{min} = 1/[\omega x_m v^2]$$
$$= 1/Kv^2$$

Here $K = \omega x_m$ is constant since base radian frequency ω and unsaturated magnetizing reactance x_m are constant. Thus, we notice that C_{min} is inversely proportional to

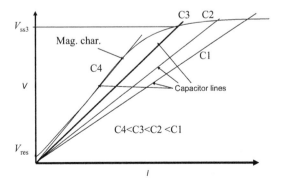

Figure 12.31 Family of capacitor curves for different capacitors intersecting the magnetization curve.

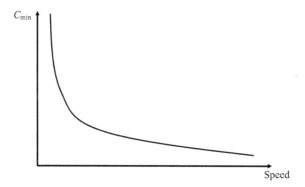

Figure 12.32 Variation of C_{min} with speed.

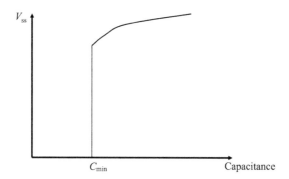

Figure 12.33 Variation of steady voltage with capacitance at a constant speed.

square of speed as in Figure 12.32 Increased speed needs a lower minimum capacitor for self-excitation and vice versa. Variation of steady voltage with capacitance at a constant speed is shown in Figure 12.33 indicating increase in voltage with connected capacitance.

12.7.3.3.2   Effect of speed
When speed of an SEIG is changed the induced pu frequency F will change almost proportionately since the negative slip $F - v$ is small, which if neglected makes $F = v$. If speed and hence frequency are increased the magnetization curve moves to the left (as voltage increases for the same current being proportional to Fx_m) and the capacitor line moves to the right (as voltage is inversely proportional to F for the same capacitance). Thus, the intersection point will be at a higher voltage. At increased speed, since the minimum capacitance line is tangential to the magnetization curve, which will have increased slope, C_{min} will be lower as proved earlier. At each speed there is one C_{min} below which self-excitation will cease. Thus, voltage increases with speed at a constant capacitance as shown in Figure 12.34. Self-excitation occurs only above a minimum speed, v_{min}, as shown in Figure 12.34.

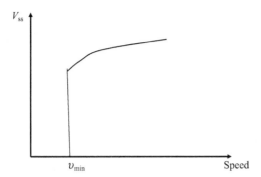

Figure 12.34 Effect of speed on voltage at constant *C*.

12.7.3.3.3 Analysis and performance of self-excited induction generators

The analysis of such generators requires knowledge of parameters of machines to determine their performance for a given capacitance, speed, and load. In the SEIG used as an isolated power source, both the terminal voltage and frequency are unknown and have to be computed for a given speed, capacitance, and load. Steady-state analysis of an SEIG is quite involved as the machine operates at varying levels of saturation decided by capacitance and speed, which alters both x_m and F. Initially the machine core is unsaturated due to low residual flux, which increases gradually achieved by the connected capacitor that drives the core to saturation as voltage builds up to the steady-state value, thus fixing the level of saturation as per the magnetization curve. Thus, the magnetization curve of the machine is central to the analysis and one needs correct data on the same obtained by design or test. "Synchronous speed" is best suited to acquire this data from measurements for a given machine. Here the magnetization curve will be the variation of air-gap voltage normalized to base frequency V_g/F with magnetizing current I_m as shown in Figure 12.35. The magnetizing reactance x_m being the ratio of the previously mentioned voltage and current will assume different values

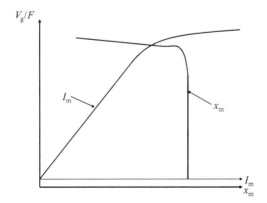

Figure 12.35 Magnetizing curve for analysis of SEIG.

based on level of saturation decided by the operating voltage (V_g/F) or flux. The corresponding variation of V_g/F with x_m is also shown in Figure 12.35.

From the equivalent circuit (Figure 12.30), we can write the loop equation as:

$$[jx_m F - jx_c/F]I = 0$$

Under self-excitation since the current $I \neq 0$:

$$jx_m F - jx_c/F = 0$$
or
$$x_m = x_c/F^2$$

As the capacitor is increased beyond C_{min}, when self-excitation occurs, x_m assumes a suitable saturated value dependent on x_c for a given speed of F. For corresponding x_m we can get the corresponding voltage V_g from Figure 12.35, which is also the intersection point as in Figure 12.29. Thus, we can get an approximate value of induced steady-state self-excited voltage on no load for a given capacitor and speed assumed equal to F.

For analysis with load on the generator we can resort to the equivalent circuit of an SEIG that includes connected load R_L (per phase) and rotor circuit as in Figure 12.36. Here all reactances refer to base frequency f while F is pu generated frequency and v is the pu speed. I_s, I_r, and I_L are stator, rotor, and load currents per phase; V_t and V_g are terminal and air-gap voltages.

Self-excitation results in the saturation of the main flux. As the value of x_m reflects the magnitude of the main flux, it is essential to consider the variation of x_m with saturation level.

The loop equation for the current I_s can be written as:

$$Z_s I_s = 0$$

Under self-excitation since the current $I_s \neq 0$, the loop impedance $Z_s = 0$, which implies that both real and imaginary parts of Z_s would be separately zero. From the equivalent circuit an expression for Z_s can be derived. By separating real and imaginary parts of the previous equation we get two nonlinear real equations in the unknowns x_m and F. We can solve these equations numerically by assuming valid initial conditions by writing a computer code. The next step is to calculate the air-gap voltage V_g and the terminal voltage V_t using the information on variation of x_m with the quantity V_g/F (Figure 12.35). From this curve of V_g/F against x_m, steady-state

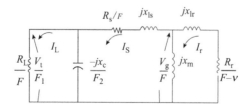

Figure 12.36 Equivalent circuit of SEIG with load.

V_g/F for the corresponding x_m can be obtained. With the knowledge of V_g, x_m, F, x_c, v, R_L, and machine parameters known, calculation of the terminal voltage V_t and the load current is straightforward using the equivalent circuit.

Based on the previous analytical procedure a general computer program can be developed, which calculates the steady-state performance of the unit for given v, C, and R_L. The program can be used to determine the steady-state operating characteristics of the generator as follows.

Load characteristics: Load characteristics of an SEIG is the variation of terminal voltage with load or output power at constant capacitor and speed. This can be obtained experimentally by varying the load resistance R_L to get variation from no load to full load. The same can be obtained by iterating R_L in the computer program. Typical characteristics for a machine for different capacitors at 1.0 pu speed are shown in Figure 12.37. Here g_c is the pu susceptance equal to $1/x_c$ or ωC and proportional to capacitance. At any fixed capacitance voltage drops with load or output power until a maximum power point beyond which we observe a bend in this curve. This is due to continuous demagnetization with load that reduces the capacitive VAR. Reduced capacitance causes reduced voltage at any load. Frequency drops slightly with load due to increased slip. If we wish to maintain constant voltage at all loads capacitance needs to be increased with load, as is apparent from this family of characteristics.

Figure 12.38 shows the family of load characteristics at different speeds. For constant voltage, the capacitance has to be increased at decreased speed. Output frequency is directly proportional to speed. By the parallel family of curves, it is apparent that both voltage and frequency regulation are almost the same at all speeds.

Excitation characteristics: This is the variation of capacitance with output power to give the desired voltage regulation or to maintain constant voltage at all loads. Figure 12.39 shows a family of excitation characteristics at different speeds to keep the load voltage constant. Here results obtained on a practical SEIG are given wherein

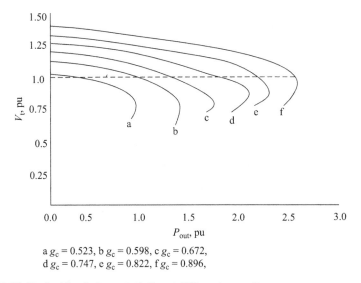

a $g_c = 0.523$, b $g_c = 0.598$, c $g_c = 0.672$,
d $g_c = 0.747$, e $g_c = 0.822$, f $g_c = 0.896$,

Figure 12.37 Typical load characteristics at different capacitors.

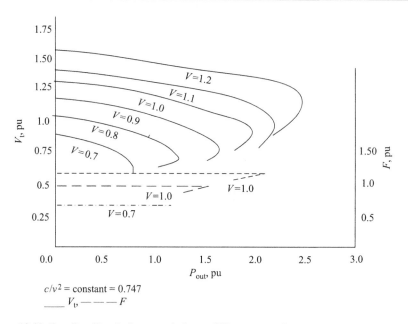

$c/v^2 = \text{constant} = 0.747$

——— V_t, — — — F

Figure 12.38 Family of load characteristics at different speeds.

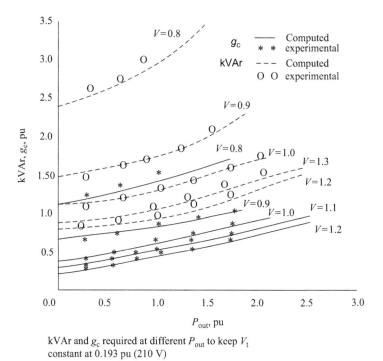

kVAr and g_c required at different P_{out} to keep V_t
constant at 0.193 pu (210 V)

Figure 12.39 Excitation characteristics.

capacitance is expressed in per unit as kVAr or g_c ($=\omega C$). Naturally the capacitor has to be increased with load at all speeds for desired constant voltage. Increased capacitor is needed at lower speeds. Any automatic voltage regulating (AVR) system must adjust the capacitor value as per these curves. With advances in power electronics different types of static capacitive VAR controllers are possible to affect variation of capacitance across the SEIG as per the previously mentioned excitation characteristics. Normally we need to vary capacitance from a low value C_{min} at no load to a high value C_{max} at full load.

12.7.3.4 Single-phase self-excited induction generators

In an earlier section, we studied a three-phase SEIG feeding three-phase isolated loads. For such off-grid applications, single-phase loads are often prevalent for domestic, commercial, and small industrial/agricultural needs. There are two possible modes in which an SEIG can be configured to supply single-phase loads: (1) a three-phase SEIG to feed single-phase loads with suitable capacitor topology, and (2) a specially designed single-phase SEIG.

12.7.3.4.1 Three-phase self-excited induction generators to feed single-phase loads

Normal three-phase SEIG will have balanced three-phase terminal capacitors and load. But a single-phase load introduces unbalance in voltage and currents that may need adjustment to capacitor topology. Several such topologies are reported in the literature. Three-phase systems operate at line voltages that are $\sqrt{3}$ times the single-phase voltages. Thus, single-phase loads cannot be directly connected across three-phase lines. Since we plan to use a standard three-phase IM to operate as an SEIG, one practical way is to take a star (Y)-connected motor and connect in delta (Δ) for a single-phase operation as an SEIG to achieve rated single-phase voltage. Figure 12.40 shows a commonly

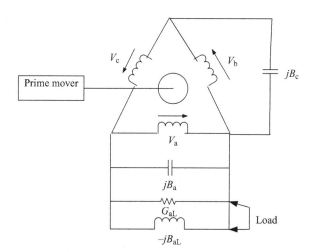

Figure 12.40 Three-phase SEIG feeding a single-phase load.

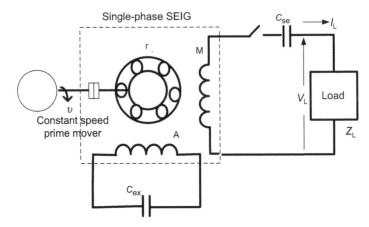

Figure 12.41 A novel two-phase SEIG feeding single-phase loads.
Indian Patent Scheme-179778.

used connection wherein *abc* are three-phase voltages connected in delta. Capacitors C_c and C_a are connected across c- and a-phases with corresponding susceptance of B_c, B_a. Normally one capacitor is twice the size of the other and results in minimum unbalance in winding voltages and currents. Single-phase (inductive) load is connected across the a-phase as shown. The analysis of such systems is involved needing use of symmetrical components to handle unbalanced systems.

12.7.3.4.2   Specially designed two-phase SEIG feeding single-phase loads
Normal single-phase IMs are found not to effectively work as single - phase SEIGs to feed single -phase loads as with a three-phase IM. A novel two-phase IM is reported in the literature to work as an SEIG to feed single -phase loads as shown in Figure 12.41. Similar to a two-phase motor, it has two-phase asymmetrical windings M (main) and A (auxiliary) in quadrature in the stator with a squirrel-cage rotor. C_{ex} is the external capacitor across the auxiliary winding that causes self-excitation when the rotor is driven at desired speed by the prime mover and causes a voltage across the main winding on no load. C_{se} is the capacitor in the main winding in series with the load. As SEIG is loaded the voltage tends to drop but C_{se} provides a compensating voltage to improve voltage regulation.

Figure 12.42 shows a test setup, measured waveform, and performance. Sinusoidal generated voltage and near constant load voltage with load can be observed.

12.8   Practical renewable energy-based power generating schemes

The foregoing sections detail different types of electric generators for both grid-fed and off-grid applications. Here we shall examine how such generators can be put to use for renewable energy (RE) applications to generate electricity.

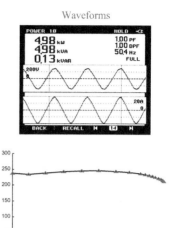

Waveforms

Test set-up

Performance graph

Figure 12.42 Test set-up, waveforms, and performance of a two-phase SEIG feeding single-phase loads.

12.8.1 Grid-fed systems

12.8.1.1 Wind energy systems

12.8.1.1.1 Squirrel-cage induction generators

SCIGs are found to be suitable for medium power RE-based power generation (typical range: 100–1000 kW) driven by wind or hydroturbines or biofuel-driven engines as they have certain inherent advantages over conventional alternators such as low unit cost, less maintenance, rugged and brushless rotor, and asynchronous operation. But system design for each of the energy sources is different. They operate in the stable part of the T–N curve in generating mode with the operating negative slip decided by the power input from the prime mover, which in turn is dependent on the energy content of the energy source. Thus, they operate in a small range of speed or nearly constant speed above the synchronous speed.

In wind systems, power varies as cube of wind speed (w) and square of the swept blade diameter (D) as per the following formula:

$$P = rC_p D^2 w^3$$

Here ρ is the air density and C_p is a constant. For each wind turbine there is a cut-in wind speed, typically 4–5 m/s below which it is infeasible to operate due to low power. Typical variation of power with wind speed is shown in Figure 12.43. Above a certain wind speed (about 12 m/s), power may go beyond rated value and hence blade

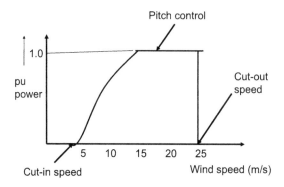

Figure 12.43 Typical variation of power with wind speed.

pitch is changed to regulate the power to be nearly constant as shown. There is also a cut-out speed, typically about 25 m/s, beyond which the machine may be discon-nected due to mechanical considerations. Thus, there is wide variation in wind power during operation due to randomly varying wind speed. Average power may be about 20% of the peak installed power depending on the wind regime, which affects the IG performance. Normally the rated IG line voltage (typically 230 V or 415 V) is lower than the grid voltage (typically 11 kV or 33 kV) and a step-up transformer between the IG and the grid is inevitable. As we learnt earlier IG feeds active power (P) to the grid and draws reactive power (Q) from the grid, which adversely affects the transmis-sion system with low voltage profiles. Therefore utilities insist on good PF (say 0.9) at the point of power evacuation needing the wind developers to restrict the VAR drain from the grid by installing terminal capacitors. A schematic of a typical wind system is shown in Figure 12.44. Such constant speed wind operated IGs have been installed in large numbers in recent decades in several countries with good wind regimes, typi-cally with a peak capacity of about 1 MW.

Advantages of this scheme are:

- Low manufacturing cost
- Robust, low maintenance cost

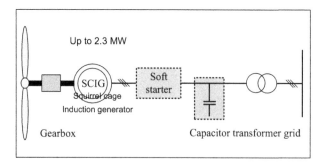

Figure 12.44 Typical wind system schematic.

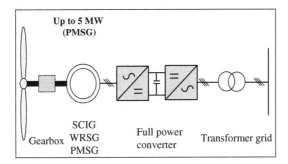

Figure 12.45 Wind energy systems with full-power converters.

Drawbacks are:

• Low conversion efficiency
• Large fluctuation in output power

However, the system can be made more efficient by introducing a pair of power electronic converters (AC–DC–AC) between the IG and the grid with a DC bus and interfacing capacitor (Figure 12.45). This facilitates variable speed operation with following features.
Advantages:

• Generator fully decoupled from the grid
• Wide speed range
• Smooth grid connection
• Reactive power compensation
• Capability to meet the strict grid code

Drawbacks:

• High system cost, reduces system efficiency

12.8.1.1.2 Doubly fed induction generators

DFIGs are widely employed for variable speed wind energy systems (WES) to adjust the rotational speed to suit the wind speed to get maximum power through maximum power point tracking based on turbine characteristics, which are a family of bell-shaped curves indicating a specific rotational speed (rpm) at which the turbine yields maximum power at each wind speed. Optimal operation must ensure that rpm for the specific wind speed. This is facilitated through DFIGs. A typical WES with DFIG is shown in Figure 12.46. At lower wind speed, due to low power in the wind, the DFIG operates in subsynchronous generating mode with rotor power being negative (taken from the grid). At higher wind speed, due to high power in the wind, the DFIG operates in supersynchronous generating mode with rotor power being positive (feeding to the grid). Now input wind power is channeled through both stator and rotor of the DFIG facilitated by the bidirectional converter.

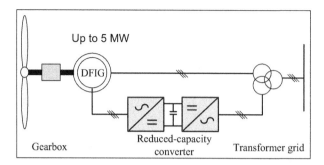

Figure 12.46 Typical WES with DFIG.

Advantages of the scheme are:

- Extended speed range
- High system efficiency and low cost
- Decoupled active and reactive power control
- Enhanced dynamic performance

Drawbacks are the limited grid-fault operation capability and regular maintenance of slip rings and brushes due to mechanical wear and tear.

12.8.1.1.3 Synchronous generators

Both wound field and permanent magnet types of SGs are widely used in WES. A recent trend has been to go for high power direct-driven low speed PMSGs. A power electronic converter to effect frequency change from generator to grid is inevitable making this scheme complicated and less reliable. The need for synchronization to the grid is a weak point unlike on IGs. The wound field synchronous generator has slip rings and brushes demanding regular maintenance. The PMSG has the advantage of being brushless at the cost of complex machine construction.

12.8.1.2 Small (or mini) hydro systems

These are discussed in detail in Chapter 5 and hence not repeated here.

12.8.1.3 Bio energy system

Bio energy sources have multiple forms needing suitable processes for energy conversion to electricity, invariably through the heat route. Two modes are possible: the "gas" route and the "steam" route. In the first mode biosources are converted to gas and fed to a combustion engine or a gas turbine, which in turn rotates the generator. Here power can be varied by adjusting the fuel intake as per demand in the range from 0 pu to 1 pu. Similar to a grid-fed hydro system both SGs and IGs can be used. With governor-controlled engine, speed can be kept nearly constant for the desired grid frequency. A typical scheme with a biomass-driven engine is shown in Figure 12.47. Biomass is fed to the gasifier whose output gas is fed to the engine that drives an IG or SG to feed converted power to the grid via a generator transformer. The controller will

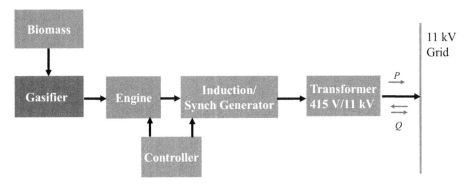

Figure 12.47 Biomass-operated engine-driven grid-fed system.

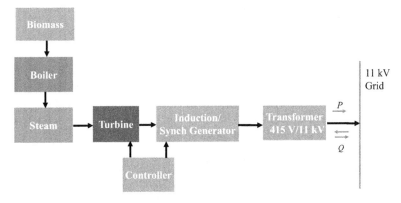

Figure 12.48 Scheme to generate electricity through the "steam" route.

comprise engine governor control and AVR or capacitor VAR control of the generator. Both P and Q control at the grid are critical to maintain good power quality. Fuel input controls P and AVR and VAR controller control Q in SG and IG, respectively.

To generate electricity through the "steam" route the scheme of Figure 12.48 can be used. Here biomass is directly fed to a boiler to produce steam to run a steam turbine, which rotates the generator to produce electricity for feeding to the grid via a step-up generator transformer to match the voltage of the grid. An SG or an IG can be used as generator with appropriate control, although SG is widely used for ease of control. P control is by adjusting steam valves to the turbine. Often a hybrid arrangement of biomass and fossil fuel such as coal is used by feeding both biomass and coal of a suitable proportion into the boiler. Increased use of biomass will reduce coal and support biomass usage.

12.8.1.4 Solar-thermal and geothermal energy

Using the scheme of the preceding section, it is possible to affect steam-based grid-fed generation by producing steam using the heat of solar-thermal or geothermal processes. The process of electricity generation and grid connection remains the same as mentioned earlier. There have recently been several innovative processes used to generate electricity by the solar-thermal process using a choice of materials to efficiently produce steam. A hybrid arrangement of using the solar-thermal process to increase the temperature of inlet water to a coal-fired boiler is another innovative experiment.

12.8.2 Off-grid systems

RE is a preferred option for off-grid systems specially to energize remote and rural areas where grid connection is uneconomical, unviable, or inaccessible. But the challenge lies in randomly varying the power of the source and the load, which needs a robust controller to match the two.

12.8.2.1 Wind energy system

As already mentioned, the power in the wind is very random (as cube of wind speed) making it nearly impossible to match with the equally random consumer load power. Thus, load power will always be unequal to wind power making it imperative to have a reliable controller that makes it equal by load adjustment, dump load, and input power augmentation. Since they operate on low power, suitable generators are SEIG and PMSG with involved control to effect power balance between source and load. They need another source to augment power under low wind speeds and a dump load to dump surplus power during high wind speeds. It is imperative that off-grid WES operates under hybrid mode with another source such as battery, diesel/bio, or solar. Figure 12.49 shows a wind battery hybrid scheme feeding a variable load. At low wind speeds and high load the battery will augment the desired load while at high wind speeds and low load the battery is charged. At all instants the power balance equation is:

$$P_w \pm P_b = P_L$$

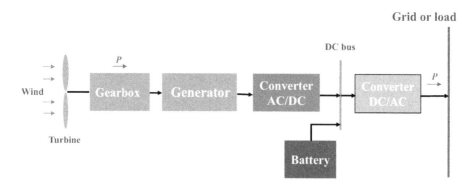

Figure 12.49 A wind battery hybrid.

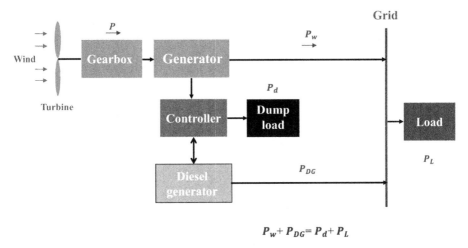

$$P_w + P_{DG} = P_d + P_L$$

Figure 12.50 A wind diesel hybrid.

where P_w, P_b, and P_L, respectively, denote power in wind, battery, and load. This needs a sophisticated controller to charge and discharge the battery. Choice of battery is critical to avoid pollution.

Figure 12.50 shows a wind diesel hybrid scheme feeding a variable load. A diesel generator (DG) acts as source of surplus power while dump load (DL) acts as sink of surplus power. At low wind speeds and high load DG supplies surplus power with an inactive DL, while at high wind speeds and low load DG is inactive and DL absorbs surplus power. At all instants the power balance equation is:

$$P_w + P_{DG} = P_d + P_L$$

where P_w, P_{DG}, P_d, and P_L, respectively, denote power in wind, diesel generator, dump load, and main load. This too needs a sophisticated controller to adjust power in DG and DL. An SEIG or PMSG can act as wind generator while an SG (with or without brushes) can act as a DG.

12.8.2.2 Small hydro energy

Small hydro energy is already covered in Chapter 5.

12.8.2.3 Bio energy

SEIGs, both three-phase and single-phase, are suited for bio energy operated engines for off-grid generation. They can also become handy for standby power driven by oil engines for domestic, commercial, camping, defense, and tourism applications. Due to the varying nature of connected loads such units will be viable only with a reliable controller that ensures good power quality across the load in terms of voltage, frequency, and waveform. For oil or bio-engine-driven SEIGs, the governor adjusts

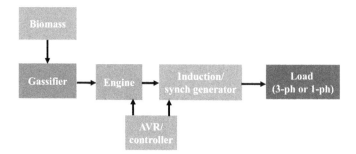

Figure 12.51 Biomass-based standalone power generation scheme.

fuel intake to maintain near constant speed, with a speed drop from no load to full
load limited to about 5%, such that the controller must project variable capacitance
as load changes for constant voltage. This is a "variable power" arrangement. An SG,
both wound field and brushless, can also be used with such engines with inherent
complexity of AVR.

A typical biomass-based power generation scheme is shown in Figure 12.51.
Biomass is converted to gas in the gasifier and fed to the engine as a fuel, which in
turn drives the generator (SG or SEIG) with a suitable controller to ensure good power
quality across the isolated load. Figure 12.52 shows the connection of a single-phase
SEIG driven by a bio-engine.

12.8.2.4 Hybrid system with microgrid

Due to random variation of RE, a combination of available local sources in hybrid
mode to supply consumer loads through a microgrid is a viable option. A general

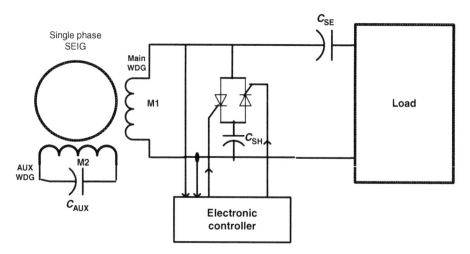

Figure 12.52 Connection of a single-phase SEIG driven by a bio-engine.

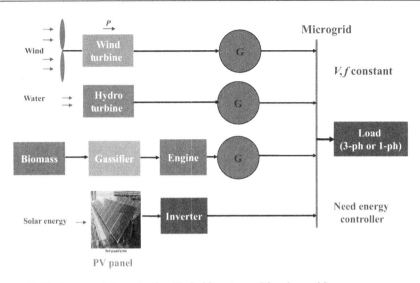

Figure 12.53 A general decentralized hybrid system with microgrid.

decentralized hybrid system with microgrid is shown in Figure 12.53, comprising bio, hydro, wind, and solar energy. Converted electricity from all sources is fed to the microgrid to which the consumer loads are connected as shown. It is imperative to operate the microgrid at constant voltage and frequency suited to the load. A sophisticated "smart" integrated energy controller is needed for the previously mentioned purpose.

12.9 Summary

This chapter dealt with different types of electrical energy conversion systems suitable for electricity generation from RE. Mechanical energy produced by bio, hydro, and wind energy sources through appropriate prime movers such as engines and turbines needs to be converted to electrical energy to feed to the grid in grid-fed systems and isolated loads in off-grid systems. Electric generators are central to this conversion process. Different types of SGs and IGs fall under this basket. Their basic construction, principles, control, and applications were detailed in the chapter. Both grid-fed and off-grid systems based on RE will have different requirements to ensure power quality at the grid and load that affects operation and control of generators. Generators being the heart of the energy system must be robust and reliable. Brushless generators are preferred to those needing brushes. Engineering challenge centers on ensuring needed power quality as regards voltage, frequency, and waveform despite the varying nature and types of sources and loads. The conversion system of each source – bio, hydro, wind – must be tailor-made both for grid-fed and off-grid cases as explained in this chapter. Smart mini- and microgrids integrating the grid, sources, loads, and energy storage units are the future.

Power semiconductor devices

Abdul R. Beig

Department of Electrical Engineering, The Petroleum Institute, Abu Dhabi, UAE

Chapter Outline

13.1   Introduction

Renewable energy systems generate electrical energy in different forms. In order to match the source and load, power conversion circuits are required. In power conversion, efficiency is the key factor. High efficiency results in low power loss, hence the compact systems. In modern power conversion circuit semiconductor devices are used as switches. Power diodes, bipolar junction transistors (BJTs), metal oxide field effect transistors (MOSFETs), insulated gate bipolar transistors (IGBT), silicon-controlled rectifiers (SCRs), gate turn-off thyristors (GTOs), and integrated gate commutated thyristors (IGCTs) are the most widely used power devices in the industry. These devices serve a broad spectrum of power levels and frequencies [1]. In high-power and medium-voltage systems (>2.2 kV) such as high voltage DC (HVDC), electric trains, medium-voltage drives, wind farms, and large solar plants, power levels range from a few megawatts to a few hundred mega watts, and switching frequency ranges from 50 Hz to 1 kHz. SCRs, GTOs, and IGCTs are used at this power level [2]. In high performance drives, electric vehicles, and standalone renewable energy sources, the switching frequency is in the range of a few kHz to 500 kHz, and the power level ranges from 10 kW to 10 MW. IGBTs are the most popular in this power range and MOSFETs are used at low voltage applications [2,3]. In communication systems, audiovisual equipment, biomedical instruments, computers, etc., the power level ranges from a few megawatt to, say, a few kilowatts, and the switching frequency ranges from hundreds of kilohertz to megahertz. In these applications the operating voltage is low and hence MOSFETs are the most suitable [3]. Diodes are required in all power levels and frequency ranges. Schottky diodes and most recently silicon carbide (SiC)-based Schottky diodes are preferred in low voltage and high-switching-frequency circuits [4].

At low power levels, the requirement is very high switching frequency and low switching losses. At high power levels, the stress is on low switching frequency but low conduction loss. Silicon-based devices are reaching the limit of voltage-blocking capacity because of high on-state resistance [4]. The wide band gap materials such as SiC and gallium nitride (GaN) promise low conduction loss, high voltage with stand capability, and high thermal stability. At a preset SiC- and GaN-based devices with low voltage rating (<1700 V) are being used in low voltage applications [5]. High manufacturing cost prevents the production of high voltage devices based on a wide bad gap of material, but recent advances in the manufacturing process ensures the availability of these devices in the near future [4–8].

The focus of research in the area of power devices is to achieve device characteristics as close as possible to ideal switches. An ideal switch should be in a position to carry current without any limit, it should be able to withstand infinite voltage in both directions, it should be fully controlled, the control circuit should not require any power, it should not have any voltage drop when the device is on, that is a zero conduction loss, it should not have leakage current when the device is off, that is a zero off-state loss, and it should turn on and off instantly that is zero switching loss [2]. The diode is an uncontrolled device, BJTs, SCRs, GTOs, and IGCTs are current controlled, and the rest of the devices are voltage-controlled devices.

The basic structure, principle of operation, v–i characteristics, and turn-on and turn-off of all the popular power devices have been discussed in this chapter. Owing to the research interest and impact of wide band-gap material-based devices, Section 13.6 has been dedicated to these emerging devices.

13.2 Power diodes

A power semiconductor diode is a two terminal P–N junction device. Power semiconductor diodes are designed to carry higher currents and withstand large voltages.

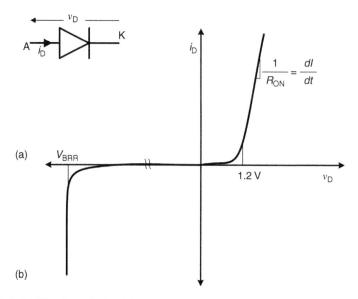

Figure 13.1 (a) Circuit symbol and (b) v–i characteristic of a diode.

The P—N junctions are formed by a masked diffusion of impurities [1–3]. The size of diffusion mask, length of diffusion, and magnitude of diffusion temperature has a direct impact on device characteristics. The power diodes have very high break-down voltages. This high breakdown voltage is achieved by controlling depletion layer boundaries, which is achieved by means of floating field plates or guard rings. Usually a silicon-based P-type semiconductor (anode) is diffused with a silicon-based N-type semiconductor (cathode). Contact terminals are formed using aluminum. The circuit symbol and v–i characteristics of a power diode are as shown in Figure 13.1.

13.2.1 Forward bias

Let $v_D = v_{AK}$ be the voltage drop across the diode. For $v_D \geq 0$, the diode is said to be in forward bias. Diode current i_D is small, as long as $v_D < V_{TD}$ (V_{TD} is in the range of 0.7–1.5 V). A diode conducts fully if $V_D > V_{TD}$. The diode current is given by,

$$I_D = I_S \left(e^{(V_D/\eta V_T)} - 1 \right)$$

(13.1)

where, V_D is the voltage drop across the anode and cathode (V). I_S is the leakage cur-rent, typically in the range of 10^{-6}–10^{-9} A. Here, η is the empirical constant in the range 1.8^{-2}. V_T is the threshold voltage ≈ 25.7 mV, for power diodes $I_D \approx I_S e^{(V_D/\eta V_T)}$. Large currents in a power diode create ohmic drop and v–i characteristics ap-pear to be linear. For large currents, the on-state resistance of the diode is given by $R_{ON} = \Delta V_D / \Delta I_D$.

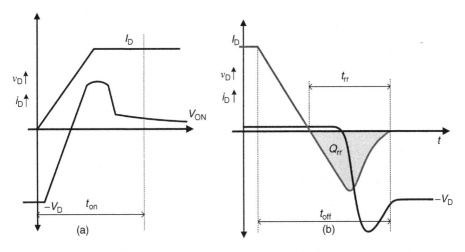

Figure 13.2 (a) Turn-on and (b) turn-off characteristics of a diode.

13.2.2 Reverse bias

For $V_{BRR} < v_D < 0$, the diode is said to be reverse biased and only a small leakage current flows through the diode. When v_D reaches the reverse bias breakdown voltage (V_{BRR}), i_D increases rapidly due to an avalanche of electrons and is limited by external circuit. The large power at breakdown will damage the device. Operation of a power diode in breakdown must be avoided.

Power diode requires finite time to go from reverse bias state to forward bias state. This time is called turn-on (t_{on}) time. The dv/dt (rate of fall of voltage across diode) is decided by the diode. The rate of rise of diode current (di/dt) during turn-on time is decided by the external circuit. Power diodes exhibit a voltage rise before attaining the on-state voltage, as shown in Figure 13.2a. The magnitude of voltage rise depends on the stray inductance in connecting leads and rate of rise of the anode current. During turn-off, the rate of fall of current (di/dt) is determined by a semiconductor diode. The rate of rise of voltage (dv/dt) is decided by the external circuit element. In Figure 13.2b, t_{rr} is the reverse recovery time and t_d is the delay time. The diode is a bipolar device. Minority carrier takes some time to become neutralized and this time is t_{rr}.

13.2.3 Types of diodes

* *Power rectifier diodes*: these diodes are used in AC to DC power rectifier circuits. These are slow, meant for low frequency circuit operations, are optimized for low conduction loss, and can withstand only moderate dynamic stresses. Typical t_{on} for a power diode is 5–20 μs and t_{off} is 20–100 μs. Voltage rating varies from a few hundred volts to 10 kV and current rating varies in the range from 1 A to 10 kA [9,10].
* *Fast diodes/fast recovery diodes*: these are usually companion diodes to fast switches like IGBTs. These diodes are optimized to accept high dynamic stress and also for switch applications. These devices have high conduction losses. Typical t_{on} time is in the range of

a few nanoseconds and typical t_{off} time is in the range of a few tens of nanoseconds to a few microseconds, depending on the rating of the diode. Voltage rating and current rating are available up to 6 kV and 3 kA, respectively [11,12].

- *Fast switching diodes*: these are optimized for high-frequency applications, such as high-frequency rectifiers in switched mode power supplies. They have a very small recovery time (1 ns to 5 μs). The power rating varies from a few hundred milliwatts to a few kilowatts [13].
- *Schottky diodes*: these diodes have very low on-state drop and very fast switching action. The on-state voltage drop can be as low as 0.1–0.7 V. Many applications such as high-frequency rectifiers in low voltage power supplies require fast diodes with low on-state drop. Schottky diode is formed by making a nonlinear contact between an N–type semiconductor (cathode) and a metal (anode), creating a Schottky barrier. The current is due to majority carriers resulting in insignificant minority carriers stored in the drift region. This reduces the turn-off time of the device significantly. The silicon-based Schottky diodes have very low (<100 V) reverse biased voltage-blocking capacity [13]. Silicon carbide (SiC)-based Schottky diodes have higher voltage-blocking capacity, say up to 3 kV [13]. Schottky diodes have low on-state resistance, very low on-state voltage drop, and low switching timings and hence are used in high-frequency resonant converters, low voltage power supplies, etc.
- *Zener diodes*: these are special purpose diodes that allow current to flow in a forward direction and also in a reverse direction. In reverse direction it is designed to operate in the breakdown region. Zener diodes are designed to have low breakdown voltage, typically a few volts up to a maximum of 1 kV. Forward current will be in the range of a few microamperes to 200 A [13].
- *Light emitting diodes*: light emitting diodes (LEDs) emit light when activated. These are used mainly as indicators and display elements. Recently they have been used for lighting purposes [2,3].

Power diodes must be protected against high di/dt and dv/dt. Snubber circuits are used for this purpose. Fast acting fuses are used to protect the power diodes against overcurrent or short circuit; however, the I^2t rating of the fuses must be far above the I^2t rating of the diode [2,3].

13.3   Bipolar junction transistors (BJT)

A power transistor is a three terminal device with emitter, collector, and base, and has two types of configuration, namely n–p–n and p–n–p. The circuit symbols of a bipolar junction transistor (BJT) are given in Figure 13.3a.

The basic structure of an n–p–n power transistor is shown in Figure 13.3b. A lightly doped N-drift region is used to achieve high blocking voltages. When collector (C) is forward biased, the collector–base junction (J_1) becomes reverse biased, no current flows. When the base–emitter junction (J_2) is forward biased, the electrons from base region sweep through the depletion region of J_1 and produce collector current (I_C). To turn off the device, J_2 should be reverse biased. The BJT has a large cross-sectional area of the collector to emitter current because the on-state resistance is kept low to reduce conduction loss. The collector-drift region width determines the breakdown voltage. In a power transistor the base thickness is significant, which results in low current gain, $\beta = I_C/I_B$. In order to achieve high β, monolithic Darlington BJTs are designed by cascading the BJTs, as shown in Figure 13.4. The overall $\beta = \beta_1\beta_2 + \beta_1 + \beta_2$.

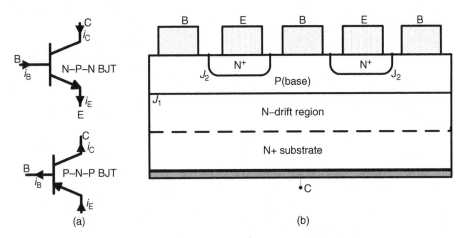

Figure 13.3 (a) Circuit symbol and (b) basic structure of a BJT.

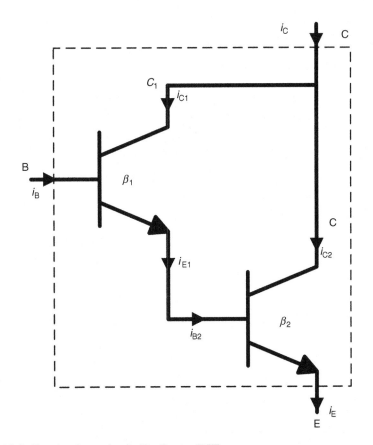

Figure 13.4 Circuit schematic of a Darlington BJT.

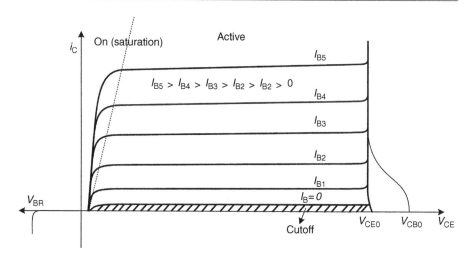

Figure 13.5 v–i **characteristics of a BJT.**

Figure 13.5 shows the v–i characteristics of an N–P–N transistor. The Darlington pair also has similar characteristics but higher current gain. There are three regions of operation for BJTs, namely saturation, active, and cut-off. In power electronic circuits, power transistors are operated either in the cut-off (OFF) or saturation (ON) region. The power BJTs are operated deep in the saturation region, known as hard saturation. Hard saturation will reduce the on-state power loss but turn-off time increases. Primary breakdown takes place in normal emitter–base junction breakdown and results in a large current. The secondary breakdown is due to thermal runaway and is associated with large power dissipation. There are two forward breakdown voltages V_{CEO} and V_{CBO}, where $V_{CBO} < V_{CEO}$.

Power BJTs are current controlled devices. Due to high power requirements, the design of the base drive circuit will be very complicated compared to voltage-controlled devices such as IGBTs and MOSFETs. The BJT has now become obsolete and has been replaced by these voltage-controlled devices.

13.4 Metal oxide semiconductor field effect transistor

Power MOSFETs entered power electronic applications in the early 1980s. Power MOSFETs are unipolar devices, in which majority carriers constitute the current. There are two types of MOSFETs, namely enhancement and depletion type. Enhancement MOSFETs are used for power applications. The power MOSFET has a four-layered structure. The $n + pn - n + (p + np - p +)$ structure forms an enhancement type n–channel (p–channel) MOSFET. Gate is insulated from the body using a SiO_2 layer, as shown in Figure 13.6.

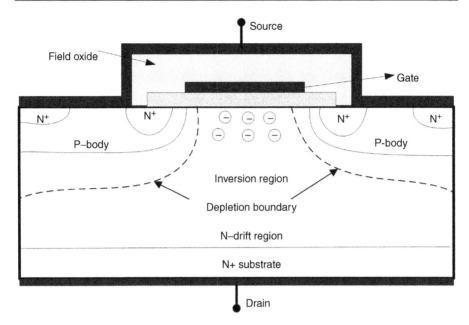

Figure 13.6 Basic structure of a power MOSFET.

13.4.1   Forward bias

For $v_{DS} > 0$ with $v_{GS} \leq 0$, no conduction takes place. So $I_D = 0$. For $v_{DS} > 0$ with a small $v_{GS} > 0$, electrons are induced into the layer below the SiO_2 layer forming a depletion region between the SiO_2 and silicon. A further increase of v_{GS} causes the depletion region to grow in thickness, thus turning the semiconductor layer between drain and source into N-type [1–3,14]. The thickness of gate oxide and width of gate decides the current for a given V_{GS}. Thus, a conduction channel is formed between the drain and source. The layer of free electron is termed an inversion layer and the value of v_{GS} is termed threshold voltage V_{GSTH}. The device will remain in the cut-off region for $v_{GS} < V_{GSTH}$. V_{GSTH} is typically a few volts for a MOSFET. The device will block voltage applied across drain and source. For $v_{DS} > B_{VDSS}$ breakdown takes place.

For $v_{GS} > V_{GSTH}$ as v_{GS} is increased, the inversion layer becomes more conductive. The MOSFET is said to be in active region, the i_D is independent of the v_{DS} and depends only on v_{GS} as in 13.2.

$$i_D = K \left(V_{GS} - V_{GSTH} \right)^2 \tag{13.2}$$

where K is constant, which depends on device geometry.

The current is said to be saturated, hence this region is called the saturation region.

As V_{GS} is increased further, such that $V_{GS} - V_{GSTH} > V_{DS} > 0$ the device is driven into the ohmic region. At the boundary of the ohmic region and saturation region, i_D is given by,

$$i_D = K\left(v_{DS}\right)^2 \tag{13.3}$$

In the ohmic region, at high currents, i_D will vary linearly with v_{DS}.

13.4.2 Reverse bias

For $V_{DS} < 0$, P- and N-type form a connected diode. So MOSFETs do not have reverse blocking capability. The diode conducts in the reverse direction. The circuit symbol and $v–i$ characteristic of a power MOSFET are given in Figure 13.7. In power electronic applications, MOSFETs are used as switches. MOSFETs are operated in cut-off (ON) and ohmic (OFF) regions. The absence of storage charges makes the turn-off time very small compared to BJTs and IGBTs.

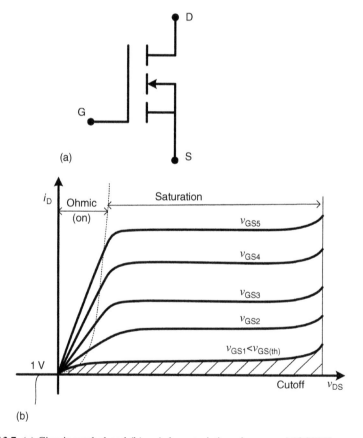

Figure 13.7 (a) Circuit symbol and (b) $v–i$ characteristics of a power MOSFET.

13.4.3 Important points

- A MOSFET is a voltage-controlled switch.
- A MOSFET is a unipolar (majority carrier) device either N- or P-type. No minority carriers contribute to the current flow. Hence, there is no storage time associated during turn-off. Hence, MOSFETs are fast devices, preferred for high switching frequency applications.
- There are two types of MOSFET namely enhancement type and depletion type. Power MOSFETs are enhancement type. For enhancement type, a device is ON for $V_{GS} > 0$ and OFF otherwise. For depletion type, a device is ON for $V_{GS} > V_T$ and OFF otherwise, where V_T is the threshold voltage and is always negative.
- In power electronic circuits, the MOSFET is used as switch. V_{GS} must be within the lower and upper limits as specified by the data sheet.
- When the MOSFET is conducting, it behaves like a resistor. So the power loss is given by $I_D^2 R_{DS\_ON}$, Where R_{DS_ON} is the on-state resistance of the MOSFET.
- The R_{DS_ON} increases with the drain current I_D. It also rises with the junction temperature. This gives rise to increased power loss during conduction and hence restricts its usage in high power circuits.
- The R_{DS_ON} increases with the drain to source breakdown voltage rating of the MOSFET. Low voltage-rating MOSFETs have low on-state resistance, hence for given voltage requirements, it is better to select power MOSFETs with minimum possible voltage rating considering the safety factor.

The protection and gate circuit requirements of MOSFETs are same as that for IGBTs and are discussed in Section 13.5.

13.5   Insulated gate bipolar transistors (IGBTs)

The IGBT is based on MOS-bipolar integration. MOSFET structure is used to give the necessary base drive for the bipolar transistor. The basic structure of the n-channel asymmetric blocking (also known as punch through – PT) IGBT is given in Figure 13.8. The N-type drift region is grown over a P$^+$ substrate. Initially, high

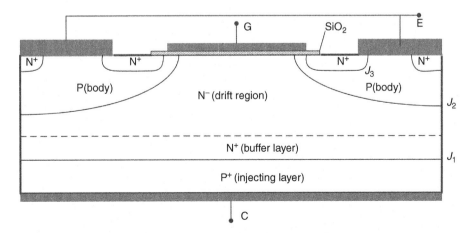

Figure 13.8 Basic structure of a power PT–IGBT.

doping is used to create a buffer layer. After this the remaining N-drift region is lightly doped to increase the voltage-blocking capability. In nonpunch through (NPT) or symmetric IGBTs, the P^+ semiconductor of collector (C) is diffused from the back side of the N-drift region. The N^+ buffer layer is absent. Then the deep P^+ body regions are formed to eliminate the effect of parasitic thyristor. Gate oxide is grown after the formation of body regions. The N^+ dopant is diffused into the body region to form an emitter (E).

13.5.1   Forward bias

For $V_{CE} > 0$, the IGBT is forward biased and device operation is controlled by gate to emitter voltage (V_{GE}). When $V_{GE} = 0$ the IGBT is capable of blocking high voltage. When the collector is forward biased, junction J_1 is forward biased whereas junction J_2 is reverse biased. When positive voltage is applied to the gate (G) terminals, that is, $V_{GE} > 0$, an inversion channel under the gate electrode is formed, which connects the N^+ emitter region to the N-drift region. The current through these two regions forms the necessary base current for the P–N–P transistor, which results in a hole injection at junction J_1. The injected holes form the emitter current of the P–N–P transistor, thus the current is established between the collector and emitter terminals. The N-drift region reduces the on-state voltage drop as the IGBT structure allows high electron injection in the drift region reducing the on-state resistance. When the V_{GS} is above the threshold voltage the device is said to be in the saturation or ON state. The rate of rise of collector current can be controlled by controlling the rate of rise of V_{GS}.

When $V_{GS} \leq 0$, the device turns off. The turn-off is associated with the removal of stored charges in the drift region. Hence, the turn-off time (t_{off}) of IGBT is larger than for MOSFETs.

13.5.2   Reverse bias

When the negative voltage is applied to the collector, that is $V_{CE} < 0$, the junction J_1 is reverse biased and J_2 is forward biased. The thickness of the N-drift region and minority carrier lifetime determines the breakdown voltage. In the case of symmetric (NPT) IGBTs, the reverse blocking voltage is equal to the collector forward bias blocking voltage. In the case of asymmetric (PT) IGBTs, the reverse blocking voltage is much lower because of the presence of the N^+ buffer layer. Asymmetric IGBTs used in inverter application where antiparallel diodes are usually connected between collector and emitter to carry reverse direction current.

The circuit symbol and the v–i characteristic of the IGBT has been shown in Figure 13.9. The IGBTs are not designed to operate in the active region due to the high on-state losses in this region. In power electronic circuits, IGBTs are operated either in the ON (saturation) or OFF (cut-off) region. Since the device is controlled by gate voltage, the gate drive is simpler and requires low power. Hence, IGBT is preferred over BJT. The IGBTs have low and constant on-state drop at very high voltage and currents compared to MOSFETs. So IGBTs are used in high-voltage and high-current

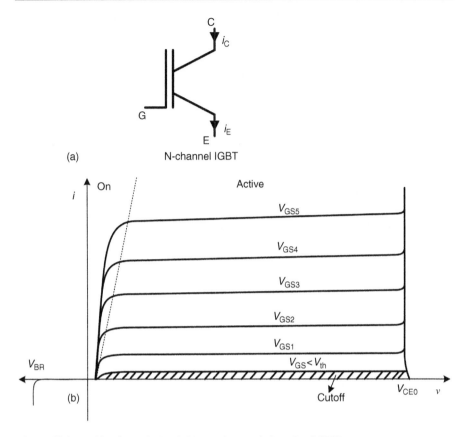

Figure 13.9 (a) Circuit symbol and (b) v–i characteristics of an IGBT.

circuits. At present, IGBTs with a voltage rating as high as 3.3 kV and current capacity as high as 3 kA, are commercially available [8–11].

13.5.3 Important points

- An IGBT is a voltage-controlled switch.
- An IGBT is a bipolar device, hence the turn-off is slower than a MOSFET due to the presence of stored charges and reverse recovery time associated with these charges.
- For n–p–n-type IGBTs, the device is ON for $V_{GE} > 0$ V and OFF otherwise.
- V_{GS} must be within the lower and upper limits, as specified by the data sheet.
- The on-state drop of an IGBT is low compared to MOSFET and remains almost constant with the device current, hence these are preferred in high-power applications.

13.5.4 Gate drive requirements for MOSFET and IGBT

- Isolation is required to isolate the gate signal from TTL/CMOS logic level to power level. The optocouplers or pulse transformers are used to isolate gate pulses.

- There exists capacitance between gate and drain. In order to charge this capacitor during turn-on, current buffers are required at the gate circuit.
- The turn-on time is decided by external series resistor, which is specified by the manufacturer data sheet. The turn-on time and/or -off time can be increased by increasing this external resistor.
- During turn-off, the gate circuit must provide a path for the gate-to-source capacitor charges to discharge.
- The gate-to-source voltage should not exceed the rated voltage. Parasitic inductance in the gate leads will result in resonance with gate-to-source capacitors and this may result in higher voltages at gate terminals. So, in order to reduce the effect of ringing in gate pulses, the gate leads will be generally short and twisted signal pair wires. Back-to-back-connected Zener diodes are connected across gate-to-source terminals to clamp the gate voltages to a safe value. A high resistance of the order of 10–100 kΩ is placed across gate-to-source terminals, in order to attenuate any spurious signals captured by these leads, thus avoiding the spurious turn-on of the devices. Also this helps in attenuating the ringing-in gate voltages. The Zener and resistors are mounted physically very close to the gate and drain terminals.
- The undervoltage sensing of the gate signal is required to ensure that gate signals are blocked when the V_{GS} or V_{GE} < minimum voltage required to keep the device in the ohmic or saturation region.

13.5.5 Precautions required while handling MOSFET and IGBT

- These devices are static-sensitive devices. So proper precautions should be taken while handling them. Gate-to-source or gate-to-emitter terminals must be shorted when the device is not in the circuit.
- The power circuit should never be energized when gate terminals are open.
- Since these devices are fast, they cannot be protected using fuse under- or overcurrent or short circuit. The preferred method for overcurrent/short-circuit protection is by current sensing or V_{CE} sensing.
- Power circuit layout is very important to reduce the stray inductance. Stray inductance in a power circuit will give rise to a large V_{CE} during turn-off. In order to limit this voltage the turn-off time should be programmed accordingly, by using proper $R_{G_{OFF}}$.

13.6 GaN- and SiC-based devices

Gallium nitride (GaN) and silicon carbide (SiC) are wide band gap materials. GaN and SiC have high critical electric filed strength (e.g., 3.2 eV for SiC and 3.4 eV for GaN) compared to silicon (e.g., 1.12 eV) [4,5,8]. For a given breakdown voltage requirement, width of the layer reduces. This results in smaller sized and reduced on-state resistance, which in turn reduces conduction loss. SiC-based Schottky diodes are now available with a voltage-blocking capacity as high as 3 kV. Other SiC-based power semiconductor devices such as SiC-based thyristors, SiC-based IGBTs, and SiC-based MOSFETs are experimentally proven but have a long way to go before they are made economically viable [4,8,14]. Hence, only a brief discussion is presented in this book. However, SiC has a promising higher-voltage rating for these devices, which was not

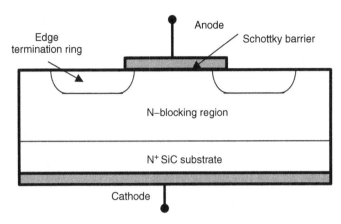

Figure 13.10 Basic structure of an SiC Schottky diode.

possible to achieve with silicon. At present SiC-based IGBTs and FETs are available in the voltage range between 600 V/50 A and 1.7 kV/50 A [12,15]. SiC has very high thermal stability compared to silicon. Experimental applications of SiC-based IGBTs in multilevel converters, inverters for drives, and grid-connected inverters are reported in [6,7]. SiC does not melt, instead it gradually sublimes at high temperatures, which makes it impossible to form large monolithic crystals. This requires a modified sublimation process, which is very expensive. However, the performance advantages offset a higher manufacturing cost in certain applications.

Figure 13.10 shows the cross-section of a typical SiC-based Schottky diode. A lightly doped N-type blocking layer is grown on a 4H–SiC substrate. The doping and thickness of this layer are chosen to achieve the desired blocking voltage. The Schottky junction on the top surface of the blocking layer is formed by implanting an edge termination ring at the surface, then depositing the Schottky metal. The edge termination rings avoid field crowding, hence avoid the reduction of voltage-blocking capacity. The SiC-based Schottky diodes are unipolar devices and have fast turn-on and -off capabilities. Due to the reduced layer thickness compared to silicon their on-state voltage drop is low, hence the conduction loss is reduced.

GaN is particularly attractive for high-voltage, high-frequency, and high-temperature applications due to its wide band gap, large critical electric field, high electron mobility, and reasonably good thermal conductivity. At present, GaN-based semiconductor material is still in the early stage concerning power applications. High manufacturing cost of GaN layers on large silicon substrates, is the main obstacle in the growth of GaN-based device technology [4,5,8]. Recently advances in the manufacturing process have enabled the growth of GaN epilayers on large Si substrates, which offer a lower cost technology.

High-electron-mobility transistors (HEMT) with a voltage rating of 600 V, for microwave applications are manufactured [16,17]. The basic structure of HEMT is shown in Figure 13.11. GaN layer is grown on a silicon substrate, an AlGaN layer is grown over a GaN layer to create a two-dimensional electron gas (2DEG) at the

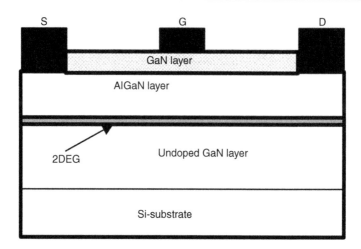

Figure 13.11 Basic structure of a HEMT.

AlGaN/GaN interface. This structure has normally on- or depletion-mode characteristics. Several modifications are proposed to achieve normally off HEMTs. These
are (1) the use of a recessed-gate structure in such a way that the AlGaN layer under the gate region is too thin for inducing a 2DEG; (2) the use of a fluorine-based
plasma treatment of the gate region; (3) combination of the gate recess together with a
fluorine-based surface treatment; (4) the selective growth of a p–n junction gate; and
(5) a cascade switch based on the series connection of a normally on GaN HEMT and
an silicon MOSFET. EPC [16] supplies normally off GaN HEMTs from 40 V/33 A
to 200 V/12 A and microGaN [17] offers normally on 600-V–170-mΩ GaN HEMTs,
and a normally off 600-V GaN HEMT. For high voltage power switching applications, lateral GaN MOSFETs show the advantage of normally off operation.

13.7   Silicon-controlled rectifiers

Silicon-controlled rectifiers (SCRs), GTOs, and IGCTs are the most popular power
semiconductor devices in the thyristor family. SCRs are the oldest semiconductor
devices, invented by General Electric in 1957. An SCR is a four-layer device with
alternating P- and N-type semiconductors, as shown in Figure 13.12a. The basic structure is constructed by starting with a lightly doped N-type silicon wafer. This forms
the drift region and determines the breakdown strength of the thyristors. The anode P$^+$
region is formed by diffusing the dopants on one side of the substrate. The p-gate and
N$^+$ region are diffused on the other side of the substrate.

13.7.1   Forward bias

Under the forward-biased conditions, junctions J_1 and J_3 are forward biased and junction J_2 (junction between the N-drift region and base P region) is reverse biased. The
device is off when the forward voltage increases to a large value, J_2 junction breaks
down, and the SCR conducts. Since V_{BO} is very large, the forward current will be large

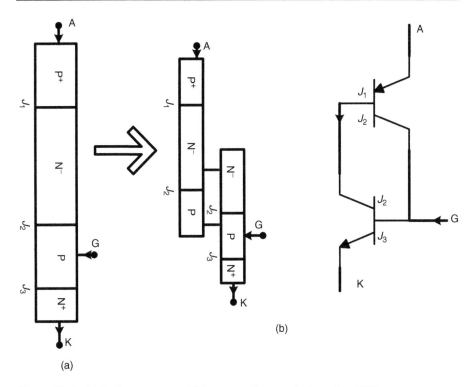

Figure 13.12 (a) Basic structure and (b) two-transistor equivalent of an SCR.

and usually this type of turning ON is not recommended. The SCR can be modeled as a back-to-back connected two-transistor equivalent as shown in Figure 13.12b. When the gate is forward biased and i_G flows into the junction, J_3 injects electrons across junction J_3. These electrons will diffuse across the base layer and be swept across junction J_2 into the n_1 base layer of the p–n–p transistor. This will result in positive feedback and the current will flow from anode to cathode [1–3].

The forward voltage, at which the SCR starts to conduct decreases with the increase of I_G. Once the SCR turns on, it behaves like a diode. The latching current I_L is the minimum on-state anode current (i_A) required to maintain the SCR in the on state immediately after the SCR is turned on and gate signal is removed. Once the device is on with $i_A > I_L$, i_G has no effect on the device. So i_G can be removed. The SCR cannot be turned off by i_G. The SCR can be turned off only by reducing the i_A to be less than I_H, where I_H is the holding current and is the minimum I_A required to maintain the thyristors in the on state. The $v–i$ characteristic of an SCR is shown in Figure 13.13.

13.7.2 Reverse bias

In the reverse bias condition, junctions J_1 and J_3 are reverse biased and the SCR behaves similar to that of a reverse-biased diode. There exists a reverse voltage of V_{BR}, which causes an avalanche of electrons and thereafter a breakdown takes place. Due to large current, the device will be damaged.

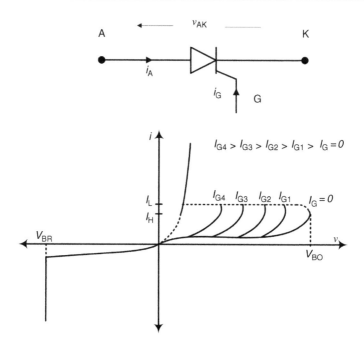

Figure 13.13 The *v–i* characteristics and circuit symbol of an SCR.

In AC-to-DC converter circuits, the SCR turns off when the current passes through zero, known as line-commutated circuits. In DC circuits, forced-commutation circuits are required to turn off the SCR. Due to the availability of other controlled devices like GTOs and IGCTs, the application of SCRs is restricted mainly to line-commutated AC-to-DC rectifier circuits.

A large di_A/dt may result in large electron flow in a small area resulting in failure of the SCR. So it is advisable to limit di_A/dt with the permissible limits of di/dt rating of the device. A small inductor is used in series with the SCR to limit the di/dt. Figure 13.14 shows the i_A and v_{AK} waveforms during turn-off. When the device is turned off from the on state a large dv_{AK}/dt may result in false turning on of the device. The dv_{AK}/dt is usually limited to a safe value by using a snubber capacitor across the SCR. A small resistor is connected in series with a capacitor to limit the discharge current during turn-on. The turn-off is slow due to the presence of minority carriers.

The thyristors (SCR and GTO) must be protected against high di/dt and dv/dt, and snubber circuits are used for this purpose.

13.8 Gate turn-off thyristors

The gate turn-off thyristor (GTO) is similar to the SCR but has the ability to turn off using gate current. The GTO is also a four-layered device like the SCR. The doping is modified and the width of the cathode region is reduced to achieve turn-off through

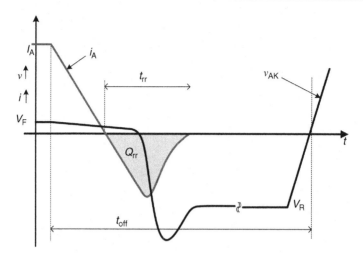

Figure 13.14 Turn-off waveforms of an SCR.

gate current [1,14]. Since the GTO has gate turn-off capability, the asymmetrical structure with the N-buffer region is preferred. The asymmetric GTOs with a shorted anode structure have a fast turn-off but their reverse voltage-blocking capacity is very low, say from 20 V to 30 V [1,14]. The GTOs without a shorted anode are slow in turn-off but have higher reverse voltage-blocking capacity. The v–i characteristic of a GTO is almost similar to that of an SCR. However, a minor difference is that, for anode currents less than a latching current, the GTO behaves like a BJT. For this operating point, the GTO returns to the turn-off position when gate current is removed. GTOs are not operated in this region.

Figure 13.15 shows the circuit symbol and turn-off waveforms of a GTO. The turn-ON process and forward v–i characteristics of a GTO are the same as that of an SCR. The GTO is turned off by applying a large negative gate current in the range of one-fifth to one-third i_A for a very short time. A large di_G/dt is used to achieve short storage time and short anode current full time. The stored charges in the n_1 and p_2 layers will result in a small anode tail current. This anode tail current will flow from anode to gate. The reverse gate current removes excess holes in the cathode–base region.

13.9   Integrated gate commutated thyristors

Integrated gate commutated thyristor (IGCT) is also a four-layer p–n–p–n device. A buffer layer is introduced at the anode to reduce thickness of the anode layer [14,18]. The electron can be extracted efficiently during turn off by the gate current. The above modification results in reduced on-state voltage drop, compared to a GTO. The IGCT has a hard drive feature $di_G/dt > 1000$ A/μs, which can be operated without snubbers. The hard drive ensures that the anode current commutates quickly [1,14].

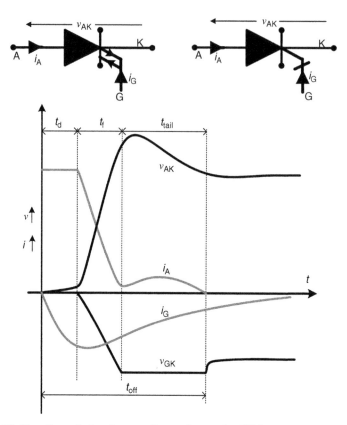

Figure 13.15 Circuit symbol and turn-off waveforms of a GTO.

Figure 13.16 shows the circuit symbol and turn-off waveforms of an IGCT. In order to turn off, the gate drive circuit should supply a fast turn-off current pulse. With this current pulse, the cathode side n–p–n transistor structure turns off within 1 μs and leaves the anode side p–n–p transistor structure base open-circuited, forcing the anode current to zero. Due to a very short gate pulse the turn-off energy at the gate is reduced drastically as compared to a GTO. Also the IGCTs have a very short tail current interval, which makes these devices much faster compared to GTOs. Due to the high-pulsed current requirements of the gate circuit, the IGCT's turn-on/off gate drive unit is supplied as an integral element of the device by the manufacturers [9,10]. IGCTs are meant for high-voltage and high-power applications, where switching frequency is quite low, hence they are optimized for low conduction losses. At present IGCTs with a forward-blocking capacity of 10 kV and a current rating as high as 6 kA are manufactured [10].

The IGCT must be protected against high di/dt and dv/dt, and snubber circuits are used for this purpose. With a well-designed gate circuit it is possible to achieve snubber-less operation with modern IGCTs.

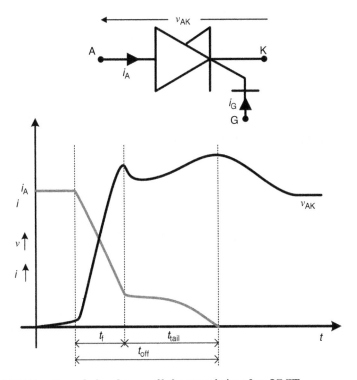

Figure 13.16 Circuit symbol and turn-off characteristics of an IGCT.

13.10   Guidelines for selecting devices

Power semiconductor devices span a wide spectrum of power levels from milliwatts to a few hundred megawatts. The switching frequency requirements range from 50 Hz at high power levels to several GHz at low power levels. Efficiency and reliability are the key factors in the power converter circuits, hence proper selection of devices is very important.

- *Switching frequency*: switching losses will increase with switching frequency. So selection of a device with proper switching frequency is important. Manufacturer data sheets will specify the maximum operating switching frequency of the device. The turn-off time and turn-on times decide the maximum switching frequency. For slow devices like rectifier power diodes, SCRs, and GTOs, the reverse recovery time (t_{rr}) decides the maximum switching frequency and t_{rr} will be specified in the data sheet. Owing to the switching losses, at high power levels, say from 500 kW to a few tens of megawatts, the switching frequency is limited to 3 kHz to power frequency that is 50/60 Hz. Modern IGCTs are capable of switching megawatt power at 3 kHz maximum and SCRs and GTOs are capable of switching in the range from 50 Hz to 1 kHz. For powers less than 500 kW, and voltage ranges from 500 V to 6.6 kV, IGBTs are used and have a maximum switching frequency of up to 30 kHz for a 600-V device, up to 20 kHz for a 1200-V device, and up to 3 kHz for a 3300-V device [11]. For low voltages (<100 V) applications and low power (<2 kW), MOSFETs are preferred as they have a switching capacity of 1 MHz maximum, at higher voltages (<1 kV), the MOSFET switching frequency is limited to 100 kHz [10]. SiC MOSFETs with a voltage rating of 1200 V and current up to 50 A can switch at 500 kHz maximum [11]. The GaN-based FETs/IGBTs with a voltage rating of 600 V and current up to 30 A can switch at 500 kHz maximum [16,17]. Low voltage Schottky rectifier diodes have a switching frequency from 1 kHz to 1 MHz [13]. Fast diodes and ultra fast recovery diodes can switch at high power levels (up to 3300 V and up to 4500 A, respectively) at frequencies in the range of 100–500 kHz [11,12]. Radio frequency and microwave grade GaN-based devices can switch at GHz, at very low power levels [10].

- *Voltage rating*: devices must be rated for maximum blocking voltage. Devices for AC applications such as rectifier diodes have repetitive peak reverse voltage (V_{RRM}) ratings [9–12]. The controlled power devices have repetitive peak off-state voltage (V_{DRM}) rating. SCRs, with a blocking voltage capacity of 12,000 V and 6,000 A, are commercially available [10]. IGBTs with a blocking voltage level of 6 kV and 6000 A [11,12], are available in the market. For power MOSFETs the voltage is limited to 500–1000 V [13].
- *Current rating*: the manufacturer's data sheet will specify the maximum value of RMS and average current rating for diodes, SCRs, GTOs, and IGCTs. The switching diodes, IGCTs, IGBTs, and MOSFETs will specify the DC current and also peak repetitive current limits.
- *I^2T rating*: manufacturer data sheets of diodes, SCRs, IGCTs, and GTOs specify the I^2T rating. The I^2T rating of these devices must be higher than I^2T rating of the protection fuse used [14].
- *Conduction loss*: it is important to estimate the on-state loss of the device. The forward on-state voltage drop and/or the on-state resistance specified in the manufacturer's data sheet, will decide the conduction loss. A proper cooling method must be used to dissipate heat and control temperature.
- *Junction temperature*: proper design of the cooling method is required to limit junction temperature of the devices within the limits specified by the manufacturer's data sheets.
- *dv/dt*: dv/dt applied to the device must be within the maximum permissible dv/dt rating.
- *di/dt*: in the case of diodes, SCRs, and GTOs di/dt should be limited to a safe value.
- *Gate drive voltage and gate resistance*: for MOSFETs and IGBTs, the gate-to-source or emitter voltage and external series resistor must be within the limits specified by the manufacturer's data sheet.

13.11 Summary

- Diodes are uncontrolled bipolar devices used in different voltage and current levels. Rectifier diodes are optimized for low-frequency operation with reduced conduction loss. Fast diodes and fast switching diodes are optimized for switching circuit applications. Schottky diodes are designed with low forward-voltage drop and used in low voltage circuits.
- IGBTs and MOSFETS are voltage-controlled devices. MOSFETs are used at high switching frequencies in low-voltage circuits. IGBTs are used in the medium power range with voltages up to 6.6 kV.
- SCRs are the only choice for controlled rectifiers.
- GTOs and IGCTs are used in high voltage (>1 kV) and high power (>1 MW) applications.
- Wide band gap material (SiC and GaN)-based devices are promising and improved devices for the future, with high switching frequency capability, low conduction and switching losses, and high temperature stability. Currently SiC- and GaN-based devices with up to 1.7 kV voltage rating, are commercially available.

Problems

1. List all the power semiconductor devices that you have studied. Classify them as,
   a. voltage-controlled and current-controlled devices
   b. uncontrolled, semicontrolled, and controlled devices
   c. unipolar and bipolar devices.
2. Explain the reverse recovery phenomenon in a power diode. How is the reverse recovery time important in circuit design?
3. Compare a Schottky diode and a power diode.
4. In power conversion circuits, the devices are not operated in the active region and are used as switches. Give reasons.
5. Darlington BJTs are used to achieve high current gain. Explain.
6. Compare a BJT and a MOSFET.
7. Compare a MOSFET and an IGBT.
8. MOSFETs with lower voltage ratings are preferred. Give reasons.
9. Power MOSFETs do not have reverse voltage-blocking capacity. Give reasons.
10. Explain the need for snubbers in thyristors.
11. Compare a SCR, an IGCT, and a GTO.
12. Explain the overcurrent protection of a MOSFET and an IGBT.
13. Explain the significance of I^2T rating of an SCR and a GTO.
14. What is the need for dv/dt protection in thyristors?
15. Explain the most common method of dv/dt protection of thyristors?
16. What is the need for di/dt protection in thyristors?
17. Explain the most common method of di/dt protection in thyristors?
18. Explain the undervoltage protection as applied to IGBT and MOSFET gate circuits.
19. Explain the significance of external ON state and OFF state gate resistance in IGBT and MOSFET gate circuits.
20. IGBTs are slower than MOSFETs. Give reasons.
21. IGBTs with higher voltage-blocking capacity are available compared to power MOSFETs. Give reasons.
22. IGBTs have lower conduction losses compared to MOSFETs at high current applications. Justify.
23. Give examples of wide band gap devices.
24. Compare wide band gap-based devices with silicon-based devices.
25. For a given voltage rating wide band gap-based devices have smaller on-state resistance. Justify.

References

[1] Baliga BJ. Fundamentals of power semiconductor devices. New York: Springer; 2008.
[2] Rashid MH. Power electronics: circuits, devices and applications. 4th ed Englewood Cliffs, NJ: Prentice-Hall; 2013.
[3] Mohan N, Undeland TM, Robbins WP. Power electronics: converters, applications and design. 3rd ed New York: John Wiley and Sons; 2008.
[4] Millán J, Godignon P, Perpina X, Perez-Tomas A, Rebollo J. A survey of wide bandgap power semiconductor devices. IEEE Trans Power Electron 2014;29(5):2155–63.

[5] Cooper JA, Agarwal A. SiC power-switching devices-the second electronics revolution? Proc IEEE 2002;90(6):956–68.

[6] Zheyu Z, Wang F, Tolbert LM, Blalock BJ, Costinett DJ. Evaluation of switching performance of SiC devices in PWM inverter-fed induction motor drives. IEEE Trans Power Electron 2015;30:5701–11.

[7] Madhusoodhanan S, Tripathi A, Patel D, Mainali K, Kadavelugu A, Hazra S, et al. Solid state transformer and MV grid tie applications enabled by 15 kV SiC IGBTs and 10 kV SiC MOSFETs based multilevel converters. IEEE Trans Ind Appl 2015;1:1626–33.

[8] Baliga BJ. Gallium semiconductor nitride devices for power electronic applications. Semicond Sci Technol 2013;28(7):411–25.

[9] Available from: http://new.abb.com/semiconductors/.

[10] Available from: http://www.mitsubishielectric.com/semiconductors/.

[11] Available from: http://www.pwrx.com/Home.aspx.

[12] Available from: http://www.semikron.com/.

[13] Available from: http://www.irf.com.

[14] Baliga BJ. Advanced high voltage power device concepts. New York: Springer; 2011.

[15] Available from: http://www.cree.com/.

[16] Available from: http://epc-co.com.

[17] Available from: http://www.microgan.com.

[18] Steimer PK, Gruning HE, Werninger J, Carroll E, Klaka S, Linder S. IGCT – a new emerging technology for high power, low cost inverters. IEEE Industry Applications Society Annual Meeting. New Orleans; 1997. p. 1592–1599.

AC–DC converters (rectifiers)   14

Ahteshamul Haque

Department of Electrical Engineering, Faculty of Engineering & Technology,
Jamia Millia Islamia University, New Delhi, India

Chapter Outline

14.1 Introduction

Renewable energy sources need power electronic converters to regulate the power generated and to convert it into a desired shape and quality. Alternating current-to-direct current (AC–DC) converters are a type of power processing converters that are mainly known as rectifiers and are used extensively in renewable energy systems (RES). One such application is the grid-connected DC microgrid [1]. In addition to RES, the use of AC–DC converters is very wide, that is, they are used in drive applications, power quality circuits, etc. It becomes necessary for a power electronics engineer to have a thorough understanding of its working mechanism. The main objective of this chapter is to provide the working of various types of AC–DC rectifiers.

Electric Renewable Energy Systems

AC–DC rectifiers are broadly classified as controlled and uncontrolled rectifiers. They are further classified, based on the AC input supply, that is, single and three phase [2]. The uncontrolled full-bridge rectifier uses diodes as switching devices in applications, where the control of power flow is not required, such as in a constant speed DC drive. However, in controlled rectifiers, switches such as thyristors, IGBT, and MOSFET, are used, which provides the control of power flow [3,4]. In this chapter the working of these two types of converters with resistive and inductive loads, along with their schematics, waveforms, and mathematical expressions have been discussed. Also, in order to evaluate the effectiveness of AC–DC converters, various performance parameters have also been discussed.

The drawbacks of the above rectifiers are poor-power factors and harmonics penetration on the AC side [5]. To overcome this negative effect pulse width modulated (PWM) rectifiers are introduced [6]. The working principle of a PWM rectifier has been explained in this chapter. Details of the filters used in AC–DC rectifiers have also been described. A few solved examples and practice problems are given at the end.

14.2 Performance parameters

Ideally, the magnitude of DC signals should be constant in amplitude and continuous with time. The rectifier is a power processing circuit that converts AC signal into DC signal and should meet the requirements, which is either ideal or close to it. At the same time, a rectifier should maintain the input AC signals that is, current as sinusoidal as possible to maintain the unity power factor. Various types of rectifier circuits used are discussed in the upcoming sections. It is important to evaluate the performance of a rectifier circuit for its closeness towards ideal performance.

The performance parameters used to evaluate the rectifier circuit have been discussed further [2].

14.2.1 Performance parameters of the output side

The *average value* of the output (load) voltage $= V_{DC}$. (14.1)

The *average value* of the output (load) current $= I_{DC}$. (14.2)

The *average value* (DC) of the output power $= V_{DC} \times I_{DC}$. (14.3)

The root-mean-square (RMS) value of the output voltage $= V_{RMS}$. (14.4)

The root-mean-square (RMS) value of the output current $= I_{RMS}$. (14.5)

The output AC power $\quad P_{AC} = V_{RMS} \times I_{RMS}$. $\hfill$ (14.6)

Rectification efficiency of a rectifier, which shows the effectiveness of the rectification process is,

$$\eta = P_{AC}/P_{AC}. \hfill (14.7)$$

The output DC voltage of a rectifier is composed of two components,

1. The DC value
2. The AC component, that is, ripple.

The effective (RMS) value of the AC component of the output voltage is,

$$V_{AC} = (V_{RMS}^2 - V_{DC}^2)\frac{1}{2}. \hfill (14.8)$$

Information about the shape of the output voltage is given by form factor (FF),

$$FF = V_{RMS}/V_{DC}. \hfill (14.9)$$

The ripple content in the output voltage is measured by ripple factor (RF),

$$RF = V_{AC}/V_{DC} \hfill (14.10)$$

or $\quad RF = ((V_{RMS}/V_{DC})^2 - 1)^{1/2} = (FF^2 - 1)^{1/2}$.

If a transformer is connected at the input side between the AC supply and the rectifier circuit, then the transformer utilization factor is,

$$TUF = P_{DC}/V_S I_S \hfill (14.11)$$

where, V_S, RMS voltage (secondary) of the transformer supply; I_S, RMS current (secondary) of the transformer supply.

14.2.2 Performance parameters of the input side

Let, I_S, RMS value of input current; I_{S1}, fundamental component of I_S.

If φ is the angle between the fundamental component of current and voltage then φ is called the displacement angle. The displacement factor is defined as,

$$DF = \cos(\varphi) \hfill (14.12)$$

The harmonic factor (HF) or total harmonic distortion (THD) of the input current is defined as,

$$HF = [(I_S^2 - I_{S1}^2)/I_{S1}^2]^{1/2} = [(I_S/I_{S1})^2 - 1]^{1/2}.$$ (14.13)

The input power factor PF is,

$$PF = (V_S I_{S1}/V_S I_S) \cos(\varphi) = I_{S1}/I_S \cos(\varphi).$$ (14.14)

The crest factor CF is,

$$CF = I_{S(peak)}/I_{S(RMS)}.$$ (14.15)

Note:

1. If the input current I_S is purely sinusoidal, $I_{S1} = I_S$, then the power factor equals the displacement factor.
2. For an ideal rectifier the values of the parameters are $\eta = 100\%$, $V_{AC} = 0$, RF = 0, TUF = 1, HF = THD = 0, PF = DF = 1.

14.3   Single-phase full-bridge rectifier circuit

The schematic of a single-phase full-bridge rectifier circuit is shown in Figure 14.1. It has four diodes D1, D2, D3, and D4 connected to an AC voltage source V_{AC}. Diodes D1 and D2 conduct through the load during the positive half cycle of AC voltage supply. Diodes D3 and D4 conduct during the negative half cycle of the AC voltage supply. The AC input voltage V_{AC} can be given directly or through an isolation transformer. The schematic shown in Figure 14.1 is commonly used in practical applications.

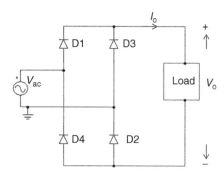

Figure 14.1 Schematic of a single-phase full-bridge rectifier circuit.

14.3.1　With resistive load

The waveforms with resistive load have been shown in Figure 14.2.
The average output voltage V_o is:

$$V_o = \frac{2}{T} \int_0^{T/2} V_m \sin wt \, dt = \frac{2V_m}{\pi} = 0.6366 \, V_m \tag{14.16}$$

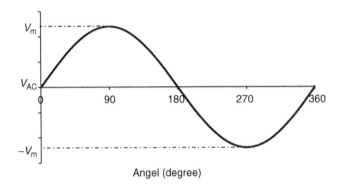

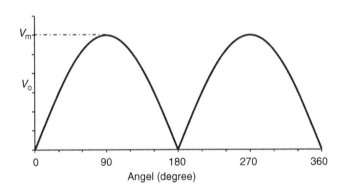

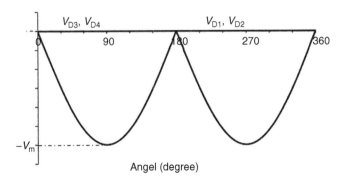

Figure 14.2　Waveforms of a single-phase full-bridge rectifier circuit with R load.

Since the output voltage is not pure DC and it has AC component in the positive half cycle, it is essential for an engineer to compute the AC component in terms of RMS value.

The RMS value of the output voltage is,

$$V_{RMS} = \left[\frac{2}{T} \int_0^{T/2} (V_m \sin wt)^2 \, dt \right]^{1/2} = \frac{V_m}{\sqrt{2}} = 0.707 V_m \tag{14.17}$$

Example 14.1

In the rectifier circuit of Figure 14.1 the load is purely resistive: $R = 100 \, \Omega$. The AC input supply voltage is $V_{AC} = 325 \sin(2\pi 60t)$. Calculate (1) the rectification efficiency, (2) the ripple factor, (3) the form factor, (4) the peak inverse voltage (PIV) of diode D1, and (5) the crest factor of the input current.

Solution

The average output voltage is,

$$V_{DC} = \frac{2V_m}{\pi} = \frac{2 \times 325}{3.14} = 207 \text{ V}.$$

The average load current is I_{DC},

$$I_{DC} = \frac{V_{DC}}{R} = \frac{207}{100} = 2.07 \text{ A}.$$

The RMS value of the output voltage is,

$$V_{RMS} = \left[\frac{2}{T} \int_0^{T/2} (V_m \sin wt)^2 \, dt \right]^{1/2}$$

$$= \frac{V_m}{\sqrt{2}} = \frac{325}{\sqrt{2}} = 229.8 \text{ V}.$$

RMS load current is,

$$I_{RMS} = \frac{V_{RMS}}{R} = \frac{229.8}{100} V = 2.298 \text{ V}$$

$$P_{DC} = V_{DC} \times I_{DC} = 207 \times 2.07 = 428.49 \text{ W}$$

$$P_{DC} = V_{RMS} \times I_{RMS} = 528.08 \text{ W}$$

1. The rectification efficiency is,

$$\eta = \frac{P_{DC}}{P_{AC}} = \frac{428.49}{528.08} \approx 81\%.$$

2. The ripple factor is,

$$= \sqrt{\left(\frac{V_{RMS}}{V_{DC}}\right)^2 - 1}$$

$$= \sqrt{\left(\frac{229.8}{207}\right)^2 - 1} = 0.482 = 48.2\%.$$

3. Form factor $= \dfrac{V_{RMS}}{V_{DC}} = \dfrac{229.8}{207} = 1.11$

4. The PIV of the diode $= V_m = 325\,V$

5. The crest factor (CF) of the input current is,

$$CF = \frac{I_{s(peak)}}{I_{s(RMS)}} = \sqrt{2} = 1.414$$

14.3.2 With battery load

Batteries are used in RES to increase reliability. In a grid-connected system, the batteries may be charged from an AC grid via a converter working as a rectifier. In this section the operation of a single-phase full-bridge converter with a battery load is discussed.

The output is connected with a battery as shown in Figure 14.3 . If $V_{AC} > V_b$ only, then the current will flow from the AC side via the rectifier circuit to the load.

$$V_{AC} = V_m \sin(wt)$$

The angle at which the current will start flowing is,

$$V_m \sin \alpha = V_b$$

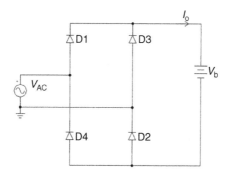

Figure 14.3 Single-phase full-bridge with battery load.

$$\alpha = \sin^{-1}\frac{V_m}{V_b}$$ (14.18)

The current will stop flowing when $V_{AC} < V_b$,

$$\gamma = \pi - \alpha$$

The charging current I_o is,

$$I_o = \frac{V_{AC} - V_b}{R} = \frac{V_m \sin wt - V_b}{R} \quad \text{where } \alpha < wt < \beta$$ (14.19)

R is the internal resistor of the battery connected in series with the battery. The waveforms are shown in Figure 14.4.

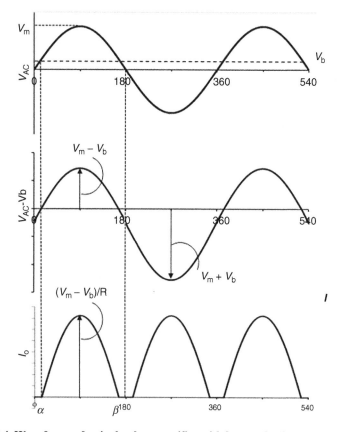

Figure 14.4 Waveforms of a single-phase rectifier with battery load.

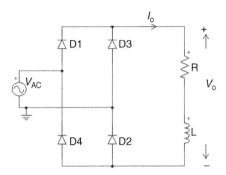

Figure 14.5 Schematic of a full-bridge rectifier circuit with R–L load.

14.3.3 With highly inductive (R–L) load

In practice, most of the load is inductive in nature. The nature of the load current depends on the value of R and L of the load. An example of highly inductive load is the armature of a DC motor. A schematic of a full-bridge rectifier with R–L load is shown in Figure 14.5.

If $V_{AC} = V_m \sin wt$ is the input voltage, then by applying KVL in the circuit of Figure 14.5,

$$L\frac{dI_o}{dt} + RI_o = V_m \sin wt \tag{14.20}$$

The waveforms with a highly inductive load are shown in Figure 14.6. Equation (14.20) has the solution of the form:

$$I_o = \frac{V_m}{Z}\sin(wt - \theta) + A_1 e^{-(R/L)t}. \tag{14.21}$$

Where load impedance $Z = (R^2 + (wL)^2)^{1/2}$ and load impedance angle $\theta = \tan^{-1}(wL/R)$.

The solution of Equation (14.21) for instantaneous load current will be,

$$I_o = \frac{V_m}{Z}\left[\sin(wt - \theta) + \frac{2}{1 - e^{-(R/L)(\Pi/\omega)}}\sin\theta e^{-(R/L)t}\right] \text{ for } 0 \le (wt - \theta) \le \Pi \text{ and } I_o \ge 0 \tag{14.22}$$

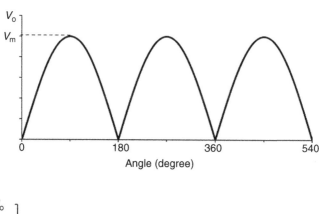

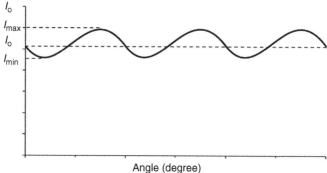

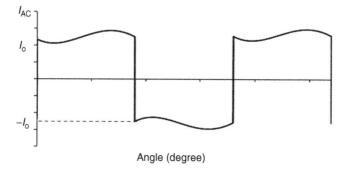

Figure 14.6 Waveforms of a single-phase rectifier with R–L load.

14.3.4   Fourier analysis

14.3.4.1   Fourier series analysis of DC output voltage

The AC–DC rectifier output voltage may be described by Fourier series analysis as,

$$V_o(t) = V_{DC} + \sum_{n=2,4,6}^{\infty} (a_n \cos nwt + b_n \sin nwt) \qquad (14.23)$$

$$V_{DC} = \frac{1}{2\pi} \int_0^{2\pi} V_0(t)d(wt) = \frac{2}{2\pi} \int_0^{\pi} V_m \sin(wt)d(wt) = \frac{2V_m}{\pi}$$

$$a_n = \frac{1}{\pi} \int_0^{2\pi} V_0 \cos nwtd(wt) = \frac{2}{\pi} \int_0^{\pi} V_m \sin(wt)\cos(nwt)d(wt)$$

$$= \frac{4V_m}{\pi} \sum_{n=2,4...}^{\infty} \frac{-1}{(n-1)(n+1)} \quad \text{for } n = 2,4 \ldots$$

$$= 0 \quad \text{for } n = 1,3,5 \ldots$$

$$b_n = \frac{1}{\pi} \int_0^{2\pi} V_0 \sin nwtd(wt) = \frac{2}{\pi} \int_0^{\pi} V_m \sin(wt)\sin(nwt)d(wt) = 0$$

Substituting the values of a_n and b_n, the expression for the output voltage is,

$$V_0(t) = \frac{2V_m}{\pi} - \frac{4V_m}{3\pi}\cos(2wt) - \frac{4V_m}{15\pi}\cos(4wt) - \frac{4V_m}{35\pi}\cos(6wt)\cdots \tag{14.24}$$

The output voltage of a single-phase AC–DC rectifier contains only even harmonics and the second harmonic is the most dominant one.

14.3.4.2 Fourier series analysis of AC input current

The input current can be expressed as a Fourier series as shown here,

$$I_s(t) = I_{DC} + \sum_{n=1,3,5...}^{\infty} (a_n \cos nwt + b_n \sin nwt) \tag{14.25}$$

Where,

$$I_{DC} = \frac{1}{2\pi} \int_0^{2\pi} i_s(t)d(wt) = \frac{1}{2\pi} \int_0^{2\pi} I_a(t)d(wt) = 0$$

$$a_n = \frac{1}{\pi} \int_0^{2\pi} i_s(t)\cos nwtd(wt) = \frac{2}{\pi} \int_0^{\pi} I_a \cos(nwt)d(wt) = 0$$

$$b_n = \frac{1}{\pi} \int_0^{2\pi} i_s(t)\sin nwtd(wt) = \frac{2}{\pi} \int_0^{\pi} I_a \sin(nwt)d(wt) = \frac{4I_a}{n\pi}$$

Substituting the values of a_n and b_n, the expression for the input current is,

$$I_s = \frac{4I_a}{n\pi}\left(\frac{\sin(wt)}{1} + \frac{\sin(3wt)}{3} + \frac{\sin(5wt)}{5} + \ldots\right). \tag{14.26}$$

14.4 Three-phase full-bridge rectifier

A three-phase full-bridge rectifier is used in high power applications. This rectifier may work with or without a transformer on the AC supply side. This rectifier gives six pulse ripples on the output voltage in one complete cycle. The numbering of diodes is done as per their sequence of conduction. One diode conducts for $120°$ duration in each cycle and a combination of two diodes conducts for $60°$ in one cycle. The combination of the two diodes conducts when the instantaneous values of AC supply voltages are higher. The load connected at the output may be of two types, that is, resistive (R) load and resistive–inductive $(R–L)$ load. The operation is analyzed in the further sections.

14.4.1 With resistive load

A schematic of a three-phase diode rectifier connected with resistive load R is shown in Figure 14.7. Six diodes, D1–D6, are connected as shown.
The instantaneous phase voltage of AC supply side is,

$$V_{an} = V_m \sin(wt) \tag{14.27}$$

$$V_{bn} = V_m \sin(wt - 120°) \tag{14.28}$$

$$V_{cn} = V_m \sin(wt) - 240° \tag{14.29}$$

The line-to-line voltages are,

$$V_{ab} = \sqrt{3} V_m \sin(wt + 30°), \tag{14.30}$$

$$V_{bc} = \sqrt{3} V_m \sin(wt - 90°), \tag{14.31}$$

$$V_{ca} = \sqrt{3} V_m \sin(wt - 210°). \tag{14.32}$$

The average output voltage is,

$$V_{DC} = \frac{3}{\pi} \int_0^{\pi/6} \sqrt{3} V_m \sin(wt - 90°) d(wt) = \frac{3\sqrt{3} V_m}{\pi}. \tag{14.33}$$

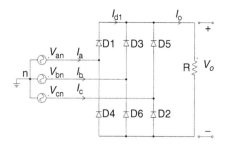

Figure 14.7 Schematic of a three-phase bridge rectifier with R load.

The RMS output voltage is,

$$V_{RMS} = \left[\frac{6}{\pi} \int 3V_m^2 \cos^2(wt)d(wt) \right]^{1/2}$$

$$= 1.6554\, V_m.$$

(14.34)

If the load is purely resistive, the peak current through the diode is,

$$I_m = \left[\frac{4}{2\pi} \int I_m^2 \cos^2(wt)d(wt) \right]^{1/2}$$

$$= 0.5518\, I_m.$$

(14.35)

Table 14.1 gives the conduction chart for diodes with line voltages, and the waveforms are given in Figure 14.8.

14.4.2 With inductive load

The waveforms of the output voltage and current for a three-phase diode bridge rectifier with R–L load are shown in Figure 14.9. The output current has negligible ripples because of its highly inductive nature.

The equation for output voltage remains the same as in the case of R load. The equations for the currents are,

The load current is:

$$i_o = \frac{\sqrt{2}V_{ab}}{Z} \left[\sin(wt - \theta) + \frac{\sin(2\pi/3 - \theta) - \sin(\pi/3 - \theta)}{1 - e^{-(R/L)(\pi/3\omega - t)}} e^{(R/L)(\pi/3\omega - t)} \right]$$

(14.36)

$$\text{for } \frac{\pi}{3} \le wt \le \frac{2\pi}{3} \text{ and } i > 0$$

$$\text{when } Z = \sqrt{R^2 + \omega L^2}$$

$$\theta = \tan^{-1}\left(\frac{\omega L}{R} \right)$$

Table 14.1 Diode conduction chart

Diodes ON	D5–D6	D1–D6	D1–D2	D3–D2	D3–D4	D5–D4	D5–D6
Line voltages	V_{cb}	V_{ab}	V_{ac}	V_{bc}	V_{ba}	V_{ca}	V_{cb}
Duration (angle)	0–30	30–90	90–150	150–210	210–270	270–330	330–360

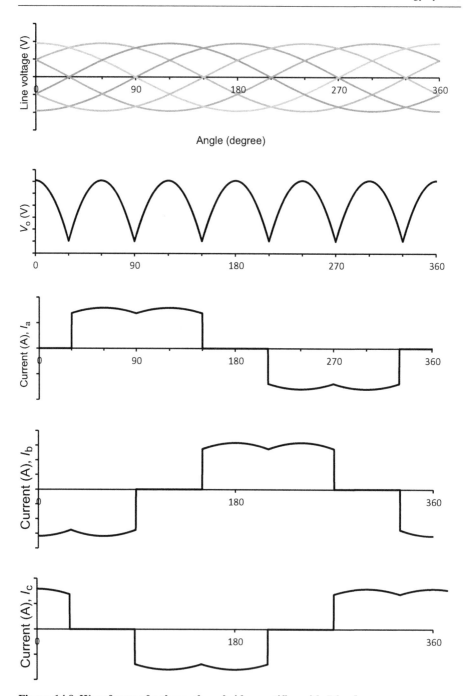

Figure 14.8 Waveforms of a three-phase bridge rectifier with R load.

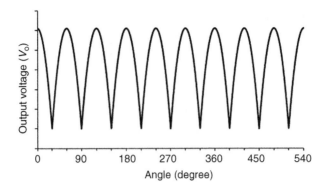

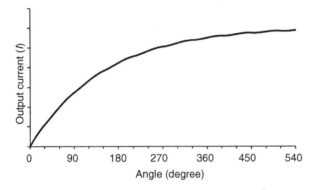

Figure 14.9 Waveforms of a three-phase bridge rectifier with R–L load.

The RMS diode current is,

$$I_{\text{RMS}} = \left(I_r^2 + I_r^2 + I_r^2 \right)^{1/2}$$
$$= \sqrt{3} I_r.$$
(14.37)

The Fourier series of instantaneous output voltage is,

$$V_o(t) = 0.9549 V_m \left(1 + \frac{2}{35} \cos(6wt) - \frac{2}{143} \cos(12wt) + ... \right).$$
(14.38)

The Fourier series expression for the input current is,

$$I_s = \sum_{n=1}^{\infty} \frac{4\sqrt{3} I_a}{2\pi} \left(\frac{\sin(wt)}{1} - \frac{\sin(5wt)}{5} - \frac{\sin(7wt)}{7} + \frac{\sin(11wt)}{11} + ... \right).$$
(14.39)

14.5   PWM rectifier

The diode- and thyristor-controlled rectifier circuits are widely used across industries. These rectifier circuits have a negative impact on the AC supply side as they deteriorate the power quality. The power factor of the AC side goes low. Also, the harmonic content on the AC side current increases. To overcome these drawbacks PWM rectifiers are developed. These rectifiers are designed to operate the switches making power factor either unity or close to unity [5,6].

In this section, a three-phase PWM rectifier is discussed. The circuit of a three-phase voltage source rectifier (VSR), along with its control block diagram, is shown in Figure 14.10.

The DC link voltage V_0 is fixed at a desired value by using a feedback control signal. It is measured and compared with a reference signal. The error signal controls the ON–OFF states of six switches. The power can flow either to or from the source, depending on the requirement.

In rectifier mode of operation, the load current I_0 is positive. The error signal demands the flow of power from the AC supply to the load. In the inverter mode the capacitor is overcharged and power flows from the load to AC supply. The control signal decides the power flow by generating an appropriate PWM gate signal.

The PWM rectifier can control both active and reactive power. It is used to correct the power factor, that is, either to unity or close to unity. The AC supply side current is also made close to sinusoidal reducing the harmonic contamination.

The waveforms of the modulation of one phase are shown in Figure 14.11. Figure 14.11a is the PWM signal and Figure 14.11b is the modulating signal.

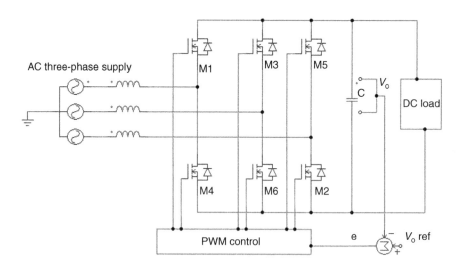

Figure 14.10 Schematic of a three-phase bridge VSR.

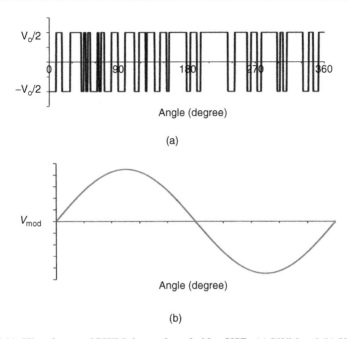

Figure 14.11 Waveforms of PWM three-phase bridge VSR. (a) PWM and (b) V_{mod}.

14.6 Single-phase full-bridge controlled rectifier

The schematic of a single-phase full-bridge rectifier with $R–L$ (highly inductive) load is shown in Figure 14.12. During the positive half cycle of the AC input supply the voltage thyristors T1 and T2 are forward biased and when both thyristors are fired simultaneously at $wt = \alpha$, load is connected to the input supply through T1 and T2. Since the load is highly inductive, switches T1 and T2 will continue to conduct even when the input voltage goes negative. During the negative half cycle of the AC supply

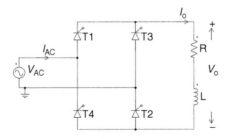

Figure 14.12 Schematic: single-phase full-bridge rectifier.

voltage, thyristors T3, T4 are forward biased, and firing T3 and T4 applies the supply voltages across T1 and T2 as a reverse blocking voltage. This will turn off T1 and T2. The load current is transferred to T3 and T4 from T1 and T2. The waveforms of this rectifier are shown in Figure 14.13. Table 14.2 gives the conduction chart for the thyristors.

The average output voltage is,

$$V_{DC} = \frac{2}{2\pi} \int_{\alpha}^{\pi+\alpha} V_m \sin wtd(wt) = \frac{2V_m}{\pi} \cos \alpha. \tag{14.40}$$

RMS value of the output voltage is,

$$V_{RMS} = \left[\frac{2}{2\pi} \int (V_m \sin wt)^2 d(wt) \right]^{1/2} = \frac{V_m}{\sqrt{2}} = V_s \tag{14.41}$$

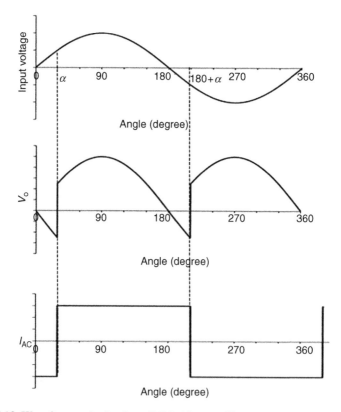

Figure 14.13 Waveforms: single-phase full-bridge rectifier.

Table 14.2 **Thyristor conduction chart**

Thyristor ON	T1–T2	T3–T4
AC voltages	Positive–negative	Negative–positive
Duration (angle)	α to $(180 + \alpha)$	$(180 + \alpha)$ to $(360 + \alpha)$

The output current is,

$$i_o = \frac{\sqrt{2}V_s}{Z}\sin(wt - \theta) + \left[I_o - \frac{\sqrt{2}V_s}{Z}\sin(\alpha - \theta)\right] e^{(R/L)(\alpha/w - t)} \text{ for } I_o > 0. \qquad (14.42)$$

Where, $Z = (R^2 + (wL)^2)^{1/2}$, $\theta = \tan^{-1}(wL/R)$.
Fourier series expansion of input current is,

$$I_{AC}(t) = \sum_{n=1,3,5\ldots}^{\infty} \sqrt{2}I_n \sin(nwt + \phi_n). \qquad (14.43)$$

Where,: $\phi_n = \tan^{-1}(a_n/b_n)$, and ϕ_n = displacement angle of the nth harmonic current.

14.7 Three-phase controlled rectifier

Three-phase rectifiers provide a higher average output voltage compared to single-phase rectifiers. Figure 14.14 shows the schematic of three-phase controlled rectifier connected with highly inductive load. The thyristors are fired at an interval of 60°. The frequency of the output ripple voltage is $6f$ (f is the frequency of the AC supply). The filtering requirement is reduced as compared to a single-phase rectifier.

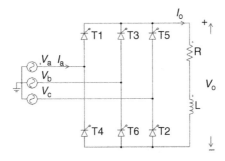

Figure 14.14 Schematic: three-phase full-bridge controlled rectifier with R–L load.

Figure 14.15 shows the waveforms of the AC phase voltages and DC output voltage. Table 14.3 gives a summary of the thyristor switch conduction.
The equations of instantaneous phase and line voltages are,

$$V_{an} = V_m \sin wt$$

$$V_{bn} = V_m \sin\left(wt - \frac{2\pi}{3}\right)$$

$$V_{cn} = V_m \sin\left(wt + \frac{2\pi}{3}\right)$$

$$V_{ab} = V_{an} - V_{bn} = \sqrt{3}V_m \sin\left(wt + \frac{\pi}{6}\right)$$ (14.44)

$$V_{bc} = V_{bn} - V_{cn} = \sqrt{3}V_m \sin\left(wt - \frac{\pi}{2}\right)$$

$$V_{ca} = V_{cn} - V_{an} = \sqrt{3}V_m \sin\left(wt + \frac{\pi}{2}\right).$$

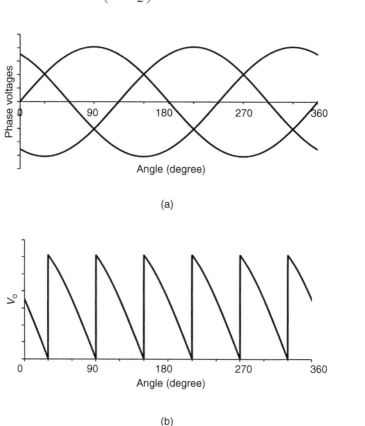

(a)

(b)

Figure 14.15 Waveforms: three-phase full-bridge controlled rectifier with R–L load.
(a) AC phase voltages and (b) DC output voltage.

Table 14.3 Summary of the thyristor switch conduction chart

Thyristors ON	T5–T6	T6–T1	T1–T2	T2–T3	T3–T4	T4–T5	T5–T6
Phase voltages	a, c, b	a, b, c	b, a, c	b, c, a	c, b, a	c, a, b	a, c, b
Duration (angle)	(0–30) + α	(30 + α) to (90 + α)	(90 + α) to (150 + α)	(150 + α) to (210 + α)	(210 + α) to (270 + α)	(270 + α) to (330 + α)	(330 + α) to 360

The average output voltage is,

$$V_{DC} = \frac{3}{\pi} \int_{\pi/6+\alpha}^{\pi/2+\alpha} V_{ab} d(wt) = \frac{3\sqrt{3}V_m}{\pi} \cos\alpha. \tag{14.45}$$

The RMS value of the output voltage is,

$$V_{RMS} = \left[\frac{3}{\pi} \int 3V_m^2 \sin^2\left(wt + \frac{\pi}{6}\right) d(wt) \right]^{1/2}$$
$$= \sqrt{3}V_m \left(\frac{1}{2} + \frac{3\sqrt{3}}{4\pi} \cos 2\alpha \right)^{1/2}. \tag{14.46}$$

Fourier series of the AC supply side current connected to a highly inductive load via a three-phase full-bridge controlled rectifier is,

$$i_{AC}(t) = \sum_{n=1,3,5...}^{\infty} \sqrt{2}I_n \sin(nwt + \Phi_n) \tag{14.47}$$

where $\Phi_n = \tan^{-1}(a_n/b_n)$.

14.8　Filters for AC to DC converters

In rectifier circuits two types of filters are used, that is, one is a DC filter connected at the output to remove the ripple of the DC voltage and other is an AC filter connected at the AC input side to remove the harmonic contents of the AC current. The schematics are shown in Figure 14.16.

In most of the applications, a capacitor is used in a DC filter (Figure 14.16a (2)). It gives satisfactory performance.

Figures 14.17 and 14.18 are waveforms of the rectified output voltage with and without a DC capacitor filter.

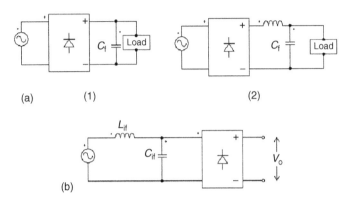

(a)   (1)   (2)

(b)

Figure 14.16 Schematic representation. (a) DC filter and (b) AC filter.

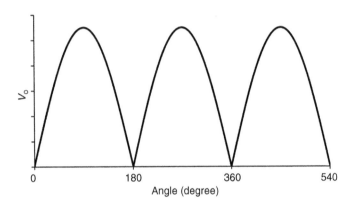

Figure 14.17 Rectified voltage without DC filter.

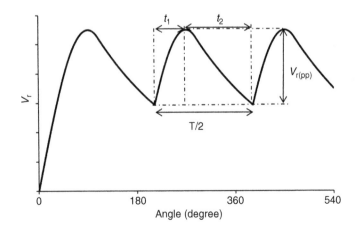

Figure 14.18 Rectified voltage with DC filter.

When the voltage across capacitor C_f of the DC filter capacitor is lower than the instantaneous input AC voltage, the C_f will be charged through the rectifier circuit. If the voltage across capacitor C_f is higher than the instantaneous AC input supply voltage then it will discharge through the resistive load. The voltage across C_f will vary. Assuming t_1 is the time for charging of C_f, it will charge to the value of the peak supply voltage V_m and t_2 is the time for discharging the capacitor C_f as shown in Figure 14.18. The capacitor C_f will be discharged through the resistive load exponentially.

The peak-to-peak ripple voltage $V_{r(pp)}$ is,

$$
\begin{aligned}
V_{r(pp)} &= V_o\left(t = t_1\right) - V_o(t = t_2) \\
&= V_m - V_m e^{-t_2/RC_f} \\
&= V_m\left(1 - e^{-t_2/RC_f}\right).
\end{aligned}
\tag{14.48}
$$

Since $e^{-x} \approx 1 - x$:

$$
\begin{aligned}
V_{r(pp)} &= V_m\left(1 - 1 + \frac{t_2}{RC_f}\right) \\
&= \frac{V_m t_2}{RC_f} = \frac{V_m}{2fRC_f}.
\end{aligned}
\tag{14.49}
$$

Average output voltage is,

$$
V_{DC} = V_m - \frac{V_{rpp}}{2} = V_m - \frac{V_m}{4fRC_f}.
\tag{14.50}
$$

RMS output ripple voltage V_{oc} is,

$$
V_{oc} = \frac{V_{r(pp)}}{2\sqrt{2}} = \frac{V_m}{4\sqrt{2}fRC_f}.
\tag{14.51}
$$

Ripple factor is,

$$
RF = \frac{V_{oc}}{V_{DC}} = \frac{1}{\sqrt{2}\left(4fRC_f - 1\right)}.
\tag{14.52}
$$

This expression can be solved to get the value of the filter capacitor C_f.

14.9 Summary

There are different types of rectifiers used in various applications depending on diode connections. These rectifiers are used in RES and in other applications. In this chapter the performance parameters of the rectifiers were defined to evaluate the quality of the rectifiers. The workings of the main types of rectifiers were discussed with schematics, waveforms, and mathematical expressions. The negative effect of the rectifier circuits is harmonics penetration. To overcome this effect PWM rectifiers are used. The other method is to use filter circuits.

Problems

Theoretical questions

1. What is the difference between a controlled and uncontrolled rectifier?
2. What is rectification efficiency? How it is different from rectifier efficiency?
3. What is the significance of the harmonic factor?
4. What is the difference between power factor and displacement factor?
5. What is the difference between AC and DC filters for rectifiers?
6. Explain the workings of battery charging using a rectifier circuit.

Numerical questions

1. A single-phase full-bridge rectifier has a purely resistive load $R = 50\ \Omega$, the peak AC supply voltage $V_m = 200$ V, and the supply frequency $f = 60$ Hz. Determine the average output voltage of the rectifier.
2. A three-phase full-bridge rectifier has a purely resistive load $R = 100\ \Omega$, the peak supply voltage $V_m = 200$ V, and the supply frequency is $f = 60$ Hz. Determine the average output voltage of the rectifier.
3. A three-phase full-bridge rectifier is required to supply an average output voltage of 400 V to a resistive load of $R = 50\ \Omega$. Determine the voltage and current ratings of the diodes.
4. A single-phase full-bridge controlled rectifier is used to charge a battery of 96 V having internal resistance of 2 Ω. The RMS value of the AC supply voltage is 230 V and frequency is 60 Hz. The SCRs are triggered by constant DC signal (Figure 14.12). If T4 is open circuited, determine the average charging current of the battery.

References

[1] Farhadi M, Mohammad A, Mohammad O. Connectivity and bidirectional energy transfer in DC microgrid featuring different voltage characteristics. IEEE Conf Green Technol 2013;244–249, vol. 1.
[2] Rashid MH. Power electronics: circuits, devices and application. India: Pearson Education Inc; 2011.
[3] Schaefer J. Rectifier circuits – theory and design. New York: John Wiley & Sons; 1975.
[4] Dixon J. Three phase controlled rectifiers. In: Rashid MH, editor. Power electronics handbook. San Diego, CA: Academic Press; 2001.
[5] Gao Y, Li J, Liang H. The simulation of three phase voltage source PWM rectifier. IEEE Conf Power Energy (APPEEC) 2012. p. 1–4, vol. 1.
[6] Rodriguez JR, Dixon JW, Espinoza JR. PWM regenerative rectifiers: state of the art. IEEE Trans Ind Electron 2005;52(1):5–22.

DC–DC converters

15

Akram Ahamd Abu-aisheh*, Majd Ghazi Batarseh**
*Department of Electrical and Computer Engineering, University of Hartford,
West Hartford, CT, USA
**Department of Electrical Engineering, Princess Sumaya University for Technology,
Amman, Jordan

Chapter Outline

15.1 Introduction

The demands for clean and sustainable energy sources have increased rapidly in the last decade, and solar energy is currently one of the most valuable, abundant, and preferred low-maintenance clean sustainable energy sources. Photovoltaic (PV) solar energy systems require the use of DC–DC converters to regulate and control the varying output of the solar panel. The three basic DC–DC converter topologies used in PV solar energy systems are buck, boost, and buck–boost converter topologies. The three basic nonisolated switch-mode DC–DC converters are also used in many industrial applications to regulate and control the amplitude of an unregulated voltage.

The single-ended primary inductance converter (SEPIC) topology is a step-up or step-down topology with no voltage polarity reversal. The isolated topology of SEPIC converters is being used more often in photovoltaic solar energy systems. A SEPIC uses two inductors and two capacitors, so it is not a basic DC–DC converter topology since basic topologies use only one inductor, one diode, one switch, and one capacitor.

The two inductors in the isolated SEPIC topology can be selected to have the same value and are wound on one core making forming a low cost 1:1 transformer with a small footprint. This makes SEPIC a more viable DC–DC converter choice.

This chapter focuses on the three basic nonisolated switch-mode DC–DC converters and SEPIC. Section 15.1 presents an introduction to nonisolated switch-mode DC–DC converters while Section 15.2 presents the three basic nonisolated switch-mode DC–DC converter structures. In Section 15.3, two application problem-based learning (PBL) projects are presented to induce in the reader the motivation to study DC–DC converters. After covering the three basic nonisolated switch-mode DC–DC converter topologies in Sections 15.4, 15.5, and 15.6, the SEPIC topology is covered in Section 15.7.

15.2 Basic nonisolated switch-mode DC–DC converters

There are three basic nonisolated switch-mode DC–DC converters: buck, boost, and buck–boost converters. Each of these basic converters consists of a MOSFET, a diode, a capacitor, and an inductor connected between the input voltage and the load in different configurations (topologies) to regulate and step down, step up, or step up or down the input voltage. Each of the three basic nonisolated switch-mode DC–DC converter topologies consists of a switching network consisting of a transistor–diode combination and an L–C filter bank consisting of only one inductor and one capacitor; hence, basic topology. Basic topologies employ only one transistor, typically a MOSFET, as the main switch and a diode. Synchronous topologies, on the other hand, employ two switches where the second switch replaces the diode. The ratio between the time the MOSFET is ON and one switching period is the duty cycle. The duty cycle is controlled using pulse width modulation to control the converter output voltage.

When selecting a DC–DC converter for any application, the designer should try first to choose the buck converter if a voltage step down is needed. The boost converter should be considered as the first choice if a voltage step up is needed for the design. If there is a need for step up or step down in the same converter, the first choice should be the buck–boost converter. If the design requirements cannot be met using any of the basic nonisolated switch-mode converters, SEPIC topology should be the next option to be considered. If higher efficiency is required, synchronous converters, with the diode replaced by a second switch to reduce converter losses, should be considered. For systems with higher isolation design requirements, isolated DC–DC converters should be considered. The preferred isolated switch-mode DC–DC converter topology is the flyback topology, which is a step-up or step-down topology with a transformer.

15.3 DC–DC converter applications

This section presents two PBL DC–DC converter application projects. The first one is an off-grid PV-powered LED street light, and the second one is a grid-connected LED street light. In addition to this chapter, senior-level power electronics textbooks [1–4] can be used to help gain a comprehensive understanding of the DC–DC converters needed for the two PBL projects presented here, and Refs [5–8] present an in-depth analysis of these two PBL application projects.

15.3.1 Off-grid PV powered LED street light

In the first PBL application project, PV panels are used to power an off-grid LED street light, so there is a need for the development of a switch-mode DC–DC converter system to power the LED street light from the solar panel. A buck, buck–boost, or SEPIC converter topology can be used to step down and regulate the solar panel output voltage from 22–26 V to the 12 V battery voltage, and a boost, buck–boost, or SEPIC topology can be used to step up the 12 V battery output voltage to 18 V needed for a set of series-connected high brightness LED loads. During the day, solar energy is delivered from the PV panel to the battery in charging mode. During the night-time, the boost converter delivers energy to the LEDs. Figure 15.1 presents this PBL project as a block diagram.

15.3.2 Grid-connected hybrid LED street light

In this BPL project, PV panels are used to power a grid-connected LED street light, and the electric grid is used as an alternative source of energy for the system when the sun is not out for a period of time beyond the capability of the batteries to serve as a backup source. The design of a high brightness (HB) LED with built-in converter that can be connected directly to the AC line requires isolated DC–DC converters, so a flyback or isolated SEPIC topology can be used here. Off-the-shelf LED lights take advantage of the high efficiency of HB LEDs but not solar energy. The hybrid HB LED-based street light design PBL presented here incorporates an automatic transfer switch, as given in Figure 15.2. Through the use of this switch, the hybrid system uses solar energy as the primary source, and it switches to the AC line only for the time when the primary source cannot supply the required power to the illumination system. In the system given in Figure 15.2, LEDs are powered from the solar panel during the daytime as the primary source, with the battery in charge mode, and they operate

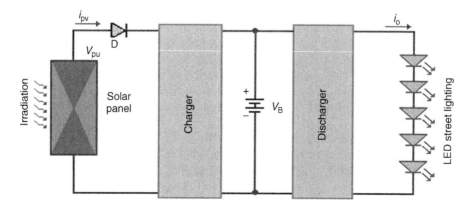

Figure 15.1 DC–DC converter application: photovoltaic powered street light.

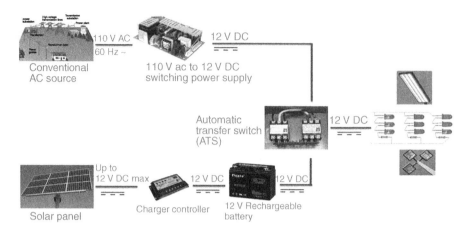

Figure 15.2 Hybrid high brightness LED illumination system.

from the battery at night. If the battery is fully discharged, the LED system operates from the AC line as a back-up source.

15.4   Buck converter

Buck converters are nonisolated switched-mode step-down DC–DC converters; the DC input voltage at the source is periodically chopped through the use of an electronic switch operating at a set frequency, and the resulting pulsated signal is filtered out and thus the load is operated at an average voltage value that is less than the input voltage. The output voltage level at the load is controlled by varying (modulating) the width of the switch chopped input pulse, which is basically controlling the duration of the time the electronic switches, MOSFETs, are ON or OFF in one cycle of the operating frequency; this is known as pulse width modulation.

Buck converters use a transistor–diode switching network; this configuration employs the transistor S_w as the main switch and the diode D_d as the freewheeling path for the inductor current. The ratio between the time the transistor is ON (pulse width) and one cycle is the duty cycle or duty ratio D. For a buck converter, the duty cycle or the gain is less than 1 making the output voltage below the input level. Figure 15.3 shows a breakdown of a conventional buck converter consisting of a DC input voltage, a switching network of a transistor–diode implementation, an L–C filtering bank, and a load resistor along with the voltage signal variation from the input side down to the load.

Buck converter efficiency can be improved by replacing the diode in the transistor–diode switching network with a second transistor; this configuration employs the transistor S_{w1} as the main switch and a second transistor S_{w2} as the freewheeling path for the inductor current. The resulting topology is known as the synchronous buck converter topology. The synchronous buck converter topology has higher efficiency

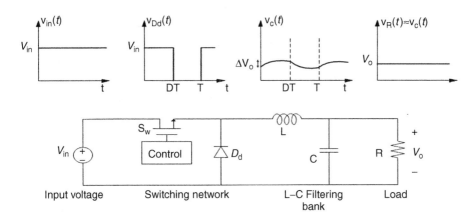

Figure 15.3 A conventional buck converter and associated voltage signals.

than the standard buck converter topology due to the reduced conduction losses resulting from replacing the diode with a transistor.

15.4.1 Steady-state analysis

The buck converter alternates between two circuit modes depending on the switch control. Assuming ideal switch operation over a period of T seconds, when the controller switches the transistor S_w ON (short circuit), the diode D_d is in the reverse biased (OFF) state and not conducting current (open circuit) as shown in Figure 15.4a. This is the charging mode of the buck where the inductor current linearly increases from an initial minimum value I_{Lmin} at steady state to a maximum value I_{Lmax} with a constant inductor voltage of $V_L = V_{in} - V_{out}$. The duration of this mode is DT_{sec} where D is the duty cycle.

After the transistor is switched off (open circuit), the built-up current in the inductor needs to discharge forcing the diode to conduct (short circuit). The converter moves from the charging mode of Figure 15.4a to the discharge mode of Figure 15.4b in which the inductor current linearly decreases from the peak value reached at the end of the charging mode back to the original initial value, during which the voltage across the inductor is $V_L = -V_o$. The duration of the discharge mode is the remainder of the cycle, which is $(1 - D)T_{sec}$. The cycle repeats and the converter enters the charging mode again by switching the transistor back ON comprising the frequency at which the converter operates of $1/T$ Hz. The signals of the voltage across and the current through the inductor are depicted in Figure 15.5.

The output voltage V_o as a function of the input voltage V_{in} and the duty cycle D can be easily derived using the volt–second balance rule applied on the inductor, which states that the average voltage over one period across the inductor has to be zero. This is implemented by taking the algebraic sum of the product of the voltage across the inductor during charging ($V_L = V_{in} - V_o$ in volts) multiplied by the time duration of the charging interval (DT in seconds), and the product of the voltage across the inductor during discharging ($V_L = -V_o$ in volts) multiplied by the time duration of the

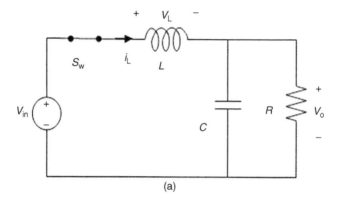

(a)

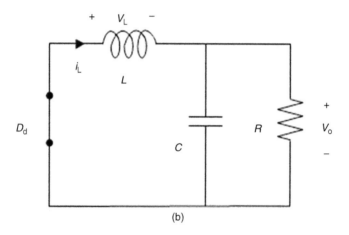

(b)

Figure 15.4 Buck converter. (a) Charging mode main switch ON and (b) discharging mode main switch OFF.

discharging interval $(1 - D)T_{sec}$ and equating them to zero. Applying the volt–second rule on the inductor of a buck converter will result in:

$$(V_{in} - V_o)DT + (-V_o)(1 - D)T = 0$$

which is simplified to give the gain of the step-down buck converter in Equation (15.1).

$$V_o = (D)V_{in} \tag{15.1}$$

The input current and the inductor and capacitor currents are all shown in Figure 15.5. The input current changes from $i_{in} = i_L$ in DT seconds during the charging mode into

$i_{in} = 0$ throughout the rest of the period of $(1 - D)T$ seconds during the discharge mode with an average value less than the inductor average current of $I_L = I_o = V_o/R$. The ripple across the inductor current varies from i_{Lmax} to i_{Lmin}. Derivation of the maximum and minimum inductor currents is well established and it is sufficient to list them in Equations (15.2) and (15.3):

$$I_{Lmin} = DV_{in}\left(\frac{1}{R} - \frac{(1-D)T}{2L}\right) \tag{15.2}$$

$$I_{Lmax} = DV_{in}\left(\frac{1}{R} + \frac{(1-D)T}{2L}\right) \tag{15.3}$$

Figure 15.5 shows the continuous conduction mode (CCM) operation of buck converters, where the inductor current is continuously present and circulating. If the energy stored in the inductor was completely discharged before switching the main power back on, then the minimum inductor current will reach zero, and the converter is said to be operating in the discontinuous conduction mode (DCM).

The boundary between CCM and DCM operations is set by a specific inductor value known as the critical inductor value $C_{critical}$ at which the minimum inductor current hits zero before charging again, this is simply found in equation (15.4) by solving for the value of the inductor that forces the minimum inductor current to zero:

$$I_{Lmin}\big|_{L=L_{critical}} = 0$$
$$I_{Lmin}\big|_{L=L_{critical}} = DV_{in}\left(\frac{1}{R} - \frac{(1-D)T}{2L}\right) \tag{15.4}$$
$$L_{critical} = \left(\frac{1-D}{2}\right)TR$$

Choosing inductors of greater values than the critical value places the converter into CCM operation, whereas inductors less than the critical value drive the operation into DCM. The large output capacitor smoothes out the variation ΔV_o on the output voltage and keeps it constant at V_o. The allowed ripple as a ratio of the desired constant output voltage $(\Delta V_o/V_o)$ determines the value of the capacitor filter needed as given in Equation (15.5):

$$C = \frac{1-D}{8L(\Delta V_o / V_o)f^2} \tag{15.5}$$

15.5 Boost converter

The DC–DC converter that powers an output load with a voltage level larger than the input source is called the boost (step-up) converter. This boost in the output voltage is achieved by placing the inductor before the switching network as in Figure 15.6.

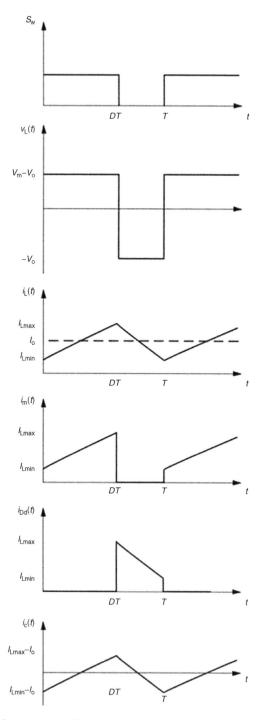

Figure 15.5 Buck converter voltage and current waveforms.

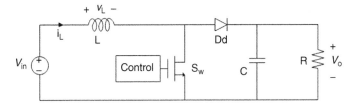

Figure 15.6 A conventional boost converter topology.

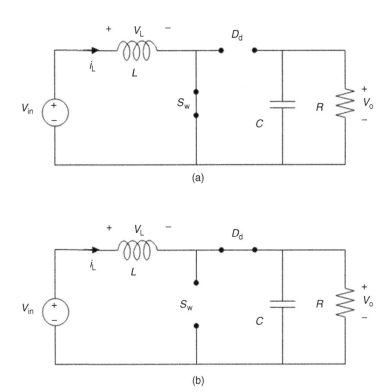

(a)

(b)

Figure 15.7 Boost converter. (a) Charging mode main switch ON and (b) discharging mode main switch OFF.

15.5.1   Steady-state analysis

Unlike the buck, during the charging mode of the boost converter the load is disconnected and the inductor is linearly charged through the constant input voltage for DT seconds. For the rest of the period of $(1 - D)T$ seconds the inductor discharges through the load while the input source is powering the circuit. The two modes of operation of the boost converter are shown in Figure 15.7a,b and the voltage and current waveforms are depicted in Figure 15.8.

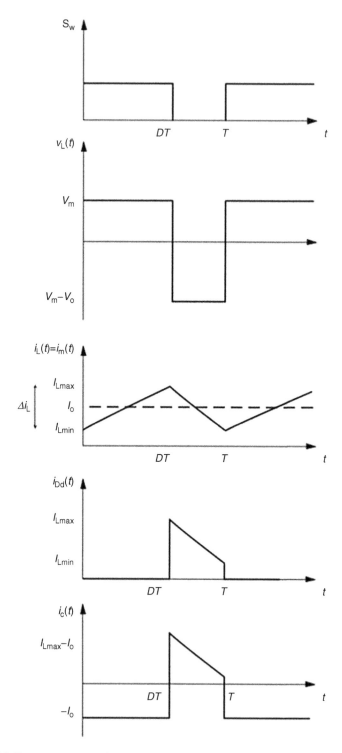

Figure 15.8 Boost converter voltage and current waveforms.

The analysis of the boost converter follows the same procedure as the buck and the results are discussed later.

Applying the volt–second balance rule on the inductor results in the gain of the boost converter given in Equation (15.6) where the output voltage exceeds the input by the factor $1/(1-D)$:

$$V_o = \frac{V_{in}}{1-D} \tag{15.6}$$

The input current, which is the same as the inductor current along with the capacitor current, is shown in Figure 15.8. In this case the average input current, which is the inductor average current, is given in Equation (15.7) and derived from equating the input and the output powers:

$$I_L = I_{in} = \frac{V_o \times I_o}{V_{in}} = \frac{I_o}{(1-D)} = \frac{V_{in}}{(1-D)^2 \times R} = \frac{V_o^2}{V_{in} \times R} \tag{15.7}$$

The ripple across the inductor current varies from I_{Lmax} to I_{Lmin} as in Equations (15.8) and (15.9):

$$I_{Lmin} = V_{in} \left(\frac{1}{R \times (1-D)^2} - \frac{DT}{2L} \right) \tag{15.8}$$

$$I_{Lmax} = V_{in} \left(\frac{1}{R \times (1-D)^2} + \frac{DT}{2L} \right) \tag{15.9}$$

The boundary between CCM and DCM is set by the critical value of the inductor given in Equation (15.10). Inductor values greater than $L_{critical}$ set the boost operation in CCM.

$$L_{critical} = \frac{D(1-D)^2 \times R}{2f} \tag{15.10}$$

The allowed ripple as a ratio of the desired constant output voltage $\left(\Delta V_o / V_o \right)$ determines the value of the capacitor filter needed as given in Equation (15.11):

$$C = \frac{D}{R \times \left(\Delta V_o / V_o \right) \times f} \tag{15.11}$$

15.6   Buck–boost converter

Buck–boost converters are nonisolated switch-mode step-up or step-down DC–DC converters; the DC input voltage at the source is stepped up or down through the use of an electronic switch, the resulting pulsated signal is filtered out, and thus the load is operated at an average voltage value that is either greater or less than the input voltage. The output voltage level at the load is controlled by varying (modulating) the width of the input pulse, which is basically controlling the duration of the time the electronic switch is ON. If the duty ratio is greater than 0.5, the input voltage is stepped up (boost mode); if the duty ratio is less than 0.5, the input voltage is stepped down (buck mode).

Buck–boost converters use a transistor–diode switching network in a configuration that employs a MOSFET as the switch and a diode as the freewheeling path for the inductor current when the MOSFET is off. The ratio between the time the transistor is ON and one period is the duty ratio. For a buck–boost converter, the gain is less than 1 when the duty ratio is less than 0.5 making the output voltage below the input level, and it is greater than 1 when the duty ratio is higher than 0.5 making the output voltage level higher than the input voltage level. Figure 15.9 shows the conventional buck–boost converter topology [3]. The diode can be replaced by a second switch resulting in the more efficient synchronous buck–boost topology.

15.6.1   Steady-state analysis

A buck–boost converter alternates between two circuit modes depending on the status of the switch. The same analysis strategy followed for the buck and boost converters applies for the buck–boost converter, and detailed analysis of this converter is presented in [1–3]. Assuming ideal switch operation over a period of T seconds, when the controller switches the MOSFET ON (short circuit), the diode is in the reverse biased (OFF) state and not conducting current (open circuit) as shown in Figure 15.10a. This is the charging mode of the buck–boost converter where the inductor current linearly increases from an initial minimum value I_{Lmin} at steady state to a maximum value I_{Lmax}.

After the transistor is switched off (open circuit), the built-up current in the inductor needs to discharge forcing the diode to conduct (short circuit). The converter moves from the charging mode of Figure 15.10a to the discharge mode of Figure 15.10b in which the inductor current linearly decreases from the peak value reached at the end

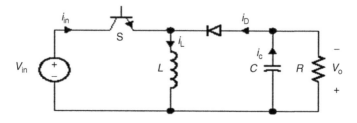

Figure 15.9 A conventional buck–boost converter topology [3].

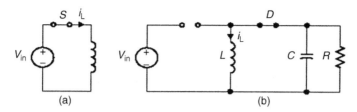

Figure 15.10 Buck–boost converter circuits. (a) Charging mode and (b) discharging mode [3].

of the charging mode back to the original initial value. The duration of the discharge mode is the remainder of the cycle, which is $(1 - D)T_{sec}$. For this analysis, the inductor current never reaches 0, so it stays in CCM.

The output voltage V_o as a function of the input voltage V_{in} and the duty cycle D can be easily derived using the volt–second balance rule applied on the inductor. This is implemented by taking the algebraic sum of the product of the voltage across the inductor during charging multiplied by the time duration of the charging interval (DT in seconds), and the product of the voltage across the inductor during discharging multiplied by the discharging interval $(1 - D)T_{sec}$ and equating them to zero. Applying the volt–second rule on the inductor of the buck–boost converter will give the gain of the buck–boost converter in Equation (15.12). For the buck–boost converter, there is a polarity reversal between the input and output voltages.

$$\frac{V_o}{V_{in}} = \frac{D}{1 - D} \tag{15.12}$$

The maximum and minimum inductor currents for the buck–boost are listed in Equations (15.13) and (15.14):

$$I_{L,max} = V_{in}\left[\frac{D}{R(1-D)^2} + \frac{DT}{2L}\right] \tag{15.13}$$

$$I_{L,min} = V_{in}\left[\frac{D}{R(1-D)^2} - \frac{DT}{2L}\right] \tag{15.14}$$

The CCM operation of the buck–boost converter is well established. In CCM operation, the inductor current is continuously present and circulating.

The allowed ripple as a ratio of the desired constant output voltage $\left(\Delta V_o/V_o\right)$ determines the value of the capacitor filter needed for the buck–boost converter and is:

$$C = \frac{D}{R \times \left(\Delta V_o/V_o\right) \times f} \tag{15.15}$$

15.7   SEPIC converter

Basic switch-mode DC–DC converters suffer from a high current ripple. This creates harmonics. In many applications, these harmonics necessitate using an LC filter. Another issue that can complicate the usage of the buck–boost converter topology is the fact that it inverts the voltage. uk converters solve the harmonics problem by using an extra capacitor and inductor. However, both uk and buck–boost converters reverse the voltage polarity of the input voltage and cause large amounts of electrical stress on the components, and this can result in device failure or overheating. SEPIC converters solve both of these problems.

SEPIC is a step-up or step-down converter. SEPIC converter topology, given in Figure 15.11, is a DC–DC converter topology that provides a low harmonic content positive regulated output voltage from an input voltage that varies above and below the output voltage. The standard SEPIC topology uses two inductors and two capacitors, so it is not considered a basic DC–DC converter topology. Two inductors of equal values can be used as a 1:1 transformer as in Figure 15.12. Coupled inductors are available in a single package at a cost only slightly more than that of the comparable single inductor. The coupled inductors not only provide a smaller footprint but also,

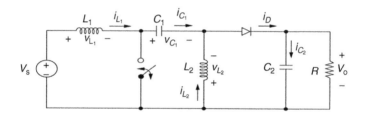

Figure 15.11 Standard nonisolated SEPIC converter topology [4].

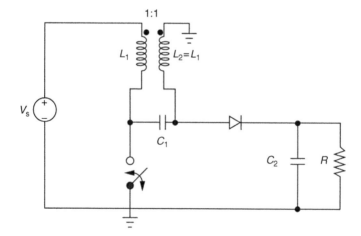

Figure 15.12 Isolated SEPIC converter topology [4].

to get the same inductor ripple current, require only half the inductance needed for a SEPIC with two separate inductors.

15.7.1 SEPIC steady-state analysis in continuous conduction mode

Applying Kirchhoff's voltage law around the path V_s, L_1, C_1, and L_2 in Figure 15.11, when the switch is open, gives $-V_s + v_{L_1} + v_{C_1} - v_{L_2} = 0$. Using the average of these voltages, the voltage across capacitor C_1 is $V_{C_1} = V_s$. When the switch is closed, the voltage across L_1 during DT is $v_{L_1} = V_s$. When the switch is open, applying KV around the outermost path gives $-V_s + v_{L_1} + v_{C_1} + V_o = 0$. Assuming the voltage across C_1 remains constant, then $v_{L_1} = -V_o$ for a time period of $(1 - D)T$. Using the voltage across the inductor of zero for a periodic operation then $V_s DT - V_o(1 - D)T = 0$ where D is duty ratio of the switch. Then $V_o = V_s(D/1 - D)$, which is expressed as $D = V_o/(V_o + V_s)$. This is similar to the buck–boost and Cuk converter equations, but with no reverse polarity.

The variation in i_{L_1} when the switch is closed is $v_{L_1} = V_s = L_1(\Delta i_{L_1}/DT)$. On solving for Δi_{L_1} we get $\Delta i_{L_1} = V_s DT/L_1 = V_s D/L_1 f$. For L_2, the average current is determined from Kirchhoff's current law at the node where L_1, C_2, and the diode are connected, $i_{L_2} = i_D - i_{C_1}$ and the diode current is $i_D = i_{C_2} + I_o$.

The output stage consisting of diode, C_2 and resistor is same as boost converter and so the output voltage ripple is: $\Delta V_o = V_{C_2} = (V_o D)/(RC_2 f)$. The voltage variation in C_1 is determined from the circuit with switch closed. Capacitor C_1 current has an average value of Io where, $\Delta V_{C_1} = (\Delta Q_{C_1}/C) = (I_o DT/C)$. Replacing with I_o with V_o/R result in $\Delta V_{C_1} = V_o D/RC_1 f$.

15.8 Summary

This chapter covered nonisolated switch-mode DC–DC converters. First, two application PBL projects were presented to show the importance of understanding the material presented in this chapter. Then, the three basic nonisolated switch-mode DC–DC converter topologies were presented.

The basic DC–DC converter topologies should be considered first when selecting a DC–DC converter for any application due to their simplicity, smaller size, and lower cost. The designer should try first to choose the buck converter if a voltage step down is needed or the boost converter if a voltage step up is needed for the design. If there is a need for a voltage step up and step down in the same converter, the first choice should be the buck–boost converter. If the design requirements cannot be met using any of the three basic nonisolated switch-mode converters, SEPIC should be the next option to be considered. If a higher efficiency is required, synchronous converters with the diode replaced by a second switch should be considered for the design implementation.

If a high isolation level is required, an isolated switch-mode DC–DC converter needs to be considered. The flyback converter is a good choice in this case. The flyback converter is a buck–boost converter with the inductor split to form a transformer to provide a higher level of electrical isolation between the input and the output. The input to output voltage gain ratio of the buck–boost topology is multiplied by the transformer turns ration to achieve the isolated switch-mode DC–DC flyback converter gain, and there is an additional advantage of input to output isolation. The analysis of the flyback converter topology is similar to the buck–boost converter topology, and Refs [1–4] present complete analysis of isolated switch-mode DC–DC converters.

Problems

1. A buck converter has the following parameters: $V_i = 15$ V, $V_o = -9$ V, $L = 10$ μH, $C = 50$ μF, and $R = 5$ Ω. The switching frequency is 150 khz. Determine the duty ratio, the maximum inductor current, and the output voltage ripple.
2. A buck converter has an input of 6 V and an output of 1.5 V. The load resistor is 3 Ω, the switching frequency is 400 kHz, $L = 5$ μH, and $C = 10$ μF. Determine the peak inductor current and the peak and average diode current.
3. A boost converter has an input of 12 V and an output of 24 V. The switching frequency is 100 kHz, and the output power to a load resistor is 125 W. Determine the duty ratio and the capacitance value to limit the output voltage ripple to 0.5%.
4. A buck–boost converter has the following parameters: $V_i = 24$ V, output voltage = -15.6 V, $L = 25$ μH, $C = 15$ μF, and $R = 10$ Ω. The switching frequency is 1 MHz. Determine the maximum inductor current and the output voltage ripple.
5. Design a buck–boost converter to produce an output voltage of –15 V across a 10-Ω load resistor. The operating frequency is 0.5 MHz and the output voltage ripple must not exceed 0.5%. The DC supply is 45 V. Design for a continuous inductor current. Calculate the value of the capacitor and the peak voltage rating of the MOSFET.
6. Design a SEPIC converter to produce an output voltage of 12 V from an input voltage of 18 V. The output power is 10 W and the operating frequency is 1 MHz. The output voltage ripple must not exceed 100 mVpp. If two 50 mH inductors are used for the converter design, calculate the value of the capacitors and the MOSFET and diode ratings.

References

[1] Mohan N, Undeland TM, Robbins WP. Power electronics converters applications and design. John Wiley and Sons, Inc. USA, 2003.
[2] Rashid MH. Power electronics: circuits, devices and applications. Pearson Education, Inc. USA, 2004.
[3] Batarseh I. Power electronic circuits. John Wiley and Sons, Inc. USA, 2004.
[4] Hart DW. Power electronics. McGraw Hill Higher Education, USA, 2010.
[5] Abu-aisheh A. Designing sustainable hybrid high brightness LED illumination systems. Int J Mod Eng 2012;12(2):35–40.
[6] Abu-aisheh A, Khader S. Hybrid MPPT-controlled LED illumination systems. ICGST-ACSE J 2012;12(2).

[7] Abu-aisheh A, Khader S, Hasan O, Hadad A. Improving the reliability of solar-powered LED illumination systems. ICGST International Conference on Recent Advances in Energy Systems. Alexandria, Egypt; April, 2012.

[8] Abu-aisheh A, Khader S, Harb A, Saleem A. Sustainable FPGA controlled hybrid LED illumination system design. The Third International Conference on Energy and Environmental Protection in Sustainable Development (ICEEP III). Palestine Polytechnic University (PPU), Hebron (Alkhaleel), West Bank; June, 2014.

DC–AC inverters

16

David (Zhiwei) Gao, Kai Sun
Department of Physics and Electrical Engineering, Faculty of Engineering and Environment, University of Northumbria, Newcastle upon Tyne, UK

Chapter Outline

16.1 Introduction 354
16.2 Single-phase voltage-source inverters 355
16.2.1 Basic operating mechanism of inverters 355
16.2.2 Single-phase half-bridge inverter 357
16.2.3 Single-phase full-bridge inverter 358
16.2.4 Phase-shift voltage control 359
16.3 Three-phase bridge voltage-source inverters 361
16.4 Multistepped Inverters 366
16.5 PWM inverters 368
16.5.1 Fundamentals of SPWM technique 369
16.5.2 Singe-phase SPWM inverter 370
16.5.2.1 Bipolar SPWM inverter 370
16.5.2.2 Unipolar SPWM inverter 372
16.5.3 Three-phase SPWM inverter 373
16.6 Current-source inverters 374
16.6.1 Single-phase current-source inverters 374
16.6.2 Three-phase current-source inverters 377
16.7 Summary 380
Problems 380
References 381

16.1 Introduction

Power converter is a kind of electronic circuits for energy conversion, which converts electrical energy of the supply into the energy suitable for the load (e.g., voltage or current with suitable frequency and/or amplitude). As one of the power converters, a DC-to-AC inverter transfers DC power to AC power. In terms of the types of power supply, inverters are categorized as voltage-source inverter and current-source inverter. As the term suggests, a voltage-source inverter utilizes a DC voltage as the supply, and the Thévenin equivalent resistance of the voltage source is ideally zero. On the other hand, a current-source inverter uses a DC current as the supply, and the Thévenin equivalent resistance of the current source is regarded ideally as infinity.

Voltage-source inverters are the second most common power converters, whose DC input voltage can be obtained either from a rectifier or a cell battery or a photovoltaics

Copyright © 2016 Elsevier Inc. All rights reserved.

array. The AC outputs after the DC-to-AC inversion can be single phase or multiphase. The single-phase and three-phase DC-to-AC converters are most common in practice. However, the development of the DC-to-AC inverters with more than three phase outputs has recently been stimulated by the construction of AC motors with more than three phases in order to improve the reliability of the motors under certain critical application scenarios. The most common waveforms of the outputs of the inverters are square wave, sinusoidal wave, or modified sinusoidal wave. PWM inverter techniques are commonly utilized, which can regulate the output AC voltage with suitable amplitudes and frequencies as well as reduce the harmonics by implementing multiple switching within the inverter with a constant DC input voltage. Voltage-source inverters have a wide scope of practical applications such as AC motor drive, AC uninterruptible power supply (UPS), active power filters, AC battery, induction heating, HVDC power transmission, etc.

Current-source inverters are typically supplied from controlled rectifiers. In order to obtain an ideal current source, a large inductor is normally connected to the DC side of the inverter for smoothing the current signal. The applications of current-source inverters can be found in high-power AC motor speed regulation, UPS systems, superconducting magnetic energy storage, etc.

There is a large number of literature addressing fundamentals of power electronic converters (e.g., see [1–9]) and their applications (e.g., see [10, 11]). In this chapter, DC–AC inverters are revisited. Specifically, the operation principles of voltage-source inverters including single-phase, three-phase, multistepped, and PWM inverters are illustrated in detail. Moreover, current-source inverters are addressed as well.

16.2 Single-phase voltage-source inverters

16.2.1 Basic operating mechanism of inverters

Here, by using the single-phase inversion circuit depicted by Figure 16.1, the operation scheme of the inverter is addressed. In Figure 16.1, V_d is the DC supply voltage, S_1–S_4 are ideal switches (S_1 and S_3 are the switches on the top, and S_2 and S_4 are the switches on the bottom).

When S_1 and S_4 are on, but S_2 and S_3 are off, the load voltage v_o is positive. On the contrary, when S_2 and S_3 are on, but S_1 and S_4 are off, the load voltage v_o is negative.

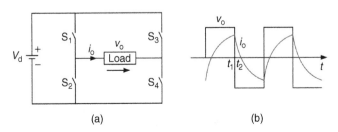

Figure 16.1 Operating scheme of an inverter. (a) Schematic diagram of an inverter, (b) waveforms of the output voltage and current.

The waveform of the load voltage is shown in Figure 16.1b, from which one can see that AC power is obtained by using the inversion circuit. In addition, when changing the switching frequency of the pair (S_1, S_4) and (S_2, S_3), the frequency of the AC power can be changed readily.

The waveform of the output current i_0 depends on the load. When the load is a pure resistive load, the current waveform and phase should be the same. However, when the load is RL load, the fundamental component of the current i_0 has a phase delay compared with the voltage v_0. In Figure 16.1, S_1 and S_4 are on before the instant t_1, and the output voltage v_0 and current i_0 are both positive. At the instant t_1, S_1 and S_4 are switched off, but S_2 and S_3 are switched on, and the output voltage v_0 becomes negative. However, due to the inductor in the load, the direction of the current i_0 cannot be changed immediately. The current i_0 flows out of the negative terminal of V_d, and flows into the positive terminal of V_d through the switches S_2, load, and S_3. As time goes by, the energy stored in the inductor reduces and the load current i_0 reduces gradually. At the instant t_2, the load current reduces to zero. After the time instant t_2, the load current i_0 turns negative and increases its amplitude. At this interval, the current i_0 flows out of the positive terminal of the source V_d, and flows back to the negative terminal through S_3, load, and S_2, and the inductor in the load stores energy.

A typical voltage-source inverter circuit is depicted by Figure 16.2 by using power semiconductor switches. In the past, SCRs were used as switches for high-power and medium-power inverters, which required commutating circuits to turn the SCRs off. At the present, fully-controlled power switches such as IGBTs (for medium-power inverters), GTOs, and IGCTs (for high-power inverters) are mostly utilized as power switches. In this chapter, fully-controlled power semiconductor switches are assumed to be employed for voltage-source inverters.

If the voltage source is not ideal, that is, the Thévenin resistance of the voltage source is not zero, a large capacitor can be connected to the voltage source in parallel. When the supply voltage is constant, the waveform of the AC output voltage is not subjected to the load; however, the waveform of AC output current and the phase are dependent on the load impedance. If the load is RL load, the AC load voltage and current may have opposite flow direction during some intervals. In each switch, there is a diode connected in parallel, which provides the path of the reactive power from the load to the DC source.

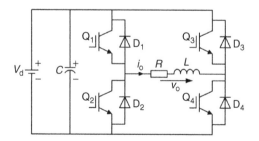

Figure 16.2 Typical DC–AC inverter.

16.2.2 Single-phase half-bridge inverter

A single-phase half-bridge inverter is one of the simplest inverters, shown by Figure 16.3, which is composed of a DC supply, two capacitors, two switches, two diodes, and load. In a switching period, denoted by T, the switches Q_1 and Q_2 are switched on in turn, respectively conducted for 180°. When Q_1 is on, but Q_2 is off, the load voltage v_o (or v_{AO}) is $0.5V_d$. However, when Q_1 is off, but Q_2 is on, the load voltage v_o (or v_{AO}) is $-0.5V_d$. For the pure resistive load, the output current i_o has the same waveform and phase as the output voltage v_o.

In the following, we look at how the circuit above works for an RL load. During the first half cycle with Q_1 on but Q_2 off, the output voltage is $0.5V_d$, and the load current i_o increases gradually. At the time instant t_2, trigger signals are sent to Q_1 and Q_2, respectively, in order to switch off Q_1 but switch on Q_2. At this instant, Q_1 is thus switched off; however, the switch Q_2 cannot be switched on immediately as the energy stored in the load inductor forces the diode D_2 to conduct so that the load current i_o maintains the same direction. At the time instant t_3, the energy stored in the load inductor is used up so that the current i_o is zero. At this instant, the diode D_2 turns off so that Q_2 can thus turn on by triggering. After the instant t_3, the current i_o changes direction and increases its amplitude as time goes by. At the time instant t_4, trigger signals are sent to Q_1 and Q_2, respectively, in order to switch on Q_1 but switch off Q_2. As a result, Q_2 is off, but Q_1 cannot be turned on immediately as the diode D_1 is forced to conduct by the energy stored in the inductor. Therefore, the current i_o flows through D_1 and goes back to the positive terminal of the supply voltage. Until t_5 the current i_o goes to zero so that Q_1 is on by triggering and the direction of the current i_o is changed readily.

When either Q_1 or Q_2 is on, the load voltage v_o and load current i_o have the same polarity, where the inverter is called on to operate in the active mode. In the active mode, the supply provides power to the load. When either D_1 or D_1 is conducted, the load voltage v_o and load current i_o have opposite polarity, where the inverter is called on to work in the feedback mode. In this mode, the power stored in the load inductor is sent back to the DC side. The fed-back reactive power is stored in the capacitors of the DC side. In order to absorb reactive power and keep the voltage constant at the "O point" in Figure 16.3, the capacitors C should be large enough. In addition, from

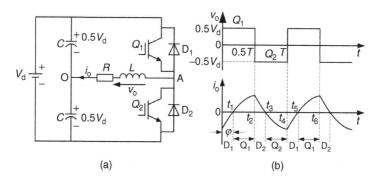

(a)   (b)

Figure 16.3 Single-phase half-bridge inverter and output waveforms. (a) Half-bridge inverter, (b) waveforms of the output voltage and current.

Figure 16.3b, one can see the fundament component of the load current i_o delays the fundament component of the load voltage v_o by $\phi°$. Actually, ϕ is the angle of the load impedance phasor.

The Fourier series of the output v_o after the half-bridge inversion can be expressed as:

$$v_o = \frac{2V_d}{\pi}\left(\sin(\omega t) + \frac{1}{3}\sin(3\omega t) + \frac{1}{5}\sin(5\omega t) + \cdots\right) \tag{16.1}$$

From Equation (16.1), the peak value and effective value (or rms value) of the fundamental component of the output v_o are $V_{o1m} = (2V_d/\pi) \approx 0.64V_d$ and $V_{o1} = (V_{o1m}/\sqrt{2}) = (\sqrt{2}V_d/\pi) \approx 0.45V_d$, respectively.

The single-phase half-bridge inverter has a simple circuit structure, which can really help understand the operation mechanism of the inverter, as more complex inverters such as single-phase full-bridge inverters and three-phase inverters can be regarded as the combination of a couple of single-phase half-bridge inverters.

16.2.3 Single-phase full-bridge inverter

A single-phase full-bridge inverter is depicted by Figure 16.4, where there are four power switches: Q_1–Q_4. The switch pairs (Q_1, Q_4) and (Q_2, Q_3) conduct in turn. The two terminals of the load are connected to the middle points of the left-hand leg and right-hand leg of the bridge circuit, respectively. The load considered is RL load with the impedance phase angle ϕ. Moreover, there are four diodes, D_1–D_4, which are employed to provide the paths for the load current driven by the stored energy in the load inductor. The period of a cycle is denoted by T.

The operation principle of the single-phase full-bridge inverter is illustrated as follows. During the interval $0 \leq t < t_1$, the switch pairs (Q_1, Q_4) and (Q_2, Q_3) are both off, but the diode pair (D_1, D_4) is forced on by the energy remaining in the load inductor.

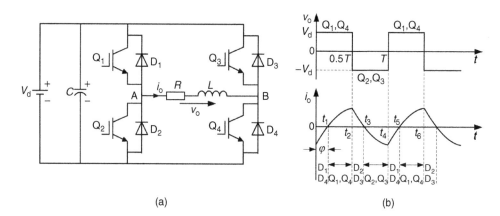

(a)   (b)

Figure 16.4 Single-phase full-bridge inverter and output waveforms. (a) Full-bridge inverter, (b) waveforms of the output voltage and current.

Therefore, at this time, the output voltage v_o is V_d, and the inductor current i_o reduces gradually in amplitude. At the time instant t_1, the load current i_o becomes zero so that the diodes D_1 and D_4 are off, but the switches Q_1 and Q_4 are switched on by triggering. Therefore, during the time interval $t_1 \leq t < t_2$, the voltage across the load is still $v_o = V_d$, but the direction of the current i_o is changed to positive. At the time instant t_2, the trigger signals are sent in order to switch off Q_1 and Q_4, but to switch on Q_2 and Q_3. The switches Q_1 and Q_4 are thus turned off immediately, but Q_2 and Q_3 cannot be turned on immediately as the energy stored in the load inductor forces the diodes D_2 and D_3 on. At this moment, the load voltage v_o becomes $-V_d$, but the load current i_o keeps the flow direction, but reduces its magnitude as times goes by. When the time reaches t_3, the load current reduces to zero so that the diodes D_2 and D_3 are off, but the switches Q_2 and Q_3 are on by triggering. Therefore, at the interval $t_3 \leq t < t_4$, the output voltage across the load is still $-V_d$, but the load current changes direction and increases its amplitude as time goes by. At the time instant t_4, the trigger signals switch off Q_2 and Q_3, but cannot turn on Q_1 and Q_4 immediately, as the energy stored in the load inductor forces the diodes D_1 and D_4 to turn on. Therefore, within the interval $t_4 \leq t < t_5$, the output voltage v_o is changed to V_d, but the load current keeps the previous direction but reduces its amplitude as time goes by. Actually, the inverter during $t_4 \leq t < t_5$ repeats the operation process of the inverter during the interval $0 \leq t < t_1$.

When either the switch pair (Q_1, Q_4) or (Q_2, Q_3) is turned on, the load voltage v_o and the load current i_o have the same polarity, which means the DC source provides the power to the load. On the other hand, when either the diode pair (D_1, D_4) or (D_2, D_3) is turned on, the load voltage v_o and the load current i_o have opposite polarity, which indicates the load feeds back the power to the DC side.

The Fourier series of the output voltage signal v_o can be given by:

$$v_o = \frac{4V_d}{\pi}\left(\sin(\omega t) + \frac{1}{3}\sin(3\omega t) + \frac{1}{5}\sin(5\omega t) + \cdots \right) \tag{16.2}$$

From (16.2), one can obtain the peak and effective values of the fundamental component of the output as $V_{o1m} = (4V_d/\pi) \approx 1.27V_d$ and $V_{o1} = (V_{o1m}/\sqrt{2}) = (2\sqrt{2}V_d/\pi) \approx 0.9V_d$.

16.2.4 Phase-shift voltage control

From (16.1) and (16.2), one can see that the output AC effective value is dependent on the DC input voltage. In other words, one can only adjust the AC output voltage by adjusting the DC input voltage, which is not convenient for the output regulation. One solution is to adjust the AC output voltage by modifying the trigger ways of the switches. This technique is called phase-shift voltage control. Here, we continue to discuss the single-phase full-bridge inverter depicted by Figure 16.4, but with shift-voltage control. The trigger signals of the switches Q_1–Q_4, denoted by $V_{G1} - V_{G4}$, are depicted by Figure 16.5. The trigger signals respectively for Q_1 and Q_2 (or for Q_3 and Q_4) have 180° phase difference, and the trigger signal for Q_3 (or Q_4) delays the trigger signal for Q_1 (or Q_2) by $\theta°$. The angle $\theta°$ is called the shift angle.

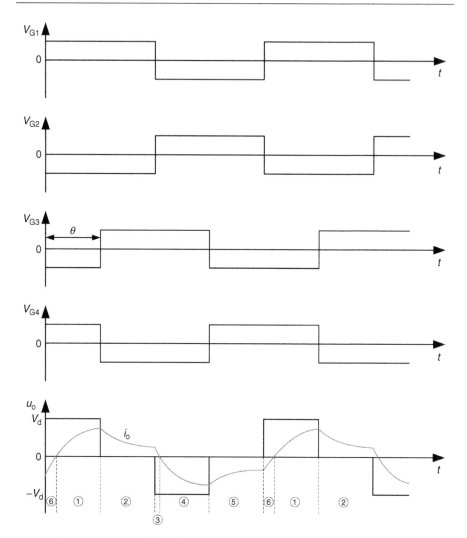

Figure 16.5 Single-phase full-bridge inversion by shift control and output waveforms .

From Figure 16.5, one can see there are six segments in a period. The operation mechanisms for these segments are summarized as follows:

1. Segment 1 (Q_1 and Q_4 conduct): Active mode with positive output voltage and positive output current. The energy from the DC side is converted to the load.
2. Segment 2 (Q_1 and D_3 conduct): Freewheeling mode with zero output voltage and positive output current.
3. Segment 3 (D_2 and D_3 conduct): Feedback mode with negative output voltage and positive output current. The energy from the load is sent back to the DC side.
4. Segment 4 (Q_2 and Q_3 conduct): Active mode with negative output voltage and negative output current. The energy from the DC side is converted to the load.

5. Segment 5 (Q_2 and D_4 conduct): Freewheeling mode with zero output voltage and negative output current.

6. Segment 6 (Q_1 and D_4 conduct): Feedback mode with positive output voltage and negative output current. The energy from the load is sent back to the DC side.

From the waveform shown in Figure 16.5, the output voltage is dependent on the phase shift $\theta°$. By regulating the control shift angle $\theta°$, one can regulate the output voltage. The Fourier series of the output voltage can be obtained as follows:

$$v_o = \sum_{n=1,3,5,\cdots} \frac{4V_d}{n\pi} \sin\left(\frac{n\theta}{2}\right) \cos(n\omega t). \tag{16.3}$$

From (16.3), one can see the peak value of the fundamental component is $a_1 = (4V_d/\pi)\sin(\theta/2)$. When $\theta = \pi$, the peak value of the fundamental component is $(4V_d/\pi) \approx 1.27V_d$, which actually reduces to the trigger way of the inverter described in Section 16.2.3.

In addition, when the load is a pure resistive load, one can obtain the same output voltage waveform as for an RL load. However, the four diodes never conduct for the resistive load. Moreover, in one period, there are only four segments:

1. Segment 1 (Q_1 and Q_4 conduct): Active mode with positive output voltage and positive output current.

2. Segment 2 (Q_1 and Q_3 conduct): The output voltage and current are both zero.

3. Segment 3 (Q_2 and Q_3 conduct): Active mode with negative output voltage and negative output current.

4. Segment 4 (Q_2 and Q_4 conduct): The output voltage and current are both zero.

16.3 Three-phase bridge voltage-source inverters

Three-phase bridge voltage-source inverters have been widely utilized for AC electric drive and general-purpose AC supplies, which are depicted by Figure 16.6. On the DC side, there is one capacitor (or a set of capacitors connected in series), which is connected to the DC source in parallel. Particularly in Figure 16.6a,b, one capacitor and two capacitors are respectively connected to the DC supply. In Figure 16.6b, the two capacitors are equal, that is, $C_1 = C_2 = 0.5C$. N', denotes the auxiliary middle-central point (or neutral point) of the DC supply voltage; therefore, the voltages across C_1 and C_2 are both $0.5V_d$.

In Figure 16.6b, there are six switches Q_1–Q_6, which constitute three bridge legs including the left-hand bridge leg (Q_1 and Q_4), middle bridge leg (Q_3 and Q_6), and right-hand bridge leg (Q_5 and Q_2). The switches at each bridge leg (or at each phase) cannot conduct at the same time, but are switched on in turn with a 180° condition period for each. The switches at the three bridge legs (or three phases) have 120° differences at the conduction time. Specifically, the condition time of Q_3 delays that of Q_1 by 120°, while the trigger time of Q_5 has 120° of delay compared with that of Q_3. Similar things happen to Q_4, Q_6, and Q_2.

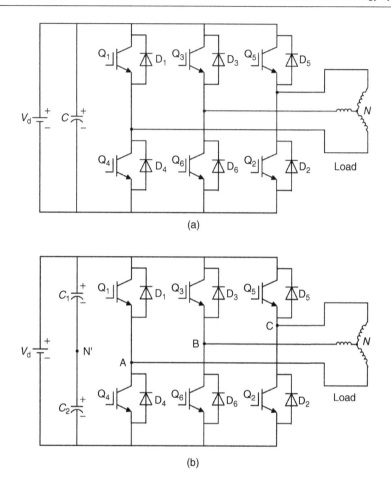

(a)

(b)

Figure 16.6 Three-phase voltage-source inverter. (a) Inverter with one capacitor in the
front end, (b) inverter with two capacitors in the front end.

When Q_1 and Q_4 are switched on in turn, the voltage of phase A, denoted by
$v_{AN'}$, is respectively $0.5V_d$ and $-0.5V_d$, leading to a square wave with the period
$360°$. When Q_3 and Q_6 are turned on in turn, the voltage of phase B, denoted by
$v_{BN'}$, has the same waveform as the $v_{AN'}$, but with $120°$ of phase delay. Similarly,
when Q_5 and Q_2 are conducted in turn, the waveform of the voltage of phase C,
denoted by $v_{CN'}$, delays the $120°$ compared with the waveform of $v_{BN'}$. The waveforms are shown in Figure 16.7a,b. It is clear that $v_{AB} = v_{AN'} - v_{BN'}$, $v_{BC} = v_{BN'}$
$- v_{CN'}$ and $v_{CA} = v_{CN'} - v_{AN'}$. Therefore, from the waveforms of $v_{AN'}$, $v_{BN'}$, and
$v_{CN'}$, the waveforms of the line voltages v_{AB}, v_{BC}, and v_{CA} can be readily obtained
(see Figure 16.7d–f).

If the neutral point N of the load is connected to the neutral point N' of the DC
supply, the three-phase load voltages are $v_{AN'}$, $v_{BN'}$, and $v_{CN'}$, respectively. However,

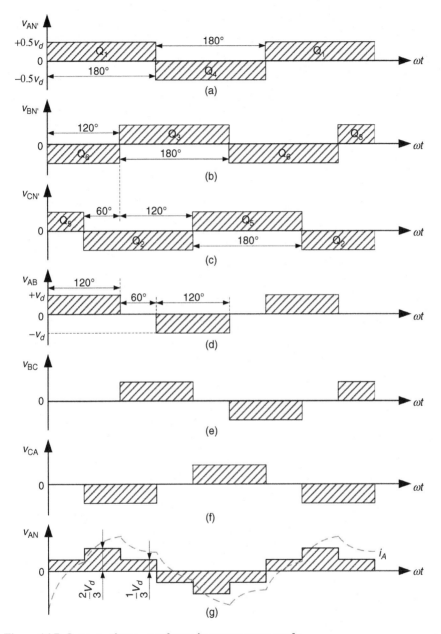

Figure 16.7 Output voltage waveforms in square-wave mode.

if the neutral points N and N' are isolated (e.g., they are isolated by the motor loads), the equivalent circuit is depicted by Figure 16.8. The triplet harmonics, that is, the zero-sequence component of the supply, appear across the points N and N'. Therefore, one has the formulae:

$$v_{AN'} = v_{AN} + v_{NN'} \tag{16.4}$$

$$v_{BN'} = v_{BN} + v_{NN'} \tag{16.5}$$

$$v_{CN'} = v_{CN} + v_{NN'}. \tag{16.6}$$

It is noted that $v_{AN} + v_{BN} + v_{CN} = 0$ for the three-phase balance loads. As a result, by adding (16.4–16.6), one can obtain:

$$v_{NN'} = \frac{1}{3}\left(v_{AN'} + v_{BN'} + v_{CN'}\right). \tag{16.7}$$

Substituting (16.7) into (16.4), (16.5), and (16.6) yields:

$$v_{AN} = \frac{2}{3}v_{AN'} - \frac{1}{3}v_{BN'} - \frac{1}{3}v_{CN'} \tag{16.8}$$

$$v_{BN} = \frac{2}{3}v_{BN'} - \frac{1}{3}v_{AN'} - \frac{1}{3}v_{CN'} \tag{16.9}$$

$$v_{CN} = \frac{2}{3}v_{CN'} - \frac{1}{3}v_{AN'} - \frac{1}{3}v_{BN'} \tag{16.10}$$

In terms of (16.8), (16.9), and (16.10), the waveforms of v_{AN}, v_{BN}, and v_{CN} can be drawn. The waveform of v_{AN} is shown in Figure 16.7g, which is a six-stepped wave. A typical current wave with an inductive load is also shown in Figure 16.7g.

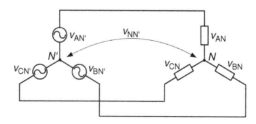

Figure 16.8 Equivalent circuit showing voltage v_{NN}, between the neural points .

The quantitative analyses of the output voltages of the three-phase bridge inverters are discussed as follows. The Fourier series of the phase output voltages with respect to the auxiliary neutral point N' can be given as:

$$v_{AN'} = \frac{2V_d}{\pi}\left(\cos\omega t - \frac{1}{3}\cos 3\omega t + \frac{1}{5}\cos 5\omega t\right) - \cdots \tag{16.11}$$

$$v_{BN'} = \frac{2V_d}{\pi}\left[\cos\left(\omega t - \frac{2\pi}{3}\right) - \frac{1}{3}\cos 3\left(\omega t - \frac{2\pi}{3}\right) + \frac{1}{5}\cos 5\left(\omega t - \frac{2\pi}{3}\right) - \cdots\right] \tag{16.12}$$

$$v_{CN'} = \frac{2V_d}{\pi}\left[\cos\left(\omega t + \frac{2\pi}{3}\right) - \frac{1}{3}\cos 3\left(\omega t + \frac{2\pi}{3}\right) + \frac{1}{5}\cos 5\left(\omega t + \frac{2\pi}{3}\right) - \cdots\right]. \tag{16.13}$$

The Fourier series of the line voltages can thus be given as follows:

$$v_{AB} = v_{AN'} - v_{BN'}$$
$$= \frac{2\sqrt{3}V_d}{\pi}\left[\cos\left(\omega t + \frac{\pi}{6}\right) + 0 - \frac{1}{5}\cos 5\left(\omega t + \frac{\pi}{6}\right) - \frac{1}{7}\cos 7\left(\omega t + \frac{\pi}{6}\right) + \cdots\right] \tag{16.14}$$

$$v_{BC} = v_{BN'} - v_{CN'}$$
$$= \frac{2\sqrt{3}V_d}{\pi}\left[\cos\left(\omega t - \frac{\pi}{2}\right) + 0 - \frac{1}{5}\cos 5\left(\omega t - \frac{\pi}{2}\right) - \frac{1}{7}\cos 7\left(\omega t - \frac{\pi}{2}\right) + \cdots\right] \tag{16.15}$$

$$v_{CA} = v_{CN'} - v_{AN'}$$
$$= \frac{2\sqrt{3}V_d}{\pi}\left[\cos\left(\omega t + \frac{5\pi}{6}\right) + 0 - \frac{1}{5}\cos 5\left(\omega t + \frac{5\pi}{6}\right) - \frac{1}{7}\cos 7\left(\omega t + \frac{5\pi}{6}\right) + \cdots\right] \tag{16.16}$$

From (16.14–16.16), the peak and effective values of the fundamental component of the line output voltage is $V_{AB1m} = ((2\sqrt{3}V_d)/\pi) = 1.1V_d$, and $V_{AB1} = (\sqrt{6}V_d/\pi) = 0.78V_d$, respectively. From (16.11–16.13), the peak and effective values of the fundamental component of the phase output voltage is $V_{AN'1m} = 2V_d/\pi = 0.637V_d$, and $V_{AN'1} = \sqrt{2}V_d/\pi = 0.45V_d$, respectively. In addition, the effective values of the line output voltages and phase output voltages are $V_{AB} = V_{BC} = V_{CA} = 0.816V_d$ and $V_{AN'} = V_{BN'} = V_{CN'} = 0.471V_d$, respectively. As a result, the line output voltages are $\sqrt{3}$ times of those of the phase output voltages. Moreover, the fundamental component of the line voltage leads the fundamental component of the phase voltage by 30°. From (16.14–16.16), one can see the characteristic harmonics in the waveform are $6n \pm 1$, where n is a nonzero integer. The three-phase fundamental as well as the harmonic components are balanced with a mutual phase difference by 120°.

If the switches at the same bridge arm (e.g., Q_1 and Q_4) are conducted at the same time, the short circuit on the DC supply side would happen. In order to avoid the unexpected situation, the trigger control system should ensure that one of the switches is off before turning on another switch. Therefore, a dead time should be set before triggering on one switch after triggering off another switch. The duration of the dead time is dependent on the switching speed. The faster the switching speed, the shorter the duration of the dead time. For the phase A bridge leg (or the left-hand bridge leg) in Figure 16.6, the trigger signals with dead time, denoted by V_{G1} and V_{G4}, respectively, for Q_1 and Q_4, are depicted by Figure 16.9. In terms of Figure 16.9, after Q_1 is triggered off, Q_4 is triggered on after a preset dead time. Similarly, after Q_4 is trigger off, Q_1 is triggered on after waiting for the duration of the dead time. Obviously, the similar trigger ways can be utilized for the other two bridge legs.

16.4   Multistepped Inverters

A multistepped inverter is an inverter with a multiple of six steps, for instance, 12, 18, 24, etc., so that the waveform of the output approaches a sine wave, which is desirable for large power applications. Since the output of the inverter can approach sinusoidal wave, the filter size can be reduced on both the DC and AC sides. It is noticed that the significant harmonics presenting in a multistepped waveform are $kn \pm 1$, where k is the number of the steps and n is an integer. For instance, the output wave of a six-stepped inverter includes 5th, 7th, 11th, 13th, 17th, 19th, 23rd, 25th, 29th, 31st, 35th, 37th, … harmonic components, while a 12-stepped inverter includes 11th, 13th, 23rd, 25th, 35th, 37th, … harmonics. As a result, some harmonics (e.g., 5th, 7th, 17th, 19th, 29th, 31st) appearing in the six-step inverters will disappear in the 12-stepped

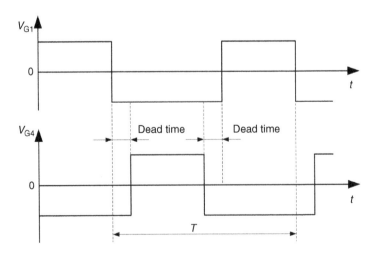

Figure 16.9 The trigger signals with dead time.

inverter. In consequence, the total harmonic distortion will be reduced significantly by using multistepped inverters.

In this section, a 12-step inverter, depicted by Figure 16.10, is discussed. In Figure 16.10, there are two three-phase inverters connected to the DC supply voltage in parallel. The second (or lower) inverter has a 30° angle delay for triggering compared with the first (or upper) inverter so that the waveform of the lower inverter has 30° phase shift. The AC output sides of the two inverters are connected to the primary windings of their respective transformers, where the winding ratios are shown in Figure 16.10. The phasor diagram of the voltages across the primary sides of the upper and lower transformers is shown in Figure 16.11a, from which one can see that the voltage across the lower transformer (e.g., $v_{d'e'}$) lags the voltage across the upper transformer (e.g., $v_{a'b'}$) by a 30° angle. The output phase voltages, obtained by the interconnection of the three secondary winding voltages, are shown by Figure 16.11b. Therefore, the phase A voltage can be given by $v_{AN} = v_{ab} + v_{de} - v_{ef}$, where $v_{de} = (v_{ab}/\sqrt{3})\angle - 30°$, and $v_{ef} = v_{de}\angle - 30° = (v_{ab}/\sqrt{3})\angle - 150°$. As a result, one has:

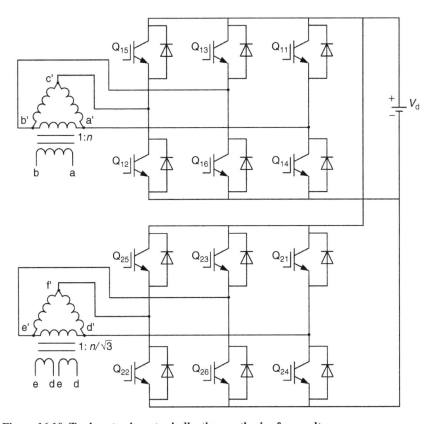

Figure 16.10 Twelve-step inverter indicating synthesis of v_{AN} voltage.

$$v_{AN} = \frac{4n\sqrt{3}V_d}{\pi}\left[\cos(\omega t)+\frac{1}{11}\cos 11(\omega t)+\cos 13(\omega t)+\cdots\right]$$

$$= \frac{4n\sqrt{3}V_d}{\pi}\left[\cos(\omega t)+\sum_{k=1}\frac{1}{k}(-1)^{12k\pm 1}\right] \tag{16.17}$$

where n is the phase turn ratio of the transformer.

The resulting waveform of the output is sketched in Figure 16.11c, which is the 12-step square wave. The effective value of the fundamental component of the output voltage is $V_{AN1} = (2n\sqrt{6}V_d/\pi)$. By adjusting the DC supply voltage V_d, one can regulate the output voltage.

Similarly, for an 18-stepped inverter, it is composed of three groups of three-phase inverters, which are connected in parallel to the DC voltage supply. The second and third inverters are phase shifted by 20° and 40°, respectively, with respect to the first inverter. The principle of an 18-stepped inverter, even a 24-stepped or a 48-stepped inverter, is similar to the 12-stepped inverter. The higher the multiple of the six-step, the lower the total harmonic distortion.

16.5   PWM inverters

The control of a three-phase six-stepped inverter is simple and the switching loss is low since only six-switching is needed during one cycle of the fundamental frequency. However, it has to be pointed out that the lower order harmonics of the six-step voltage wave create a large distortion of the current wave unless it is filtered by low-pass

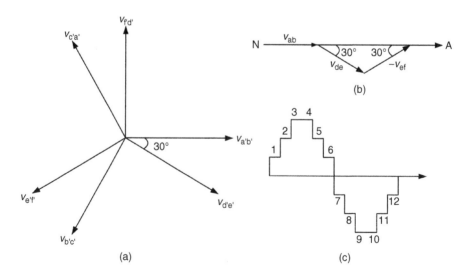

Figure 16.11 Phasors and the output waveform. (a) Phasor of transformer primary fundamental voltages, (b) phasor of three-phase output synthesis, and (c) voltage output.

filters or more than two six-stepped inverters are utilized to form multi-inverters. Extra filters and inverters will add extra economic cost unavoidably. Moreover, voltage regulation by controlling the line-side rectifier has the usual disadvantages. As a result, PWM inverter techniques are motivated and developed to control the output AC voltage as well as reduce the harmonics by implementing multiple switching within the inverter with a constant DC input voltage. There are a variety of PWM techniques such as sinusoidal PWM (SPWM), hysteresis band current control PWM, random PWM, space-vector PWM, etc. In this section, we will focus on the SPWM technique, since SPWM is very popular for industrial inverters.

16.5.1 Fundamentals of SPWM technique

As is well known, most of the DC–AC inverters are required to provide a clean sinusoidal voltage supply with a fixed or varying frequency, which is normally much lower than the switching frequency. The SPWM technique can provide a solution. The principle of the generation of the SPWM signal is depicted by Figure 16.12. Unlike the conventional PWM, the modulation signal, denoted by v_m is a sinusoidal signal rather than a DC constant signal. The modulation signal v_m and the high-frequency sawtooth carrier signal v_{cr} are sent to the comparator, and the points of intersection determine the switching frequency of power inverters. When $v_m < v_{cr}$, the output of the comparator is low DC constant voltage, while the comparator produces a high DC constant voltage when $v_m < v_{cr}$. As a result, a square-wave pulse signal is generated. The notch and pulse widths of the generated square wave vary in a sinusoidal manner so that the frequency of the fundamental component of the generated pulse signal is the same as the frequency of the modulation waveform. Therefore, one can regulate the amplitude and frequency of the output by changing the modulation waveform.

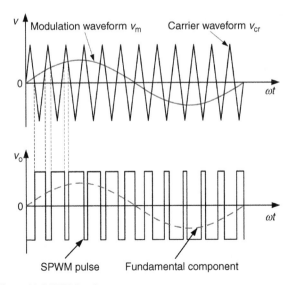

Figure 16.12 Sinusoidal PWM pulses.

The modulation index is defined as:

$$M = \frac{V_m}{V_{cr}}$$ (16.18)

where V_m and V_{cr} are the peak voltages of the modulation waveform and carrier waveform, respectively. Ideally, the modulation index M varies between 0 and 1. When $0 < M < 1$, one has the linear relationship $V_{m1} = MV_{DIN}$, where V_{m1} is the fundamental component of the output voltage and V_{DIN} is the DC supply voltage. Therefore, when M is high, the fundamental output voltage (sinusoidal wave) is high. The modulation ratio is defined as:

$$p = \frac{f_{cr}}{f_m}$$ (16.19)

where f_{cr} and f_m are the frequencies of the carrier and modulation waveforms, respectively. The output harmonics for SPWM are around the multiple of the carrier frequency, that is:

$$f = kf_{cr} = kpf_m$$ (16.20)

where k is a nonzero integer.

Figure 16.13 shows the harmonic amplitudes of the SPWM, which is redrawing of the PWM spectra figure of [7]. As the modulation index M (or depth of modulation) decreases, the amplitudes of fundamental output decreases, which is agreeable with the formula $V_{m1} = MV_{DIN}$. It is also noticed that the harmonics appear in clusters with main components at frequencies of $kpf_m, k = 1, 2, 3, \cdots$. The harmonics amplitudes vary as the modulation index M varies.

16.5.2   Single-phase SPWM inverter

The SPWM techniques can be categorized as bipolar SPWM and unipolar SPWM. First, let us look at the bipolar SPWM inverter.

16.5.2.1   Bipolar SPWM inverter

Consider the single-phase inverter with an RL load, which is triggered by the bipolar SPWM pulse signals shown by Figure 16.14. When $v_m > v_{cr}$, the positive pulse signals are generated, which are sent for triggering Q_1 and Q_2 on, but triggering Q_3 and Q_4 off. At this moment, if the load current i_o is in a positive direction, Q_1 and Q_2 will be turned on; otherwise, the diodes D_1 and D_2 will turn on. In both cases, the output voltage v_o equals V_d. When $v_{cr} > v_m$, the negative pulse signals are sent for triggering Q_3 and Q_4 on, but triggering Q_1 and Q_2 off. At this time instance, if the load current is in the reversed direction, Q_3 and Q_4 will be switched on; otherwise, the diodes D_3 and D_4 conduct. The load voltage v_o equals $-V_d$. The output waveform of the voltage is shown in Figure 16.15.

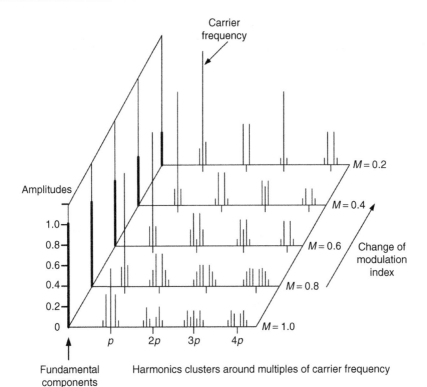

Figure 16.13 Harmonic amplitudes for SPWM. *M*, Modulation index and *p*, modulation ratio.

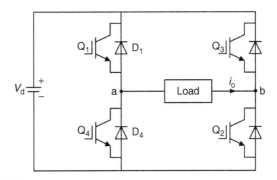

Figure 16.14 SPWM single-phase inverter.

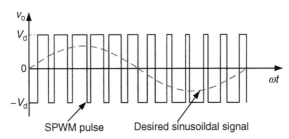

Figure 16.15 Output voltage of SPWM single-phase inverter.

16.5.2.2   Unipolar SPWM inverter

Different from the bipolar SPWM, the unipolar SPWM produces impulse signals with different polarities in positive and negative half cycles, respectively, depicted by Figure 16.16. v_m is the modulation signal with sinusoidal waveform, and v_{cr} is the sawtooth carrier signal with positive amplitude during the positive half cycle of the modulation signal, but negative amplitude during the negative half cycle of the sinusoidal modulation signal. During the positive half cycle, when $v_m > v_{cr}$, positive DC voltages are outputted; otherwise, zero outputs are given. During the negative half cycle, when $v_m > v_{cr}$, zero outputs are given; otherwise, negative DC constant outputs are given. The fundamental frequency of the generated pulse signal is the same as the frequency of the modulation signal.

Now let us look at how to use unipolar SPWM to regulate the single-phase inverter depicted by Figure 16.14. The switches Q_1 and Q_4 on the left-hand leg can be driven by a square wave synchronized with the modulation signal v_m and the switches Q_2 and Q_3 on the other leg are driven by the unipolar SPWM signal shown by Figure 16.16. During the positive half cycle, Q_1 is kept on, but Q_4 is off, while Q_3 and Q_2 are switched alternatively triggered by the SPWM impulses. Specifically, when $v_m > v_{cr}$ Q_2 is triggered on, but Q_3 is triggered off, so that the load voltage v_o equals V_d. When $v_m < v_{cr}$ Q_2 is triggered off, but Q_3 is triggered on, leading to $v_o = 0$. During the negative half cycle, Q_4 is kept on, but Q_1 is off, and the switches Q_2 and Q_3 are switched in turn. When $v_m < v_{cr}$ Q_3 is triggered on, but Q_2 is switched off, so that the load voltage v_o equals $-V_d$. When $v_m > v_{cr}$, Q_3 is turned off, but Q_2 turns on, therefore one

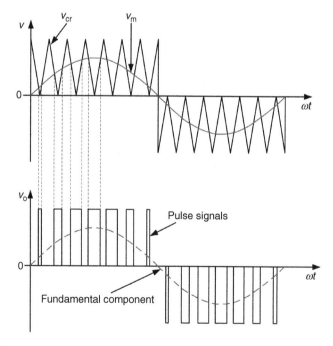

Figure 16.16 Unipolar SPWM pulse signals.

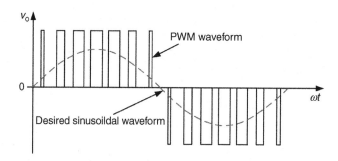

Figure 16.17 Output voltage of unipolar SPWM single-phase inverter.

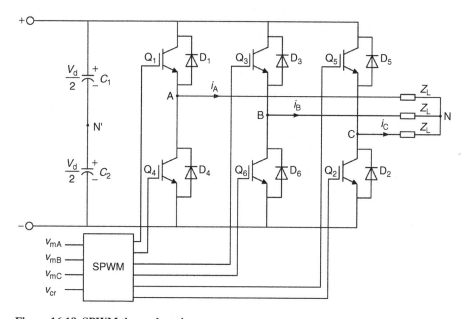

Figure 16.18 SPWM three-phase inverter.

has $v_o = 0$. Consequently one can have the waveform of the output voltage depicted by Figure 16.17.

16.5.3   Three-phase SPWM inverter

Three-phase SPWM inverters normally utilize bipolar SPWM control manners, which can be depicted by Figure 16.18. There are three sinusoidal modulation waveforms v_{mA}, v_{mB}, and v_{mC}, which displace each other by 120°. The common carrier waveform is a sawtooth signal, denoted by v_{cr}. For each leg, the trigger way is the same. Now, we take phase A to illustrate the operation principle. When $u_{mA} > v_{cr}$, Q_4 is triggered off, and Q_1 is sent a trigger-on signal. If the load current is negative, D_1 is on; otherwise, Q_1 is conducted. In both cases, one has $v_{AN'} = V_d/2$. Similarly, when $u_{mA} < v_{cr}$, Q_1

is trigged off, Q_4 (or D_4) is on, leading to $v_{AN'} = -(V_d/2)$. As a result, one can obtain the waveforms $v_{AN'}$, $v_{BN'}$, and $v_{CN'}$, as shown in Figure 16.19. In terms of $v_{AB} = v_{AN'} - v_{BN'}$, and $v_{AN} = v_{AN'} - (v_{AN'} + v_{BN'} + v_{CN'})/3$, one can sketch the waveform of the line voltage and phase voltage, respectively. One can see the phase voltage v_{AN} has five levels of voltage, composed of 0, $\pm V_d/3$, and $\pm 2V_d/3$; and the line voltage v_{AB} has three levels of voltage, including 0 and $\pm V_d$.

16.6 Current-source inverters

Ideally, DC supplies of current-source inverters are constant current sources with infinite Thévenin impedances. However, ideal current sources do not commonly exist in practice. Generally, a controlled rectifier with feedback loop and a DC link with sufficiently large inductance are utilized to produce an approximate ideal DC current source. Under the constant current supply, the waveforms of the AC output current are not affected by the load conditions. A typical current-source inverter is depicted by Figure 16.20, where the DC voltage can be regarded as the output after the AC-to-DC rectification. The large inductor L_d on the DC side is employed to smooth the current ripples. The freewheeling diodes become redundant for the current-source inverters, therefore, the current entering any leg of the inverter cannot change its polarity. As power semiconductor devices in current-source inverters must withstand reverse voltages, standard asymmetric voltage blocking devices such as power BJTs, power MOSFETs, IGBTs, MCTs, IGCTs, and GTOs cannot be utilized. Therefore, symmetric voltage blocking power semiconductor devices such as GTOs and SCRs (also called thyristors) should be used in the current-source inverters. Before analyzing single-phase and three-phase current-source inverter circuits with details, characteristics of the current-source inverters are given as follows:

1. A large inductor is connected to the DC side in series for smoothing current ripples. The current on the DC side is almost constant and the impedance of the DC side is high.
2. The waveform of the AC output current is square, which is independent of the load angle. However, the waveform of the AC output voltage and phase are subjected to the load angle.
3. For the RL load on the AC side, the DC source provides reactive power to the load. The DC side current does not change flow direction when the reactive power is fed back to the supply. Therefore, the freewheeling diodes become redundant in current-source inverters so that the size and weight are reduced, and reliability is improved.

16.6.1 Single-phase current-source inverters

A single-phase current-source inverter is depicted by Figure 16.21, where the DC source is provided by a bridge rectifier and is connected by a DC link inductor in series, and the RL load is connected by a capacitor in parallel. The purpose of the capacitor is to make the current of the effective load lead that of the voltage so that the load commutation of the thyristors can be realized. The circuit is generally utilized for high-frequency induction heating applications. It is assumed that the DC

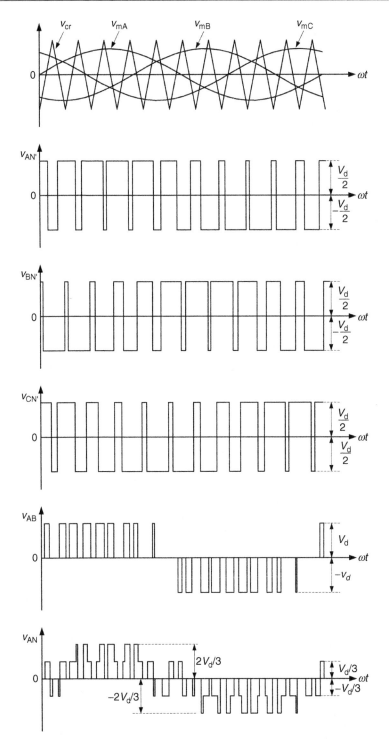

Figure 16.19 Waveforms of the SPWM three-phase inverter.

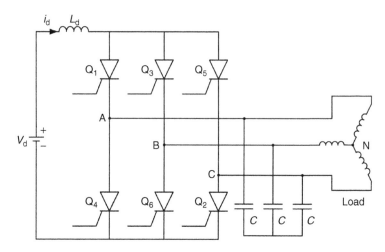

Figure 16.20 Typical current-source inverter.

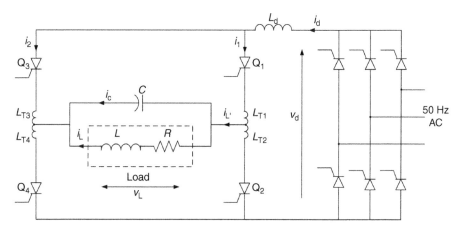

Figure 16.21 Single-phase current-source inverter.

link inductance is sufficiently large to smooth the DC current-source ripples and the capacitor has near perfect filtering of harmonic currents. The waveforms of load voltage and load current are shown by Figure 16.22a. The switch pairs (Q_1, Q_4) and (Q_2, Q_3) are switched alternately for an 180° angle to produce a square-wave current at the output. The fundamental component of the load current leads the nearly sinusoidal wave output by $\beta°$. When the pair (Q_1, Q_4) is switched on, the outgoing pair (Q_2, Q_3) is impressed with a negative voltage for the duration $\beta°$, leading to load commutation. As $\beta = \omega t_q$, the minimum value of β is sufficient to ensure that the thyristors are switched off during the time t_q. The phasor of the load voltage and current is depicted by Figure 16.22b. The RL load current I_L lags the load voltage V_L by the angle φ, and the load current I_L can be resolved into the active component I_P and the reactive

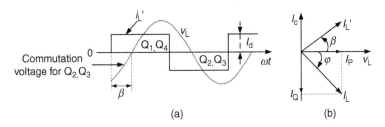

Figure 16.22 (a) Load voltage and current waveforms and (b) phasor.

Figure 16.23 The phase-locked-loop control.

component I_Q. The capacitor current I_C overcomes the reactive current I_Q so that the current of the effective load I_L' leads the load voltage V_L by β.

In order to obtain a desired β, one method is to adjust the capacitance of the capacitor C, while the alternative is to adjust the frequency ω of the inverter. Obviously, the latter is easiest to implement. One can make the inverter frequency ω slightly higher than the resonant frequency ω_r, where the effective load is considered as a parallel resonant circuit, so that the power factor $\cos\beta$ of the effective load becomes leading. The phase-locked-loop control can be utilized to regulate the frequency of the inverter. From Figure 16.23, one can see the β (the function of the inverter frequency) forms a feedback loop to compare with the expected β, denoted by β^*, so that the actual β can track β^*. For an induction heating-type load, a variation in frequency is of no concern. On the other hand, a constant marginal time t_β is desirable in a variable-frequency operation, instead of a constant marginal angle β. For a constant β angle, the marginal time t_β will increase when the frequency decreases. A marginal time t_β larger than necessary can cause unnecessary reactive power loading of the inverter.

16.6.2 Three-phase current-source inverters

The three-phase autosequential current-source inverter (ASCI) is depicted by Figure 16.24.

In Figure 16.24, the load is an induction motor, which can be approximately described by a per-phase equivalent circuit, consisting of a sinusoidal counterelectromotive force (CEMF) in series with an effective leakage inductance L. Under the stalled condition of the induction machine, the CEMF becomes zero so that the motor can ideally be considered as an inductive load. The thyristors Q_1–Q_6 are the principal switches, each of which conducts in sequence preferably for 120° angle, leading to the usual six-stepped current waveform. It is noticed that each thyristor is connected to a diode in series, and a bank of capacitors with equal values is connected to the upper

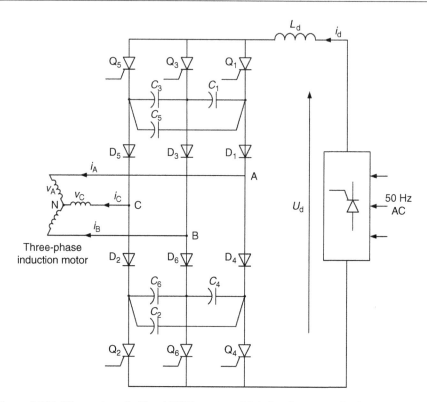

Figure 16.24 Three-phase bridge ASCI inverter with induction motor load.

and lower groups of the thyristors in delta forms, respectively. The diodes and capacitor banks constitute the forced commutation elements, where the capacitors store a charge with the correct polarity for commutation and the series diodes isolate the capacitors from the load. During normal operation, the upper and lower group devices operate independently, and six-time commutations are carried out per cycle of the fundamental frequency.

The equivalent circuit during the commutation from Q_2 to Q_4 is shown in Figure 16.25. All the other commutations are similar. When the incoming thyristor Q_4 is fired, the outgoing thyristor Q_2 is impressed with the reversed voltage across the capacitor bank so that the thyristor Q_2 is switched off almost instantaneously with the reverse voltage. Therefore, the DC current I_d flows through Q_3 and D_3, phases b and c of the machine, the diode D_2, the capacitor bank of the lower group, and Q_4 to the negative supply. The capacitor bank of the lower group is charged. During the charge for the capacitor bank by the constant current, there is no voltage drops across the load inductances. The charge continues until the voltages across the capacitors equal the line voltage v_{ca}. As a result, the diode D_4 conducts so that the current I_d is transferred to the D_4 completely and terminates the commutation process.

In terms of the formula $L(di_L/dt)$, a large voltage spike will be induced across each inductor during the current transfer duration, and the spike voltage will be added with

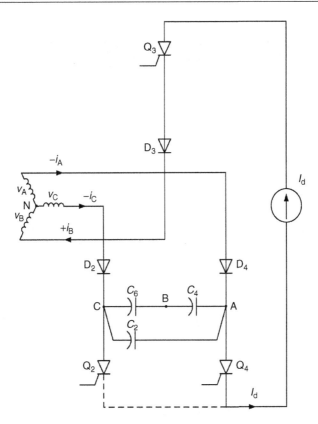

Figure 16.25 Equivalent ASCI circuit during commutation from Q₂ to Q₄.

the CEMF of the machine. Machine spike voltage is a serious problem, which can be attenuated by equipping a diode bridge circuit at the machine terminal with a Zener diode load. In addition, a low leakage inductance may be selected during the machine design so that the spike voltage can be reduced readily.

ASCI inverter-fed induction motor drives have medium or large capacity, which has been widely applied in a variety of industrial systems.

Multistepped current-source inverters can solve harmonic heating and torque pulsation problems caused by a six-stepped current-source inverter. Multistepped current-source inverters can be realized by connecting more than two six-stepped current-source inverters in series where there is a preset phase displacement between these six-stepped inverters. Moreover, PWM inverters can provide a powerful alternative to overcome the disadvantages of a six-step current wave such as harmonic heating, torque pulsation, and acoustic noise. Furthermore, the PWM current waves with reduced harmonic content can be further filtered by using a commuting capacitor bank to make the load current nearly sinusoidal. The PWM techniques for the current-source inverters are somewhat different from those for voltage-source inverters. The commonly used PWM techniques for current-source inverters include trapezoidal PWM and selected harmonic elimination PWM, which are omitted due to space limitations. This sub-chapter (current-resource inverter) was reviewed by referring to [1–11], particularly [2]. The readers can refer to [1–11] for more detailed materials about a variety of techniques of current-source inverters and their applications.

16.7 Summary

In terms of the types of DC sources, the DC–AC inverters can be classified into voltage-source inverters and current-source inverters. It is noticed that voltage-source inverters possess freewheeling diodes that provide pathways of reactive power from the load to the supply when the current and voltage across the RL load have opposite polarity. However, the freewheeling diodes are absent in the current-source inverters, which reduce the size and weight of the power circuit and improve the reliability of the inverter. Inverters can be made to operate either in square-wave mode or PWM mode. The operation of the square-wave mode inverter is simple with low power loss but high harmonic distortion. On the other hand, the operation of the PWM mode inverter can produce higher quality output by reducing harmonic distortion. The quality of output can be further improved by using filtering techniques such as an active harmonic filter (a static VAR compensator when the PWM frequency is sufficiently high), LC filters, capacitor banks, etc. DC–AC inverters have found broad applications in direct DC-to-AC conversion or indirect AC-to-AC conversion schemes, which play an important role in current renewable energy conversion.

In this chapter, the operation principles of voltage-source inverters, including single-phase half-bridge inverters, single-phase full-bridge inverters, three-phase bridge inverters, multistepped inverters, and PWM inverters, were reviewed in detail. Moreover, the current-source inverters including single and three-phase inverters were also reviewed.

Problems

1. Given a single-phase half-bridge voltage-source inverter with an RL load, sketch the waveforms of the load voltage and current, and explain the operation principle of the circuit. If the supply DC voltage is 100 V, calculate the peak and effective voltages of the fundamental component of the output voltage.

2. With the aid of the sketched output waveforms, explain the operation principle of a single-phase full-bridge voltage-source inverter with an RL load.

3. Explain briefly the concept of the phase-shift voltage control and the function of the freewheeling diodes in single-phase full-bridge voltage-source inverters.

4. Explain briefly the concept of the phase-shift voltage control and the function of the freewheeling diodes in single-phase full-bridge voltage-source inverters.

5. With the aid of the sketched output waveforms in square-wave mode, explain the operation principle of a three-phase full-bridge voltage-source inverter.

6. Calculate the effective values of the fundamental components of the phase and line output voltages of a three-phase bridge voltage-source inverter with 200 V DC supply. Calculate the effective voltages of the 5th and 7th harmonics of the phase output voltage in the inverter.

7. What would happen if the switches on the same bridge leg are turned on simultaneously? What measures can be taken to avoid this situation?

8. Explain why the total harmonic distortion of the output will be reduced significantly by using multistepped inverters (e.g., 12-stepped inverters).

9. Explain the operation principle of a 12-stepped inverter with the aid of the circuit, phasor, and output waveform.

10. Explain the merits and disadvantages of the PWM inverters compared with six-step inverters and multistepped inverters.

11. Describe the principle of the sinusoidal PWM with the aid of diagrams.

12. Briefly explain the concepts of the modulation index and modulate ratio and their relationships with the harmonic amplitudes by using the SPWM. Describe how to regulate the frequency and amplitude of the output voltage of the SPWM inverter.

13. Explain the operation principle of a single-phase full-bridge inverter with an RL load by using the bipolar SPWM technique.

14. Explain the operation principle of a single-phase full-bridge inverter by using the unipolar SPWM technique.

15. Explain the operation principle of a three-phase SPWM inverter with aid of the sketched waveforms of modulation signals, carrier signal, output phase, and line voltages.

16. Explain the load commutation technique utilized in the single-phase current-source inverter shown in Figure 16.21.

17. Explain the operating principle of the three-phase ASCI shown in Figure 16.24.

18. Explain how to reduce the machine spike voltage in a three-phase current-source inverter-driven motor system.

19. What is the function of the diodes that are connected to the switches in series in a current-source inverter?

20. Compare the characteristics of voltage-source inverters and current-source inverters.

References

[1] Bird B, King K, Pedder D. An introduction to power electronics. John Wiley & Sons: New York, USA; 1993.

[2] Bose BK. Modern power electronics and AC drives. Prentice Hall PTR: New Jersey, USA; 2002.

[3] Erickson R, Maksimovic D. Fundamentals of power electronics. Kluwer Academic Plenum Publisher: New York, USA; 2004.

[4] Lander C. Power electronics. McGraw-Hill Higher Education: London, UK; 1994.

[5] Mohan N, Robbins W, Undeland T. Power electronics: converters, applications and design. John Wiley & Sons: New York, USA; 2003.

[6] Rashid M. Power electronics: circuits, devices and applications. Prentice Hall: New Jersey, USA; 2014.

[7] Salam Z. Power electronics and drives. Johor Bahru: UTM; 2003.

[8] Trzynadloswski A. Introduction to modern power electronics. John Wiley & Sons: New York, USA; 2010.

[9] Wang ZA, Liu J. Power electronics technology. China Machine Press: Beijing; 2009.

[10] Zhong Q, Hornic T. Control of power inverters in renewable energy and smart grid integration. Wiley: Chichester, UK; 2013.

[11] Abu-Rub H, Malinowski M, Al-Haddad K. Power electronics for renewable energy systems, transportation and industrial applications. Wiley: Chichester, UK; 2014.

Electric power transmission

17

Miszaina Osman, Izham Zainal Abidin,
Tuan Ab Rashid Tuan Abdullah, Marayati Marsadek
College of Engineering, Universiti Tenaga Nasional, Jalan IKRAM-UNITEN,
Selangor Darul Ehsan, Malaysia

Chapter Outline

17.1   Introduction

The purpose of the transmission system in a power grid is to transmit electrical energy from the generating stations to the distribution networks. The transmission network also plays a role as an interconnection system with neighboring countries' power grids, which allows economic dispatch of power within regions during normal and emergency conditions.

Because of AC flowing in the conductors, transmission lines exhibit the electrical properties of resistance, inductance, capacitance, and conductance. Inductance and capacitance is a result of magnetic and electric fields around the current carrying conductors. Transmission line models can be developed with these two parameters. As for leakage currents flowing across the insulators and ionized pathways in the air, these are represented as shunt conductance in the circuit.

17.2 Overhead transmission lines

A transmission line consists of conductors, support structures, and other equipment such as insulators, spacers, jumpers, etc. A few examples of typical structures of a transmission line are poles, lattice structures, and H-frame structures. The conductors are hung from these towers or structures, which are usually made of steel, wood, or reinforced concrete. Most transmission lines are also installed with shield wires and lightning protection devices such as the surge arrestors. Voltages above 110 kV are usually considered to be the transmission level voltages. Voltages less than 33 kV are usually used for distribution, voltages above 230 kV are considered extra-high voltage, and voltages above 765 kV are usually referred to as ultra-high voltage.

The most commonly used conductor materials for high voltage transmission lines are ACSR (aluminum conductor steel-reinforced), AAC (all-aluminum conductor), AAAC (all-aluminum alloy conductor), and ACAR (aluminum conductor alloy reinforced). These types of conductors are widely used in transmission and some distribution systems due to their relatively low cost and high strength to weight ratio. Detail characteristics of these conductors can be found in manufacturers' data sheets. The conductors are stranded to increase their stability and flexibility and consist of a center core of steel strands surrounded by layers of aluminum strands. For purposes of heat dissipation, overhead power line conductors are bare (no insulating cover). ACSR conductors have long been widely used as overhead high tension power lines and have an established reputation for economy and dependability. Typical ACSR conductors and their standard sizes are shown in Figure 17.1.

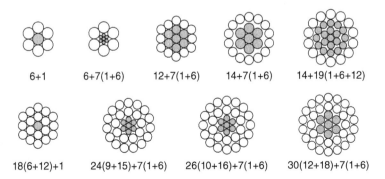

6+1   6+7(1+6)   12+7(1+6)   14+7(1+6)   14+19(1+6+12)

18(6+12)+1   24(9+15)+7(1+6)   26(10+16)+7(1+6)   30(12+18)+7(1+6)

Figure 17.1 Cross-sectional view of typical standard sizes and stranding pattern for ACSR conductors.

17.3 Transmission line parameters

For power system analysis, a particular transmission line can be represented by its resistance, inductance or inductive reactance, capacitance or capacitive reactance, and leakage resistance.

17.3.1 Line resistance

Resistance of the line will determine the performance of the line in terms of its efficiency and costing. The DC resistance of a solid round conductor at a specified temperature is given by:

$$R_{DC} = \frac{\rho l}{A} \; \Omega \tag{17.1}$$

where ρ is the conductor resistivity, l is the conductor length, and A is the conductor cross sectional area.

The conductor resistance is affected by three factors, which are frequency, spiraling, and temperature. When AC flows in a conductor, the phenomenon of skin effect happens. This is where the current distribution is not uniformly distributed over the conductor cross-sectional area, and the current density is greatest at the surface of the conductor. Thus, AC resistance is somewhat higher than DC resistance. Spiraling a stranded conductor will make the strand longer than the finished conductor, thus creating a higher AC resistance.

The conductor resistance increases as temperature increases. This change can be considered linear over the range of temperatures normally encountered and is calculated as follows:

$$R_2 = R_1 \frac{T + t_2}{T + t_1} \tag{17.2}$$

where R_2 and R_1 are conductor resistances at t_2 and t_1, respectively. T is a temperature constant and depends on the conductor material. For example, for hard-drawn copper T is 241, and for annealed copper T is 234.5.

17.3.2 Line inductance

17.3.2.1 Single-phase overhead lines

Figure 17.2 shows a single-phase overhead line, consisting of two solid round conductors with radius r and separated by a distance D.

Assume that the current flows out from conductor X and returns in conductor Y. The currents will result in magnetic field lines that link between the conductors. Thus, the inductance of the conductor is expressed as:

$$L = 0.2 \ln \frac{D}{D_s} \; \text{mH/km} \tag{17.3}$$

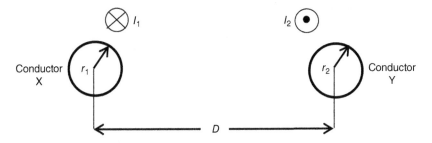

Figure 17.2 Single-phase with two wire line.

where D is the distance between the two conductors and D_s is the geometric mean radius (GMR).

The GMR for stranded conductors is usually provided by the manufacturers, as calculations for conductors with a large number of strands can be very tedious. For a solid cylindrical conductor, GMR is calculated as $re^{-1/4}$ or $0.7788r$.

17.3.2.2 Three-phase overhead lines

Practically, transmission lines are not able to maintain a symmetrical spacing between their conductors due to construction constraints. A three-phase line conductor with asymmetrical spacing is shown in Figure 17.3.

For a given conductor configuration, the average values of the inductance and capacitance can be found by representing them with equivalent equilateral spacing. Geometric mean distance (GMD) is the equivalent conductor spacing and it is calculated as:

$$\text{GMD} = \sqrt[3]{D_{ab}D_{bc}D_{ca}} \tag{17.4}$$

In practice, the conductors of a transmission line are transposed. The transposition is an operation of exchanging the conductor positions, and is usually carried out at the switching stations. Therefore, the average inductance per phase is:

$$L = 0.2\ln\frac{\text{GMD}}{D_s}\,\text{mH/km} \tag{17.5}$$

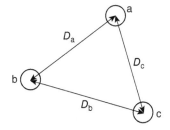

Figure 17.3 Three-phase line with asymmetrical spacing.

Typically, high voltage transmission lines are constructed with bundled conductors and consist of two, three, or four subconductors. The subconductors within the bundle are separated using spacer-dampers. The GMR calculations for the bundled conductors are as follows:

$$\text{GMR}_L = D_s^b = \sqrt{D_s \times d} \ (\text{for two-subconductor bundle}) \tag{17.6}$$

$$\text{GMR}_L = D_s^b = \sqrt[3]{D_s \times d^2} \ (\text{for three-subconductor bundle}) \tag{17.7}$$

$$\text{GMR}_L = D_s^b = 1.09 \sqrt[4]{D_s \times d^3} \ (\text{for four-subconductor bundle}) \tag{17.8}$$

17.3.3 Line capacitance

17.3.3.1 Single-phase overhead lines

In Figure 17.2, the conductors of the single-phase line have a radius r and are separated by a distance D. If conductors X and Y are carrying charges, the presence of the second conductor and ground will disturb the field of the first conductor. However, the charge is assumed to be uniformly distributed since the separation distance D is larger with respect to the radius r and the height of the conductors from ground is larger compared with D. The line-to-neutral capacitance is given by:

$$C = \frac{0.0556}{\ln \dfrac{D}{r}} \mu\text{F/km} \tag{17.9}$$

where D is the distance between the two conductors and r is the radius of the conductor.

17.3.3.2 Three-phase overhead lines

Similar to the inductance calculation, for capacitance of a three-phase line, the value of the capacitance will take into account the GMD of the three phases as follows:

$$C = \frac{0.0556}{\ln \dfrac{\text{GMD}}{r}} \mu\text{F/km} \tag{17.10}$$

In capacitance calculation, the effect of bundling will introduce an equivalent radius r^b. The r^b calculations for the bundled conductors are as follows:

$$r^b = \sqrt{r \times d} \ (\text{for two-subconductor bundle}) \tag{17.11}$$

$$r^b = \sqrt[3]{r \times d^2} \; \text{(for three-subconductor bundle)} \qquad (17.12)$$

$$r^b = 1.09\sqrt[4]{r \times d^3} \; \text{(for four-subconductor bundle)} \qquad (17.13)$$

Example 17.1

A 500-kV three-phase transposed transmission line is 100 km long and consists of one ACSR 1,272,000-cmil, 45/7 Bittern conductor per phase in a flat horizontal configuration, with a spacing of 10 m. The conductors have a diameter of 3.4160 cm and a GMR of 1.3560 cm. Calculate the inductance per phase in mH/km and the capacitance per phase in µF/km for the configuration.

Solution
From (17.4),

$$GMD = \sqrt[3]{D_{12}D_{23}D_{13}} = \sqrt[3]{(10)(10)(20)} = 12.5992\,\text{m}$$
$$r = \frac{3.4160}{2} = 1.708\,\text{cm} = 0.01708\,\text{m}$$

Using (17.5):

$$L = 0.2 \ln \frac{GMD}{D_s}\,\text{mH/km} = 0.2 \ln \frac{12.5992}{0.013560} = 1.3669\,\text{mH/km}$$

Using (17.10):

$$C = \frac{0.0556}{\ln \dfrac{GMD}{r}}\,\mu\text{F/km} = \frac{0.0556}{\ln \dfrac{12.5992}{0.01708}}\,\mu\text{F/km} = 0.00842\,\mu\text{F/km}$$

Example 17.2

The line in Example 17.1 is now replaced with a two-conductor bundle of ACSR 636,000-cmil, 24/7 Rook conductors per phase in a flat horizontal configuration, with a spacing of 10 m measured from the center of the bundles. The spacing between the conductors in the bundle is 45 cm. The conductors have a diameter of 2.4816 cm and a GMR of 1.0028 cm. Calculate the inductance per phase in mH/km and the capacitance per phase in µF/km for the configuration.

Solution

$$r = \frac{2.4816}{2} = 1.2408\,\text{cm} = 0.012408\,\text{m}$$

For inductance calculation from (17.6), $GMR_L = \sqrt{D_s \times d} = \sqrt{0.010028 \times (0.45)} = 0.0672\,\text{m}$

$$L = 0.2 \ln \frac{GMD}{GMR_L}\,\text{mH/km} = 0.2 \ln \frac{12.5992}{0.0672} = 1.0467\,\text{mH/km}$$

For inductance calculation from (17.11),

$$GMR_c = r^b = \sqrt{r \times d} = \sqrt{0.012408 \times 0.45} = 0.0747\,m$$

$$C = \frac{0.0556}{\ln\dfrac{GMD}{GMR_c}}\,\mu F/km = \frac{0.0556}{\ln\dfrac{12.5992}{0.0747}}\,\mu F/km = 0.0108\,\mu F/km$$

17.4 Transmission line representation

This section presents a transmission line representation using a π model. An alternative representation of the line using a T model is briefly discussed. The model is based on formulae to calculate the voltage, current, and power along a transmission line. In a transmission line, a set of values such as voltage and current are usually known at one end of the line. Those values are required to calculate other unknown electrical quantities at another point in the line. The model may be used to evaluate the design and operational performance of a power system such as line efficiency, losses, and limits of power flow over a line both under steady-state and transient conditions. It can also be used to study the effect of the parameters of the line on bus voltages and the flow of the power.

For discussion purposes, consider the representation of an electric system consisting of a generator supplying a balanced three-phase load Z_L through a power transmission line with a lump parameter $Z = R + j\omega L$. The three-phase electrical connection of this system is Y-connected. A schematic of the generator, transmission line, and load is shown in Figure 17.4. The line parameters are represented by lump parameters and the line capacitor is omitted.

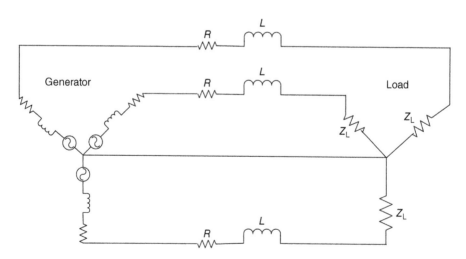

Figure 17.4 Generator supplying a balance three-phase load.

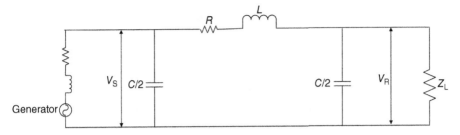

Figure 17.5 A single-phase equivalent of the circuit with capacitance to neutral added.

Normally, the transmission lines are operated in a balanced three-phase load. A single-phase equivalent circuit of the three-phase schematic can be used to simplify the discussion and calculation presented in this section. A single-phase equivalent circuit with shunt capacitors added at the sending end and receiving end of the transmission line is shown in Figure 17.5. The total capacitance of the line, C, is divided equally at the sending and receiving end and labeled as $C/2$. To differentiate the total series impedances of a line and the series impedance per unit length, the following symbols are used:

z = series impedance per unit length per phase,
y = shunt admittance per unit length per phase to neutral,
l = length of line,
$Z = zl$ = total series admittance per phase,
$Y = yl$ = total shunt admittance per phase to neutral,
$\gamma = \alpha + j\beta$ = a propagation constant. Its real part is called attenuation constant α measured in nepers per unit length. Its quadrature part is called phase constant β measured in radians per unit length,
Z_c = characteristic impedance of a line.

An equivalent circuit of a short-length transmission line is depicted in Figure 17.6, the sending and receiving end currents I_S and I_R are the same since there is no shunt arm. The V_S and V_R are the sending and receiving end line-to-neutral voltages. The line impedance is modeled as a lumped parameter impedance $Z = R + j\omega L$.

The formulae to calculate the currents and voltages for the short power transmission line are:

$$I_S = I_R \tag{17.14}$$

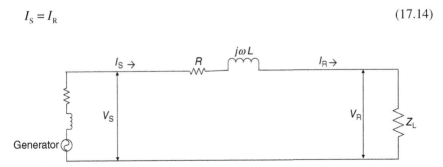

Figure 17.6 Equivalent circuit of a short transmission line.

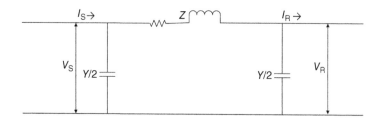

Figure 17.7 Nominal circuit of a medium-length transmission line.

$$V_S = V_R + I_R Z \tag{17.15}$$

Given a set of currents and voltages at one end of the line and the line parameters, the value of the currents and voltages at the other end can be calculated from these two equations, which modeled the short-length power transmission line.

An equivalent circuit of a medium-length transmission line is depicted in Figure 17.7. The sending and receiving end currents I_S and I_R are not necessarily the same since there are two shunt arms. The shunts arms represent the total admittance of the line divided into two equal parts and placed at the sending and receiving ends of the line. The circuit is called a nominal π. An expression for V_S can be represented by noting the capacitance at the receiving end $V_R Y/2$ and the current in the series arm is $I_R + V_R Y/2$. The formulae to compute the current and voltage are:

$$V_S = \left(V_R \frac{Y}{2} + I_R \right) Z + V_R \tag{17.16}$$

$$V_S = \left(\frac{ZY}{2} + 1 \right) V_R + Z I_R \tag{17.17}$$

Noting that the current in the shunt capacitance at the sending end is $V_S(Y/2)$ added to the current in the series arm gives I_S, thus:

$$I_S = V_S \frac{Y}{2} + V_R \frac{Y}{2} + I_R \tag{17.18}$$

Substituting V_S given by equation (17.17) into equation (17.18) gives:

$$I_S = V_R Y \left(1 + \frac{ZY}{4} \right) + \left(\frac{ZY}{2} + 1 \right) I_R \tag{17.19}$$

Corresponding equations can also be derived for the nominal T. In the latter case, all the shunt admittances of the line are lumped in the shunt arm of the T and the series impedance is divided equally between the two series arms.

In the previous two cases, lump parameters were used to represent the line constants in the short-length and medium-length power transmission lines. However, the nominal π model may not represent a long-length power transmission line exactly since the nominal model does not take into account the uniformly distributed parameters of the long-length line. Despite that, an equivalent circuit of a long line can be established by modifying the model for the medium-length transmission line.

For discussion purposes, denote the series equivalent π circuit Z' and the shunt arms $Y'/2$ to distinguish them from the nominal π parameters. Then, replace the respective Z and $Y/2$ in equation (17.17), thus obtaining the sending voltage as:

$$V_S = \left(\frac{Z'Y'}{2} + 1 \right) V_R + Z'I_R \tag{17.20}$$

We note that the generalized circuit constant for a long-length power transmission line is:

$$A = \cosh \gamma l \tag{17.21}$$

$$B = Z_c \sinh \gamma l \tag{17.22}$$

$$C = \frac{\sinh \gamma l}{Z_c} \tag{17.23}$$

$$D = \cosh \gamma l \tag{17.24}$$

And the line equations are:

$$V_S = AV_R + BI_R \tag{17.25}$$

$$I_S = CV_R + DI_R \tag{17.26}$$

Inspecting and equating the coefficients of Equations (17.20) and (17.25), we can deduce that:

$$Z' = Z_c \sinh \gamma l \tag{17.27}$$

$$Z' = \sqrt{\frac{z}{y}} \sinh \gamma l = zl \frac{\sinh \gamma l}{\sqrt{zl}\, l} \tag{17.28}$$

$$Z' = Z \frac{\sinh \gamma l}{\gamma l} \tag{17.29}$$

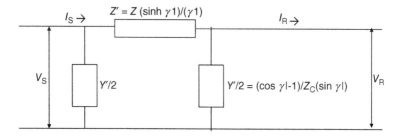

Figure 17.8 Equivalent π circuit of a long-length transmission line.

Recall that Z is equal to the zl, the total series impedance of the line. For the shunt arm of the equivalent π circuit, equate the coefficient of V_R in Equation (17.20) with circuit constant A in Equation (17.21), thus:

$$\frac{Z'Y'}{2} + 1 = \cosh \gamma l \qquad (17.30)$$

Use Equation (17.27) to eliminate Z' in Equation (17.30) and obtain:

$$\frac{Y'Z_c \sinh \gamma l}{2} + 1 = \cosh \gamma l \qquad (17.31)$$

Rearrange Equation (17.31) and obtain:

$$\frac{Y'}{2} = \frac{1}{Z_c} \frac{\cosh \gamma l - 1}{\sinh \gamma l} \qquad (17.32)$$

From series impedance Equation (17.29) and the shunts arms Equation (17.32), we can deduce the equivalent π circuit for the long-length power transmission line as shown in Figure 17.8.

17.5 Transmission line as a two-port network and power flow

A transmission line is very important in ensuring energy is transported from energy sources to loads. From a power system analysis point of view, transmission line models are very important to ensure better representation of transmission lines. In the previous section, detailed line models were derived. Since a detailed model is assumed to be the best, however, from a power system simulation point of view, simplification of models is recommended. In general, there are three different types of transmission line models based upon line distance. They are as follows:

1. Short line model (below 80 km)
2. Medium line model (80–250 km)
3. Long line model (above 250 km)

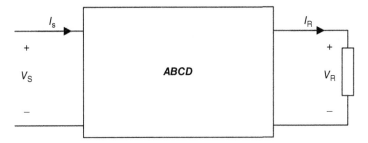

Figure 17.9 Two-port model.

17.5.1 The two-port model

The two-port model is a very simplified approach to determine the correct model parameters to be used for different types of transmission lines. The two-port model is shown in Figure 17.9.

The two-port model assumes that the transmission line model is a black box with sending end current (I_S) and voltage (V_S) at the input and the receiving end current (I_R) and voltage (V_R) at the output. Based upon Figure 17.9, the derived equations are as follows:

$$V_S = AV_R + BI_R \tag{17.33}$$

$$I_S = CV_R + DI_R \tag{17.34}$$

In matrix form:

$$\begin{bmatrix} V_S \\ I_S \end{bmatrix} = \begin{bmatrix} A & B \\ C & D \end{bmatrix} \begin{bmatrix} V_R \\ I_R \end{bmatrix} \tag{17.35}$$

17.5.2 Short line model

The short line model is reserved for lines that are less than 80 km in distance. This is because the impact of line capacitance is considered to be insignificant. Therefore the short line model is simply modeled with a unit series impedance (per km) multiplied by the total line distance. This is shown in Equation (17.36), and the circuit diagram representing the short line model is shown in Figure 17.6.

$$Z_{line} = (R + j\omega L)l_{line} = R + jX \ \Omega \tag{17.36}$$

where R is the per phase resistance per unit distance, L is the per phase inductance per unit distance, and l_{line} is the total line distance (usually in km).

The two-port model representation of a short line is as follows:

$$A = 1, B = Z, C = 0, D = 1 \tag{17.37}$$

17.5.3 Medium line model

For lines that are between 80 km and 250 km, the medium line model is used. This is because at this distance, shunt capacitance will have some impact on the overall impedance of the line. The circuit representation is shown in Figure 17.7. The two-port $ABCD$ model for medium lines is as follows:

$$A = \left(1 + \frac{ZY}{2}\right), \quad B = Z, C = Y\left(1 + \frac{ZY}{4}\right), \quad D = \left(1 + \frac{ZY}{2}\right) \tag{17.38}$$

17.5.4 Long line model

If the lines are longer than 250 km, distributed line parameters must be considered in order to determine the overall line impedance. The circuit representation is shown in Figure 17.10.

The two-port $ABCD$ model for long lines is as follows:

$$A = \cosh(\gamma l), \quad B = Z_c \sinh(\gamma l), \quad C = \frac{1}{Z_c}\sinh(\gamma l), \quad D = \cosh(\gamma l) \tag{17.39}$$

where $\gamma = \sqrt{zy}$ (propogation constant), $Z_c = \sqrt{\frac{z}{y}}$ (characteristic impedance), and $l =$ line length.

The previously mentioned model can be further simplified by assuming a lossless model where the two-port $ABCD$ model for this is as follows:

$$A = \cos(\beta l), \quad B = jZ_c \sin(\beta l), \quad C = j\frac{1}{Z_c}\sin(\beta l), \quad D = \cos(\beta l) \tag{17.40}$$

where $l =$ line length, $\beta = \omega\sqrt{LC}$ (phase constant), and $Z_c = \sqrt{\frac{z}{y}}$ (characteristic impedance).

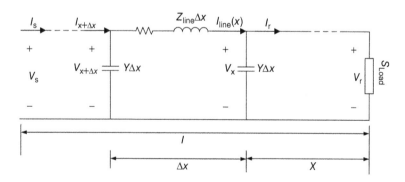

Figure 17.10 Long line model.

Example 17.3

A 500-kV, 60-Hz, 3-Ø line has a length of 180 km and delivers 1600 MW at 475 kV and at 0.95 power factor leading to the receiving end at full load. The series impedance (z) is given as 0.0201 + j0.335 Ω/Ø/km and the shunt admittance (y) is $j4.807 \times 10^{-6}$ S/Ø/km. Using the nominal π-circuit for the problem stated, determine the *ABCD* parameters for this model, sending-end voltage and current, sending-end power and power factor, full-load line losses and efficiency, and voltage regulation.

Solution
Using the medium line model:

$$A = D = 1 + \frac{YZ}{Z} = 1 + \frac{1}{2}\left(0.336 \times 180 \angle 86.6°\right)\left(4.807 \times 10^{-6} 180 \angle 90°\right)$$
$$= 0.9739 \angle 0.0912° \text{ pu}$$
$$B = Z = z\ell = 0.336(180) \angle 86.6° = 60.48 \angle 86.6° \, \Omega$$
$$C = Y\left(1 + \frac{YZ}{4}\right) = \left(4.307 \times 10^{-6} \times 180 \angle 90°\right)\left(1 + 0.0131 \angle 176.6°\right) = 8.54 \times 10^{-4} \angle 90.05° \text{ S}$$

To calculate the sending-end voltage and current:

$$V_R = \frac{475}{\sqrt{3}} \angle 0° = 274.24 \angle 0° \text{ kV}, \quad I_R = \frac{P_R \angle \cos^{-1}(pf)}{\sqrt{3} \times V_{R\text{-LL}(pf)}} = \frac{1600 \, \text{MVA} \angle \cos^{-1}(0.95)}{\sqrt{3} \times 475 \, \text{kV}(0.95)}$$
$$= 2.047 \angle 18.19° \text{ kA}$$
$$V_S = AV_R + BI_R = \left(0.9739 \angle 0.0912°\right)(274.24) + \left(60.48 \angle 86.6°\right)(2.047 \angle 18.19°)$$
$$= 264.4 \angle 27.02° \text{ kV}_{3\varnothing}$$
So, $V_S = \sqrt{3} \times 264.4 = 457.9 \text{ kV}_{3\varnothing}$
$$I_S = CV_R + DI_R = \left(8.54 \times 10^{-4} \angle 90.05°\right)(274.24) + \left(0.9739 \angle 0.0912°\right)(2.047 \angle 18.19° \text{ kA})$$
$$= 2.079 \angle 24.42° \text{ kA}$$

To calculate sending-end power and power factor:

$$P_S = \sqrt{3} \times V_{S\text{-}3\varnothing} I_S (pf) = \sqrt{3} \times 457.9(2.079)\cos(27.02° - 24.42°) = 1647 \, \text{MW}$$
$$pf = \cos(27.02° - 24.42°) = 0.999 \, \text{lagging}$$

Calculating full-load line losses and efficiency:
Full load line losses = $P_S - P_R = 1647 - 1600 = 47$ MW
$$\text{Efficiency} = \left(\frac{P_R}{P_S}\right)100 = \left(\frac{1600}{1647}\right)100 = 97.1\%$$
Voltage regulation is as follows:

$$V_{R\text{-NL}} = \frac{V_S}{A} = \frac{457.9}{0.9739} = 470.2 \text{ kV}_{3\varnothing}$$
$$\%V_R = \frac{V_{R\text{-NL}} - V_{R\text{-FL}}}{V_{R\text{-FL}}} = \frac{470.2 - 475}{475} \times 100\% = -1\%$$

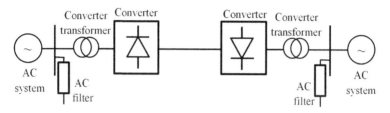

Figure 17.11 HVDC station.

17.6 High voltage DC transmission

The improvement and enhancement of power transfer capability is made possible with the advent of power electronics technologies. High voltage DC (HVDC) transmission connects two AC systems via a DC link and utilizes a power electronics-based converter. Figure 17.11 shows the typical arrangement of an HVDC system. Power transfer using HVDC offers the following advantages:

1. Fewer line losses
2. Less expensive for long distance transmission
3. Reduced system disturbance
4. Enables power transmission between two asynchronous AC systems
5. Power flow can be controlled

17.6.1 High voltage DC transmission converter

A converter is one of the main components in an HVDC system. The converter used in an HVDC station can either be a 6-pulse or 12-pulse converter. The typical circuit arrangement for a six-pulse converter is shown in Figure 17.12. In a six-pulse converter, the thyristor valve is fired every 60°. The average output voltage of the converter (i.e., V_d) can be varied by varying the firing delay angle α of the thyristor. The mode of operation of the converter is determined by α, which is summarized in Table 17.1. Figure 17.13 shows the output voltage of the converter when α is 0°, 30°, and 120°, where curves ab, ba, ac, ca, bc, and cb represent the line-to-line voltage. The average output voltage of the converter can be calculated as follows:

$$V_d = V_{do}\cos(\alpha) \tag{17.41}$$

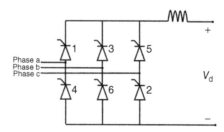

Figure 17.12 Six-pulse converter.

Table 17.1 **Mode of operation of a converter**

Range of α	Mode of operation	Remarks
$0° < \alpha < 90°$	Rectification	V_d has positive value
$\alpha = 90°$	–	V_d is zero
$90° < \alpha < 180°$	Inversion	V_d has negative value

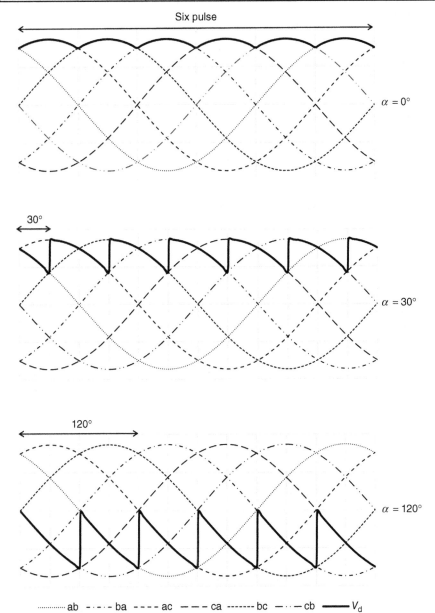

Figure 17.13 Output voltage of converter.

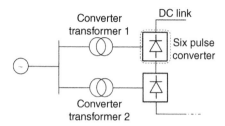

Figure 17.14 Simplified diagram of a 12-pulse converter.

where:

$$V_{do} = \frac{3\sqrt{3}}{\pi} V_m$$

In a 12-pulse converter, the thyristor valve is fired every 30° and it is obtained by arranging two 6-pulse converters in series. Figure 17.14 shows a simplified diagram of a 12-pulse converter.

17.6.2   AC and DC filter

The converters in an HVDC system introduce harmonics to the system. The harmonics generated by the converter are fed into the AC system and DC link. Harmonics must be sufficiently filtered otherwise they may cause problems like excessive power losses, over-voltage, and heating in the machine. From an AC system point of view, the converter acts as a source of current harmonics while in the DC link the converter acts as a source of voltage harmonics. AC and DC filter arms are installed on the AC system and DC link, respectively, in order to reduce the harmonics order that penetrates into the system. The functions and order of harmonics that exist in the HVDC system are summarized in Table 17.2.

17.6.3   Control of HVDC

The power flow in an HVDC system can be inherently controlled. The control strategies in an HVDC system should fulfill the following criteria:

1. The maximum DC current should be limited to 1.2 pu.
2. The maximum DC voltage should be maintained in order to reduce line losses.

Table 17.2 Filter in HVDC system

Type of filter	Function	Order of harmonics, n
AC filter	• To suppress the harmonics generated by the converter • To supply reactive power to the converter	$n = 6k \pm 1$ (for six-pulse converter) $n = 12k \pm 1$ (for 12-pulse converter) where $k = 1, 2, \ldots$
DC filter	• To suppress the harmonics generated by the converter	$n = 6k$ (for six-pulse converter) $n = 12k$ (for 12-pulse converter) where $k = 1, 2, \ldots$

Table 17.3 Typical HVDC control mode

Converter operation mode	Control mode	Description
Rectification	CC	• Rectifier maintains constant current by changing α
	CIA	• Rectifier operates with minimum firing delay angle (i.e., α_{min}) and no further voltage increase is possible
Inversion	CEA	• Inverter maintains constant extinction angle (i.e., γ)
	CC	• The CC control mode in the inverter is provided when there is a reduction in voltage that may cause the current to drop to zero

3. The converter should minimize reactive power consumption.
4. The minimum extinction angle should be maintained so as to avoid commutation failure.

The general control mode of a converter is summarized in Table 17.3. During normal voltage, the converter in the rectification process will operate in CC (constant current) control mode while the converter in the inversion process operates in CEA (constant extinction angle) control mode. In the event of disturbance where voltage at the rectifier side drops to a level at which the minimum firing delay angle has been reached and no further voltage increase is possible, the control mode of the converter in the rectification process switches from CC to CIA (constant ignition angle) control mode. When operating at this lower voltage, the DC current will reduce to a value within the current margin I_m and the converter in the inversion process operates in CC control mode. Figure 17.15 shows the $V_d - I_d$ characteristic of converters in an HVDC system. The intersection of control mode between rectifier and inverter signifies the operating point. The analysis of HVDC control

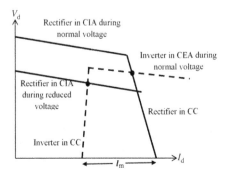

Figure 17.15 $V_d - I_d$ **characteristic of converters.**

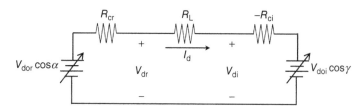

Figure 17.16 Simplified converter diagram.

mode can be carried out by considering the simplified converter circuit shown in Figure 17.16.

17.7 Summary

An electric transmission line has important parameters, namely resistance, inductance, and capacitance. These parameters will affect the electrical design and performance of the line. The line parameters are uniformly distributed along the whole line and are functions of line geometry, construction material, and operational frequency. These parameters and the load current and power factor will determine the electrical performance of the line.

Problems

1. A three-phase, 500-kV, 50-Hz, transposed line is composed of four ACSR 1,033,525, 54/7 Curlew conductors per phase with flat horizontal spacing of 14 m. The Curlew conductors have a diameter of 3.162 cm and a GMR of 0.5715 cm. The bundle spacing is 45 cm and the line is 260 km long. For the purpose of this problem, a lossless line is assumed.
   a. Based on the parameters given, calculate the *ABCD* constant.
   b. If the line delivers a load of 1500 MVA at 0.7 lagging power factor at 500 kV, determine the sending end quantities (V_S, I_S, and S_S (three-phase)) and voltage regulation.
   c. If the line now delivers a purely resistive load of 238 Ω, discuss what will happen to the voltage regulation and the receiving end current (I_R) as compared to load and condition in (b). Give reasons to your answer.
2. A three-phase transposed transmission line with voltage rating of 765 kV, 60 Hz is composed of four ACSR 1,431,000 cimil, 45/7 Bobolink-type conductors per phase. This configuration has a flat horizontal spacing of 14 m. The conductors have a diameter of 3.625 cm and a GMR of 1.439 cm. The bundle has a spacing of 45 cm. The length of the line is 400 km.
   a. From the structure of the transmission line conductors described earlier, calculate the inductance (L) in mH/km and the capacitance (C) in μF/km of the line.
   b. Using the long line lossless model, determine the transmission line *ABCD* constants.
   c. Determine the sending-end quantities, sending-end current (I_S), sending-end line voltage ($V_{S(LL)}$), sending-end power ($S_{S(3\emptyset)}$), and voltage regulation if this line delivered 2000 MVA at 0.8 lagging power factor at 735 kV.

3. What is the fraction in error on the calculation of the series impedance of the long-length power transmission line using the equivalent π model for the medium-length power transmission line as compared to the equivalent π model for the long-length power transmission line?

4. Show that the shunt arm of admittance for the equivalent π model for the long-length power transmission line is:

$$\frac{Y'}{2} = \frac{Y}{2}\frac{\tanh\left(\dfrac{\gamma l}{2}\right)}{\left(\dfrac{\gamma l}{2}\right)}$$

Hint: $\tanh\left(\dfrac{\gamma l}{2}\right) = \dfrac{\cosh\gamma l - 1}{\sinh\gamma l}$

5. Derive an equivalent T circuit model for the long-length power transmission line. Note: An equivalent T circuit model for the long-length power transmission line has all the shunt admittance of the line lumped in the shunt arm of the T and the series impedance is divided equally between the two series arms.

6. Figures 17.17 and 17.18 show the $V_d - I_d$ characteristics and equivalent circuit of converters in an HVDC station, respectively. The rectifier and inverter are initially operated with ignition

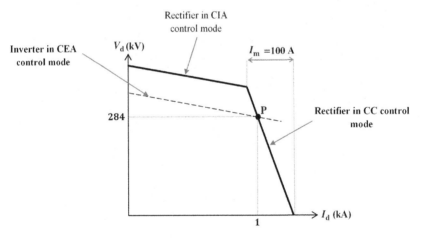

Figure 17.17 $V_d - I_d$ characteristic of converters used in Problem 6.

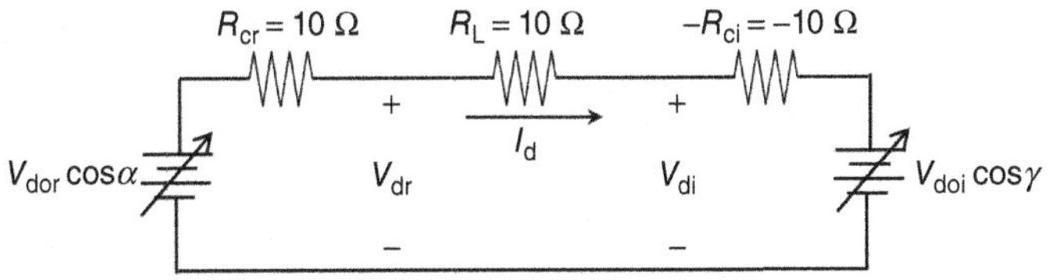

Figure 17.18 Simplified converter diagram.

delay angle and extinction angle of 20° and 15°, respectively. Point "P" is the converter's operating point at normal voltage. Answer the following questions:

a. If the converters operate at a normal voltage, determine the value of the rectifier no-load voltage V_{dor} and rectifier end voltage V_{dr} required to transmit a load at 1000 A. The rectifier ignition angle and inverter extinction angle is given as 15° and 18°, respectively.

b. What is the maximum reduction of voltage (in %) on the rectifier side in order for the converters to remain in the existing control mode if the minimum firing delay angle is given as 5°?

c. Describe what will happen to the control mode of each converter if further voltage reduction on the rectifier side occurred due to some disturbances.

Suggested readings

[1] Gonen T. Modern power system analysis. 2nd ed. Boca Raton, United States of America: CRC Press; 2013.

[2] Saadat H. Power system analysis. 2nd ed. International Edition. Singapore: McGraw Hill; 2004.

[3] Grainger J, Stevenson W. Power system analysis. International Edition. Singapore: McGraw Hill; 1994.

[4] Chapman SJ. Electric machinery and power system fundamentals. New York, United States of America: McGraw Hill; 2002.

[5] Electric Power Research Institute. Transmission line reference book – 345 kV and above. Palo Alto, CA: EPRI; 1979.

[6] Stevenson WD Jr. Elements of power system analysis. 4th ed. New York, United States of America: McGraw-Hill; 1982.

[7] Sluis LVD. Transients in power system. Chichester, United Kingdom UK: John Wiley & Sons Ltd; 2001.

[8] Arrillaga J, Smith BC, Watson NR, Wood AR. Power system harmonic analysis. Chichester, United Kingdom: John Wiley & Sons Ltd; 1997.

[9] Pavella M, Murthy PG. Transient stability of power systems – theory and practices. Chichester, United Kingdom: John Wiley & Sons Ltd; 1994.

[10] Arillaga J. High voltage direct current transmission. 2nd ed. London, United Kingdom: The Institution of Electrical Engineers; 1998.

Electric power systems

S. Vasantharathna

Department of Electrical and Electronics Engineering, Coimbatore Institute
of Technology, Coimbatore, Tamil Nadu, India

18

Chapter Outline

Electric Renewable Energy Systems

18.1   Introduction

Basic components of a power system are generators, transformers, transmission lines, and loads. The interconnection of the components discussed in section 18.2 in a power system is represented in a *one-line diagram* or *single-line diagram*. The advantage of a one-line diagram is its simplicity. One line in the single line diagram represents single phase or/and all three phases of the balanced system. Equivalent circuits of the components are replaced by their standard symbols and completion of the circuit neutral is omitted. Figure 18.1 shows the one line diagram of a typical power system network.

Figure 18.2 [1] shows different configurations of the single-phase transformers that are commonly used. But in a one-line diagram, it is always represented symbolically as in Figure 18.3 irrespective of whether it is a single-phase or a three-phase transformer. A generating station may have one or more generators and a pool of

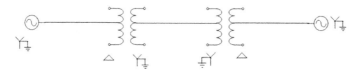

Figure 18.1 One-line diagram.

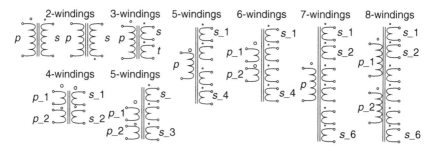

Figure 18.2 Winding representations of single-phase transformers.

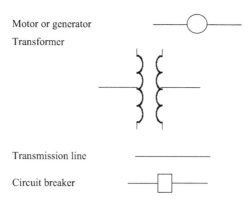

Figure 18.3 Symbolic representation of components in a single-line diagram.

generating stations synchronized as represented by a single circle in a one-line diagram. The generator, load, transmission line, and circuit breakers are represented in a single line diagram as shown in Figure 18.3. Figure 18.4 shows a model of a single line diagram of the power system network that can be modeled using a simulation tool (Figures 18.5 to 18.7).

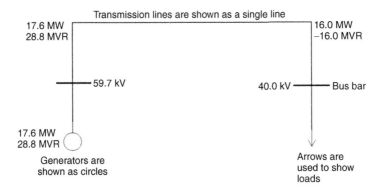

Figure 18.4 One-line diagram.

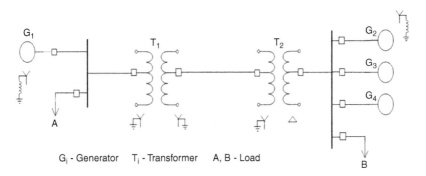

Figure 18.5 Single-line diagram.

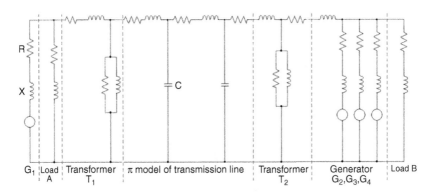

Figure 18.6 Impedance diagram.

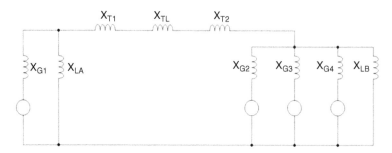

Figure 18.7 Reactance diagram.

An *impedance diagram* [4] is derived from the one-line diagram, representing the equivalent circuits of power system components. If resistances, static loads, and transmission line capacitances are neglected, it is known as a *reactance diagram*. The impedance diagram and reactance diagram is much helpful in load flow studies, fault studies and stability analysis of a power system network. The location of circuit breakers are not required for load flow studies.

The following assumptions are made while drawing the impedance and reactance diagram.

1. A generator can be represented by a voltage source in series with an inductive reactance. The internal resistance of the generator is negligible compared to the reactance.
2. The loads are inductive.
3. The transformer core is ideal and may be represented by a reactance.
4. The transmission line is a medium length line and can be denoted by a T or π circuit.

Problem 18.1

Generating station 1 is connected to a load *A* and is transmitting power through a transmission line. The receiving end of the transmission line is connected to load *B* and three other generating stations. Draw the one line diagram, impedance diagram, and reactance diagram.

Solution

The single line diagram of the given power system network is shown in Figure 18.5. The small squares represent the location of circuit breakers. The vertical lines are the bus. The impedance diagram shown in Figure 18.6 is derived by drawing the respective equivalent circuits. The resistance (R), inductive reactance (X_l) and capacitive reactance (X_c) are calculated based on standard formulae. The reactance diagram shown in Figure 18.7 is drawn by removing resistances and capacitive reactances as their impact on performance analysis is negligible.

18.1.1 Per unit quantities

Per unit quantities, like percentage quantities, are fractional quantities of a reference quantity used to reduce the computational complexity. Per unit values are written with "pu" after the value. For power, voltage, current, and impedance, a per unit quantity may be obtained by dividing the respective base or reference of that quantity.

$$pu \text{ quantity} = \frac{\text{Actual quantity}}{\text{Base quantity}}.$$

The pu representation of the quantities viz., complex power, voltage, current, and impedance, respectively are given as follows.

$$S_{pu} = \frac{S}{S_{base}}, \quad V_{pu} = \frac{V}{V_{base}}, \quad I_{pu} = \frac{I}{I_{base}}, \quad \text{and} \quad Z_{pu} = \frac{Z}{Z_{base}}.$$

Only two base or reference quantities need to be independently defined, because voltage, current, impedance, and power are related. The base quantities for the other two can be derived therefrom. Since power and voltage are most often specified, they are usually chosen to define the independent base quantities.

If VA_{base} and V_{base} are the selected base quantities of power (complex, active, or reactive) and voltage, respectively, then,

$$\text{Base current } I_{base} = \frac{V_{base}I_{base}}{V_{base}} = \frac{VA_{base}}{V_{base}}$$

$$\text{Base impedance } Z_{base} = Z_{base} = \frac{V_{base}}{I_{base}} = \frac{V_{base}^2}{I_{base}V_{base}} = \frac{V_{base}^2}{VA_{base}}.$$

In a power system, voltages and power are usually expressed in kilovolts (kV) and megavolt amperes (MVA), thus it is usual to select an MVA_{base} and a kV_{base} and to express them as,

$$\text{Base current } I_{base} = \frac{MVA_{base}}{kV_{base}} \text{ in kA}$$

$$\text{Base impedance } Z_{base} = \frac{kV_{base}^2}{MVA_{base}} \text{ in } \Omega.$$

In these expressions, all the quantities are single-phase quantities. In three-phase systems the line voltage and total power are usually used rather than the single-phase quantities. It is thus usual to express base quantities in terms of these.

If $VA_{3\Phi base}$ and V_{LLbase} are base three-phase power and line-to-line voltage, respectively, then

$$\text{Base current } I_{base} = \frac{MVA_{3\Phi base}}{\sqrt{3} \, kV_{LLbase}} \text{ in kA}$$

$$\text{Base impedance } Z_{base} = \frac{kV_{LLbase}^2}{MVA_{3\Phi base}} \text{ in } \Omega.$$

Problem 18.2

Given the actual and base quantities, express the following quantities in pu form.
Actual quantities are 20 A, 0.2 A, 50 V, 1000 V, and 2 Ω.
Base quantities are 10 A, 200 V, and 20 Ω.

$$I_{pu} = \frac{20}{10} = 2 \, pu.$$

$$I_{pu} = \frac{0.2}{10} = 0.02 \, pu.$$

$$V_{pu} = \frac{50}{200} = 0.25 \, pu.$$

$$V_{pu} = \frac{1000}{200} = 5 \, pu.$$

$$Z_{pu} = \frac{2}{20} = 0.1 \, pu.$$

Problem 18.3

In the circuit shown in Figure 18.8, consider the base quantities of voltage and impedance as $V_b = 100$ V; $Z_b = 0.01 \, \Omega$. Find I_b, I_{pu}, V_{pu}, Z_{pu}, and I.

$$Z = 0.01 + j0.01 \, \Omega$$
$$I_b = V_b/Z_b = 100/0.01 = 10^4 \, A$$
$$V_{pu} = 100/100 = 1 \, pu$$
$$Z_{pu} = 0.01 + j0.1/0.01 = 1 + j1 \, pu$$
$$I_{pu} = 1/1 + j1 = 0.5 - 0.5 \, pu.$$

Problem 18.4

Choosing a base MVA of 50 and a base kV of 33, find the pu value of 10 Ω resistance.

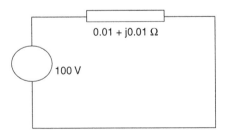

0.01 + j0.01 Ω

100 V

Figure 18.8 Figure for Problem 18.3.

$Z_b = 332/50 = 21.78\,\Omega$
$Z_{pu} = 10/21.78 = 0.45914\,\text{pu.}$

Problem 18.5

A three-phase, 13 kV transmission line delivers 8 MVA load. The per phase impedance of the line is 0.01 + j0.05 pu. What is the voltage drop across the line, when it is referred to a 13 kV, 8 MVA$_{base}$?
The given base quantities yield,

Base kVA = 8000 = 1 pu
Base kV = 13 = 1 pu.

Then the other base quantities are,

Base current = 8000/13
$\text{Base current} = \dfrac{8000}{13\sqrt{3}} = 355.292\,\text{A}$
$\text{Base impedance} = \dfrac{13000}{355.292} = 36.59\,\Omega$
Impedance = 36.59(0.01 + j0.05) = 0.3659 + j1.8295 Ω
Voltage drop = 355.292(0.3659 + j1.8295) = 130.001 + j650.01 = 662.88 V.

Conversions from one base to another: normally the per unit value is defined to its own rating. In a power system network, different components can have different ratings, and may be different from the system rating, therefore it is necessary to convert all quantities to a common base to perform numeric computations. Also if a new station is added/removed to/from a network, the reference quantities might get changed. Instead of recalculating the pu quantities based on new reference values, for all the systems, change of base is preferred. The conversion from one base to another in a system is as follows,

$$Z_{pu} = Z_{old}\,\frac{MVA_{base\,new}}{MVA_{base\,old}}\,\frac{kV^2_{base\,old}}{kV^2_{base\,new}}.$$

Problem 18.6

A 11 kV, 15 MVA generator has a reactance of 0.15 pu referred to its own ratings as base. The new bases chosen are 110 kV and 30 MVA. Calculate the new pu reactance:

$$Z_{pu} = 0.15 \times \frac{30}{15} \times \frac{11^2}{110^2} = 0.003\,\text{pu}$$

Problem 18.7

Three generators are rated as follows. Draw the reactance diagram.

$G1 = 100\,\text{MVA}, 33\,\text{kV}, X'' = 10\%$
$G2 = 150\,\text{MVA}, 32\,\text{kV}, X'' = 8\%$
$G3 = 110\,\text{MVA}, 30\,\text{kV}, X'' = 12\%$
Base $= 200\,\text{MVA}, 35\,\text{kV}$

$$X_{G1}\,\text{pu} = 0.1 \times \frac{200}{100} \times \frac{33^2}{35^2} = 0.1773\,\text{pu}$$

$$X_{G2}\,\text{pu} = 0.08 \times \frac{200}{150} \times \frac{32^2}{35^2} = 0.0892\,\text{pu}$$

$$X_{G3}\,\text{pu} = 0.12 \times \frac{200}{110} \times \frac{30^2}{35^2} = 0.1603\,\text{pu}.$$

The reactance diagram is shown in Figure 18.9.

Per unit representation of a transformer: Consider the equivalent circuit of the transformer shown in Figure 18.10.

Here, Z_p, leakage reactance on the primary side; Z_s, leakage reactance on the secondary side.

Transformation ratio $= 1:a$.

Choose VA_{base} and V_{base} on two sides of a transformer such that,

$$\frac{V_{1b}}{V_{2b}} = 1/a \quad \frac{I_{1b}}{I_{2b}} = a \quad Z_{1b} = \frac{V_{1b}}{I_{1b}} \quad Z_{2b} = \frac{V_{2b}}{I_{2b}}.$$

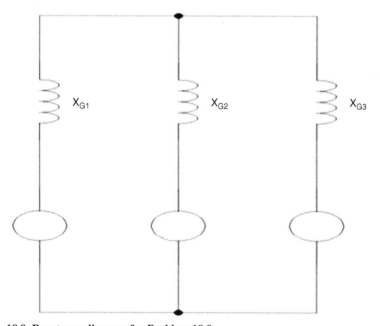

Figure 18.9 Reactance diagram for Problem 18.9.

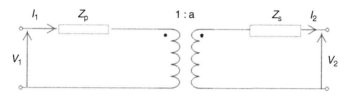

Figure 18.10 Equivalent circuit.

From Figure 18.10, it is written as $V_2 = (V_1 - I_1 Z_p)a - (I_2 Z_s)$.

In pu form, $V_{2pu}V_{2b} = [V_{1pu}V_{1b} - I_{1pu}I_{1b}Z_{ppu}Z_{1b}]a - I_{2pu}I_{2b}Z_{spu}Z_{2b}$.

Divide by V_{2b} throughout using the base relation $V_{2pu} = V_{1pu} - I_{1pu}Z_{ppu} - I_{2pu}Z_{spu}$.

Using the relations $\dfrac{I_1}{I_2} = \dfrac{I_{1b}}{I_{2b}} = a, \dfrac{I_1}{I_{1b}} = \dfrac{I_2}{I_{2b}} \rightarrow I_{1pu} = I_{2pu} = I_{pu}$, it is rewritten as,

$$V_{2pu} = V_{1pu} - I_{pu}Z_{pu}.$$

Where, $Z_{pu} = Z_{ppu} + Z_{spu}$.

Even from the primary or secondary side pu Z can be calculated.

On the primary side,

$$Z_1 = Z_p + \frac{Z_s}{a^2}$$

$$Z_{1pu} = \frac{Z_1}{Z_{1b}} = \frac{Z_p}{Z_{1b}} + \frac{Z_s/a^2}{Z_{1b}} = \frac{Z_p}{Z_{1b}} + \frac{Z_s}{Z_{1b}\,a^2}$$

$$\therefore Z_{1pu} = Z_{ppu} + Z_{spu} = Z_{pu}.$$

On the secondary side:

$$Z_2 = Z_s + a^2 Z_p$$

$$Z_{2pu} = \frac{Z_2}{Z_{2b}} = \frac{Z_s}{Z_{2b}} + a^2 \frac{Z_p}{Z_{2b}}$$

$$\therefore Z_{2pu} = Z_{spu} + Z_{ppu} = Z_{pu}.$$

Therefore, pu impedance of a transformer is the same whether computed from the primary or secondary side, as long as the voltage bases on two sides are the ratio of transformation.

Problem 18.8

A generating station is supplying power to a distant village 50 km away. The transmission is done on 110 kV transmission line. The generator is rated at 400 MVA giving the output at 11 kV and has a subtransient reactance of 20%. The load consists of motors running at 11 kV and rated for 60, 80, and 100 MVA. The subtransient reactance of motors is 18%. The transformer at the generating station is rated for 300 MVA with a leakage reactance of 10% and a voltage rating

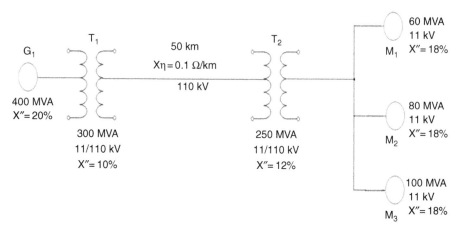

Figure 18.11 One-line diagram.

of 11/110 kV. The transformer at the village is rated for 250 MVA with a leakage reactance of 12% and a voltage rating of 110/11 kV. The reactance of the line is 0.1 Ω/km. Draw the reactance diagram of the system.

The one line diagram of the descripted power system network is shown in Figure 18.11. G1: 400 MVA, 11 kV selected as base.

Base value in the transmission line = 11 $\times$ (110/11) = 110 kV.

Base value in the motor = 110 $\times$ (11/110) = 11 kV.

$$Z_{pu} = Z_{old} \frac{MVA_{base\,new}}{MVA_{base\,old}} \frac{kV^2_{base\,old}}{kV^2_{base\,new}}.$$

Reactance of a transformer based on its own rating, converted to the common base quantity is,

$$X\,T1\,pu = 0.1 \times \frac{400}{300} \times \frac{11^2}{11^2} = 0.1333\,pu$$

$$X\,T2\,pu = 0.12 \times \frac{400}{250} \times \frac{110^2}{110^2} = 0.1920\,pu$$

$$X\,TL\,pu = 0.1 \times 50 \times \frac{400}{110^2} = 0.1652\,pu$$

$$X\,m1\,pu = 0.18 \times \frac{400}{60} \times \frac{11^2}{11^2} = 1.2\,pu$$

$$X\,m2\,pu = 0.18 \times \frac{400}{80} \times \frac{11^2}{11^2} = 0.9\,pu$$

$$X\,m1\,pu = 0.18 \times \frac{400}{100} \times \frac{11^2}{11^2} = 0.72\,pu$$

The reactance diagram is shown in Figure 18.12.

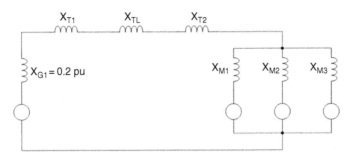

Figure 18.12 Reactance diagram.

18.2 Phases of power system engineering

The power system network includes generation, transmission, distribution, and utilization systems as shown in Figure 18.13. In the generation subsystem, the power plant produces the electricity. The transmission subsystem transmits the electricity to the

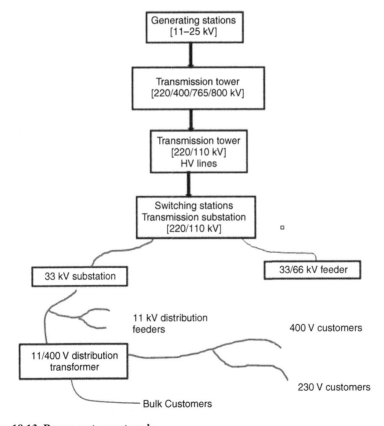

Figure 18.13 Power system network.

load centers. The distribution subsystem distributes the power to the customers. The utilization system is concerned with the different uses of electrical energy.

The *grid* is an electrical network that connects a variety of electric generators in different generating stations located remotely to the users of electric power. In high voltage transmission lines step-up transformers are used to reduce transmission power loss. Substations convert power to higher voltages before transmission and to lower voltages suitable for appliances after transmission [5].

18.3   Interconnected systems

Power generating stations are interconnected to improve the reliability of the supply, to reduce the reserve capacity required, and to improve the load factor, diversity factor, and the overall efficiency. More efficient plants are operated as base load plants and less efficient plants are operated as peak load plants, such that the generation cost and capital cost per kW are reduced. In the deregulated environment, interconnection of generating stations is being supported to enable the exchange of power among regions or countries and transport cheaper energy over long distances to the load centers. For this exchange of power through the interconnected grid, it must be ensured that all the generators run not only at the same frequency, but also at the same phase. A large failure in one part of the grid, if not quickly compensated, may lead to cascaded failures. Load dispatch centers facilitate communication between generating stations to maintain a stable grid. HVDC or variable frequency transformers can be used to connect two alternating current interconnection networks, which are not synchronized with each other, without the need to synchronize the wider area. However, there are technical and economical limitations in the interconnections, if the power has to be transmitted over extremely long distances.

18.3.1   Microgrids [2]

The need for a sustainable and reliable power supply and the availability of renewable energy sources lead to the development of the microgrid. Combinations of different but complementary energy generation systems based on renewable or mixed energy (renewable energy with a backup biofuel/biodiesel generator) are known as *renewable energy hybrid systems*. The grid formed by this system is known as a microgrid due to its size compared to the main grid thereby reducing the transmission and distribution losses.

A typical microgrid configuration is shown in Figure 18.14. It consists of electrical/ heat loads and microsources connected through a low-voltage distribution network. The microgrid is provided with systems to implement the control, metering, and protection functions during standalone and grid-connected modes of operation. In Figure 18.14, the microgrid consists of radial feeders to supply the electrical and heat loads, which can also be categorized under priority (that requires uninterrupted power supply) and nonpriority loads. Microsources and storage devices are connected to feeders through microsource controllers (MCs). The microgrid is coupled with the main medium voltage (MV) utility grid through the point of common coupling (PCC). A circuit breaker is operated to

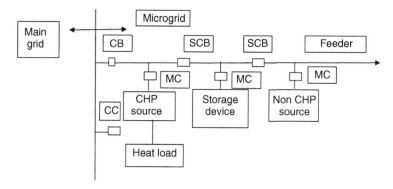

Figure 18.14 Typical microgrid configuration.

connect and disconnect the entire microgrid from the main grid, as per the selected mode of operation, such as, (1) grid connected and (2) standalone.

In *grid-connected mode*, the microgrid remains connected to the main grid either totally or partially, and imports or exports power from or to the main grid. The operation and management of the microgrid in different modes is controlled and coordinated through a local MC and the central controller (CC).

1. MC – The main function of the MC is to independently control the power flow and load end voltage profile of the microsource in response to any disturbance and load changes. The MC also participates in economic generation scheduling, load tracking/management and demand-side management by controlling the storage devices. The built-in control features of the MCs are:
   a. Active and reactive power control.
   b. Voltage control.
   c. Storage requirement for fast load tracking.
   d. Load sharing through power–frequency (*P–f*) control.
2. CC – Two main functional modules of CC are the energy management module (EMM) and the protection coordination module (PCM). The objectives of the CC are,
   a. to maintain a specified voltage and frequency at the load end through *P–f* and voltage control,
   b. to ensure energy optimization for the microgrid,
   c. to perform protection coordination, and
   d. to provide the power dispatch and voltage set points for all the MCs.
      - EMM provides the set points for active and reactive power output, voltage, and frequency to each MC.
      - PCM responds to the microgrid and main grid faults and loss of grid scenarios in a way so as to ensure correct protection coordination of the microgrid. It also adapts to the change in fault current levels during changeover from grid-connected to standalone mode.

The functions of CC in grid-connected mode are as follows.

1. Collects information from the microsources and loads connected to the microgrid and monitors system diagnostics.
2. Performs state estimation and security assessment evaluation, economic generation scheduling, active and reactive power control of the microsources, and demand-side management functions, by using collected information.

3. Ensures synchronized operation with the main grid maintaining the power exchange at prior contract points.

The functions of the CC in standalone mode are as follows.

1. Performs active and reactive power control of the microsources in order to maintain stable voltage and frequency at load ends.
2. Adopts load interruption/shedding strategies using demand-side management with storage device support to maintain power balance and bus voltage.
3. Initiates a local black start to ensure improved reliability and continuity of service.
4. Switches over the microgrid to grid-connected mode after main grid supply is restored, without hampering the stability of either grid.

Supervisory control design philosophy includes optimal dispatch control to provide P&Q set point with a slow time constant of 5–10 min and timeline control to manage voltage and power at the point of interconnection and to provide control set points for local control with a faster time constant of 10–100 ms. For the load flow analysis, bus admittance matrix need to be formulated. For fault studies, Bus impedance matrix need to be formulated.

18.3.2   Bus admittance matrix [5]

The bus admittance matrix is formulated to load flow analysis.
  Generator power: $S_{Gi} = P_{Gi} + jQ_{Gi}$ and load power: $S_{Li} = P_{Li} + jQ_{Li}$.
  Complex power: $S_i = S_{Gi} - S_{Li} = (P_{Gi} - P_{Li}) + j(Q_{Gi} - Q_{Li}) = P_{Li} + jQ_{Li}$.
  Consider the one-line diagram and its equivalent impedance diagram of a four-bus system shown in Figure 18.15. The circuit with common reference at ground potential is redrawn as shown in Figure 18.16.

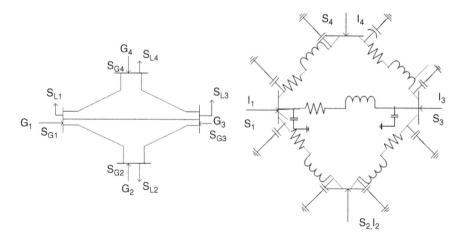

Figure 18.15 (a) One line diagram of a four-bus system. (b) Impedance diagram of a four-bus system.

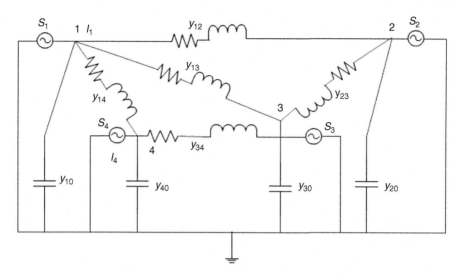

Figure 18.16 Equivalent circuit.

Applying KCL to four nodes:

$$I_1 = V_1 y_{10} + (V_1 - V_2)y_{12} + (V_1 - V_3)y_{12} + (V_1 - V_4)y_{14}$$
$$I_2 = V_2 y_{20} + (V_2 - V_1)y_{12} + (V_2 - V_3)y_{23}$$
$$I_3 = V_3 y_{20} + (V_3 - V_1)y_{13} + (V_3 - V_2)y_{23} + (V_3 - V_4)y_{34}$$
$$I_4 = V_4 y_{40} + (V_4 - V_1)y_{14} + (V_4 - V_3)y_{34}.$$

In matrix form, it is written as,

$$
\begin{bmatrix} I_1 \\ I_2 \\ I_3 \\ I_4 \end{bmatrix} =
\begin{bmatrix}
y_{10} + y_{12} + y_{13} + y_{14} & -y_{12} & -y_{13} & -y_{14} \\
-y_{12} & y_{20} + y_{12} + y_{23} & -y_{23} & 0 \\
-y_{13} & -y_{23} & y_{30} + y_{13} + y_{23} + y_{34} & -y_{34} \\
-y_{14} & 0 & -y_{34} & y_{40} + y_{14} + y_{34}
\end{bmatrix}
\begin{bmatrix} V_1 \\ V_2 \\ V_3 \\ V_4 \end{bmatrix}
$$

$$
\begin{bmatrix} I_1 \\ I_2 \\ I_3 \\ I_4 \end{bmatrix} =
\begin{bmatrix}
y_{11} & y_{12} & y_{13} & y_{14} \\
y_{21} & y_{22} & y_{32} & y_{42} \\
y_{31} & y_{32} & y_{33} & y_{34} \\
y_{41} & y_{42} & y_{43} & y_{44}
\end{bmatrix}
\begin{bmatrix} V_1 \\ V_2 \\ V_3 \\ V_4 \end{bmatrix}
\rightarrow I_{bus} = Y_{bus} \cdot V_{bus}
$$

where, y_{ii}, self-admittance or driving point – summation of all the admittances connected to the bus, y_{ip}, off diagonal – mutual or transfer admittance – negative of the connected admittance.

$$V_{bus} = Z_{bus} \cdot I_{bus} \rightarrow Z_{bus} = Y_{bus}^{-1}$$

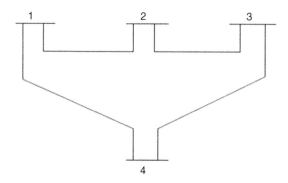

Figure 18.17 A fourbus system.

Problem 18.9

Formulate Y_{bus} for the four-bus system shown in Figure 18.17. Consider bus 4 as a reference bus. The shunt admittance at the buses is negligible. The line impedances are as shown here.

Line bus to bus	1–2	2–3	3–4	1–4
R in pu	0.025	0.02	0.05	0.04
X in pu	0.1	0.08	0.2	0.16

Solution

$$G_{bus} = \frac{R}{R^2 + X^2} \quad B = \frac{-X}{R^2 + X^2} \quad Y = G + jB$$

Line bus to bus	1–2	2–3	3–4	1–4
G in pu	2.35	2.94	1.176	1.47
B in pu	−9.41	−11.76	−4.706	−5.88

$$Y_{11} = Y_{12} + Y_{14} \quad Y_{22} = Y_{12} + Y_{23} \quad Y_{33} = Y_{23} + Y_{34}$$
$$Y_{12} = Y_{21} = -Y_{12} \quad Y_{23} = Y_{32} = -Y_{23}$$

$Y_{13} = Y_{31} = -Y_{13} = 0$ as there is no connection.

$$Y_{bus} = \begin{bmatrix} 3.82 - j15.29 & -2.35 + j9.41 & 0 \\ -2.35 + j9.41 & 5.29 - j21.17 & -2.94 + j11.76 \\ 0 & -2.94 + j11.76 & 4.116 - j16.466 \end{bmatrix}.$$

Problem 18.10

The parameters of a four-bus system are as shown below. Draw the network and find the bus admittance matrix.

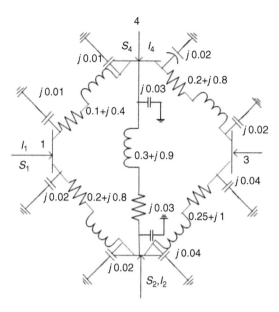

Figure 18.18 Impedance diagram of the four-bus system defined in Problem 18.11.

Bus code	Line impedance (pu)	Charging admittance (pu)
1–2	$0.2 + j0.8$	$j0.02$
2–3	$0.3 + j0.9$	$j0.03$
2–4	$0.25 + j1$	$j0.04$
3–4	$0.2 + j0.8$	$j0.02$
1–3	$0.1 + j0.4$	$j0.01$

Solution
The Figure 18.18 shows the network defined.

$$Y_{12} = \frac{1}{Z_{12}} = \frac{1}{0.2 + j0.8} = 0.294 - j1.176 \, \text{pu}$$

Shunt admittance: $y_{10} = j0.03$, $y_{20} = j0.09$, $y_{30} = j0.06$, $y_{40} = j0.06$ pu.

$$Y_{bus} = \begin{bmatrix} 0.882 - j3.498 & -0.294 + j1.176 & -0.588 + j2.352 & 0 \\ -0.294 + j1.176 & 0.862 - j3.026 & 0.333 + j1 & -0.235 + j0.94 \\ -0.588 + j2.352 & -0.333 + j1 & 1.215 - j4.468 & -0.294 + j1.1.76 \\ 0 & -0.235 + j0.94 & -0.294 + j1.176 & 0.529 - j2.056 \end{bmatrix}.$$

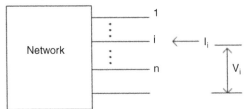

Figure 18.19 N-port network.

18.3.3   Bus impedance matrix [6]

Fault studies use an impedance matrix. Islanding can be detected by monitoring impedance changes. Consider the N-port network shown in Figure 18.19. Z_{ij} represents impedance between bus i and bus j.

$$\begin{bmatrix} V_1 \\ V_2 \\ V_i \\ V_n \end{bmatrix} = \begin{bmatrix} Z_{11} & Z_{12} & \cdots & Z_{1n} \\ Z_{21} & Z_{22} & \cdots & Z_{2n} \\ Z_{i1} & Z_{i2} & \cdots & Z_{in} \\ Z_{n1} & Z_{n2} & \cdots & Z_{nn} \end{bmatrix} \begin{bmatrix} I_1 \\ I_2 \\ I_i \\ I_n \end{bmatrix} \rightarrow V_{\text{bus}} = Z_{\text{bus}} \cdot I_{\text{bus}} \text{ and } Z_{ik} = Z_k \quad \text{diagonally symmetric.}$$

A step-by-step algorithm to build the Z_{bus} has the benefit of avoiding recomputation of Z_{bus}, even if there is an addition/removal of a bus in a network. The four modifications that are possible with the existing Z_{bus} are as follows.

1. Adding self-impedance Z_s from a new bus to reference.
2. Adding Z_s from a new bus to an old bus.
3. Adding Z_s from an old bus to reference.
4. Adding Z_s between old buses.

Type 1 modification – Addition of a tree branch Z_s from a new bus q to reference as shown in Figure 18.20:

$$\begin{bmatrix} V_1 \\ V_2 \\ V_i \\ V_n \\ V_q \end{bmatrix} = \begin{bmatrix} z_{11} & z_{12} & \cdots & z_{1n} \\ z_{21} & z_{22} & \cdots & z_{2n} \\ z_{i1} & z_{i2} & \cdots & z_{in} \\ z_{n1} & z_{n2} & \cdots & z_{nn} \\ 0 & 0 & 0 & z_s \end{bmatrix} \begin{bmatrix} I_1 \\ I_2 \\ I_i \\ I_n \\ I_q \end{bmatrix}$$

$$Z_{iq} = Z_{qi} = 0, Z_{qq} = Z_s$$

$$[z_{\text{bus}}]_{\text{new}} = \begin{bmatrix} [z_{\text{bus}}]_{\text{old}} & 0 \\ 0 & z_s \end{bmatrix}$$

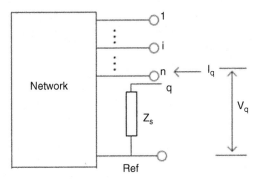

Figure 18.20 Type I modification.

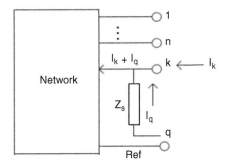

Figure 18.21 Type II modification.

Type 2 modification – Addition of a tree branch Z_s from a new bus q to an old bus k as shown in Figure 18.21:

$$V_q = Z_s I_q + V_k$$
$$= Z_s I_q + Z_{k1} I_1 + Z_{k2} I_2 + \cdots + Z_{kk}(I_k + I_q) + \cdots + Z_{kn} I_n$$
$$= Z_{k1} I_1 + Z_{k2} I_2 + \cdots + Z_{kk} I_k + \cdots + Z_{kn} I_n + (Z_{kk} + Z_s) I_q.$$

Similarly,

$$V_1 = Z_{11} I_1 + Z_{12} I_2 + Z_{13} I_3 + \cdots + Z_{1k}(I_k + I_q) + \cdots + Z_{1n} I_n$$
$$= Z_{11} I_1 + Z_{12} I_2 + \cdots + Z_{1k} I_k + Z_{1n} I_n + Z_{1k} I_q.$$

Similarly,

$$V_2, V_3, \ldots V_n$$

can be expanded and in matrix form.

$$
\begin{bmatrix} V_1 \\ V_2 \\ \vdots \\ V_k \\ \vdots \\ V_n \\ V_q \end{bmatrix} = \begin{bmatrix} z_{11} & z_{12} & \cdots & z_{1n} \\ z_{21} & z_{22} & \cdots & z_{2n} \\ \vdots & \vdots & \vdots & \vdots \\ z_{i1} & z_{i2} & \cdots & z_{in} \\ \vdots & \vdots & \vdots & \vdots \\ z_{n1} & z_{n2} & \cdots & z_{nn} \\ 0 & 0 & 0 & z_s \end{bmatrix} \begin{bmatrix} I_1 \\ I_2 \\ \vdots \\ I_k \\ \vdots \\ I_n \\ I_q \end{bmatrix}
$$

$$
[z_{bus}]_{old} = \begin{bmatrix} & & & & & z_{1k} \\ & & [z_{bus}]_{old} & & & z_{2k} \\ & & & & & z_{kk} \\ & & & & & z_{nk} \\ \hline z_{k1} & z_{k2} & z_{kk} & z_{kn} & & z_{1k+}z_s \end{bmatrix}
$$

Type 3 modification – Addition of a link Z_s between an old bus k and reference, has been shown in Figure 18.22. This is an extension of type 2 modification. If node q is eliminated, it results in type 3 modification.

$$
\begin{bmatrix} V_1 \\ V_2 \\ \vdots \\ \dfrac{V_n}{0} \end{bmatrix} = \begin{bmatrix} & & & z_{1k} \\ & [z_{bus}]_{old} & & z_{2k} \\ & & & \vdots \\ & & & z_{nk} \\ \hline z_{k1} & z_{k2} & \cdots & z_{kn} & z_{kk+}z_s \end{bmatrix} \begin{bmatrix} I_1 \\ I_2 \\ \vdots \\ \dfrac{In}{I_q} \end{bmatrix}
$$

If I_q is eliminated the last row and last column are eliminated resulting in,

$$
[z_{bus}]_{new} = [z_{bus}]_{old} - \frac{1}{z_{kk} + z_s} \begin{bmatrix} z_{1k} \\ z_{2k} \\ \vdots \\ z_{nk} \end{bmatrix} \begin{bmatrix} z_{k1} & z_{k2} & \cdots & z_{kn} \end{bmatrix}
$$

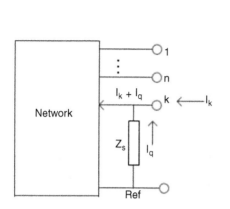

Figure 18.22 Type III modification.

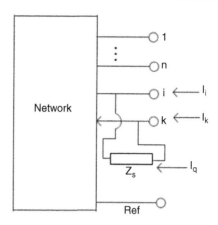

Figure 18.23 Type IV modification.

Type 4 modification – Addition of a link Z_s between two old buses i and k are shown in Figure 18.23:

$$V_1 = Z_{11}I_1 + Z_{12}I_2 + Z_{1i}(I_i + I_q) + Z_{1j}I_j + Z_{1k}(I_k - I_q) + \dots Z_{1n}I_n$$
$$V_1 = Z_{11}I_1 + Z_{12}I_2 + \dots + Z_{1n}I_n + I_q(Z_{1i} - Z_{1k})$$
$$V_k = Z_sI_q + V_i.$$

Expanding the equation for V_k:

$$Z_{k1}I_1 + Z_{k2}I_2 + \dots + Z_{ki}(I_i + I_q) + Z_{kj}I_j + Z_{kk}(I_k - I_q) + \dots$$
$$= Z_sI_q + Z_{i1}I_1 + Z_{i2}I_2 + \dots + Z_{ii}(I_i + I_q) + Z_{ij}I_j + Z_{ik}(I_k - I_q) + \dots$$

Similarly for all the buses rearranging,

$$0 = (Z_{i1} - Z_{k1})I_1 + \dots + (Z_{ii} - Z_{ki})I_i + (Z_{ij} - Z_{kj})I_j + (Z_{ik} - Z_{kk})I_k + \dots$$
$$+ (Z_s + Z_{ii} - Z_{ik} - Z_{ki} + Z_{kk})I_q$$

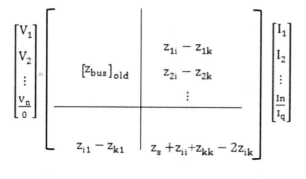

Eliminating I_q,

$$[z_{bus}]_{new} = [z_{bus}]_{old} - \frac{1}{z_s + z_{ii} + z_{kk} - 2z_{ik}} \begin{bmatrix} z_{1i} - z_{1k} \\ z_{2i} - z_{2k} \\ \vdots \\ z_{ni} - z_{nk} \end{bmatrix} [(z_{i1} - z_{k1}) \quad \cdots \quad (z_{in} - z_{kn})]$$

Problem 18.11

Figure 18.24 shows a four-bus system. Treating bus four as the reference bus, obtain Z_{bus}.

Step I. Bus four is considered as the reference bus: add new bus one to reference: type 1 modification $[z_{bus}] = [1]$.

Step II. Connect new bus two to reference: type 1 modification,

$$Z_{bus} = \begin{bmatrix} 1 & 0 \\ 0 & 1 \end{bmatrix}$$

Step III. Connect bus three to reference: type 1 modification,

$$Z_{bus} = \begin{bmatrix} 1 & 0 & 0 \\ 0 & 1 & 0 \\ 0 & 0 & 1 \end{bmatrix}$$

Step IV. Connect bus one to bus two: type 4 modification,

$$Z_{bus} = \begin{bmatrix} 1 & 0 & 0 \\ 0 & 1 & 0 \\ 0 & 0 & 1 \end{bmatrix} - \frac{1}{1+1+1} \begin{bmatrix} 1 \\ -1 \\ 0 \end{bmatrix} \begin{bmatrix} 1 & -1 & 0 \end{bmatrix} \rightarrow Z_{bus} = \begin{bmatrix} \frac{2}{3} & \frac{1}{3} & 0 \\ \frac{1}{3} & \frac{2}{3} & 0 \\ 0 & 0 & 1 \end{bmatrix}.$$

Step V. Connect bus one to bus three: type 4 modification,

$$Z_{bus} = \begin{bmatrix} \frac{2}{3} & \frac{1}{3} & 0 \\ \frac{1}{3} & \frac{2}{3} & 0 \\ 0 & 0 & 1 \end{bmatrix} - \frac{1}{1+\frac{2}{3}+1} \begin{bmatrix} \frac{2}{3} \\ \frac{1}{3} \\ -1 \end{bmatrix} \begin{bmatrix} \frac{2}{3} & \frac{1}{3} & -1 \end{bmatrix} = \begin{bmatrix} \frac{1}{2} & \frac{1}{4} & \frac{1}{4} \\ \frac{1}{4} & \frac{5}{3} & \frac{1}{8} \\ \frac{1}{4} & \frac{1}{8} & \frac{5}{8} \end{bmatrix}.$$

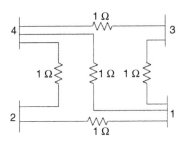

Figure 18.24 Four-bus system.

18.4 Fault analysis

Fault analysis in power systems is required to design the rating of protective devices and to study the stability of a system. Fault studies need to be routinely performed, as the power system network is dynamic with the addition or removal of generators, transmission lines, and loads.

Faults usually occur in a power system, due to insulation failure, flashover, physical damage, or human errors. These faults may be either three phases in nature involving all three phases in a symmetrical manner, or asymmetrical, where usually only one or two phases may be involved. Faults may also be caused by either short circuits to earth or between live conductors, or may be caused by broken conductors (open circuit faults) in one or more phases. Sometimes simultaneous faults may occur involving both short circuit and broken conductor faults.

Balanced three-phase faults may be analyzed using an equivalent single-phase circuit. Symmetrical components are used to reduce the computational complexity with the study of asymmetrical three-phase faults.

18.4.1 Symmetrical component analysis

According to Fortscue's theorem, unbalanced three-phase systems can be expressed in terms of three balanced components called symmetrical components, such as, positive sequence (balanced and having the same phase sequence as the unbalanced supply), negative sequence (balanced and having the opposite phase sequence to the unbalanced supply), and zero sequence (balanced but having the same phase and hence no phase sequence).

The phase components are the addition of symmetrical components, as shown in Figure 18.25 and are mathematically expressed as follows, such that it is possible to convert from either sequence components to phase components or vice versa.

$$a = a_1 + a_2 + a_0$$
$$b = b_1 + b_2 + b_0$$
$$c = c_1 + c_2 + c_0$$

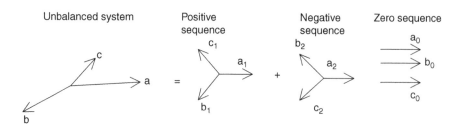

Figure 18.25 Phase component in terms of symmetrical components.

Where a, b, and c are phase components and subscript 1 represents positive sequence, subscript 2 represents negative sequence and subscript 0 represents zero sequence components in an unbalanced systems.

18.4.2 Definition of the operator α

With the balanced components, the angle is 120°. Complex operator j is defined as $\sqrt{-1}$, that is, a vector of unit magnitude rotated anticlockwise by an angle of 90°.

that is, $j = \sqrt{-1} = 1 \angle 90°$

Similarly a new complex operator α is defined with a magnitude of unity and when operated on any complex number rotates it anticlockwise by an angle of 120°:

that is $\alpha = 1 \angle 120° = -0.500 + j\,0.866$

Some properties of α:

$\alpha = 1 \angle 120°$ in anticlockwise direction
$\alpha^2 = 1 \angle 240°$ or $\quad 1 \angle -120°$
$\alpha^3 = 1 \angle 360°$ or 1
that is, $\alpha^3 - 1 = (\alpha - 1)(\alpha^2 + \alpha + 1) = 0$.

Since α is complex, it cannot be equal to 1, so that $\alpha - 1$ cannot be zero.

$\therefore \alpha^2 + \alpha + 1 = 0$

The sequence components of the unbalanced quantity, with each of the components written in terms of phase a components and the operator α, is given in Figure 18.26.

It is inferred that even in the unbalanced systems, phase b and phase c requires no measurements and can be computed with reference to the quantities measured in phase a itself. All the sequence components in terms of the quantities for the a phase using the properties of rotation can be expressed as,

$a = a_0 + a_1 + a_2$
$b = a_0 + \alpha^2 a_1 + \alpha a_2$
$c = a_0 + \alpha a_1 + \alpha^2 a_2.$

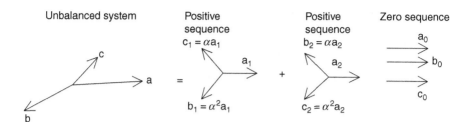

Figure 18.26 Expressing components in terms of phase a.

This can be written in matrix form.

$$\begin{bmatrix} a \\ b \\ c \end{bmatrix} = \begin{bmatrix} 1 & 1 & 1 \\ 1 & \alpha^2 & \alpha \\ 1 & \alpha & \alpha^2 \end{bmatrix} \begin{bmatrix} a_0 \\ a_1 \\ a_2 \end{bmatrix}$$

$$V_a = a_{11}V_1 + a_{12}V_2 + a_{13}V_3 = V_{a_1} + V_{a_2} + V_{a_0}$$

$$V_b = a_{21}V_1 + a_{22}V_2 + a_{23}V_3 = V_{b_1} + V_{b_2} + V_{b_0}$$

$$V_c = a_{31}V_1 + a_{32}V_2 + a_{33}V_3 = V_{c_1} + V_{c_2} + V_{c_0}$$

From Figure 18.26:

$$V_{b_1} = \alpha^2 V_{a_1}$$
$$V_{b_2} = \alpha V_{a_2}$$
$$V_{c_1} = \alpha V_{a_1}$$
$$V_{c_2} = \alpha^2 V_{a_2}$$
$$V_{b_0} = V_{c_0} = V_{a_0}$$

∴ The phase components of an unbalanced system can be represented using the symmetrical components.

$$V_a = V_{a_1} + V_{a_2} + V_{a_0}$$
$$\therefore V_b = \alpha^2 V_{a_1} + \alpha V_{a_2} + V_{a_0}$$
$$\therefore V_c = \alpha V_{a_1} + \alpha^2 V_{a_2} + V_{a_0}.$$

Similarly,

$$\therefore I_a = I_{a_1} + I_{a_2} + I_{a_0}$$
$$\therefore I_b = \alpha^2 I_{a_1} + \alpha I_{a_2} + I_{a_0}$$
$$\therefore I_c = \alpha I_{a_1} + \alpha^2 I_{a_2} + I_{a_0}.$$

Problem 18.12

Given the phase voltages V_a, V_b, and V_c find the symmetrical components V_{a_0}, V_{a_1}, and V_{a_2} and find the average three phase power in terms of symmetrical components.

Solution

$$V_a + V_b + V_c = 3V_{a_0}$$

$$\therefore V_{a_0} = \frac{1}{3}(V_a + V_b + V_c).$$

To get V_{a_1}, multiply by 1, α, and α^2:

$$\alpha^2 V_c + \alpha V_b + V_a = V_{a_1} + \alpha V_b + \alpha^2 V_c = V_{a_1}(1 + \alpha^3 + \alpha^3) + V_{a_2}(1 + \alpha^2 + \alpha^4) + V_{a_0}(1 + \alpha + \alpha^2) = 3V_{a_1}$$

$$\therefore V_{a_1} = \frac{1}{3}(V_a + \alpha V_b + \alpha^2 V_c).$$

Similarly to get V_{a_2}, multiply by 1, α^2, and α,

$$\alpha V_c + \alpha^2 V_b + V_a = 3V_{a_2}$$

$$\therefore V_{a_2} = \frac{1}{3}(V_a + \alpha^2 V_b + \alpha V_c).$$

Average three-phase power in terms of symmetrical components is,

$$P + jQ = V_a I_a^* + V_b I_b^* + V_c I_c^* = \begin{bmatrix} V_a & V_b & V_c \end{bmatrix} \begin{bmatrix} I_a \\ I_b \\ I_c \end{bmatrix}^*$$

$$\begin{bmatrix} V_a \\ V_b \\ V_c \end{bmatrix} = \begin{bmatrix} 1 & 1 & 1 \\ 1 & \alpha^2 & \alpha \\ 1 & \alpha & \alpha^2 \end{bmatrix} \begin{bmatrix} V_{a_0} \\ V_{a_1} \\ V_{a_2} \end{bmatrix} = AV$$

$$\begin{bmatrix} V_a \\ V_b \\ V_c \end{bmatrix}^T = (AV)^T = V^T A^T.$$

But $A^T = A$ as α and α^2 are conjugate.

$$\begin{bmatrix} I_a \\ I_b \\ I_c \end{bmatrix}^* = \left(\begin{bmatrix} 1 & 1 & 1 \\ 1 & \alpha^2 & \alpha \\ 1 & \alpha & \alpha^2 \end{bmatrix} \begin{bmatrix} I_{a_0} \\ I_{a_1} \\ I_{a_2} \end{bmatrix} \right)^*$$

$$P + jQ = \begin{bmatrix} V_{a_0} & V_{a_1} & V_{a_2} \end{bmatrix} \begin{bmatrix} 1 & 1 & 1 \\ 1 & \alpha^2 & \alpha \\ 1 & \alpha & \alpha^2 \end{bmatrix} \begin{bmatrix} 1 & 1 & 1 \\ 1 & \alpha & \alpha^2 \\ 1 & \alpha^2 & \alpha \end{bmatrix} \begin{bmatrix} I_{a_0} \\ I_{a_1} \\ I_{a_2} \end{bmatrix}$$

$$P + jQ = 3\begin{bmatrix} V_{a_0} I_{a_0}^* & V_{a_1} I_{a_1}^* & V_{a_2} I_{a_2}^* \end{bmatrix}.$$

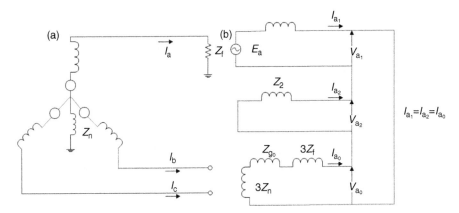

Figure 18.27 (a) A three-phase unloaded alternator with neutral grounded through impedance Z_n and fault impedance Z_f, L–G fault and (b) interconnection of sequence network for L–G fault.

18.4.3   Line-to-ground fault with Z_f [3]

Consider a three-phase system with a fault impedance of Z_f and the neutral impedance Z_n, as shown in Figure 18.27a.

From Figure 18.27a,

$$V_a = I_a Z_f \qquad I_b = 0 \qquad I_c = 0.$$

The sequence network equations are,

$$V_{a_0} = -I_{a_0} Z_0 \qquad\qquad V_{a_1} = E_a - I_{a_1} Z_1 \qquad\qquad V_{a_2} = -I_{a_2} Z_2$$

Substituting I_b and I_c values, the solution of these equations gives the unknown quantities,

$$I_{a_1} = \frac{1}{3}(I_a + \alpha I_b + \alpha^2 I_c) = \frac{1}{3} I_a$$

$$I_{a_2} = \frac{1}{3}(I_a + \alpha^2 I_b + \alpha I_c) = \frac{1}{3} I_a$$

$$I_{a_0} = \frac{1}{3}(I_a + I_b + I_c) = \frac{1}{3} I_a$$

$$I_{a_1} = I_{a_2} = I_{a_0} = \frac{I_a}{3} \rightarrow I_a = 3I_{a_1}.$$

$V_a = I_a Z_f \rightarrow I_a = V_a/Z_f$, written in terms of symmetrical components:

$$V_{a_1} + V_{a_2} + V_{a_0} = V_a = 3I_{a_1} Z_f$$

$$E_a - I_{a_1} Z_1 - I_{a_2} Z_2 - I_{a_0} Z_0 = 3I_{a_1} Z_f$$

$$E_a = I_{a_1}[Z_1 + Z_2 + Z_0 + 3Z_f]$$

$$I_{a_1} = \frac{E_a}{Z_1 + Z_2 + Z_0 + 3Z_f}.$$

Since I_{a_1}, I_{a_2}, and I_{a_0} are known, $V_{a_1}, V_{a_2}, V_{a_0}$ can be calculated from the sequence network equations. Thereafter the fault current I_a can be calculated. The sequence network interconnections are shown in Figure 18.27b.

18.4.4 Line-to-line fault with Z_f

From Figure 18.28a,

$$I_a = 0 \qquad I_b + I_c = 0 \rightarrow I_b = -I_c \qquad V_b = V_c + I_b Z_f$$

and the sequence network equations are,

$$V_{a_1} = E_a - I_{a_1} Z_1 \qquad V_{a_2} = -I_{a_2} Z_2 \qquad V_{a_0} = -I_{a_0} Z_0$$

$$I_{a_1} = \frac{1}{3}(I_a + \alpha \ I_b + \alpha^2 I_c) = I_b$$

$$I_{a_2} = \frac{1}{3}(I_a + \alpha^2 I_b + \alpha \ I_c) = -I_b$$

$$I_{a_0} = \frac{1}{3}(I_a + I_b + I_c) = 0$$

$$I_{a_0} = 0, I_{a_2} = -I_{a_1}$$

$$V_b = V_b + I_b Z_f$$

$$V_{a_0} + \alpha^2 V_{a_1} + \alpha V_{a_2} = V_{a_0} + \alpha V_{a_1} + \alpha^2 V_{a_2} + (I_{a_0} + \alpha^2 I_{a_1} + \alpha I_{a_2})Z_f$$

$$(\alpha^2 - \alpha)V_{a_1} = (\alpha^2 - \alpha)V_{a_2} + (\alpha^2 - \alpha)I_{a_1}Z_f \quad \text{as } I_{a_0} = 0 \rightarrow V_{a_0} = 0 \text{ and } I_{a_2} = -I_{a_1}$$

$$V_{a_1} = V_{a_2} + I_{a_1}Z_f.$$

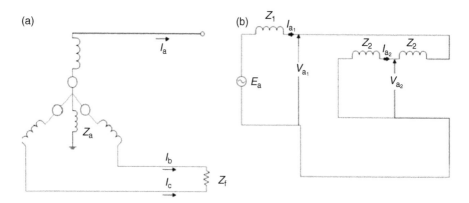

Figure 18.28 (a) A three-phase unloaded alternator with neutral grounded through impedance Z_n and fault impedance Z_f, and (b) interconnection of sequence network, fault impedance Z_f, L–L fault.

Now substituting for V_{a_1} and V_{a_2} from the sequence network equations we get,

$$E_a - I_{a_1} Z_1 = -I_{a_2} Z_2 + I_{a_1} Z_f$$
$$E_a - I_{a_1} Z_1 = I_{a_1} (Z_2 + Z_f)$$
$$I_{a_1} = \frac{E_a}{Z_1 + (Z_2 + Z_f)}.$$

The interconnections of the sequence network, are shown in Figure 18.28b.

18.4.5 Double line-to-ground fault through Z_f

Figure 18.29 shows the double line-to-ground (DLG) fault through impedance Z_f. From Figure 18.29,

$$I_a = 0 \quad = I_{a_1} + I_{a_2} + I_{a_0} \qquad V_b = V_c = (I_b + I_c) Z_f$$

and the sequence network equations are,

$$V_{a_1} = E_a - I_{a_1} Z_1 \qquad V_{a_2} = -I_{a_2} Z_2 \qquad V_{a_0} = -I_{a_0} Z_0$$
$$V_b = V_c \;\rightarrow\; V_{a_0} + \alpha^2 V_{a_1} + \alpha V_{a_2} = V_{a_0} + \alpha V_{a_1} + \alpha^2 V_{a_2} \;\rightarrow\; V_{a_1} = V_{a_2}$$
$$V_b = (I_b + I_c) Z_f$$
$$V_{a_0} + \alpha^2 V_{a_1} + \alpha V_{a_2} = [I_{a_0} + \alpha^2 I_{a_1} + \alpha I_{a_2} + I_{a_0} + \alpha I_{a_1} + \alpha^2 I_{a_2}] Z_f$$
$$V_{a_0} + (\alpha^2 - \alpha) V_{a_1} = 2(I_{a_0} + (\alpha^2 + \alpha)(I_{a_1} + I_{a_2}) Z_f$$
$$V_{a_0} - V_{a_1} = [2I_{a_0} (\alpha^2 + \alpha)(-I_{a_0})] Z_f$$
$$V_{a_0} - V_{a_1} = 3 I_{a_0} Z_f$$
$$-I_{a_0} Z_0 - E_a - I_{a_1} Z_1 = I_{a_0} Z_f$$
$$I_{a_0} = -\frac{E_a - I_{a_1} Z_1}{Z_0 + 3Z_f}.$$

Similarly making use of the relation $V_{a_1} = V_{a_2}$, expressing I_{a_2} in terms of I_{a_1},

$$E_a - I_{a_1} Z_1 = -I_{a_2} Z_2$$
$$I_{a_2} = -\frac{E_a - I_{a_1} Z_1}{Z_2}.$$

Now substituting the values of I_{a_2} and I_{a_0} in the equation,

$$I_a = I_{a_1} + I_{a_2} + I_{a_0} = 0$$
$$I_{a_1} = -\frac{E_a - I_{a_1} Z_1}{Z_2} - \frac{E_a - I_{a_1} Z_1}{Z_0 + 3Z_f} = 0$$
$$I_{a_1} = \frac{E_a}{Z_1 + (Z_2 (Z_0 + 3Z_f) / Z_2 + Z_0 + 3Z_f)}.$$

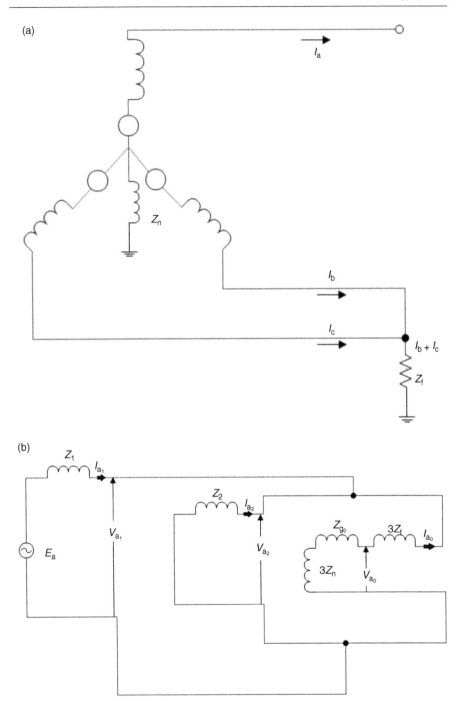

Figure 18.29 (a) L–L–G faultthrough a fault impedance Z_f and neutral impedance Z_n and (b) interconnection of sequence networks for (a).

The interconnection of the sequence network is shown in Figure 13.29b. Based on the above fault analysis discussions, it is inferred that

1. Normally if the system is balanced, the analysis is done in any one of the phase, which can be extended for the rest of the phases. Similarly, even if the system is unbalanced, the fault current can be calculated in phase a and can be extended for all other phases.
2. With the presence of circulating sequence currents, it is possible to discriminate between fault types. For example, it is proved that positive sequence currents flow irrespective of the fault types. Negative sequence currents flow only if the fault involves more than one phase. Zero sequence currents flow if the fault involves ground.

18.4.6 Sequence networks

In all respect, the positive sequence network is identical to the usual networks considered. Each synchronous machine must be considered as a source of EMF, which may vary in magnitude and phase position depending upon the distribution of power and reactive volt amperes, just prior to the occurrence of the fault. The positive sequence voltage at the point of fault will drop, the amount being dependent upon the type of faults; for three-phase faults it will be zero; for DLG fault, line-to-line fault, and single line-to-ground fault, it will be higher in the order stated.

The negative sequence network is in general quite similar to the positive sequence network except that since no negative sequence voltages are generated, the source of EMF is absent.

The zero sequence networks will be free of internal voltages, flow of current resulting from the voltage at the point of fault. The impedances to zero sequence current are very frequently different from positive or negative sequence currents. Transformer and generator impedances depend upon the type of star or delta connections.

The zero sequence equivalent circuits of three-phase transformers require special attention because of the possibility of various combinations. A general circuit for any combination is given in Figure 18.30a. Z_0 is the zero sequence impedance of the windings of the transformer. These are two series and two shunt switches. One series and one shunt switch are for both the sides separately. The series switch of a particular side is closed if it is star grounded and the shunt switch is closed if that side is delta connected, otherwise they are left open. For example, consider the transformer Δ/Y is connected with star ungrounded, as shown in Figure 18.30b. The switching arrangement shown in Figure 18.30c, since the primary is delta connected, the shunt switch of the primary side is closed and series is left open. The secondary is star ungrounded, therefore the series switch is left open and the shunt switch is also left open. The zero sequence network is shown in Figure 18.30d.

The zero sequence equivalent circuits for a few more combinations using this arrangement are shown in Figure 18.31.

18.4.7 Systematic fault analysis using Z_{bus} matrix

For large networks, this method is applicable. Consider the network shown in Figure 18.32. Assume a three-phase fault at bus k has occurred through fault impedance Z_f.

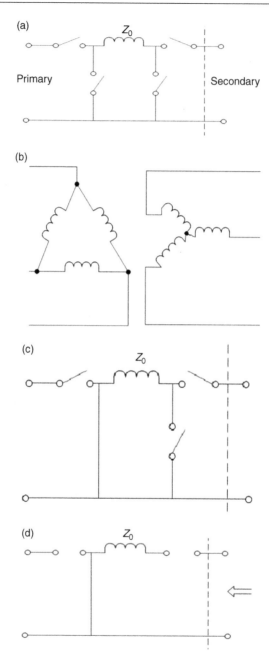

Figure 18.30 (a) General switching combination, (b) star–delta transformer, (c) its switch arrangements for zero sequence network of a Δ/Y transformer, (d) zero sequence equivalent of a Δ/Y transformer.

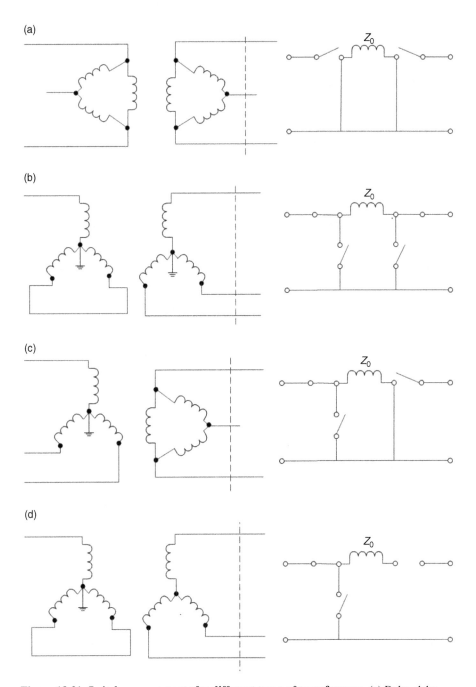

Figure 18.31 Switch arrangements for different types of transformers. (a) Delta–delta transformer and its switch arrangements, (b) star grounded–star grounded transformer and its switch arrangements, (c) star grounded–delta transformer and its switch arrangements, (d) star grounded–star ungrounded transformer and its switch arrangements, (e) star grounded through an impedance–star grounded transformer and its switch arrangements, and (f) star ungrounded–delta transformer and its switch arrangements.

(e)

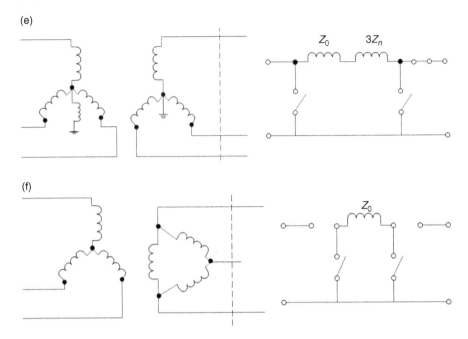

(f)

Figure 18.31 *(cont.)*

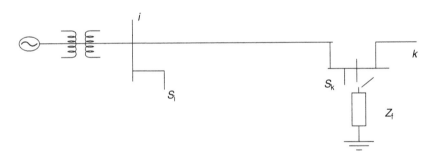

Figure 18.32 Fault at bus k.

Prefault-bus voltages are obtained from load-flow solution and are a column vector.

$$V_{\text{bus}}(0) = \begin{bmatrix} V_1(0) \\ \vdots \\ V_k(0) \\ \vdots \\ V_n(0) \end{bmatrix}$$

Short circuit currents are larger than steady-state values and hence can be neglected. Represent the bus load by a constant impedance evaluated at the prefault bus voltage, that is,

$$Z_{iL} = \frac{|V_i(0)|^2}{S_L^2}.$$

The change in network voltage caused by fault in impedance Z_f, is equivalent to those caused by the added voltage $V_k(0)$ with all other sources short circuited. The bus voltage changes caused by the fault in this circuit are represented by:

$$\underline{\Delta V}_{bus}(0) = \begin{bmatrix} \underline{\Delta V}_1(0) \\ \vdots \\ \underline{\Delta V}_k(0) \\ \vdots \\ \underline{\Delta V}_n(0) \end{bmatrix}.$$

Using Thévenin's theorem, bus voltages during the fault are obtained by superposition of the prefault bus voltages and changes in bus voltages are given by,

$$V_{bus}(F) = V_{bus}(0) + \underline{\Delta V}_{bus}$$

It is known that $I_{bus} = Y_{bus}V_{bus}$.

Current entering every bus is zero except at the faulty bus. As the current leaves the faulty bus, it is taken as negative current entering bus k:

$$\begin{bmatrix} 0 \\ \vdots \\ -I_k(F) \\ \vdots \\ 0 \end{bmatrix} = \begin{bmatrix} Y_{11} & \cdots & Y_{1k} & & Y_{1n} \\ \vdots & & \vdots & & \vdots \\ Y_{k1} & \cdots & Y_{kk} & \cdots & Y_{kn} \\ \vdots & & \vdots & & \vdots \\ Y_{n1} & \cdots & Y_{n1} & \cdots & Y_{nn} \end{bmatrix} \begin{bmatrix} \underline{\Delta V}_1 \\ \vdots \\ \underline{\Delta V}_k \\ \vdots \\ \underline{\Delta V}_n \end{bmatrix}$$

$$I_{bus}(F) = Y_{bus}\Delta V_{bus}.$$

Solving for ΔV_{bus},

$$\Delta V_{bus} = Z_{bus} I_{bus}(F)$$

therefore, $V_{bus}(F) = V_{bus}(0) + Z_{bus} I_{bus}(F)$

for fault in bus k, the bus voltage, $V_k(F) = V_k(0) - Z_{kk}I_k(F)$ and fault current is

$$I_k(F) = \frac{V_k(0)}{Z_{kk} + Z_f}.$$

18.5   Power flow study

The load flow analysis is required to be performed to decide upon addition/removal of a substation, distribution transformers, tap changers, reactive power controlling devices, etc. From the conservation of energy, real power supplied by the source is equal to the sum of real powers absorbed by the load and real losses in the system. Reactive power must also be balanced between the sum of leading and sum of lagging reactive power-producing elements. The total complex power delivered to the loads in parallel is the sum of the complex powers delivered to each,

$$0 = \sum P_{\text{gen}} - \sum P_{\text{loads}} - \sum P_{\text{losses}}$$
$$0 = \sum Q_{\text{leading}} + \sum Q_{\text{caps}} - \sum Q_{\text{lagging}} - \sum Q_{\text{ind}}$$
$$0 = \sum S_{\text{gen}} - \sum S_{\text{loads}} - \sum S_{\text{losses}}.$$

Complex power injected into the ith bus is $S_i = P_i + jQ_i = V_i I_i^*$ and $i = 1, 2, \ldots, n$ where, V_i is the voltage at the ith bus and I_i is the source current injected into the bus,

Hence $S_i = P_i - jQ_i = V_i^* I_i = V_i^* \sum_{k=1}^{n} Y_{ik} V_k$.

Therefore, $P_i = \text{real}\left[V_i^* \sum_{k=1}^{n} Y_{ik} V_k \right]$.

$$Q_i = -i_m\left[V_i^* \sum_{k=1}^{n} Y_{ik} V_k \right]$$

In polar form $V_i = |V_i| e^{j\delta i}$ and $Y_i = |Y_{ik}| e^{jq_{ik}}$.

$$P_i = |V_i| \sum_{k=1}^{n} |Y_{ik}||V_k| \cos\left(\theta_{ik} - \delta_i + \delta_k\right)$$
$$Q_i = -|V_i| \sum_{k=1}^{n} |Y_{ik}||V_k| \sin\left(\theta_{ik} - \delta_i + \delta_k\right).$$

The P and Q equations are known as static load flow equations. To solve the static load flow equations, no explicit solutions are possible with the variables: P_i, Q_i, $|V_i|$, δ_I, as it involves trigonometry and only iterative solutions.

18.5.1   Load flow equations and methods of solution

Load flow solution is a solution of the network under steady-state condition subject to certain inequality constraints, under which the system operates. These constraints can be in the form of load nodal voltages, reactive power generation of the generators, the tap settings of a tap changing under load transformer, etc. To solve the load flow equation, the buses are classified as in Table 18.1:

$$0 = \sum P_{\text{gen}} - \sum P_{\text{loads}} - \sum P_{\text{losses}}$$

But loss remains unknown until the load flow equation is solved. Hence, one of the generator buses is made to take up additional real and reactive power to supply

Table 18.1 **Bus classification**

Bus type	Quantities specified	To be obtained
Load bus	P, Q	$\lvert V \rvert, \delta$
Generator bus or voltage controlled bus	$P, \lvert V \rvert$	Q, δ
Slack bus/swing or reference bus	$\lvert V \rvert, \delta$	P, Q

transmission loss and is named as swing bus. It is known that $I_{bus} = [Y_{bus}]V_{bus}$, where V and Y are obtained from the prefault-bus voltage and bus-admittance matrix, as discussed in the previous topics.

$$I_i = Y_{i_1}V_1 + Y_{i_2}V_2 + \cdots + Y_{i_i}V_i + \cdots + Y_{i_n}V_n$$

$$= \sum_{p=1}^{n} \left| Y_{i_p} V_p \right|$$

$$= \sum_{p=1}^{n} \left| V_p \right| \left| Y_{i_p} \right| \underline{\left| \delta_p + \gamma_{i_p} \right.}$$

where $V_p \big| = \lvert V_p \rvert \underline{\lvert \delta_p}$ and $Y_{i_p} = \left| Y_{i_p} \right| \underline{\lvert \gamma_{i_p}}$.

Complex power injected into the ith bus is,

$$S_i = P_i + jQ_i = V_i I_i^*$$

$$S_i^* = P_i - jQ_i = V_i^* I_i$$

$$S_i^* = P_i - jQ_i = V_i^* \sum_{p=1}^{n} Y_{i_p} V_p$$

$$V_i = \lvert V_i \rvert \underline{\lvert \delta_i}; \underline{\lvert} \; V_i^* = \lvert V_i \rvert \underline{\lvert \delta_i} \qquad Y_{i_p} = Y_{p_i} = \left| Y_{i_p} \right| \underline{\lvert \gamma_{i_p}}$$

$$P_i - jQ_i = \lvert V_i \rvert \sum_{p=1}^{n} \left| V_p \right| \left| Y_{i_p} \right| \underline{\left| \delta_p + \gamma_{i_p} - \delta_i \right.}$$

$$= \lvert V_i \rvert \sum_{p=1}^{n} \left| V_p \right| \left| Y_{i_p} \right| \underline{\left| -\left(\delta_i - \delta_p - \gamma_{i_p} \right). \right.}$$

The static load-flow equation can be rewritten as,

$$\text{Real part} = P_i = \lvert V_i \rvert \sum_{p=1}^{n} \left| V_p \right| \left| Y_{i_p} \right| \cos\left(\delta_i - \gamma_{i_p} - \delta_p \right)$$

$$\text{Imaginary part} = Q_i = \lvert V_i \rvert \sum_{p=1}^{n} \left| V_p \right| \left| Y_{i_p} \right| \sin\left(\delta_i - \gamma_{i_p} - \delta_p \right).$$

18.5.2   *Solution of load flow equation*

$$X_i^k = f_i(X_1^k, X_2^k, \dots X_{i-1}^k, X_i^{k-1}, X_{i+1}^{k-1}, X_n^{k-1})$$

$$\Delta X_i = X_i^k - X_i^{k-1}$$

$$X_i^k = X_i^{k-1} + a\Delta X_i.$$

Its advantages are,

1. Simplicity of the technique.
2. Small computer memory requirement.
3. Less computation time per iteration.

Its disadvantages are,

1. Slow rate of convergence, therefore large number of iterations.
2. Increase in number of iterations directly with increase in number of buses.
3. Effect on convergence due to choice of slack bus.

Hence, Gauss–Seidel (G–S) method is used only for systems having a small number of buses.

18.5.2.1 Using Gauss-Seidel (G-S) method when PV buses are absent

Out of n buses, 1 is the slack bus, so $n - 1$ PQ buses.

$$I_i = Y_{i_i} V_i \sum_{p=1}^{n} Y_{i_p} V_p$$

$$V_i = \frac{1}{Y_{i_i}} \left[I_i - \sum_{p=1}^{n} Y_{i_p} V_p \right].$$

Similarly,

$$S_i^* = P_i - jQ_i = V_i^* I_i$$

$$I_i = \frac{P_i - jQ_i}{V_i^*}$$

$$V_i = \frac{1}{Y_{i_i}} \left[\frac{P_i - jQ_i}{V_i^*} - \sum_{p=1}^{n} Y_{i_p} V_p \right] i = 2, 3, \ldots n, \text{ since 1 is a slack bus.}$$

Solve for $V_2, V_3, \ldots, V_n$,

$$K_i = \frac{P_i - jQ_i}{Y_{i_i}} \quad L_{i_p} = \frac{Y_{i_p}}{Y_{i_i}} \quad i = 2, 3, \ldots n \quad p = 1, 2, \ldots n \quad p \neq n$$

$$V_i^{k+1} = \frac{K_i}{(V_i^k)^*} - \sum_{p=1}^{n} L_{i_p} V_p^{k+1} - \sum_{p=i+1}^{n} L_{i_p} V_p^k \quad i = 2, 3, \ldots n.$$

Problem 18.13

The following is the system data for a load flow solutions. The line admittances are,

Bus code	Admittance
1–2	$2-j8.0$
1–3	$1-j4.0$
2–3	$0.666-j2.664$
2–4	$1-j4.0$
3–4	$2-j8.0$

The schedule of active and reactive powers,

Bus code	P	Q	V	Remarks
1	–	–	1.06	Slack
2	0.5	0.2	$1 + j0$	PQ
3	0.4	0.3	$1 + j0$	PQ
4	0.3	0.1	$1 + j0$	PQ

Determine the voltages at the end of first iteration using G–S method. Take $\propto = 1.6$,

$$Y_{bus} = \begin{bmatrix} 3 - j12 & -2 + j8 & -1 + j4 & 0 \\ -2 + j8 & 3.666 - j4.664 & -0.666 + j2.664 & -1 + j4 \\ -1 + j4 & -0.666 + j2.664 & 3.666 - j14.664 & -2 + j8 \\ 0 & -1 + j4 & -2 + j8 & 3 - j12 \end{bmatrix}.$$

Power for load buses is negative, generation buses is positive,

$$V_2^1 = \frac{1}{Y_{22}} \left[\frac{P_i - JQ_i}{V_{ii}} - Y_{21}V_1^0 - Y_{23}V_3^0 - Y_{24}V_4^0 \right]$$

$$= \frac{1}{3.666 - j14.664} \left[\frac{-0.5 + j0.2}{1 - j0} - 1.06(-2 + j8) - 1(0.666 - j2.664) - 1(-1 + j4) \right]$$

$$= 1.01187 - j0.02888$$

$$V'_{2acc} = (1 + j0) + 1.6(1.01187 - j0.02888 - 1 - j0)$$

$$V'_3 = \frac{1}{Y_{33}} \left[\frac{P_3 - JQ_3}{V_3^*} - Y_{31}V_1 - Y_{32}V'_2 - Y_{34}V_4^0 \right]$$

$$= \frac{1}{3.666 - j14.664} \left[\begin{array}{l} \frac{-0.4 + j0.3}{1 - j0} - (-1 + j4)1.06 - (-0.666 + j2.664) \\ (1.01899 - j0.046208) - (-2 + j8)(1 + j0) \end{array} \right]$$

$$= 0.994119 - j0.029248$$

$$V'_{3acc} = 0.99059 - j0.0467968$$

$$V'_4 = \frac{1}{Y_{44}} \left[\frac{P_4 - JQ_4}{V_4^*} - Y_{41}V_1 - Y_{42}V'_2 - Y_{43}V'_3 \right]$$

$$= 0.9716032 - j0.064684$$

$$V'_{4acc} = 0.954565 - j0.1034944$$

18.5.2.2 Modification of G–S method when PV buses are present

Now, $i = 1$ slack bus, $i = 2, 3, \ldots, m$ PV bus, and $i = m + 1, \ldots, n$ PQ bus.
Conditions to be met by PV buses are:

1. $|V_i| = |V_i|_{specified}$ for $i = 2, 3, \ldots, m$
2. $Q_{i,min} < Q_i < Q_{i,max}$ for $i = 2, 3, \ldots, m$.

The second requirement can be violated if the specified bus voltage $|V_i|_{specified}$ is either too high or too low. It is possible to control $|V_i|$ only by controlling Q_i. Hence, if Q constraint is violated, treat it as a PQ bus with Q equal to maximum or minimum values.

Steps:

1. Calculate $Q_i = |V_i| \sum_{p=1}^{n} |V_p||Y_{ip}| \sin(\delta_i - \gamma_{ip} - \delta_p)$.

2. For every iteration $|V_i|$ must be set equal to $|V_i|_{\text{specified}}$.

3. $Q_i^{k+1} = |V_i|_{\text{specified}} \sum_{p=1}^{i-1} |V_p^{k+1}||Y_{ip}| \sin(\delta_i^k - \gamma_{ip} - \delta_p^{k+1}) + |V_i|_{\text{specified}} \sum_{p=1}^{n} |V_p^k||Y_{ip}| \sin(\delta_i^k - \gamma_{ip} - \delta_p^k)$.

4. Check for constraints, if violated, treat as PQ bus.

Problem 18.14

If in Problem 18.13, bus two is taken as a generator bus with $|V_2| = 1.04$ and reactive power constraint is $0.1 \leq Q_2 \leq 1.0$, determine the voltages starting with a flat voltage profile and assuming the accelerating factor as 1.

Since bus two is a PV bus, Q is not specified. Hence, to find V_2', Q_2 must be calculated, with $V_2 = 1.04$ and the phase angle of voltage = 0.

$$P_2 - jQ_2 = V_2^* \sum_{q=1}^{4} Y_{2q} V_q$$
$$= V_2^* [Y_{21}V_1 + Y_{22}V_2 + Y_{23}V_3 + Y_{24}V_4]$$
$$Q_2 = |1.04|[Y_{21}V_1 \sin(\delta_2 - \gamma_{21} - \delta_1) + Y_{22}V_2 \sin(\delta_2 - \gamma_{22} - \delta_2) + Y_{23}V_3 \sin(\delta_2 - \gamma_{23} - \delta_3)$$
$$+ Y_{24}V_4 \sin(\delta_2 - \gamma_{24} - \delta_4)]$$
$$= 0.1108.$$

Since Q_2 lies within limits, $V_2 = V_{2\text{specified}}$:

$$V_2 = \frac{1}{Y_{22}} \left[\frac{0.5 - j0.1108}{1.04 - j0} - Y_{22}V_1 - Y_{23}V_3 - Y_{24}V_4 \right]$$
$$V_2' = 1.0472846 + j0.0291476, \delta = 1.59°$$
$$V_2 = 1.04\underline{|1.59°} = 1.0395985 + j0.02891159$$
$$V_{2acc}' = \text{negative}$$
$$V_3' = \frac{1}{Y_{33}} \left[\frac{R_3 - jQ_3}{V_3} - Y_{31}V_1 - Y_{32}V_2' - Y_{34}V_4 \right]$$
$$= 0.9978866 - j0.015607057$$
$$V_4' = 0.998065 - j0.022336.$$

Problem 18.15

For the same problem, if the reactive power constraint on generator two is $0.2 \leq Q_2 \leq 1$, solve the problem for voltages at the end of the first iteration.

$Q_2 = 0.1108$. Hence, violated. So assume as PQ bus with $P_2 = 0.5$, $Q_2 = Q_{2\text{min}} = 0.2$, and $V_2° = 1 + j0$ and as usual procedure, but $P_2 + jQ_2$ is positive, though it is assumed to be a PQ bus.

$$V_2' = \frac{1}{Y_{22}} \left[\frac{0.5 - j0.2}{1 - j0} - Y_{22}V_1 - Y_{23}V_3° - Y_{24}V_4° \right]$$
$$= 1.098221 + j0.030105.$$

Similarly V_3' and V_4' are calculated.

18.5.2.3   Newton–Raphson method

It is suitable for large systems.
Its advantages are:

1. Increased accuracy and surety of convergence.
2. Only about three iterations are required compared to more than 25 or so required by the G–S method.
3. Number of iterations is independent of system size.
4. This method is insensitive to factors like slack bus selection, regulating transformers, etc.

Its disadvantages are:

1. Solution technique is difficult.
2. More calculations are involved, hence more computation time/iteration.
3. Memory requirement is large.

This method can be used with both rectangular and polar coordinates. But the rectangular coordinate requires more number of equations as compared to the polar form. Hence, the polar form is preferred.

N–R method using rectangular coordinates is,

$$V_p = |V_p| \underline{|\delta_p} = e_p + jQ_p$$
$$P_i = u_1(e, f), Q_i = u_2(e, f).$$

Changes in active and reactive power with bus as slack,

$$\Delta P_i = \sum_{p=2}^{n} \frac{\partial pi}{\partial 1_p} \cdot \Delta 1_p + \sum_{p=2}^{n} \frac{\partial pi}{\partial 1_p} \cdot \Delta f_p$$

$$\Delta Q_i = \sum_{p=2}^{n} \frac{\partial Qi}{\partial 1_p} \cdot \Delta 1_p + \sum_{p=2}^{n} \frac{\partial Qi}{\partial f_p} \cdot \Delta f_p.$$

$$\begin{bmatrix} \Delta P \\ \Delta Q \end{bmatrix} = \begin{bmatrix} j_1 & j_2 \\ j_3 & j_4 \end{bmatrix} \begin{bmatrix} \Delta e \\ \Delta f \end{bmatrix} \quad \text{for PV bus } |V_i|^2 = e_i^2 + f_i^2.$$

$$\Delta V_i^2 = \frac{\partial |V_i|2}{\partial \rho i} \Delta \rho i + \frac{\partial |V_i|2}{\partial fi} \Delta fi, \text{ with PV bus (instead of } Q_i \text{ and } V_i).$$

Total number of equations = $(n-1)^2$.
N–R method using polar coordinates is,

$$P_i = f_1(\delta, |V|), Q_i = f_2(\delta, |V|).$$

$$\Delta P_i = \sum_{p=2}^{n} \frac{\partial pi}{\partial \delta_p} \cdot \Delta \delta_p + \sum_{p=2}^{n} \frac{\partial pi}{\partial |V_p|} \cdot \Delta |V_p|. \tag{18.1}$$

$$\Delta Q_i = \sum_{p=2}^{n} \frac{\partial Qi}{\partial \delta_p} \cdot \Delta \delta_p + \sum_{p=2}^{n} \frac{\partial Qi}{\partial |V_p|} \cdot \Delta |V_p|. \tag{18.2}$$

but for PV bus (18.2) does not exist.

The n bus system has 1 slack bus and g PV buses.
Total number of equations = $(2n - 2 - g)$.
Therefore,

$$\frac{\partial P_i}{\partial \delta_p} - \frac{\partial Q_i}{\partial \delta_p} = j(e_i - jf_i)(G_{ip} + jB_{ip})(e_p + jf_p)$$

Replace ΔV_p by, $\dfrac{\Delta|V_p|}{|V_p|}$

$$\Delta P_i = \sum_{p=2}^{n} \frac{\partial pi}{\partial \delta_p} \cdot \Delta \delta_p + \sum_{p=2}^{n} \frac{\partial pi}{\partial |V_p|} \cdot |V_p| \frac{\Delta|V_p|}{|V_p|}$$

$$\Delta Q_i = \sum_{p=2}^{n} \frac{\partial Qi}{\partial \delta_p} \cdot \Delta \delta_p + \sum_{p=2}^{n} \frac{\partial Qi}{\partial |V_p|} \cdot |V_p| \frac{\Delta|V_p|}{|V_p|}$$

$$\begin{bmatrix} \Delta P \\ \Delta Q \end{bmatrix} = \begin{bmatrix} H & N \\ J & L \end{bmatrix} \begin{bmatrix} \Delta \delta \\ \dfrac{\Delta|V|}{|V|} \end{bmatrix}$$

$$H_{ip} = \frac{\partial pi}{\partial \delta_p} \quad N_{ip} = \frac{\partial pi}{\partial |V_p|}|V_p| \quad J_{ip} = \frac{\partial Qi}{\partial \delta_p} \quad L_{ip} = \frac{\partial Qi}{\partial |V_p|}|V_p|$$

$$P_i - jQ_i = V_i^* \sum_{p=1}^{n} Y_{ip} V_p \ldots (*).$$

Hence, a Jacobian matrix can be formed from Y_{bus}:

$$Y = G + JB \quad V_i = e_i + jf_i$$
$$a_p - jb_p = (G_{ip} + jB_{ip})(e_p + jf_p)$$
$$\frac{\partial p_i}{\partial \delta_p} = H_{ip} = a_p f_i - b_p e_i \qquad J_{ip} = -(a_p e_i + b_p f_i)$$
$$H_{ii} = -Q_i - |V_i|^2 B_{ii} \qquad J_{ii} = P_i - |V_i|^2 G_{ii}$$
$$N_{ip} = -J_{ip} \quad \text{and} \quad L_{ip} = H_{ip}$$
$$N_{ii} = P_i + |V_i|^2 G_{ii} \quad \text{and} \quad L_{ii} = Q_i - |V_i|^2 B_{ii}$$
$$P_i - jQ_i = |V_i|e^{-j\delta_i} \sum_{p=1}^{n} |Y_{ip}|e^{j\gamma L_p} |V_p|e^{j\delta_p}$$
$$\frac{\partial p_i}{\partial \delta_p} - j\frac{\partial Q_i}{\partial \delta_p} = j(|V_i|e^{-j\delta_i})(|Y_{ip}|e^{j\gamma L_p})(|V_p|e^{j\delta_p})$$
$$|V_i|e^{-j\delta_1} = e_i - jf_i \qquad |Y_{ip}|e^{j\gamma L_p} = G_{ip} + jB_{ip} \qquad |V_p|e^{j\delta_p} = e_p + jf_p.$$

Problem 18.16

Figure 18.33 shows a six-bus system. Assuming bus one is a slack bus, formulate the Jacobian matrix for this system.

Using the following conventions, the Jacobian matrix is formulated.

If bus i and bus m are both PQ buses, then all the Jacobian components are present.

If bus i is a PQ bus and bus m is a PV bus, then the Jacobian components present are H_{im} and J_{im}.

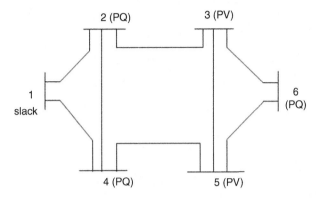

Figure 18.33 Six-bus system for Problem 18.16.

If bus *i* is a PV bus and bus *m* is a PQ bus, then the Jacobian components present are H_{im} and N_{im}. If bus *i* and bus *m* are both PV buses, then the Jacobian component present is only H_{im} as $\Delta|V_m| = 0$.

$$
\begin{array}{c}
\Delta P_2 \\
\Delta Q_2 \\
\Delta P_3 \\
\Delta P_4 \\
\Delta Q_4 \\
\Delta P_5 \\
\Delta P_6 \\
\Delta Q_6
\end{array}
=
\begin{bmatrix}
H_{22} & N_{22} & H_{23} & H_{24} & H_{24} & & & \\
J_{22} & L_{22} & J_{23} & J_{24} & L_{24} & & & \\
H_{23} & N_{23} & H_{33} & & & H_{35} & H_{36} & N_{36} \\
H_{42} & N_{42} & & H_{44} & N_{44} & N_{45} & & \\
J_{42} & L_{42} & & J_{44} & L_{44} & J_{45} & & \\
& & H_{53} & H_{54} & N_{54} & H_{55} & H_{56} & H_{56} \\
& & H_{63} & & & H_{65} & H_{66} & N_{66} \\
& & J_{63} & & & J_{65} & J_{66} & L_{66}
\end{bmatrix}
$$

Problem 18.17

Figure 18.34 shows a five-bus system. Assuming bus one is a slack bus, formulate the Jacobian matrix for this system:

$$
\begin{bmatrix}
H_{22} & N_{22} & H_{23} & H_{24} & H_{24} & 0 \\
J_{22} & L_{22} & J_{23} & J_{24} & L_{24} & 0 \\
H_{32} & N_{32} & H_{33} & 0 & 0 & H_{35} \\
H_{42} & N_{42} & 0 & H_{44} & N_{44} & H_{45} \\
J_{42} & L_{42} & 0 & J_{44} & L_{44} & J_{45} \\
0 & 0 & H_{53} & H_{54} & N_{54} & H_{55}
\end{bmatrix}
$$

Problem 18.18

Determine the set of load flow equations at the end of first iteration using NR method. The load flow data for the given power system are given next. The voltage magnitude of bus 2 is to be

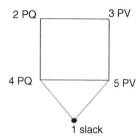

Figure 18.34 Five-bus system for Problem 18.17.

maintained at 1.04 pu. The maximum and minimum reactive power limits of the generator at bus 2 are 0.35 and 0 pu, respectively.

Bus code	Assumed voltages	Generation		Load	
		MW	**MVAR**	**MW**	**MVAR**
1	$1.06 + j0$	0	0	0	0
2	$1 + j0$	0.2	0	0	0
3	$1 + j0$	0	0	0.6	0.25

The admittance matrix is given as:

$$Y_{bus} = \begin{bmatrix} 6.25 - j18.75 & -1.25 + j3.75 & -5 + j15 \\ -1.25 + j3.75 & 2.916 - j8.75 & -1.666 + j5 \\ -5 + j15 & -1.666 + j5 & 6.666 - j20 \end{bmatrix}$$

Solution

From the admittance matrix, it is observed that:

$G_{11} = 6.25$	$G_{12} = -1.25$	$G_{13} = -5$	$G_{22} = 2.916$	$G_{23} = -1.666$	$G_{33} = 6.666$
$B_{11} = 18.75$	$B_{12} = -15$	$B_{13} = -15$	$B_{22} = 8.75$	$B_{23} = -5$	$B_{33} = 20$

The voltage amplitude and phase angle of the bus are specified as:

$$e_1 = 1.06 \qquad e_2 = 1 \qquad e_3 = 1 \qquad\qquad f_1 = 0 \qquad f_2 = 0 \qquad f_3 = 0$$

From the static load flow equations, active power in bus 2 can be calculated as follows. θ values are taken from the admittance matrix written in polar coordinates form.

$$P_2 = |V_2|\,|V_1|\,|Y_{21}|\cos(\theta_{21} + \delta_1 - \delta_2) + |V_2|^2\,|Y_{22}|\cos(\theta_{22}) + |V_2|\,|V_3|\,|Y_{23}|$$
$$\cos(\theta_{23} + \delta_3 - \delta_2) = -0.075$$

Alternately, the active and reactive power can also be calculated as follows.

$$P_i - jQ_i = (e_i + jf_i)^* \sum_{p=1}^{n} (G_{ip} - jB_{ip})(e_p + jf_p)$$
$$= (e_i - jf_i)^* \sum_{p=1}^{n} (G_{ip} - jB_{ip})(e_p + jf_p)$$

$$P_i = \sum_{p=1}^{n} e_i (e_p G_{ip} + f_p B_{ip}) + f_i (f_p G_{ip} - e_p B_{ip})$$

$$Q_i = \sum_{p=1}^{n} f_i (e_p G_{ip} + f_p B_{ip}) - e_i (f_p G_{ip} - e_p B_{ip})$$

$$P_2 = e_2 (e_1 G_{21} + f_1 B_{21}) + f_2 (f_1 G_{21} - e_1 B_{21}) + e_2 (e_2 G_{22} + f_2 B_{22}) + f_2 (f_2 G_{22} - e_2 B_{22})$$
$$+ e_2 (e_3 G_{23} + f_3 B_{23}) + f_2 (f_3 G_{23} - e_3 B_{23})$$
$$= -0.075$$

Similarly the active and reactive powers of bus 2 and 3 are calculated.

$$P_3 = -0.3 \qquad\qquad Q_2 = -0.225 \qquad\qquad Q_3 = -0.9$$

The Δ values are the difference between specified values and the calculated values. Hence,

$$\Delta P_2 = 0.275 \qquad \Delta P_3 = -0.3 \qquad \Delta Q_2 = 0.225 \qquad \Delta Q_3 = 0.65$$

Q_2 violates the limits specified, therefore bus 2 is treated as load bus with $Q_{2spec} = 0$.

$$\frac{\partial P_p}{\partial e_p} = 2 e_p G_{pp} + \sum_{\substack{q=1 \\ q \neq p}}^{n} \left(e_p G_{pq} + f_q B_{pq} \right)$$

$$\frac{\partial P_2}{\partial e_2} = 2 e_2 G_{22} + \sum_{\substack{q=1 \\ q \neq 2}}^{3} \left(e_2 G_{2q} + f_q B_{2q} \right)$$

$$\frac{\partial P_2}{\partial e_2} = 2 e_2 G_{22} + [e_2 G_{21} + f_2 B_{21} + e_2 G_{23} + f_3 B_{23}$$
$$= 2 \times 1 \times 2.916 + 1.06(-1.25) + 0(-3.75) + 1(-1.66) + 0(-5) = 2.848$$

Similarly, $\dfrac{\partial P_3}{\partial e_3} = 6.367$

$$\frac{\partial P_p}{\partial f_p} = 2 f_p G_{pp} + \sum_{\substack{q=1 \\ q \neq p}}^{n} \left(f_q G_{pq} + e_q B_{pq} \right)$$

$$\frac{\partial P_2}{\partial f_2} = 2 f_2 G_{22} + [f_1 G_{21} - e_1 B_{21} + f_3 G_{23} - e_3 B_{23}] = 8.975$$

Similarly, $\dfrac{\partial P_3}{\partial f_3} = 20.9$

$$\frac{\partial P_p}{\partial e_q} = e_p G_{pq} - f_p B_{pq}$$

$$\frac{\partial P_2}{\partial e_3} = -1.666$$

$$\frac{\partial P_3}{\partial e_2} = -1.666$$

$$\frac{\partial P_p}{\partial f_q} = e_p B_{pq} + f_p G_{pq}$$

$$\frac{\partial P_2}{\partial f_3} = -5.0$$

$$\frac{\partial P_3}{\partial f_2} = -5.0$$

$$\frac{\partial Q_p}{\partial e_p} = 2 e_p B_{pp} + \sum_{\substack{q=1 \\ q \neq p}}^{n} (f_q G_{pq} - e_q B_{pq})$$

$$\frac{\partial Q_2}{\partial e_2} = 8.525$$

$$\frac{\partial Q_3}{\partial e_3} = 19.1$$

$$\frac{\partial Q_p}{\partial f_p} = 2 f_p B_{pp} + \sum_{\substack{q=1 \\ q \neq p}}^{n} (e_q G_{pq} + f_q B_{pq})$$

$$\frac{\partial Q_2}{\partial f_2} = -2.991$$

$$\frac{\partial Q_3}{\partial f_3} = -6.966$$

At the end of first iteration, the load flow equations are:

$$
\begin{bmatrix} 0.275 \\ -0.3 \\ 0.225 \\ 0.65 \end{bmatrix} =
\begin{bmatrix}
2.846 & -1.666 & 8.975 & -5.0 \\
-1.666 & 6.366 & -5.0 & 20.90 \\
8.525 & -5.0 & -2.991 & 1.666 \\
-5.0 & 19.1 & 1.666 & -6.966
\end{bmatrix}
\begin{bmatrix} \Delta e_2 \\ \Delta e_3 \\ \Delta f_2 \\ \Delta f_3 \end{bmatrix}
$$

Problem 18.19

Consider the three-bus system. Each of the three lines has a series impedance of 0.02 + j0.08 pu and a total shunt admittance of j0.02 pu. The specified quantities at the buses are tabulated below.

Bus	Real load demand (P_D)	Reactive load demand (Q_D)	Real power generation (P_G)	Reactive power generation	Voltage specified
1	2	1			$1.04 + j0$
2	0	0	0.5	1	(PQ)
3	1.5	0.6	0	?	$V_3 = 1.04$ (PV)

A controllable reactive power source is available at bus three with the constraint $0 \leq Q_{G3} \leq$ 1.5 pu. Find the load flow solution using the N–R method; use a tolerance of 0.01 for power mismatch.

1.

$$Y_{bus} = \begin{bmatrix} 24.23\angle-75.95 & 12.13\angle104.04 & 12.13\angle104.04 \\ 12.13\angle104.04 & 24.13\angle-75.95 & 12.13\angle104.04 \\ 12.13\angle104.04 & 12.13\angle104.04 & 24.23\angle-75.95 \end{bmatrix}$$

$$P_2 = V_2 V_1 Y_{12} \cos(\theta_{21} + \delta_1 - \delta_2) + V_2^2 Y_{22} \cos(\theta_{22}) + V_2 V_3 Y_{23} \cos(\theta_{23} + \delta_3 - \delta_2)$$

$$P_2 = 0.063\,pu, P_3 = -0.122\,pu$$

$$Q_2 = 0.224\,pu, Q_3 = -0.557\,pu$$

$$J = \begin{bmatrix} H_{22} & H_{23} & N_{23} \\ H_{32} & H_{33} & N_{33} \\ J_{32} & J_{33} & L_{33} \end{bmatrix}$$

$$H_{22} = -Q_2 - |V_2|^2 B_{22} = -0.224 - |1.03|^2(-17.18) = 18.002$$

$$a_3 + jb_3 = (G_{23} + jB_{23})(a_3 + jb_3) = -2.035 + j8.61$$

$$H_{23} = a_3 f_2 - b_3 e_2 = -8.868$$

$$J = \begin{bmatrix} 18.002 & -8.868 & -2.035 \\ -8.868 & 17.736 & 3.948 \\ 2.096 & -4.192 & 16.623 \end{bmatrix}$$

2. $\Delta P_2 = P_{2spec} - P_{2calc} = 1.5 - 0.063 = 1.437$ pu

$\Delta P_3 = -1.2 - (-0.122) = -1.078$ pu

$\Delta Q_3 = -0.5 - (-0.557) = 0.057$ pu

$$\begin{bmatrix} 1.437 \\ -1.078 \\ 0.057 \end{bmatrix} = \begin{bmatrix} 18.002 & -8.868 & -2.035 \\ -8.868 & 17.736 & 3.948 \\ 2.096 & -4.192 & 16.623 \end{bmatrix} \begin{bmatrix} \Delta\delta_2 \\ \Delta\delta_2 \\ \dfrac{\Delta|V_3|}{|V_3|} \end{bmatrix}$$

18.5.2.4 Fast decoupled load flow method

Sparsity of Y_{Bus} and loose physical interactions between MW and MVAR flows are taken to make load-flow studies faster and more efficient. $P \rightarrow \delta$ and $Q \rightarrow V$ are strong whereas $P{\rightarrow}V$ and $Q{\rightarrow}\delta$ are weak. Therefore, MW–δ MVAR–V calculations are decoupled and hence N and J are neglected in the Jacobian matrix:

$$[\Delta P] = [H][\Delta \delta]$$
$$[\Delta Q] = [L]\left[\frac{\Delta |V|}{|V|}\right].$$

- Voltage angle corrected using P
- Voltage magnitude corrected using Q

$$\left|y_{ip}\right|e^{jyL_p} = G_{ip} + jB_{ip}$$
$$\frac{\partial p_i}{\partial \delta_p} - j\frac{\partial Q_i}{\partial \delta_p} = j|V_i||V_p|e^{j(\delta_p - \delta_i)} \cdot (G_{ip} + jB_{ip})$$

But $\delta_p - \delta_i$ is very small. Hence, $e^{-j(\delta p - \delta i)} \approx 1$.
1. $\cos(\delta p - \delta i) \approx 1$ and $\sin(\delta p - \delta i) = (\delta p - \delta i)$.
Therefore,

$$\frac{\partial p_i}{\partial \delta_p} - j\frac{\partial Q_i}{\partial \delta_p} = j|V_i||V_p|(G_{ip} + jB_{ip})$$

Separating real and imaginary terms where $I \neq p$:

2. $H_{ip} = \dfrac{\partial p_i}{\partial \delta_p} = -|V_i||V_p||B_{ip}|$ but $L = H$.

$$L_{ip} = H_{ip} = -|V_i||V_p||B_{ip}|$$

In the expressions for L_{ii} and H_{ii}, Q_i is generally very small compared to $|V_i|^2 B_{ii}$:

$$Q_i \ll |V_i|^2 B_{ii}$$
$$H_{ii} = L_{ii} = -|V_i|^2 B_{ii}$$
$$[\Delta P] = [|V||B'||V|][\Delta \delta]$$
$$[\Delta Q] = [|V||B''||V|]\left[\frac{\Delta |V|}{|V|}\right]$$

B' and B'' elements $[-B_{ip}]$.
In the PV bus, Q is not specified and $\Delta |V| = 0$.

Therefore such rows can be simply neglected.
The final algorithm is obtained by making the following approximations.

1. Omit from $[B']$ the representation of those network elements, which predominantly affect MVAR flow only and do not affect MW flow significantly, that is, shunt reactance, off nominal X^r.
2. Omit from $[B'']$ the angle shifting effects of phase shifters.
3. Neglect the series resistance is calculating the elements of $[B]$. With these modifications:

$$\left[\frac{\Delta P}{|V|}\right] = |B'|[\Delta\delta]$$

$$\left[\frac{\Delta Q}{|V|}\right] = \|B''\|[\Delta|V|].$$

B' and B'' are real and sparse.

18.5.2.5   Fixed slope decoupled Newton–Raphson

Load flow computations in a large system are,

1. sparsity: 97% sparse,
2. slope only diagonal and off diagonal + row, column matrix,
3. Gauss elimination, triangular factorization,
4. optimal ordering – least zero, most nonzero number, hence row to be eliminated is with least zero elements.

18.5.2.6   Load flow solution for microgrids

Load flow is the procedure used for obtaining the steady-state voltages of electric power systems at fundamental frequency. An efficient power flow solution looks for fast convergence, minimum usage of memory (computationally efficient), and a numerically robust solution for all the scenarios. Load flow studies on transmission networks are well developed using G–S and N–R methods and their decoupled versions. Because of the some of the following special features the distribution networks fall in the category of ill-conditioned power systems for these conventional load flow methods.

• Radial or weakly meshed networks.
• High R/X ratios.
• Multiphase, unbalanced operation.
• Unbalanced distributed load.
• Distributed generation.

A single-phase representation of three-phase system is used for power flow studies on a transmission system that is assumed as a balanced network in most cases. But the unbalanced loads, radial structure of the network, and untransposed conductors make the distribution system as an unbalanced system. Hence, three-phase power flow analysis needs to be used for distribution systems. The three-phase power flow analysis can be carried out in two different reference frames, namely, phase frame and sequence frame. Phase frame deals directly with unbalanced quantities and sequence frame deals with three separate positive, negative, and zero sequence systems to solve the unbalanced load flow conditions in the circuit. Load flow analysis in distributed generation is carried out using the forward and backward sweep method, compensation methods, implicit Z bus method, direct method (bus injection to branch current and branch current to bus voltage matrices method), or modified Newton methods.

18.6   Power system stability

The dynamics of power systems with the continuous load variations and varying power generation capacity impacts on the stability of the system. Power system stability is the ability of the system to bring back its operation to a steady-state condition within the minimum possible time if undergoing a transient or other disturbance in the system. In power plants, synchronous generators with different voltage ratings are connected to the bus terminals having the same frequency and phase sequence. For example, consider the case defined in the Problem 18.19. After careful analysis, it is found that synchronization depends upon load sharing based on the droop characteristics of the generator. The same is also true for microgrids.

Problem 18.20

Two generators rated 200 and 400 MW are operating in parallel. The droop characteristics shown in Figure 18.35, of their governors are 4 and 5%, respectively, from no load to full load. Assuming that the generators are operating at 50 Hz at no load, how would a load of 600 MW be shared between them? What will be the system frequency at this load?

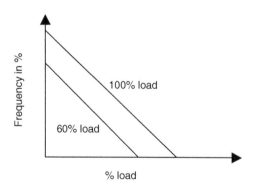

Figure 18.35 Droop characteristics.

As the generators run in parallel, they operate at the same frequency. Let load on G1 (200 MW) = x MW and load on G2 (400 MW) = (600 − x) MW. Reduction in frequency = Δf,

$$\frac{\Delta f}{x} = \frac{0.04 \times 50}{200}$$

$$\frac{\Delta f}{600 - x} = \frac{0.05 \times 50}{400}.$$

Equating the Δf and solving, x = 231 MW. Therefore the load on G1 is 231 MW (overloaded) and that of generator two is 369 MW (underloaded). The system frequency will be,

$$\text{System frequency} = 50 - \frac{0.04 \times 50}{200} \times 231 = 47.69\,\text{Hz}.$$

As the droop characteristics are different, G1 is overloaded and G2 is underloaded. If both the governors are 4% droop then they will share the load as 200 and 600 MW, respectively.

Apart from this, stable and quality power is required for ICT applications and therefore stability analysis is important. The steady-state power limit defines the maximum power permissible to flow through a part of the system when subjected to a fault or disturbances. Stability analysis is done under various types of disturbances,

1. Steady-state stability – ability of a system to maintain its stability following a small disturbance like normal load fluctuations, action of automatic voltage regulator, etc., where the variations are gradual and infinitely small power change.
2. Transient state stability – ability of the system to maintain its synchronism following a large disturbance like sudden addition or removal of a large load, switching operations, faults, or loss of excitation, which exists for a reasonably longer period of time.

Power system stability studies as shown in Figure 18.36 are classified based on the duration, such as, short-term or long-term disturbance, and the factors, such as, voltage, frequency, and rotor angle disturbing the stability of the power system.

Basic phenomena associated with rotor angle stability are,

1. imbalance between accelerating and decelerating generator torque,
2. temporary surplus energy stored in the rotating masses,
3. synchronizing torque limited by pullout torque,
4. loss of the synchronization.

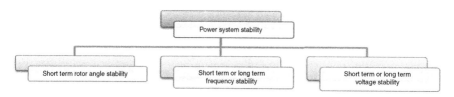

Figure 18.36 Classification of power system stability.

Basic phenomena associated with voltage stability are,

1. increased reactive load reducing the voltage magnitude,
2. temporary load reduction,
3. reduction in transfer capability between areas,
4. if there is no solution to load flow, the voltage collapses.

Basic phenomena associated with frequency stability are,

1. connected load,
2. speed of the generators,
3. prime mover.

18.6.1 Power angle curve and the swing equation [6]

The steady-state power limit is defined by the equation,

$$P = \frac{|E||V|}{X}$$

where E is the generated voltage, V is the terminal voltage, and X is the transfer reactance.

The power $P = P_m \sin \delta$ and the power angle curve are drawn as shown in Figure 18.37.

The dynamics of a generator depend on the inertia constant designated by the manufacturer and the kinetic energy developed while running.

$$\text{Kinetic energy} = \frac{1}{2} M \omega_s$$

where M is the moment of inertia in MJs/rad and ω_s is the rotor speed in rad/s.

The kinetic energy is also defined as:

$$GH = \text{Kinetic energy} = \frac{1}{2} M \omega_s$$

where G is the machine rating in MVA_{base} and H is the inertia constant in MJ/MVA.

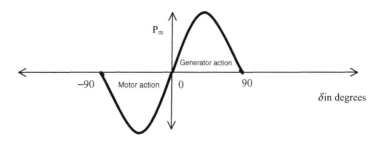

Figure 18.37 Power angle curve.

Considering the rotor angle (δ) torque, speed, flow of mechanical (P_m) and electrical (P_e) powers in a synchronous machine, the rotor dynamics are defined by the swing equation:

$$M \frac{d^2 \delta}{dt^2} = P_m - P_e$$

18.6.2 Solutions of swing equation [7]

For steady-state and transient stability analysis, swing equations can be solved using,

1. *Equal area criterion method*: in a power angle curve, the accelerating area should be equal to the decelerating area for synchronism to be achieved and stability to be regained following the disturbance.
2. *Numerical solution of swing equation*: modified Euler's method.

18.6.3 Stability analysis in microgrids [2]

Apart from active and reactive power control using FACTS controllers, a microgrid's overall stability depends on voltage control. With a large number of microsources connected, microgrids suffer from reactive power oscillations without proper voltage control resulting in circulating currents. This circulating control is controlled by using voltage-reactive power (V–Q) droop controllers. V–Q droop controllers will increase the voltage set point if the microsource reactive current becomes predominantly inductive and decreases the set point if the reactive currents are predominantly capacitive. The reactive power limits are set by the apparent power rating of an inverter and active power output of the microsource.

18.7 Summary

This chapter introduced basic methods and algorithms to study the performance of power systems. Based on the results, solutions may be provided to improve the reliability and security of a power-system network.

Problems

1. A 300 MVA, 20 kV, 3 ϕ generator has a subtransient reactance of 20%. The generator supplies a number of motors over a 64-km transmission line having transformers at both ends. The motors are all rated at 13.2 kV and are represented by two equivalent motors. The rated inputs to the motors are 200 MVA and 100 MVA, respectively. Both have a reactance of 20%. The 3 ϕ transformer T1 is rated 350 MVA, 230/20 kV with a leakage reactance of 10%. T2 is composed of three numbers of 1 ϕ transformers each rated 127/13.2 kV, 100 MVA with a leakage reactance of 10.1. Reactance of the transmission line is 0.5 Ω/km. Draw the one-line diagram, impedance diagram, and reactance diagram.

References

[1] Kothari DP, Nagrath IJ. Modern power system analysis. 4th ed. New Delhi: Tata McGraw Hill Education Private Limited; 2011.

[2] Chowdhury S, Chowdhury SP, Crossley P. Microgrids and active distribution networks. London, UK: The Institution of Engineering and Technology; 2009.

[3] Wadhwa CL. Electrical power systems. New Delhi, India: New Age International (P) Ltd; 2007.

[4] Grainger JJ, Stevenson WD. Elements of power system analysis. Noida, India: Tata McGraw Hill; 2007.

[5] Gupta BR. Power system analysis and design. 3rd ed. Waterville, ME: Wheeler Publishers; 2003.

[6] Saadat H. Power system analysis. 3rd ed. Noida, India: Tata McGraw Hill; 2004. reprint.

[7] Weedy BM. Electric power systems. New York: John Wiley; 1987.

Control of photovoltaic technology

Sukumar Mishra, Dushyant Sharma
Department of Electrical Engineering, Indian Institute of Technology Delhi, New Delhi, India

Chapter Outline

19.1 Introduction to semiconductor physics

A material or device is called a Photovoltaic (PV) if it is capable of converting the energy contained in light (photons) to electrical power. The term *photo* represents photons or light and the term *volt* represents (electrical) voltage. The solar energy contained in the photons is directly converted to electricity, ignoring loss mechanisms.

Silicon is widely used as a semiconductor material. The conductivity of silicon depends on the number of mobile carriers present in the active semiconductor layer.

The band gap energy, denoted by E_g, is measured in electron-volts (eV); where 1 eV = 1.6×10^{-19} J. The band gap energy is 1.12 eV for silicon.

The energy needed for carrier generation can be provided in different forms. In PVs, the energy is provided by the photons in light. If a photon with energy greater than 1.12 eV is absorbed by a solar cell, an electron-hole pair is formed, which can be harvested with some energy lost as heat. If the photon energy is less than 1.12 eV it just passes through the solar cell and an electron-hole pair is not formed. If the energy of the photon is higher than 1.12 eV, the excess energy is dissipated as heat.

Solar radiation consists of electromagnetic radiations of different wavelengths. The energy contained in a photon is related to the wavelength of the electromagnetic radiation and is given by (19.1):

$$E = \frac{hc}{\lambda} \tag{19.1}$$

where E is the energy of the photon (J), c is the speed of light (3×10^8 m/s), λ is the wavelength (m), and h is the Planck's constant (6.626×10^{-34} Js).

Example 19.1

Let 1.12 eV energy be needed to promote an electron to the conduction band. Calculate the minimum frequency of the photon required.

Solution
Band gap energy, E_g = 1.12 eV
1 eV = 1.6×10^{-19} J
So E_g = 1.12 eV = $1.12 \times 1.6 \times 10^{-19}$ J = 1.792×10^{-19} J

The maximum wavelength (λ_{max}) that contains energy just equal to band gap energy is calculated as $E = \dfrac{hc}{\lambda_{max}} = E_g$

Thus $\dfrac{6.626 \times 10^{-34} \text{ Js} \times 3 \times 10^8 \text{ m/s}}{\lambda_{max}} = 1.792 \times 10^{-19}$ J

or $\lambda_{max} = \dfrac{6.626 \times 10^{-34} \text{ Js} \times 3 \times 10^8 \text{ m/s}}{1.792 \times 10^{-19} \text{ J}} = 1.109 \times 10^{-6}$ m and the corresponding frequency is

$\upsilon_{min} = \dfrac{c}{\lambda_{max}} = \dfrac{3 \times 10^8 \text{ m/s}}{1.109 \times 10^{-6} \text{ m}} = 2.705 \times 10^{14}$ Hz.

The electron-hole pair thus created needs to be separated to avoid recombination of the free electron and the hole. This can be achieved by employing a p–n junction.

The simplest p–n junction is the p–n junction diode (shown in Figure 19.1). The current that can flow through the diode is given in (19.2):

$$I_d = I_{rs} \left\{ \exp\left(\frac{qV_d}{kT_c\,A} \right) - 1 \right\} \tag{19.2}$$

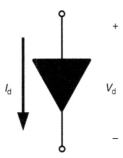

Figure 19.1 Symbol of a diode.

where I_d is the diode current (A), I_{rs} is the reverse saturation current (A), q is the charge of an electron (1.6×10^{-19} C), V_d is the voltage across the diode (V), k is the Boltzmann constant (1.38×10^{-23} J/ K), T_c is the temperature of the diode (K), and A is a constant known as the ideality factor, which depends on the mechanism by which the charge carriers move across the p–n junction. $A = 1$ represents diffusion and $A = 2$ represents recombination.

The reverse saturation current is dependent on temperature and the relationship can be approximated by (19.3):

$$I_{rs(T_c)} = I_{rs(T_{ref})} \left[\exp\left\{ K_1 \left(T_c - T_{ref} \right) \right\} \right] \tag{19.3}$$

where $I_{rs(T_c)}$ is the cell's reverse saturation current at working temperature T_c, $I_{rs(T_{ref})}$ is the cell's reverse saturation current at reference temperature T_{ref}, and K_1 is the temperature coefficient of the reverse saturation current.

19.2 Basics of a photovoltaic cell

A p–n junction diode forms the most elementary archetype of a PV cell. When energy from photons is absorbed by the solar cell, electron-hole pairs are formed, which are confined to the n side and p side, respectively, because of the depletion region. A voltage equal to the reduction in barrier potential is developed at the terminals of a cell, which can appear across an external load leading to current flow.

19.2.1 Structure of a simple PV cell

A simple PV cell is essentially a p–n junction. A cell is provided with a glass cover, adhesive, and an antireflection layer. The antireflection layer is provided to enhance the absorption of solar radiation and the glass cover is provided for mechanical protection [1].

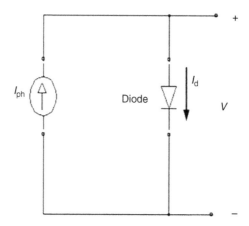

Figure 19.2 Equivalent circuit of an ideal PV cell.

19.2.2   Equivalent circuit of an ideal PV cell

An ideal PV cell can be treated as a diode in parallel to a constant current source (Figure 19.2). The constant current source represents the photogenerated current (I_{ph}). The forward biasing of the p–n junction is represented in the diode. Considering these, the PV current is obtained by (19.4):

$$I = I_{ph} - I_d = I_{ph} - I_{rs}\left\{ \exp\left(\frac{qV}{kT_cA} \right) - 1 \right\}$$ (19.4)

The photogenerated current is dependent on the solar irradiation and temperature and is a constant for a particular temperature and insolation level. The photogenerated current is related to the cell's short circuit current and the cell temperature and irradiation by (19.5):

$$I_{ph} = \left\{ I_{sc} + K_I\left(T_c - T_{ref} \right) \right\} S$$ (19.5)

where I_{sc} is the cell's short circuit current at reference temperature T_{ref}, K_I is the short circuit temperature coefficient, and S is the solar irradiation level normalized to 1000 W/m^2 [2]. The photogenerated current can also be called the short circuit current of a cell at a particular temperature and solar irradiation.

Example 19.2

The short circuit current of a PV cell is 8.03 A at a reference temperature of 300 K. The short circuit current coefficient is 0.0017. Find the photogenerated current of the cell at 310 K and 60% solar irradiation.

Solution

$S = 0.6$
$I_{sc} = 8.03$ A
$K_1 = 0.0017$
$T_c = 310$ K
$T_{ref} = 300$ K

From (19.5) $I_{ph} = \{I_{sc} + K_1(T_c - T_{ref})\}S = \{8.03 + 0.0017(310 - 300)\}0.6$

or $I_{ph} = \{8.03 + 0.0017(10)\}0.6 = \{8.03 + 0.017\}0.6 = 8.047 \times 0.6 = 4.828$ A

Example 19.3

Find the open circuit voltage of a PV cell whose short circuit current is 8.03 A at a reference temperature of 300 K. The short circuit current coefficient is 0.0017. The cell's working temperature is 300 K and is under full Sun. Reverse saturation current of the p–n junction diode at reference temperature is 1.2×10^{-7} A, and the ideality factor is 1.92.

Solution

$S = 1$
$I_{sc} = 8.03$ A
$K_1 = 0.0017$
$T_c = 300$ K
$T_{ref} = 300$ K
$I_{rs} = 1.2 \times 10^{-7}$ A
$A = 1.92$

From (19.5), the photogenerated current of the cell $I_{ph} = 8.03$ A.
At open circuit, PV current is zero ($I = 0$). The corresponding voltage is known as the open circuit voltage (V_{oc}).

Thus, from (19.4) $0 = I_{ph} - I_{rs}\left\{\exp\left(\dfrac{qV_{oc}}{kT_cA}\right) - 1\right\}$ or $I_{rs}\left\{\exp\left(\dfrac{qV_{oc}}{kT_cA}\right) - 1\right\} = I_{ph}$

$$\Rightarrow \exp\left(\frac{qV_{oc}}{kT_cA}\right) - 1 = \frac{I_{ph}}{I_{rs}}$$

$$\Rightarrow \exp\left(\frac{qV_{oc}}{kT_cA}\right) = \frac{I_{ph}}{I_{rs}} + 1$$

$$\Rightarrow \frac{qV_{oc}}{kT_cA} = \ln\left(\frac{I_{ph}}{I_{rs}} + 1\right)$$

$$\Rightarrow V_{oc} = \left(\frac{kT_cA}{q}\right) \times \ln\left(\frac{I_{ph}}{I_{rs}} + 1\right) \tag{19.6}$$

$$\Rightarrow V_{oc} = \left(\frac{1.38 \times 10^{-23} \times 300 \times 1.92}{1.6 \times 10^{-19}}\right) \times \ln\left(\frac{8.03}{1.2 \times 10^{-7}} + 1\right) = 0.895 \text{ V}$$

Example 19.4

For the same PV cell as used in Example 19.3, find the open circuit voltage if the cell's temperature (1) increases to 315 K and (2) decreases to 290 K and the temperature coefficient of the reverse saturation current is 0.15/°C.

Solution

$$S = 1$$
$$I_{sc} = 8.03 \text{ A}$$
$$K_1 = 0.0017$$
$$T_{ref} = 300 \text{ K}$$
$$I_{rs(300)} = 1.2 \times 10^{-7} \text{ A}$$
$$A = 1.92$$
$$K_1 = 0.15$$

1. $T_C = 315$ K

From (19.4) $I_{ph} = \{8.03 + 0.0017(315 - 300)\}1 = \{8.03 + 0.0017(15)\} = 8.0555$ A

From (19.3) $I_{rs(315)} = I_{rs(300)}\left[\exp\{0.15(315 - 300)\}\right] = 1.2 \times 10^{-7} \times \left[\exp\{0.15(15)\}\right] = 1.138 \times 10^{-6}$ A

From (19.6) obtained in Example 19.3

$$V_{oc} = \left(\frac{1.38 \times 10^{-23} \times 315 \times 1.92}{1.6 \times 10^{-19}}\right) \times \ln\left(\frac{8.0555}{1.138 \times 10^{-6}} + 1\right) = 0.822 \text{ V}$$

2. $T_C = 290$ K

From (19.4) $I_{ph} = \{8.03 + 0.0017(290 - 300)\}1 = \{8.03 + 0.0017(-10)\} = 8.013$ A

From (19.3) $I_{rs(290)} = I_{rs(300)}\left[\exp\{0.15(290 - 300)\}\right] = 1.2 \times 10^{-7} \times \left[\exp\{0.15(-10)\}\right] = 2.677 \times 10^{-8}$ A

From (19.6) obtained in Example 19.3

$$V_{oc} = \left(\frac{1.38 \times 10^{-23} \times 290 \times 1.92}{1.6 \times 10^{-19}}\right) \times \ln\left(\frac{8.013}{2.677 \times 10^{-8}} + 1\right) = 0.937 \text{ V}$$

The previously mentioned examples illustrate that the photogenerated current depends mainly on solar irradiation and is slightly dependent on the temperature while the open circuit voltage depends mainly on temperature and is slightly dependent on the solar irradiation.

These relationships are further discussed in the current–voltage (*I–V*) and power–voltage (*P–V*) characteristics of a solar cell.

19.2.3 Characteristics of an ideal PV cell

An ideal PV cell described in the previous section can be treated as a constant current source in parallel with a diode in opposition. The opposition is due to the difference between the number of charge carriers crossing the junction by diffusion and the number of charge carriers crossing the junction by drift as the direction of charge carrier diffusion (similar to a current in a p–n junction diode) is opposite to the direction of charge carrier drift. When energy from photons creates electron-hole pairs, the charge carriers are drifted away from the junction. If the external terminals are short circuited, the amount of current flowing due to charge carrier drift is very large as compared to the amount of current flowing due to diffusion of charge carriers as the potential barrier does not allow diffusion of charger carriers. Thus, the net PV current is a constant depending on the amount of electron-hole pair created by the photons. If the voltage across the PV terminals increases and approaches toward open circuit, the charger carrier concentration (electrons in the n-type side and holes in the p-type side) increases. This generates an electric field at the junction, which is opposite to the barrier potential, and results in reduced barrier potential and diffusion current increases. At open circuit, an equilibrium point is reached such that the diffusion current is exactly equal (in magnitude) to the drift current and no net current flows through the PV cell.

Thus, an ideal PV cell can be treated as a constant current source under low voltage (as almost constant current flows through the PV cell) and can be treated as a constant voltage source (as the voltage across the PV cell remains almost constant) around the open circuit point. This current characteristic is shown in Figure 19.3. The characteristics of an ideal PV cell at different solar irradiation and temperature are shown in Figures 19.4 and 19.5, respectively.

The power obtained from a PV cell initially increases with voltage to a maximum value (called the maximum power point (MPP) of the cell) and then reduces sharply.

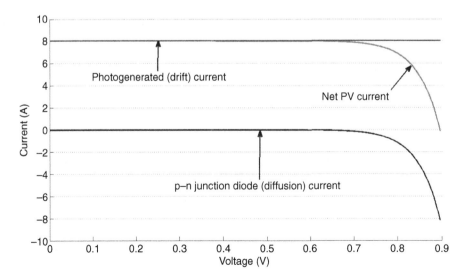

Figure 19.3 Current–voltage (*I–V*) characteristics of a PV cell.

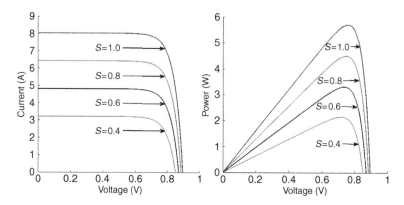

Figure 19.4 *I–V* and *P–V* characteristics of an ideal PV cell at different solar irradiations.

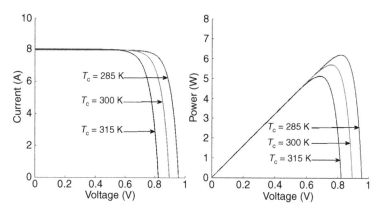

Figure 19.5 *I–V* and *P–V* characteristics of an ideal PV cell at different cell temperatures.

Example 19.5

A load resistance of 0.1 Ω is connected across the PV cell used in previous examples operating at the reference temperature and full irradiation. Plot the PV cell characteristics and the load characteristics to obtain the operating point and find the current and the voltage.

Solution
$R_L = 0.1 \ \Omega$
The characteristics of the cell and the load are plotted in Figure 19.6. The point of intersection gives the operating point.
From Figure 19.6, $V = 0.755$ V and $I = 7.55$ A.

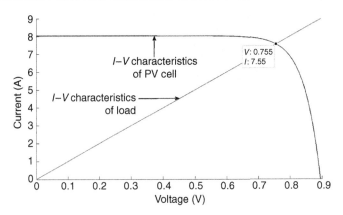

Figure 19.6 *I–V* characteristics of an ideal PV cell and resistive load.

19.2.4 Equivalent circuit of an actual PV cell

An actual PV cell has a series and a parallel resistance. The parallel resistance represents the leakage effect while the series resistance shows the effect of resistance of the semi-conductor and the resistance of the bond between the cell and wire leads. The equivalent circuit is shown in Figure 19.7. Usually the parallel leakage resistance is very large (infinite in an ideal cell) and series resistance is very small (zero in an ideal cell).

When the parallel resistance (R_P) is considered and the series resistance (R_S) is neglected, that is, R_S is zero, the PV cell current is modified as:

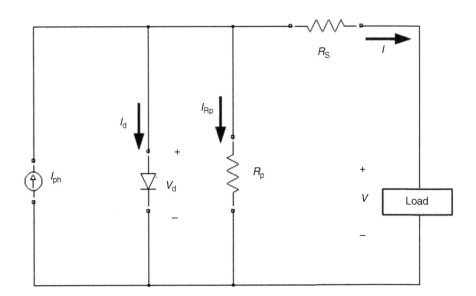

Figure 19.7 Equivalent circuit of an actual PV cell.

$$I = I_{ph} - I_d - I_{Rp}$$

$$\text{or } I = I_{ph} - I_{rs}\left\{\exp\left(\frac{qV}{kT_cA}\right) - 1\right\} - \frac{V}{R_P} \tag{19.7}$$

When the parallel resistance (R_P) is neglected, that is, R_P is infinite and the series resistance (R_S) is considered, the PV cell current is modified as:

$$I = I_{ph} - I_{rs}\left[\exp\left\{\frac{q(V + IR_S)}{kT_cA}\right\} - 1\right] \tag{19.8}$$

When both the parallel resistance (R_P) and the series resistance (R_S) are considered, the PV cell current is modified as:

$$I = I_{ph} - I_{rs}\left[\exp\left\{\frac{q(V + IR_S)}{kT_cA}\right\} - 1\right] - \frac{V + IR_S}{R_P} \tag{19.9}$$

19.2.5   Characteristics of an actual PV cell

The impact of solar irradiation and temperature on an actual cell remains the same but the characteristics are modified due to the impact of the series and the parallel resistance as shown in Figure 19.8.

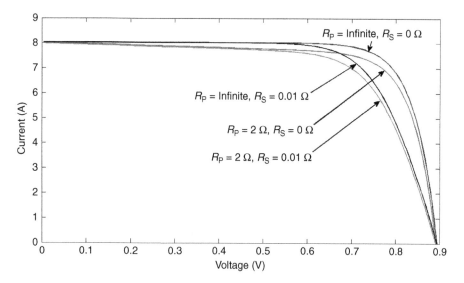

Figure 19.8 Characteristics of an actual PV cell considering the effect of series and parallel resistance.

19.2.6   From cell to module and array

A single cell can produce a maximum voltage equal to a diode turn-on voltage (around 0.5–0.8 V depending on the type of cell). A very low voltage of this range is not sufficient to supply to electrical loads. Thus, a number of cells are connected in series to form a module and modules are connected in series or parallel to form an array providing the required voltage and current rating. When N_S number of cells are connected in series and N_P number of cells are connected in parallel while both the parallel resistance (R_P) and the series resistance (R_S) are considered, the PV cell current is modified as:

$$I = N_P I_{ph} - N_P I_{rs} \left[\exp \left\{ \frac{q\left(V/N_S + IR_S/N_P\right)}{kT_c A} \right\} - 1 \right] - N_P \left(\frac{V/N_S + IR_S/N_P}{R_P} \right) \quad (19.10)$$

For an ideal cell the current equation can be written as [3]:

$$I = N_P I_{ph} - N_P I_{rs} \left[\exp \left\{ \frac{q(V/N_S)}{kT_c A} \right\} - 1 \right] \quad\quad (19.11)$$

19.3   Maximum power point tracking

The power obtained from a PV module is maximum at a particular voltage and current. To achieve maximum efficiency, it is desired to operate the PV at maximum power. Many maximum power point tracking (MPPT) algorithms are discussed in the literature [4–6]; the most accepted of these are the perturb and observe, and the incremental conductance methods.

19.3.1   Perturb and Observe MPPT algorithm

The Perturb and Observe (P&O) method is like climbing a hill in which the DC voltage is changed in steps ($= \Delta V$) to operate around the MPP. After each step the change in power is observed. If the power increases, the next voltage change (increase/decrease) should have same sign as the previous one and if the power reduces, that is, the change in power is negative, the next perturbation is done contrary to the previous. It can be mathematically related to the sign of ratio of change in power to change in voltage, that is, the voltage change in the next perturbation should have the same sign as that of the ratio of change in power to change in voltage in the current step.

The algorithm is described in Figure 19.9.

As the DC voltage reference is changed at each time step, the reference voltage always oscillates around the MPP voltage (Figure 19.10).

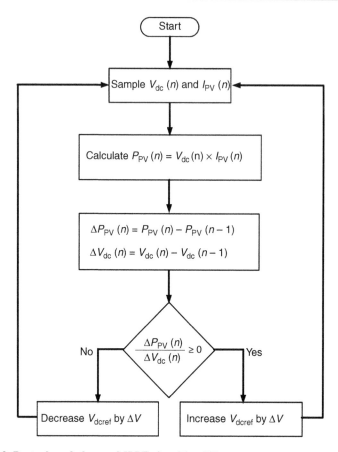

Figure 19.9 Perturb and observe MPPT algorithm [2].

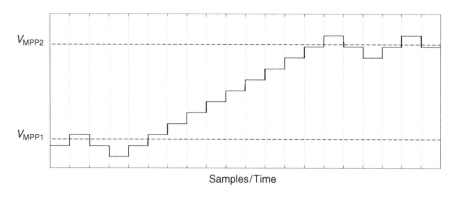

Figure 19.10 Voltage reference variation in perturb and observe MPPT algorithm [2].

19.3.2 Incremental Conductance MPPT algorithm

At the point of maximum power the rate of change of power with respect to voltage (dP/dV) is zero; it is positive for voltage less than MPP voltage and negative for voltage greater than MPP voltage.

$$\text{At MPPT, } \frac{dP}{dV} = 0 \tag{19.12}$$

$$\frac{dP}{dV} = \frac{d(V \times I)}{dV} = V \times \frac{dI}{dV} + I = 0 \tag{19.13}$$

$$\frac{dI}{dV} = -\frac{I}{V} \tag{19.14}$$

The term on the left-hand side of (19.14) denotes the incremental conductance while the term on the right-hand side denotes the actual conductance.

Similar to the P&O algorithm, voltage and current are sensed and voltage is varied in steps to satisfy the condition given in (19.14) to track MPP.

19.4 Shading impact on PV characteristics

19.4.1 Shading impact on series-connected cells and shade mitigation

Few cells or modules in an array undergo partial or full shading due to movement of clouds, shadows of trees, etc. Shading creates many problems like reduction in output voltage and power, heating of cells, and damage to devices because of fluctuations [7].

The impact of shading will depend on the number of cells under shading, amount of shading, and parallel leakage resistance, and it can be understood by reference to Figure 19.11.

If n number of cells are connected in series and one of the cells is shaded while the remaining are under full Sun, the characteristics will depend on the amount of shading, that is, the photogenerated current that the shaded cell can provide as per (19.5).

Two scenarios can arise. The first is when the photogenerated current is greater than the load current (the current supplied by the remaining $n - 1$ cells). This scenario is shown in Figure 19.11a. As the photogenerated current is sufficient to match the current supplied by the remaining cells, the series string behaves like a normal series-connected string operating under uniform solar irradiation. However, in the second scenario (Figure 19.11b), the photogenerated current is not sufficient to match the current supplied by the remaining cells. In this case, the excess current flows through the parallel leakage resistance leading to a drop in voltage across the cell. The diode becomes reverse biased and does not carry any current. The cell produces negative

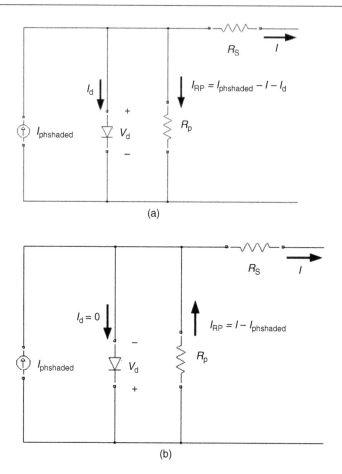

Figure 19.11 Shading impact. (a) When shaded, cell's photogenerated current is greater than the current supplied by the remaining cells and (b) when shaded, cell's photogenerated current is less than the current supplied by the remaining cells.

voltage, that is, reduces the overall voltage rather than increasing it. Moreover, as current is flowing through the high parallel resistance, heating of the cell may occur leading to hot spot formation in the series string.

The reduction in output voltage is obtained by subtracting the output voltage during shading from the original output voltage obtained under normal operation, that is, $\Delta V = V - V_{SH}$; where ΔV is the reduction in the output voltage, V is the output voltage under uniform irradiation, and V_{SH} is the output voltage under shaded conditions.

The voltage V_{SH} is obtained by subtracting the drop across the parallel resistance from the voltage developed by the remaining cells (V_{n-1}). In normal operation, the voltage across the string is V. As all the cells are under uniform irradiation, the voltage across each cell is $V_{cell} = V/n$. Thus, the voltage across the $n-1$ cells is $V_{n-1} = ((n-1)/n)V$.

So the output voltage during shading is $V_{SH} = V_{n-1} - \left(I - I_{\text{phshaded}}\right)\left(R_P + R_S\right)$. The series resistance is generally very small as compared to the shunt resistance, and so it can be neglected. Thus, the output voltage due to shading of one cell in a series-connected string is:

$$V_{SH} = V_{n-1} - \left(I - I_{\text{phshaded}}\right)\left(R_P\right) = \left(\frac{n-1}{n}\right)V - \left(I - I_{\text{phshaded}}\right)\left(R_P\right). \tag{19.15}$$

The drop in output voltage is:

$$\Delta V = V - V_{SH} = V - \left(\frac{n-1}{n}\right)V + \left(I - I_{\text{phshaded}}\right)\left(R_P\right) = \frac{V}{n} + \left(I - I_{\text{phshaded}}\right)\left(R_P\right) \tag{19.16}$$

where I_{phshaded} is the photogenerated current of the shaded cell.

The problems arising due to shading are generally taken care of by the use of a bypass diode. A bypass diode is connected across the cell antiparallel to the diode of the cell. During uniform irradiation the bypass diode has no role to play as it is reverse biased, so there is no impact on the bypass diode, but during shading scenarios due to the drop across the resistance, the bypass diode becomes forward biased and all the current flows through it. Thus, no current flows through the parallel resistance and hot spot formation is avoided. Moreover, the reduction of voltage in this case is just around 0.5–0.8 V depending upon the type of diode rather than the larger drop occurring without it.

The circuit topology with a bypass diode is shown in Figure 19.12. The I–V characteristics under shading, when one cell is 50% shaded and the impact of a bypass diode, is shown in Figure 19.13.

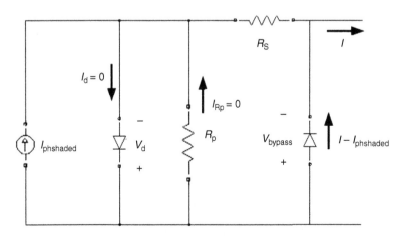

Figure 19.12 Use of bypass diode for shade mitigation.

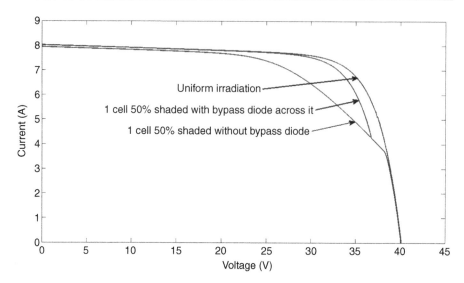

Figure 19.13 *I–V* characteristics under partial shading.

In practice, a bypass diode is not used across each cell but rather across a group of cells in a module or across each module.

When a bypass diode is used, the *I–V* curve shows two knee points, suggesting that there can be multiple maximum power points. A normal P&O technique may end up with operation at local maxima if the initial point is not around the global MPP. Thus, for partial shading scenarios advanced MPPT techniques are required.

Example 19.6

A PV module having 60 cells, each having the parallel resistance of 10 Ω, is operating at reference temperature. The short circuit current at a reference temperature of 300 K and full insolation is 8.03 A. The reverse saturation current of the p–n junction diode of each cell at reference temperature is 1.2×10^{-7} A, and the ideality factor is 1.92. All the cells are receiving uniform irradiation of 700 W/m². One of the cells undergoes partial shading such that the solar irradiation drops to 300 W/m². Find out what current the module can feed to a load without being heated up (assume bypass diode is absent).

Solution

$$R_p = 10$$
$$I_{sc} = 8.03 \text{ A}$$

As the module is operating at reference temperature, photogenerated current at 300 W/m² is

$$I_{ph300} = 8.03 \times 0.3 \text{ A} = 2.409 \text{ A}.$$

Thus, the module can provide 2.409 A without being heated up.

Example 19.7

A PV module having 60 cells, each having a parallel resistance of 10 Ω, is operating at reference temperature. The short circuit current at reference temperature of 300 K and full insolation is 8.03 A. The reverse saturation current of the p–n junction diode of each cell at reference temperature is 1.2×10^{-7} A, and the ideality factor is 1.92. All the cells are receiving uniform irradiation of 700 W/m². The module is operating at its maximum power. The MPP voltage is 80% of the open circuit voltage. Find the module output voltage and current. One of the cells undergoes partial shading such that the solar irradiation drops to 300 W/m². Find the new output voltage and the reduction in the output voltage if the module is still supplying the same current to the load (assume bypass diode is absent).

Solution

$R_p = 10\,\Omega$

$I_{sc} = 8.03\,A$

As the module is operating at reference temperature, photogenerated current at 300 W/m² is

$I_{phshaded} = I_{ph\,300} = 8.03 \times 0.3\,A = 2.409$ A and at 700 W/m² it is $I_{ph\,700} = 8.03 \times 0.7\,A = 5.621$ A.

From (19.6), the open circuit voltage of each cell is

$$V_{oc} = \left(\frac{1.38 \times 10^{-23} \times 300 \times 1.92}{1.6 \times 10^{-19}} \right) \times \ln\left(\frac{5.621}{1.2 \times 10^{-7}} + 1 \right) = 0.895 \text{ V}$$

Thus, the voltage across each cell at MPPT is $0.895 \times 0.8 = 0.716$ V.
Thus, the output voltage (voltage across the module) is $V = 60 \times 0.716 = 42.96$ V.
The output current is obtained using (19.4)

$$I = 5.621 - 1.2 \times 10^{-7} \left\{ \exp\left(\frac{1.6 \times 10^{-19} \times 0.716}{1.38 \times 10^{-23} \times 300 \times 1.92} \right) - 1 \right\} = 5.403 \text{ A}.$$

Under partial shading the module is still supplying the same current, so $I = 5.403$ A.
From (19.15), the module output voltage during shading is
$V_{SH} = (59/60)42.96 - (5.403 - 2.409)(10) = 12.304$ V and the drop in voltage is 42.96 − 12.304 = 30.656 V.
Alternatively the reduction in output voltage can also be obtained using (19.16)

$$\Delta V = 0.716 + (5.403 - 2.409)(10) = 30.656 \text{ V}.$$

19.4.2 Shading impact on parallel-connected cells and shade mitigation

If a few of the parallel-connected strings are under shading, the shaded string may withdraw current from the remaining strings rather than supply current. Thus, the current that can be supplied to load reduces and so does the output power [1]. Moreover, when PV is not producing power (e.g., at night), power may flow into the PV system from other energy sources like a battery.

This is avoided by using a blocking diode (also called isolation diode) at the top of each string such that it allows current from the PV to pass through it while it blocks any reverse current flowing into the PV during shading. The scheme of using blocking diodes is shown in Figure 19.14.

19.5 Mode of operation of a PV system

A PV system can be operated in grid-connected mode as well as isolated mode. A general PV system consists of a DC/DC converter, battery storage, charge controller, inverter, and transformer. Different schemes use different configurations using all or a few of these equipments. The converters are controlled to achieve desired operating conditions.

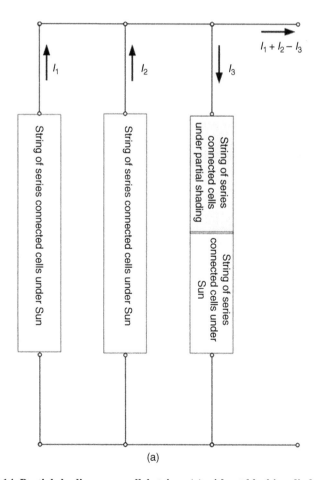

(a)

Figure 19.14 Partial shading on parallel strings (a) without blocking diode and

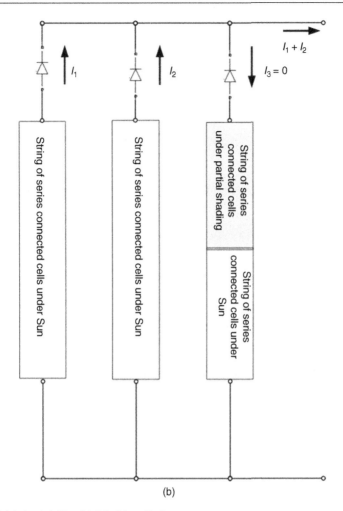

(b)

Figure 19.14 *(cont.)* **(b) with blocking diode.**

Generally in a *single-stage PV system*, a PV array is connected to the AC system through an inverter and filter. The inverter is controlled to achieve MPP as well as to achieve desired AC system parameters. In a *double-stage PV system*, a DC/DC converter is used between PV array and inverter. The DC/DC converter is controlled to ensure MPPT and the inverter is controlled to achieve desired AC system parameters. Schemes for both single-phase and three-phase AC systems are available. A schematic for a grid-connected PV system is shown in Figure 19.15. A PV may also feed to a DC grid through a DC/DC converter.

In isolated operation a battery is generally used as a backup. It can supply power when PV power is reduced and can consume power when there is surplus PV power. The battery is controlled through a charge controller.

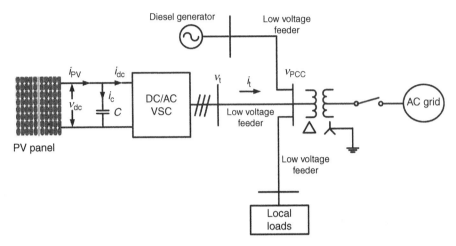

Figure 19.15 Typical grid-connected PV system with a diesel generator and local loads [8].

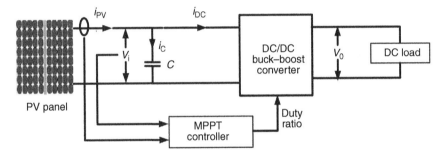

Figure 19.16 PV system supplying a DC load.

19.5.1 PV supplying a DC system

Generally, a boost or buck–boost converter is used to increase the DC voltage and maintain MPPT; a scheme is shown in Figure 19.16. The output and input DC voltages are related by the duty cycle (D) [9] as:

$$\frac{V_o}{V_i} = -\left(\frac{D}{1-D}\right) \tag{19.17}$$

The minus sign shows the reverse polarity of the output voltage.

Example 19.8

A resistive load of 15 Ω is fed through a buck–boost converter being supplied by the PV module given in Example 19.7. All the cells are receiving uniform irradiation of 700 W/m^2. Find the duty cycle at which the converter is operated so that maximum power is extracted from the PV module.

Solution

$R_L = 15\ \Omega$

As per Example 19.7, the voltage at maximum power is 42.96 V and the corresponding current is 5.403 A. Thus, the maximum power the PV module can provide is $P_{max} = 42.96 \times 5.403$ W = 232.112 W.

The power delivered to the load is $P_L = V_o^2 / R_L$. To extract maximum power, $P_L = P_{max}$.

Thus, $\dfrac{V_o^2}{R_L} = 232.112$ W

$\Rightarrow V_o = \sqrt{232.112 \times R_L} = \sqrt{232.112 \times 15} = 59$ V

From (19.17), the average output voltage $V_0 = \left(\dfrac{D}{1-D}\right)V_i$.

Thus, $V_0 = DV_i + DV_0$ or $D = \dfrac{V_0}{V_0 + V_i} = \dfrac{59}{42.96 + 59} = \dfrac{59}{101.96} = 0.578$.

19.5.2 PV supplying an AC system

As discussed earlier, an inverter is used in an AC system. There are various methods of inverter control like decoupled control, sliding mode control, vector control, neural network, fuzzy-based controls, etc. [10–13].

In a grid-connected system, the inverter is operated to obtain desired active and reactive power while in an isolated system the inverter is operated to obtain desired voltage and frequency. For a three-phase PWM inverter the DC and AC voltage are related by the modulation index (m) [14] as:

$$V_{dc} = \frac{2\sqrt{2}V_{iphase}}{m}. \tag{19.18}$$

where V_{iphase} is the inverter root mean square (RMS) phase voltage.

Example 19.9

Thirty PV modules, each having an MPP voltage of 30.7 V and an MPP current of 8.08 A, are connected in series to form an array and have MPPT controllers to ensure operation at maximum power. The PV array supplies power to a three-phase 415 V (l–l), 50 Hz AC system through a PWM inverter operating at a switching frequency of 10 kHz. Each phase has a filter inductor of 7.50 mH. Find the modulation index at which the inverter will be operated. Assume the inverter to be lossless, unity power factor operation, all the cells under uniform irradiation of 1000 W/m², and operating at standard conditions. Ignore parallel current path.

Solution

$V_{l-l} = 415$ V

$V_{phase} = \dfrac{415}{\sqrt{3}} = 239.6$ V

$V_{MPP} = 30.7 \times 30 = 921$ V

$I_{MPP} = 8.08$ A

Maximum power $P_{MPP} = 921 \times 8.08 = 7441.68\,\text{W}$.

As the inverter is lossless, the power generated by the PV is completely supplied to the AC system.

The AC system RMS phase current $I_{phase} = \dfrac{P_{MPP}}{3 \times (415/\sqrt{3})} = \dfrac{7441.68}{\sqrt{3} \times 415} = 10.353\,\text{A}$.

Per phase reactance between the inverter and the AC system

$X_L = 2 \times \pi \times 50 \times L = 100 \times \pi \times 0.0075 = 2.356\,\Omega$.

Per phase RMS voltage at the inverter terminals is

$V_{iphase} = \sqrt{239.6^2 + (10.353 \times 2.356)^2} = 240.838\,\text{V}$.

The inverter terminal phase voltage is related to the DC voltage by (19.18) as $V_{dc} = \dfrac{2\sqrt{2}V_{iphase}}{m}$.

Thus, $m = \dfrac{2\sqrt{2}V_{iphase}}{V_{dc}} = \dfrac{2\sqrt{2} \times 240.838}{921} = 0.739$.

19.5.3 Application of transformer in grid-connected PV system

A grid-connected inverter-based system may or may not have a transformer. The topology depends on voltage and isolation requirements. A grid-connected system having a transformer is shown in Figure 19.15. A transformer may be needed to provide galvanic isolation from the AC grid. A transformer is also required to amplify the voltage if the inverter voltage is less than the grid voltage.

In general, a transformer is connected in delta–star with neutral grounding (Δ–Y) configuration. The PV side is Δ connected while the grid side is Y connected with neutral grounding. This is required to provide a zero sequence current path.

19.5.4 Decoupled d–q control structure

Three-phase time varying AC quantities can be converted to synchronously rotating d–q quantities using Park's transformation. Balanced three-phase quantities when transformed to d and q axis become constant quantities and are easy to analyze and control. Park's transformation is given in (19.19):

$$\begin{bmatrix} u_d \\ u_q \\ u_0 \end{bmatrix} = \sqrt{\frac{2}{3}} \begin{bmatrix} \cos(\theta) & \cos\left(\theta - \dfrac{2\pi}{3}\right) & \cos\left(\theta + \dfrac{2\pi}{3}\right) \\ -\sin(\theta) & -\sin\left(\theta - \dfrac{2\pi}{3}\right) & -\sin\left(\theta + \dfrac{2\pi}{3}\right) \\ \dfrac{1}{\sqrt{2}} & \dfrac{1}{\sqrt{2}} & \dfrac{1}{\sqrt{2}} \end{bmatrix} \begin{bmatrix} u_a \\ u_b \\ u_c \end{bmatrix} \tag{19.19}$$

The θ in the previous equation is the reference phase angle obtained from a *phase locked loop* (PLL). A PLL generates an output whose phase is related to the phase of the input signal. Generally in a grid-connected system (as shown in Figure 19.17) the

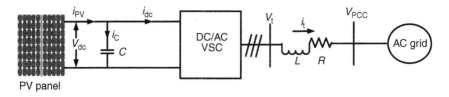

Figure 19.17 Grid-connected system.

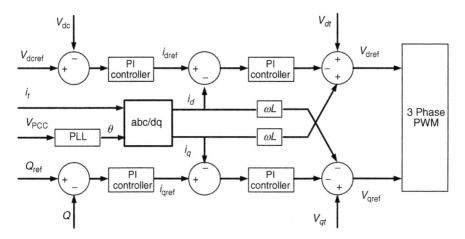

Figure 19.18 Decoupled control.

phase of the AC system at the point of common coupling (PCC) is used as reference and all other quantities are converted to d and q axis using the reference obtained from a PLL (Figure 19.18).

The d and q axis current controls the active and the reactive power being fed from the inverter. The active power feed affects the DC link voltage behind the inverter. Thus, the reference d and q axis current is obtained from a voltage and reactive power controller. Generally the controller output is limited to generate the reference currents within limits that are needed for safe operation of an inverter. High currents may damage the power electronics devices used in the inverter. The reactive power reference is made zero for unity power factor operation. The voltage reference for the voltage controller is obtained from the MPPT algorithm. Assuming the inverter terminal voltage as v_t, PCC voltage as v_{PCC}, current fed from the inverter as i_t, and the filter inductance and resistance as L and R, respectively. The d and q axis reference currents provided by the voltage and reactive power controller are controlled as shown in Figure 19.18. The differential equations relating current and voltage are given in (19.20) and (19.21) [10,11,15]:

$$L\frac{di_{dt}}{dt} = -Ri_{dt} + \omega Li_{qt} + \left(v_{dt} - v_{d\,PCC}\right)$$

(19.20)

$$L\frac{di_{qt}}{dt} = -Ri_{qt} - \omega Li_{dt} + \left(v_{qt} - v_{q\text{PCC}}\right) \tag{19.21}$$

Equations (19.20) and (19.21) can be simplified to (19.22) and (19.23), respectively:

$$L\frac{di_d}{dt} = -Ri_d + u_d \tag{19.22}$$

$$L\frac{di_q}{dt} = -Ri_q + u_q \tag{19.23}$$

where $u_d = v_{dt} - v_{d\text{PCC}} + \omega Li_{qt}$ and $u_q = v_{qt} - v_{q\text{PCC}} - \omega Li_{dt}$. The control signals u_d and u_q are derived from the current control loop.

The controllers are designed such that the current control loop is faster than the voltage and reactive power control loop. The time constant of the current control loop depends on the PWM switching frequency and the time constant of the voltage control loop depends on the settling time of the current control loop.

Current control loop design is defined as follows. From (19.22) the transfer function relating control u_d and current i_d can be written as $\dfrac{i_d}{u_d} = \dfrac{1}{Ls + R}$. As the control signal is obtained from a proportional integral (PI) controller the resultant open loop transfer function is $\left(K_p + \dfrac{K_i}{s}\right) \times \left(\dfrac{1}{Ls + R}\right) = \dfrac{K_p}{Ls} \times \left(s + \dfrac{K_i}{K_p}\right) \times \left(\dfrac{1}{s + (R/L)}\right)$. The closed loop current control is shown in Figure 19.19.

The controller gains can be chosen by various methods. Using pole zero cancellation methods, the gains can be taken as $K_p = L/\tau$ and $K_i = R/\tau$; where τ is the desired time constant. Using these gains the open loop transfer function becomes $1/\tau s$ and the closed loop transfer function becomes $1/(\tau s + 1)$. The resultant current control loop is shown in Figure 19.20.

Time constant is generally taken in the range of 2–4 ms. The outer (voltage control) loop can be designed in a similar way. The time constant of the voltage control loop should be greater than the settling time of the current control loop. When the inner

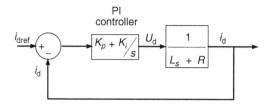

Figure 19.19 Current control loop.

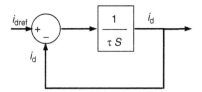

Figure 19.20 Resultant current control loop.

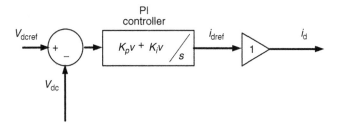

Figure 19.21 Voltage control loop when the current loop has settled.

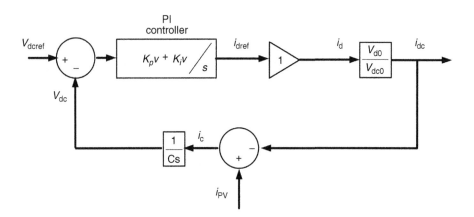

Figure 19.22 Closed loop voltage control.

loop has settled the current can be assumed to be the same as the current reference. Thus, while designing the voltage loop the overall current control loop may be treated as a gain (= 1), shown in Figure 19.21.

The DC voltage (V_{dc}) is related to DC current (i_{dc}) by a transfer function $V_{dc}(s) = (I_{pv}(s) - I_{dc}(s))/Cs$, and the DC current is related to d axis current as $P = i_{dc} \times V_{dc} = v_d \times i_d$. Thus, $i_{dc} = (v_d / V_{dc}) \times i_d$. Using these relations a closed loop transfer function of the voltage control loop can be obtained at the initial operating condition, shown in Figure 19.22.

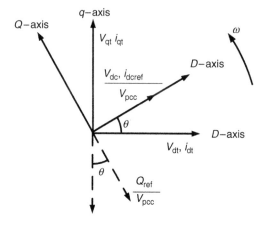

Figure 19.23 Phasor diagram of a grid-connected system [16].

The d and q axis reference currents (i_{dref}) and (i_{qtref}) are related to the DC current reference (i_{dcref}) by (19.24). The corresponding phasor diagram is shown in Figure 19.23.

$$\begin{bmatrix} i_{dt\,ref} \\ i_{qt\,ref} \end{bmatrix} = \begin{bmatrix} \cos\theta & \sin\theta \\ \sin\theta & -\cos\theta \end{bmatrix} \begin{bmatrix} \dfrac{V_{dc}i_{dcref}}{V_{PCC}} \\ \dfrac{Q_{ref}}{V_{PCC}} \end{bmatrix} \tag{19.24}$$

$$v_{PCC} = \sqrt{v_{d\,PCC}^2 + v_{q\,PCC}^2} \tag{19.25}$$

$$\theta = \tan^{-1}\left(\frac{v_{q\,PCC}}{v_{d\,PCC}} \right) \tag{19.26}$$

Example 19.10

Using the pole zero cancellation method in decoupled d–q control, find the minimum time constant of the outer loops if the time constant of the current control loop is 4 ms and settling time is considered as three times the time constant. Also find the PI controller gains if the filter inductor is 7.50 mH and resistance is 0.01 Ω.

Solution

$L = 7.5$ mH
$R = 0.01\ \Omega$
$\tau = 0.004$ s

Settling time $= 3 \times \tau = 3 \times 0.004 = 0.012$ s.

Thus, the time constant of the voltage loop must be greater than 0.012 s. Using the pole zero cancellation method, the control gains are

$K_p = L/\tau = 0.0075/0.004 = 1.875$ and $K_i = R/\tau = 0.01/0.004 = 2.5$.

19.5.5 Derated operation of PV system

A PV system may be operated under derating mode, which means operation at a power lower than the maximum power. A PV may be operated under derated mode under load following mode in a microgrid or under some control strategy.

A PV under derated mode can be treated as a reserve like a battery. A PV under derating can be used for frequency control of an associated AC system. Such a control scheme is shown in Figure 19.24.

The scheme shown in Figure 19.24 has a proportional control. In this kind of scheme the PV voltage (or power) is varied only during transients, that is, whenever frequency is deviated from the nominal reference frequency. In this scheme, in steady state any load change is met by a change in power generation of other source(s) in the power system. In steady state the PV operates at the predefined derated power. Another possible scheme is to use a PI controller, which varies the steady-state PV power. In this scheme, in the steady state any load change is met by a change in PV power generation along with the other source(s) having a secondary controller. Such a control scheme is shown in Figure 19.25.

Using derated PV for frequency control better frequency regulation is observed as shown in Figure 19.26. There is considerable improvement in both settling time and peak deviation. In Figure 19.26a, the settling time from the instant of load change is 40 s and the range of frequency oscillation is from 0.97 (pu) to 1.015 (pu) while the settling time and the range of oscillation in Figure 19.26b is 50 s and 0.95 (pu) to 1.03 (pu), respectively.

The secondary frequency regulation shown in Figure 19.25 does not consider the amount of reserve available with multiple PVs. When this type of control is used all PVs having a secondary controller share the load requirement equally irrespective of the reserve they have. Under this control, a PV with less reserve may become exhausted. Thus, it is necessary to consider the reserve available with each PV. The control

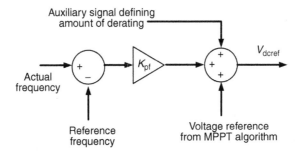

Figure 19.24 Control of a PV system during derating mode of operation with proportional controller.

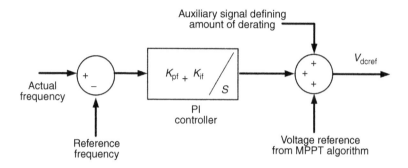

Figure 19.25 Control of a PV system during derating mode of operation with a PI controller (secondary frequency control).

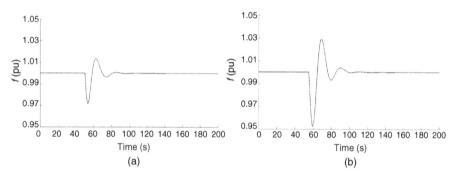

Figure 19.26 Frequency response of AC system for step load change when (a) PV is participating in frequency control and (b) PV is not participating in the frequency control.

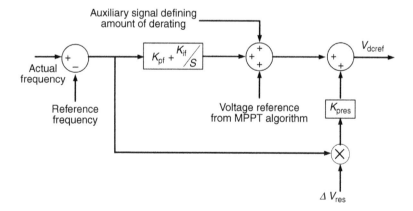

Figure 19.27 Secondary frequency regulation using a PV considering the reserve available.

scheme is modified considering the reserves in Figure 19.27. The PV with maximum reserve will share the maximum load change. This scheme ensures that none of the PVs are exhausted [17].

19.5.6 Droop control for power sharing

In a synchronous generator-based power system, all generators share power change as per their droop coefficients. When multiple PV systems are operating in parallel in a microgrid, they need to share power as per their rating.

To meet these requirements droop control is also introduced in inverter control and the reference voltage and frequency is generated as per the droop characteristics:

$$f_{ref} = f_0 - K_{fdelP}\Delta P \tag{19.27}$$

$$V_{ref} = V_0 - K_{VdelQ}\Delta Q \tag{19.28}$$

where P_0 is the active power at nominal frequency f_0, K_{fdelP} is the frequency droop coefficient, ΔP is the change in active power, f_{ref} is the reference active power at the new frequency, V_0 is the nominal voltage corresponding to the reactive power Q_0, K_{VdelQ} is the voltage droop coefficient, ΔQ is the change in reactive power, and V_{ref} is the reference voltage corresponding to new reactive power.

The droop coefficients are decided on the basis of ratings of each inverter and are proportional to the individual ratings. The droop philosophy is shown in Figures 19.28 and 19.29 [15].

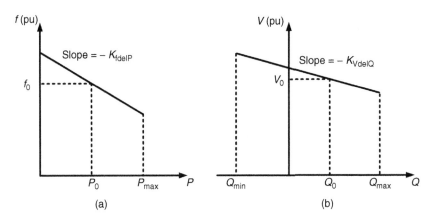

Figure 19.28 Droop characteristics for load sharing. (a) P–f and (b) Q–V.

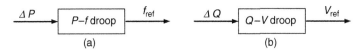

Figure 19.29 Droop control. (a) Active power and (b) reactive power.

References

[1] Masters GM. Renewable and efficient electric power systems. John Wiley & Sons, Inc., Hoboken, New Jersey; 2004.

[2] Sekhar PC, Mishra S. Takagi–Sugeno fuzzy-based incremental conductance algorithm for maximum power point tracking of a photovoltaic generating system. IET Renew Power Gen 2014;8(8):900–14.

[3] Yazdani A, Dash PP. A control methodology and characterization of dynamics for a photovoltaic (PV) system interfaced with a distribution network. IEEE Trans Power Deliv 2009;24(3):1538–51.

[4] Esram T, Chapman PL. Comparison of photovoltaic array maximum power point tracking techniques. IEEE Trans Energy Conv 2007;22(2):439–49.

[5] Subudhi B, Pradhan R. A comparative study on maximum power point tracking techniques for photovoltaic power systems. IEEE Trans Sust Energy 2013;4(1):89–98.

[6] de Brito MAG, Galotto L, Sampaio LP, de Azevedo e Melo G, Canesin CA. Evaluation of the main MPPT techniques for photovoltaic applications. IEEE Trans Ind Electron 2013;60(3):1156–67.

[7] Jewell W, Ramakumar R. The effects of moving clouds on electric utilities with dispersed photovoltaic generation. IEEE Trans Energy Conv 1987;EC-2(4):570–6.

[8] Mishra S, Ramasubramanian D, Sekhar PC. A seamless control methodology for a grid connected and isolated PV-diesel microgrid. IEEE Trans Power Syst 2013;28(4): 4393–404.

[9] Mohan N. First course on power electronics and drives. MNPERE: Minneapolis, United States of America; 2003.

[10] Vahedi H, Noroozian R, Jalilvand A, Gharehpetian GB. A new method for islanding detection of inverter-based distributed generation using DC-link voltage control. IEEE Trans Power Deliv 2011;26(2):1176–86.

[11] Bajracharya C, Molinas M, Suul JA, Undeland TM. Understanding of tuning techniques of converter controllers for VSC-HVDC. Nordic workshop on power and industrial electronics. June 9–11, 2008.

[12] Mishra S, Sekhar PC. Sliding mode based feedback linearizing controller for a PV system to improve the performance under grid frequency variation. International Conference on Energy, Automation and Signal, ICEAS – 2011. p. 106–112.

[13] Mahmood H, Jiang J. Modeling and control system design of a grid connected VSC considering the effect of the interface transformer type. IEEE Trans Smart Grid 2012;3(1):122–34.

[14] Rashid MH. Power electronics: circuits, devices, and applications. 3rd ed. Upper Saddle River, NJ: Pearson, Prentice Hall; 2004.

[15] Katiraei F, Iravani R, Hatziargyriou N, Dimeas A. Microgrids management. IEEE Power Energy Mag 2008;6(3):54–65.

[16] Tiwari A, Boukherroub R, Sharon M, editors. Solar cell nanotechnology. Hoboken, NJ, USA: John Wiley & Sons, Inc.; 2013.

[17] Zarina PP, Mishra S, Sekhar PC. Exploring frequency control capability of a PV system in a hybrid PV-rotating machine-without storage system. Int J Elec Power Energy Syst 2014;60:258–67.

Integration of distributed renewable energy systems into the smart grid

Ghanim Putrus, Edward Bentley
Electrical Power Engineering at Northumbria University Newcastle upon Tyne,
United Kingdom

Chapter Outline

20.1   Introduction

About 30% of all primary energy resources worldwide are used to generate electrical energy. Since the invention of the electric incandescent light bulb in 1879, the growth of electric power systems progressed at an exponential rate, particularly after the development of AC power generation and the transformer. Transforming AC power from one voltage level to much higher levels meant that losses and voltage drops in the supply lines could be kept at acceptable values.

The contribution of different renewable energy sources to the electricity generation mix varies from one country to another, but generally this is currently a small proportion of the total installed capacity. However, energy policy targets for 2050 are very ambitious, setting the commitment of some countries to 80% reduction of greenhouse gas emissions below 1990 levels, by 2050 [1]. Renewable energy obligations such as the European 20/20/20 targets are driving a fast growth of renewable energy installations. For example, in the United Kingdom, renewable electricity installed capacity at the third quarter of 2014 was 23.1 GW, which is over 25% of the total installed capacity (around 85 GW). The contribution of renewable energy to electricity generation in this period was about 17.8% of total generation [2,3]. This represents an increase of 24% compared to the same quarter in 2013. The growth of renewable energy generation capacity in the United Kingdom during 2009–2013 is shown in Figure 20.1. As can be noticed, growth of renewable energy in the last few years is steadily increasing.

The increase in renewable energy generation involves significant challenges in establishing cost-effective and reliable renewable energy systems (RES) in addition to solving the technical problems associated with their connection to the grid. In order to understand the impact of increased connection of RES on the grid and the need for smart grid solutions, it is important to first understand how electricity is currently generated, the characteristics of generation from RES, and some aspects of grid control.

20.2   Conventional power generation

Conventional power plant is the general term applied to the production of electrical energy from coal, oil, or natural gas using the intermediary of steam. The generator is usually a synchronous machine having a small number of poles (two or four) and running at high speeds (1500–3600 rpm). The overall efficiency of energy conversion from fuel to electrical is greatly influenced by the poor efficiency of the turbine and condenser. Typical overall efficiency ranges from 30% to 40%. The main features of these conventional plants are their low capital cost per kilowatt installed as compared to other plants and virtually no limit on their size.

The combined-cycle power plant is relatively more efficient and environmentally friendly. It operates in two stages with an overall efficiency of up to 55%. The first stage includes a gas turbine that drives the first alternator and the second stage uses hot exhaust gases from the gas turbine to produce steam through a heat exchanger, which operates a steam turbine coupled to a second alternator. Another way of increasing the

(a)

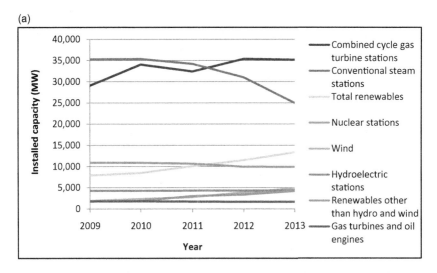

(b)

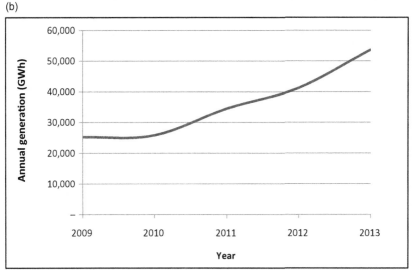

Figure 20.1 Growth of renewable energy generation capacity in the United Kingdom during 2009–2013.

overall efficiency of a conventional power plant is to utilize the remaining energy in the steam, after leaving the turbine for industrial processing or central heating (e.g., large areas of towns). In this arrangement, usually referred to as combined heat and power, the plant overall efficiency is increased to around 90%.

These plants are usually centrally controlled at national level, taking into consideration their capacity and dynamic characteristics, in order to match the overall demand, which continually varies but follows a fairly predictable pattern. This match is necessary in order to keep the frequency of supply constant. To meet a particular demand, the national grid control center selects a number of generating units with sufficient capacities to meet the demand and some spare capacity (generation dispatch). Units selected (committed) must be at appropriate locations in order to reduce transmission losses. To minimize generation costs, units committed must have the lowest production costs. Once a unit has been committed, it must be allowed to deliver at least the minimum power it can handle. The levels of units loading (generation schedule) is determined and thus controlled by the turbines' governors. Also, minimizing the cost of starting up and shutting down units is an additional constraint. Therefore, generating stations are normally listed in the order of economic merit, which is used to determine when each unit is to be committed and its output power at any loading conditions (generation scheduling). The incremental fuel cost, defined as the additional cost to increase the output of a generating plant by 1 MWh, is usually used as the benchmark for unit loading. For optimum economical operation, any demand for extra power should be met by a unit with the lowest incremental fuel cost, until this exceeds that of another unit [4].

A power generation plant may be assigned as a PQ plant, where the plant is assigned a fixed real and reactive power generation, or as a PV plant, where the plant is assigned a fixed real power generation and reactive power is varied to keep the voltage at the plant constant. A plant may be assigned to provide a balancing service to control frequency and voltage, where the active and reactive power outputs are variable.

20.3 Electricity generation from renewable energy resources

Renewable resources such as water, wind, biomass, tidal, wave, and solar energy are available at zero cost, are pollution free, and inexhaustible. At present, the cost of generation per unit of energy (MWh) from these resources (with the exception of hydroelectric and wind power plants) is relatively high compared to fossil-fuel sources. However, advances in technology and economy of scale are bringing down the cost of generation from RES.

Hydroelectric power plants have been used for a long time. However, the cost of construction of these plants is very high, in particular the civil engineering works. The pumped-storage plant is similar to a normal hydro plant except that two (upper and lower) reservoirs are used and the turbine–generator set is designed to also operate as a motor–pump set. When electrical energy is most needed, during peak load, the plant operates in generating mode where water flows from the upper reservoir to the lower one and electric power is generated and fed to the system network. Water stored in the lower reservoir is pumped back to the upper reservoir during off-peak periods when electricity is cheap and the plant is made ready for the next peak load. The overall efficiency of the pumped-storage plant (defined as electrical energy output/electrical energy input) is about 67%.

In further sections, main characteristics and grid interface controllers for RES are described, with emphasis on photovoltaic (PV) and wind energy conversion systems (WECSs). Although the characteristics of other types of RES (such as biomass, tidal, and wave) may be different, the way they generate electricity, and their connection and impact on the grid are similar.

20.3.1 Photovoltaic systems

The performance of a PV module is usually determined by its I–V characteristics, shown in Figure 20.2. These characteristics define the power output at any load point, giving the power curve shown in this figure. As can be seen, maximum power is only generated at a certain voltage level. This voltage varies with incident radiation, shading, temperature, and module soiling conditions. Therefore, it is important that a PV module is operated at (or around) this value in order to maximize energy capture.

The current (I) generated by a PV cell at any given voltage (V) may be expressed as [5]:

$$I = I_{SC}\left[1 - \exp\left\{\frac{\ln\left(1 - \dfrac{I_{MPP}}{I_{SC}}\right)(V - V_{OC})}{V_{MPP} - V_{OC}}\right\}\right] \tag{20.1}$$

where,

I_{SC} is the short circuit current,
V_{OC} is the open circuit voltage,
V_{MPP} and I_{MPP} are the cell voltage and current at maximum power point (MPP), which may be found from manufacturer data sheets.

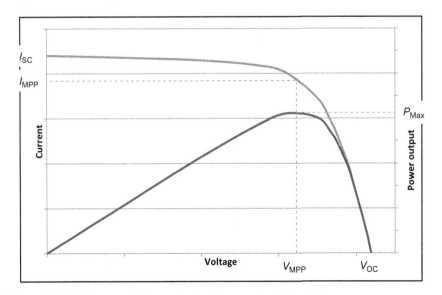

Figure 20.2 I–V characteristics and output power of a PV module.

A maximum power point tracker (MPPT) controller is usually used to determine the voltage, at which a PV module is able to produce maximum power and shifts the operating point to that voltage, regardless of load voltage. The MPPT is a high efficiency DC-to-DC converter that presents an optimal electrical load to a solar module (or array) and produces a voltage suitable for the load, irrespective of voltage variation on the PV module side (due to variations in irradiance, temperature, etc.). The MPPT decouples the PV module voltage from the load voltage, that is, it allows the PV to operate at MPP voltage (at the controller input) while delivering the required load voltage at the output of the converter. In addition to providing MPPT and "matching" functions, the converter is usually of boost type, which steps up the PV module voltage to allow operation at higher voltage levels, thus reducing power losses. The converter also provides flexibility in selecting rated voltages of the modules and batteries.

There are different approaches to control the DC-to-DC converter to track the MPP. They vary from a simple current controlled system where the PV or load voltage is assumed constant and the MPPT tries to maximize the current to a more sophisticated system, where the output power generated is optimized, such as incremental conductance, perturbation, and observation (P and O) and hill climbing MPPT controllers, which may employ digital signal processing techniques [6]. The principle for controlling the DC-to-DC converter is to change the duty ratio (mark to space ratio) of the converter such that the PV module voltage is controlled to occur at a value that produces maximum output power.

In standalone PV systems, the operating point is largely determined by the battery voltage, which is usually close to the MPP. As the battery voltage varies (depending on its charging state), the operating point also varies and may move away from the MPP. In practice, the battery is not allowed to fully discharge, hence the voltage is kept reasonably constant and close to the MPP. Therefore, for standalone systems, MPPT is desirable (to maximize the PV energy output), but not essential, as the PV module operates most of the time around the MPP.

20.3.2   Wind energy conversion systems

An alternative source of electrical energy that is becoming increasingly popular is the wind power plant. Due to their commercial viability, WECSs are currently the fastest growing renewable energy resources. A single wind-powered generator produces a limited amount of electrical power and hence a large group of generators is normally deployed and is called a "wind farm." As with many other renewable energy resources, wind energy has the disadvantage that degree and period of its availability are uncertain. Also, the balance between the capital costs of a wind turbine (or a farm) and the revenue from the electricity generated over the lifetime of the turbine must be carefully looked at.

In WECSs, the wind pressure rotates turbine blades attached to a shaft, which is coupled to the rotor of a generator. Large wind turbines usually employ a mechanical gearbox to increase the speed of the generator (usually induction type). Smaller wind turbines (below 10 kW) usually use permanent magnet generators.

The mechanical power output of a wind turbine (and hence electrical power output) is proportional to the cube of wind speed and may be expressed as in Ref. [7].

$$P_a = \frac{1}{2}\rho\pi R^2 v^3 C_p \tag{20.2}$$

where P_a is captured power by the wind rotor, R is the radius of the rotor in m, ρ is the air density in kg/m^3, v is the speed of the incident wind in m/s, C_p is the power coefficient, which for a given wind rotor depends on the pitch angle of the wind rotor blades and on the tip speed ratio (λ) defined as,

$$\lambda = \frac{\omega R}{v} \tag{20.3}$$

where ω is the rotational speed of the rotor in rad/s.

Equations (20.2) and (20.3) show that the output power of a wind rotor is a function of the wind speed (v) and rotational speed (ω) of the rotor and this relationship may be represented, as shown in Figure 20.3 [8]. Therefore, in order to maximize energy capture from the wind, it is necessary to regulate the wind turbine speed to match the optimal speed (ω_1 or ω_2) for any particular wind speed (v_1 or v_2), as shown in Figure 20.3.

There are two main types of wind turbines, namely, fixed speed and variable speed. Fixed-speed wind turbines employ a complex variable-pitch control mechanism of the blades and gearbox to optimize the energy captured. Consequently they have lower efficiency and require regular maintenance, as compared to variable-speed wind turbines, which employ power electronics converters to optimize the output power performance (without the need for aerodynamic controllers). In fixed-pitch variable-speed wind turbines, the electrical torque of the generator is usually controlled to maintain optimum wind rotor speed at a particular wind speed for maximum power output. A power electronic MPPT controller, similar to that used for PV systems, is usually used for this purpose.

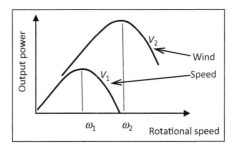

Figure 20.3 Wind turbine characteristics.

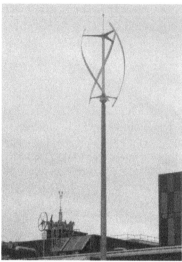

Figure 20.4 Horizontal and vertical axes small wind turbines.

In general, variable-speed wind turbines are characterized as having higher efficiency and better reliability. Hence, they are becoming more popular, particularly for small-scale applications.

The rotor in large wind turbines is usually a horizontal-axis, which provides better wind energy capturing capability. However, in turbulent wind environments, control of a horizontal-axis rotor to continuously follow the wind direction and operate at the MPP becomes difficult due to the fluctuation of wind speed. Hence, in such areas it is better to employ vertical-axis turbines, as shown in Figure 20.4.

20.3.3   Other renewable energy resources

Other renewable energy resources such as water, biomass, tidal, and wave are also available. Hydroelectric, pumped-storage, and biomass plants have been in use for a long time. These are usually large plants and centrally controlled, similar to conventional power plants. Tidal energy is generated by the rotation of the earth within the gravitational attractions of the Moon and the Sun, which causes the sea level to change. Power generation involves extracting the kinetic energy from moving water, set up by tidal flows. The output power is periodic and reasonably predictable. Waves are produced by the action of wind on the surface of the seas. Power is generated when the main mechanical structure of the turbine moves with the waves, causing oscillations or driving hydraulic motors, which drives an asynchronous generator. Wave energy can be considerable, up to 70 MW/km of shoreline coastal processes [9]. However, one major problem is that wave energy is not consistent, and therefore the output power is variable and largely unpredictable.

20.4 Grid connection of distributed RES

PV panels generate DC output and hence they need inverters to connect to the grid. Generators used with wind turbines can be asynchronous or synchronous. An asynchronous (induction) generator may be connected directly to the grid (usually through a soft starter to reduce transient currents) while a synchronous generator is connected through power electronics converters. Converters are also used for power generation from wave and tide. Generally, these converters generate harmonics, which can be harmful to the grid (e.g., extra losses, overheating, etc.) and therefore they have to meet applicable harmonic emission standards, as explained in Section 20.7.

20.4.1 Grid interface of PV systems

There are two types of PV systems, namely standalone- and grid-connected systems. In the former, PV modules are used to supply individual loads in isolation from other sources of electricity (the grid). In the latter, PV system is connected to and operated as a part of the main grid. As a PV system generates DC voltage, a DC/AC inverter is required if the system is to supply AC loads or is to be connected to the grid.

The main components of a PV system, which is suitable for both standalone and grid-connected operation are, PV modules, rechargeable batteries, a control unit, and an inverter (if AC output is required), as shown in Figure 20.5 [6].

An inverter is the interfacing device between the PV modules and the grid. Line-commutated inverters usually employ thyristors as the power electronic switching devices. These devices are not self-commutated and rely on the grid voltage to operate (commutate). Therefore, line-commutated inverters are not suitable for standalone PV systems. They draw reactive power from the grid (for thyristors commutation), hence they increase losses and reduce overall efficiency.

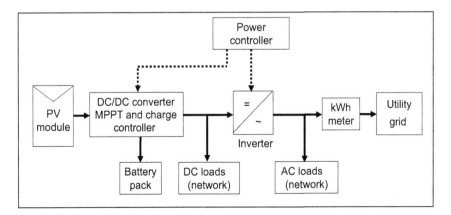

Figure 20.5 Main components of a PV system.

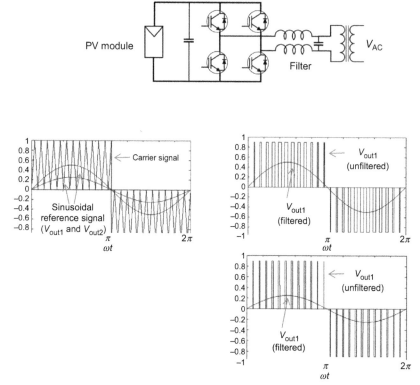

Figure 20.6 Typical self-commutated PV inverter with low frequency transformer and (PWM).

Self-commutated inverters employ self-commutated power electronic switching devices (usually MOSFETs, for low power applications and IGBTs, for medium to high power applications), as shown in Figure 20.6. These devices are switched on or off by a gate signal, thus the grid voltage is not necessary. Therefore, these inverters are suitable for both grid-connected and standalone PV systems. They do not need reactive power to operate, hence fewer power losses and better overall efficiency. In fact, the inverter may be used to generate reactive power, if required.

Voltage source inverters (VSI) are usually used in PV systems and these may be voltage controlled, where the output voltage is controlled such that it follows the desired reference signal while the current is load dependent. Alternatively, the VSI can be current controlled, where the output current is controlled to follow the desired reference signal. The current controlled VSI allows control of the power factor (i.e., unity or leading power-factor operation) and fault current (i.e., low fault currents) and therefore it is widely used for grid-connected systems. In standalone systems, the output current is determined by the load, so a voltage controlled VSI is essential.

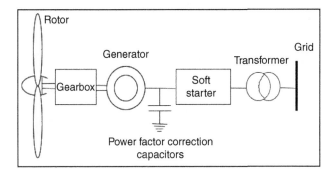

Figure 20.7 Fixed-speed wind turbine.

20.4.2 Grid interface of WECS

20.4.2.1 Fixed-speed wind turbine

In fixed-speed wind turbines, a variable-pitch fixed-speed wind turbine drives (through a gearbox) an asynchronous (induction) generator, which is directly connected to the grid, as shown in Figure 20.7. The gearbox is needed to increase the speed and together with the variable pitch they control the speed and optimize the energy captured from the wind. The gearbox reduces the overall efficiency of the WECS and requires maintenance. A soft starter is usually used to energize the generator and reduces transient currents while starting. Also, switched capacitors are needed to supply some of the reactive power demand of the induction generator.

When using an induction generator, the machine generates AC power by running at a small slip speed above synchronous speed. Hence, the turbine speed variations are also small, since any increase in wind speed increases the torque. Typical generator voltage is ~690 V, so a three-phase transformer is needed to step up the voltage. This is a simple (electrical) and relatively cheap system. This system tends to overspeed during faults in the grid and network operators expect the wind turbine to meet their "fault ride-through" requirements (i.e., the turbine system continues to work for a short time during faults on the grid) [10].

20.4.2.2 Variable-speed wind turbines

An alternative scheme employs power electronics converters to connect the generator to the grid. The converter decouples the generator from the grid, maintaining synchronization while allowing the turbine speed to vary depending on wind conditions. Thus, there is no need for pitch control or it can be very simple. Compared to fixed-speed systems, variable-speed systems have improved system efficiency, energy capture, and reduced mechanical stresses. Hence, they are becoming more popular. Two versions of this system are common, and are briefly described as follows.

1. *Wide-range variable-speed wind turbines with synchronous generator:* in this system, a multipole-synchronous generator, designed for low-speed operation, is usually used in order

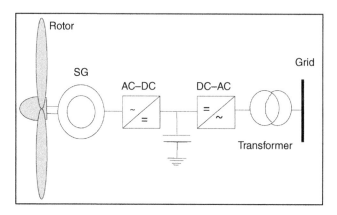

Figure 20.8 Wide-range variable-speed WECS.

to eliminate the need for the gearbox. Connection to the grid is through a power electronics converter, as shown in Figure 20.8, which completely decouples the generator speed from grid frequency [11]. The rating of a power converter corresponds to the rated power of the generator plus losses. The generator can be electrically excited through a field winding or a permanent magnet machine. Small-scale wind turbines usually employ permanent magnet generators. The generator produces variable frequency power, which is first rectified (AC-to-DC) and then converted to 50 Hz by using a self-commutated inverter (DC-to-AC).

2. *Doubly fed induction generator* (*DFIG*): in this case, a wound rotor induction (asynchronous) generator is used and variable-speed operation is obtained by injecting controllable voltage into the rotor at slip frequency [11]. The rotor winding current is fed through slip rings from a variable frequency power electronics converter, which is supplied from the grid, as shown in Figure 20.9. The power converter decouples the grid frequency from the rotor mechanical frequency, enabling variable-speed operation of the wind turbine. The power rating of this converter is determined by the rotor power, which in turn is determined by the operating slip of the generator, which is usually very small, resulting in a converter rating, which is a fraction of the full power (usually about one-third of the generator's rated power).

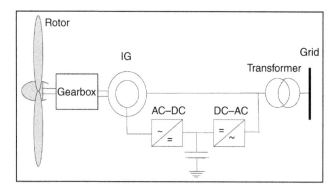

Figure 20.9 DFIG.

20.4.3 Grid interface of other RES

Similar to conventional power plants, synchronous generators are used in hydroelectric, pumped-storage, and biomass plants. The plants are usually centrally controlled and "dispatchable." Therefore, no special grid connection issues will normally arise unless the power plant, for logistical reasons, is located in a weak part of the grid.

Power generation from tidal and wave energy is not steady and therefore not centrally "dispatchable." However, it is worth noting that tidal power is periodic and reasonably predictable. Power is usually extracted from wave energy by using linear or asynchronous generators operating at variable speeds. Similar to WECSs, power electronics converters are usually needed to connect the generators (driven by tidal and wave turbines) to the grid. Therefore, grid connection considerations are similar to those applicable to WECS.

20.5 Distributed renewable energy sources

Electric power networks have evolved over the years to have large centrally controlled generators connected to the high voltage side of the network. Consequently, power flow is from the high voltage side (where generators are connected) to the low voltage side of the network (where medium and small size loads are connected).

Renewable energy sources connected to the grid can be either relatively very large plants (e.g., hydro and wind farms) with hundreds of MW power capacity or small plants less than 50 MW. The former is usually connected to the transmission system and is centrally controlled. The latter is relatively small, compared to central generation and is neither centrally planned, nor dispatched. Distributed RESs are relatively small and connected to the distribution network, typically below 33 kV (close to the point of use). This usually relates to noncentralized generators, ranging in size from around 1 kW to around 50 MW, and are referred to as distributed, embedded, or dispersed generation. Examples of distributed RESs are large wind turbines (or farms) up to 50 MW and small-scale wind and PV systems (or microgenerators) [12].

20.5.1 Benefits of distributed RES

Potential benefits of distributed RESs may be summarized as follows.

1. Energy efficient, reduced CO_2 emissions, and environment friendly.
2. Reduced network (infrastructure) and central generation capacities.
3. Generation closer to loads, thus reduced power losses in the network.
4. Diversity in terms of energy supply and location, thus a more secure and reliable supply.

To encourage the uptake of low carbon technologies, several countries have introduced incentives to support generation from distributed RES up to 5 MW. This is usually referred to as the feed-in tariff (FIT), which is a payment made to customers for both generation and export of produced renewable energy.

20.5.2 Impact of distributed RES on the grid

Currently, distributed RES amounts to only a small proportion of the total network generating capacity. Hence, the impact on network performance is not significant. However, with the increased number of new and renewable energy resources being connected to the grid, these will start to have an impact and create potential problems in existing power networks (which were not designed to accommodate RES).

Existing power networks rely on central generation and a national transmission grid that enables central control of the system at a national level by means of accurate statistical prediction of overall demand (allowing for diversity). Loads vary with respect to size and time but are fairly predictable. To ensure continuity of supply, each supply utility maintains its own generation margin, typically about 10% of peak demand or the size of the largest generator. Large centrally controlled generators are connected to the transmission system, which is highly automated and centrally controlled. Loads are connected at the distribution network, which has limited automation, for example, transformer tap changers and autoreclosing of circuit breakers. This arrangement results in power flow from generating plants via the transmission and distribution networks to medium and smaller sized loads, that is, directional power flow, as shown in Figure 20.10. As power flow is directional, the quality of power supply (frequency and voltage control) is maintained by a central control, and utilities have developed the necessary skills and procedures to do this.

Significant connection of RES at the distribution level will change the structure and power flow in future power networks. Figure 20.11 shows a possible structure of future power networks. Studies [13–16] have shown that significant deployment of distributed RES could lead to a bidirectional power flow in distribution networks and, unless adequately controlled, this may have significant impacts on the quality of power supply. Understanding these effects is very important, as they will influence the way future power networks will be designed and operated.

Potential problems for distribution networks from deployment of distributed RES have been identified as follows [13–16].

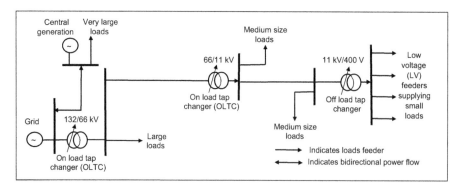

Figure 20.10 Existing power distribution networks.

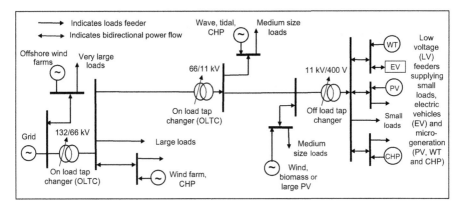

Figure 20.11 Future power distribution networks.

1. *Change in voltage level and violation of statutory limits*: these limits are set up according to the distribution code in each country, for example, in Europe these are –6% to +10% for the low voltage (400 V) level and ±5% for the higher voltage levels.
2. *Difficulties in power flow control and potential reverse power flow*: distributed RES will influence power flow in the network and this may lead to overloads on network feeders and equipment, depending on demand and generation profiles. Transformer tap changers normally have "line drop compensation" and this will malfunction with reverse power flow.
3. *Increase in short circuit contribution and fault levels*: network equipment has a defined fault handling capacity, usually for a short time (until protection operates and isolates the faulty part, which usually takes 0.2–1.0 s). Increase of fault level depends on the type of RES. When using rotating machines, for example, wind turbines, the increase in fault level can be significant (depending on the penetration level). When using power electronics converters to interface RES to the grid, the contribution to fault current can be made negligible, as the converter can be switched off almost immediately as soon as a fault is detected.
4. *Issues with protection management*: distributed RES will cause additional "uncontrolled" power flow in the network during faults, and this will affect existing relay settings and may also jeopardize network protection. It will also make it difficult to set relays, as fault current seen by the relay keeps changing depending on the state of generation of the distributed generator (DG) (whether on or off). Another issue to consider is that of islanding operation. As described in Section 20.5.3, according to current legislations, distributed RES can only operate when connected to the grid, that is, it must shut down if connection to the grid is lost (likely due to faults on the grid). Therefore, the DG must have anti-islanding protection or a loss-of-mains protection installed.
5. Potential voltage imbalance, which is specific to single-phase small-scale distributed generation.

The anticipated effects of large deployments of distributed RES need to be properly analyzed and appropriate measures should be taken in order to avoid potential problems in future power networks. Factors to consider include the following:

1. Type and capacity of the generator and voltage level for connection to the grid (point of common coupling, (PCC)).
2. Technology and type of grid interface.
3. Availability of power generated and reliability of plant.

4. Variation of power generated with time and predictability of variation.

5. Regulations and grid code for connection.

20.5.3   Islanded operation and microgrids

The distributed generation owner needs to ensure that the system is safe for installers, operators, and users (e.g., earthing, compliance with standards, labeling, isolation, etc.). Current regulations for power systems operation require that distributed generation be disconnected in the event of a fault on the grid or if the grid voltage drops [12,17]. This is usually done if any of the following exceeds a preset level: frequency, rate of change of frequency (ROCOF), or voltage level. The requirement for disconnection is mainly due to issues related to safety and control of the grid and the island. Therefore, it is not permitted to operate an isolated local power area (an "island") electrically separated from the grid, even though this would permit DGs to carry on supplying local demand. Under IEEE Standard 1547 [16], DGs rated below 10 MVA must be disconnected in the event of a major fault occurring on the grid. There is no provision for setting up a power "island" in this situation, however desirable this might be from the point of view of continuity of supply. The reasons for this are related to safety and technical issues. If an isolated area continued to be powered from distributed RES internal to the area, repair workers will be at risk from unexpected dangerous voltages and there is also the risk of damage to distribution equipment during reconnection. The other safety issue is that the grid operator cannot guarantee adequate earthing in the island, if the grid connection is lost. The technical issue is with regard to balancing supply and demand to maintain constant voltage and frequency in the island.

The opportunity of setting up "islands" was not thought important enough when IEEE 1547 was drafted to be worth special provision, since distributed generation was in its infancy. IEEE Standard 1547.4 [18], written more recently, considers the deliberate retention of distributed generation within an "island" when a system fault is present elsewhere. Sources of distributed generation if connected to the grid in conformance to the standard must be capable of operating in "island mode," thus benefiting adjacent consumers. Clearly for "islanding" to work, distributed generation within the island must be capable of supplying the load requirements, including those for reactive power. The standard calls for provision of distributed control and monitoring apparatus to ensure that power supply and demand are kept in equilibrium, and that frequency and voltage limits are observed. The cost of the additional equipment is considerable, restricting the degree to which "islanding" has so far been adopted. Autonomous control of voltage and frequency is only permitted during operation as an "island", and is not needed when the area is reconnected to the remainder of the grid.

Microgrids are power "islands" that are designed to operate as separate entities separated from the grid for extended lengths of time. Typically a microgrid would contain its own sources of generation plus power storage and may be uninterruptible power supplies [19,20]. Where consumers require very high levels of supply reliability such a scheme may be attractive. Possible users may include hospitals and defense establishments. Microgrids operating as "islands" would ensure continuity of supply even if there were failures in the grid [21]. At present, developments in microgrids are in their early stages. As of 2011 there were only 160 ongoing microgrid projects worldwide, involving some 1.2 GW [22]. Microgrids suffer from the same cost drawbacks as "islanding" systems conformable with IEEE 1547.4 [18]. The benefits, in terms of reliability of supply, may be obtained using conventional backup generators, and this approach tends to be simpler and cheaper.

20.6   Voltage control in power networks

As explained in Section 20.5.2, the flow of power in existing power distribution networks is from higher to lower voltage levels. Therefore, utilities estimate the voltage drop in the feeders based on the network configuration and assume that minimum and maximum loading on the feeders will remain relatively constant. Accordingly, a transformer's off-load tap position and on-load tap changer's (OLTC) initial position are set to raise the voltage to a level that will compensate for the voltage drops.

20.6.1   Voltage drop in feeders

It is usual practice in power systems to represent a part of a network by its Thévenin equivalent circuit. A Thévenin circuit comprises of a voltage source (e.g., sending-end voltage of a feeder) and a series impedance (e.g., impedance of the feeder), as shown in Figure 20.12, where a line represented by series impedance, $R + jX$, is supplying a load drawing a power, $P + jQ$.

The phasor diagram of the circuit is shown in Figure 20.13 and the line voltage drop may be derived as follows.

From the phasor diagram,

$$V_S^2 = (V_R + \Delta V)^2 + \delta V^2 \tag{20.4}$$

$$= (V_R + RI\cos\phi + XI\sin\phi)^2 + (XI\cos\phi - RI\sin\phi)^2 \tag{20.5}$$

Since $P = V_R I \cos\phi$ and $Q = V_R I \sin\phi$,

$$\therefore \quad V_S^2 = \left(V_R + \frac{RP}{V_R} + \frac{XQ}{V_R} \right)^2 + \left(\frac{XP}{V_R} - \frac{RQ}{V_R} \right)^2. \tag{20.6}$$

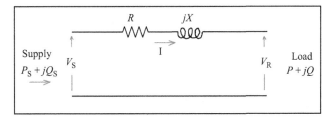

Figure 20.12 Equivalent circuit of a power network.

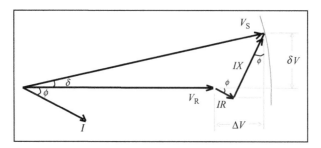

Figure 20.13 Phasor diagram.

Normally, $\delta V \ll V_R + \Delta V$

$$\therefore \; V_S^2 \approx \left(V_R + \frac{RP}{V_R} + \frac{XQ}{V_R} \right)^2 . \tag{20.7}$$

Hence, magnitude of the line voltage drop is approximately given by,

$$V_S - V_R = \Delta V \approx \frac{RP + XQ}{V_R} \tag{20.8}$$

and the angular shift is determined by,

$$\delta V = \frac{XP - RQ}{V_R} \tag{20.9}$$

where ΔV is the voltage drop, R and X are the resistance and reactance of the feeder, P and Q are the active and reactive power flow, and V is the grid voltage.

The effect of the line resistance (R) is often significant in low voltage (400 V) distribution networks. However, on the high voltage side of the network (11 kV and higher), the X/R ratio is normally high and the effect of the feeder resistance may be ignored.

Therefore, normal variations in the power generation of RES may cause voltage variations, which will be experienced by other consumers connected to the PCC (a substation that supplies several points is referred to as the, PCC).

20.6.2 Voltage control using tap changers on distribution transformers

Control of transformer's tap changer is the most popular form of voltage control at all voltage levels in a power system. This is based on changing the turns ratio of a transformer, hence the voltage in the secondary circuit is varied and voltage control

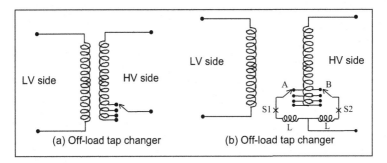

Figure 20.14 Distribution transformers.

is obtained. The tap changer is usually placed on the high voltage winding as the current is lower and this minimizes the current handling requirements and stress during operation of the tap changer.

Figure 20.14a shows a schematic diagram of an off-load tap changer that requires disconnection of the transformer when the tap setting is to be changed. Most transformers have an OLTC, which is shown in basic form in Figure 20.14b. In the position shown, the voltage at the LV side is at its maximum and the current divides equally in to the two halves of the coil L, resulting in zero resultant flux and minimum impedance. To reduce the voltage at the LV side, S1 opens and the total current passes through the other half of the reactor L. Selector switch A then moves to the next contact and S1 closes. A circulating current now flows in L superimposed on the load current. Then, S2 opens and B moves to the next tapping; S2 then closes and the operation is complete.

To avoid large voltage disturbances, the voltage change between taps is normally small, about 1.25% of the nominal voltage. The total range of tapping varies with transformer usage; a typical figure for generator transformers is +2% to −16% in 18 steps.

Voltage control using OLTCs is usually based on voltage measurement at the transformer location. Some OLTCs also measure the current through the transformer to adjust for the variation in load currents (line compensation). Existing controllers for line compensation are designed to measure the current in one direction only, as power flow in existing power networks is in one direction.

As explained in Section 20.5.2, with significant penetration of distributed RES, power flow in the network may be altered, with potential reverse power flow and voltage rise, which falls beyond the control limits of the OLTC, as shown in Figure 20.15. As can be seen, the voltage rise is most significant at the remote end of the feeder where the generator is connected. The amount of voltage rise depends on the level of RES penetration level, operating p–f, loading conditions (peak or off peak) and distance from the main transformer. Violation of statutory upper voltage limit is likely to occur at minimum loading condition and this worsens when the generator is operating at lagging p–f (supplying reactive power).

This rise may be felt at the point of the RES connection and possibly at the PCC with other sensitive loads. Too many distributed RES connected to the network increase the

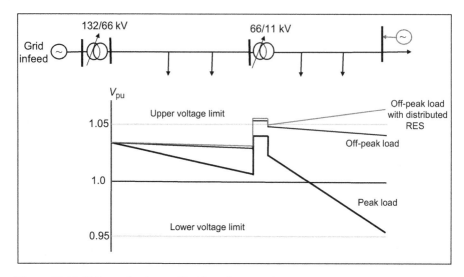

Figure 20.15 Voltage rise due to distributed generation.

risk of voltage rises. Therefore, the philosophy of voltage control in future distribution networks will need to change in order to allow for large penetration of distributed RES.

20.6.3   Voltage fluctuation (flicker)

When connecting distributed RES to the grid, one needs to ensure that the system does not interfere with other users of the power network (flicker, harmonics, transients, fault in-feed, etc.). Voltage fluctuations or "flicker" are rapid changes in voltage magnitude within the statutory limits of the usual slow variations of voltage (±5% of nominal value). These fluctuations can cause noticeable variations in lighting (flicker) and interrupt the operations of some electronic controllers. Research has shown that some loads are very sensitive to certain fluctuations of voltage magnitude such that a fluctuation of 0.5% at a frequency of 5–6 Hz will cause annoying flickers, as shown in Figure 20.16 [23].

Voltage fluctuations can occur due to rapid change in current flow in the grid. Power generated from distributed RES (e.g., wind and wave) can be variable and intermittent with frequent and repetitive changes. Also, clouds passing by a PV system may produce a rapid change in its power output. Such generation results in a large change in active and reactive power exchanged with the grid and may give rise to voltage fluctuations at the PCC.

Figure 20.17 shows two alternative ways of supplying a 1 pu purely inductive load X_L. This load may be connected to busbar A or B, both of which are connected to supply point C via transformers T_2 and T_3. The effect of switching on this load on the voltage level at busbar A where another (sensitive) load (load 1) is already connected is analyzed. To simplify the analysis, load 1 is assumed to be very small relative to the inductive load under consideration.

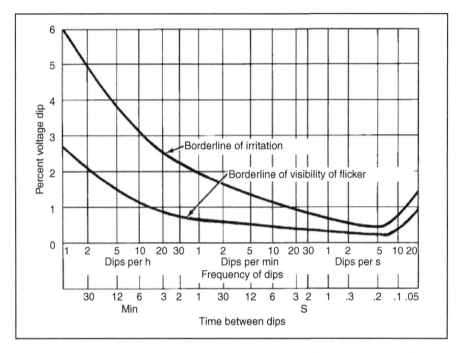

Figure 20.16 Range of observable and objectionable voltage flicker versus time [23].

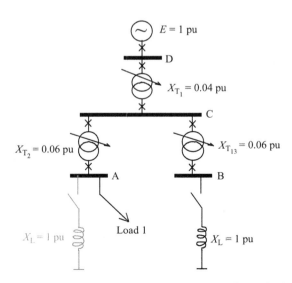

Figure 20.17 Two alternative ways of supplying a 1 pu purely inductive load.

If the inductive load X_L is connected to busbar A and is switched on, the voltage at busbar A will be,

$$V_A = \frac{E}{X_{T_1} + X_{T_2} + X_L} X_L$$
$$= \frac{1}{1.1} \times 1 = 0.9091 \text{ pu}$$

(20.10)

That is, when the inductive load X_L is switched on, a voltage drop of approximately 9% is produced at busbar A. Note that as load 1 is assumed to be very small, the voltage at busbar A before connecting the inductive load is equal to 1.0 pu (grid voltage). Clearly, this voltage drop will also be experienced by other consumers on this busbar (load 1). However, if load X_L is connected to busbar B, the same voltage drop will be produced at busbar B, but only a fraction of this voltage drop will be seen at busbar A, through the voltage changes produced at busbar C (the PCC). The voltage at busbar C is:

$$V_C = \frac{E}{X_{T_1} + X_{T_3} + X_L}(X_{T_3} + X_L)$$
$$= \frac{1}{1.1} \times 1.06 = 0.9636 \text{ pu}.$$

(20.11)

Therefore, when load X_L is connected at busbar B and switched on, it will produce less than 4% voltage drop at busbar A.

These two alternatives represent a change in the PCC for the loads (load 1 and load X_L) from busbar A to busbar C. That is, moving the PCC to a higher voltage level will significantly improve the quality of supply in the presence of disturbing loads.

Variations in voltage level at the PCC can be minimized by providing a "strong" system, that is, one with lower source impedance X_{T_1}, X_{T_2}, and X_{T_3}, or by using voltage control devices. Slow acting devices such as automatic voltage regulators for generators, transformers, OLTCs, and conventional reactive power compensators are relatively slow and may not provide smooth control. Hence, they are inadequate for compensation of rapid fluctuations in voltage level (flicker). To deal with such fluctuations fast-acting reactive power compensators are necessary, such as those that employ power electronics (FACTS) technology [24].

20.7 Power quality and harmonics

20.7.1 Definitions and impact of RES

Power quality (PQ) is a generic term often used in relation to unwanted disturbances of the electricity supply. A PQ event (disturbance) is defined as any deviation (steady state or transient) of voltage or current waveforms from a pure sinusoidal form of a specified magnitude [25]. PQ events, as defined by IEEE Standards 1159–1995, include steady-state events (long-term) abnormalities in the voltage/current waveform as well as transient, short and long duration events that are sudden abnormalities with time scales ranging from a few nanoseconds to several minutes [26,27]. These standards define harmonics as "sinusoidal voltages or currents having frequencies that are integer multiples of the frequency at which the supply system is designed to operate (termed the fundamental frequency; usually 50 or 60 Hz)" [27]. Harmonics are superimposed on the fundamental component and cause waveform distortion.

In the recent years, there has been an increased number of PQ related problems. This is mainly due to the rapid growth in the use of nonlinear loads, mainly power electronic equipment that employs switching devices such as switched mode power supplies, variable speed drives, converters for connection of distributed RES, etc. This equipment usually includes sophisticated control, which is also sensitive to PQ disturbances. The most common PQ problem associated with the use of power electronics converters used for interfacing distributed RES to the grid is to do with harmonics, which is described in this section.

Power electronics converters used for grid connection and control of distributed RES are usually designed and tested to comply with existing standards and therefore should not produce significant PQ events. However, the interaction of different converters during normal operation may result in amplification of certain harmonics or events, which will be experienced by other consumers connected to the PCC. The consumer has a right to receive a supply free from significant distortion. Equally, the user must not cause distortion, which will reflect upon other consumers.

20.7.2   Harmonics effects and solutions

The effects of harmonics and the degree of susceptibility to harmonics depend upon the characteristics of the equipment (those that generate harmonics and those that are affected by them) and also on the specifications of the supply network. In general, possible effects of harmonics can be summarized as follows [28]:

1. Extra losses (I^2R), heating and overloading of power system equipment, for example, cables and transformers. In three-phase four conductor systems, triplen harmonic components add in the neutral conductor, causing overloading of the neutral conductor unless it is oversized.
2. Harmonics giving rise to resonance between the inductive and capacitive impedances in the system at harmonic frequencies. Consequently, high current or voltage may appear at points in the system causing damage to equipment.
3. Maloperation of control and protection equipment normally designed to function with a sinusoidal wave, for example, protective relays and controllers of adjustable speed drives or any electronics equipment that use the mains zero crossing point as a timing signal.
4. Interference with communication systems caused by induced noise from harmonic currents and voltages. Telephone systems may be subject to interference where the lines run adjacent to power lines carrying harmonic currents.
5. Metering and instrumentation affected by the presence of harmonics, particularly if resonant conditions occur.

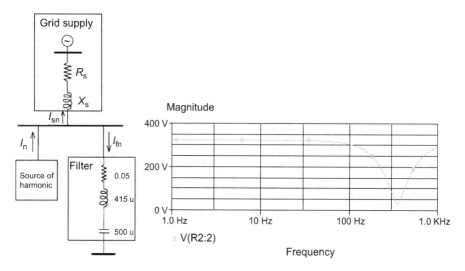

Figure 20.18 Harmonic current flow. (a) Power circuit and (b) harmonic spectrum.

6. Rotating machines experiencing overheating, noise, and torque pulsation in the presence of excessive harmonic distortion. The possible shaft torsional-vibration problems can be harmful to the motor loads, particularly for critical processes. Harmonics can also set up resonant conditions if the natural frequency of the mechanical system is excited by them.

The best way to deal with harmonics is not to generate them. Power electronics converters may be designed to produce low or no harmonics by using appropriate control, for example, using high pulse number, pulse width modulation (PWM) control, etc. If this proves difficult or expensive then filters may be used. Filtering of harmonics involves the provision of a low impedance path to ground for the harmonic frequencies of interest, while providing high impedance for the basic mains frequency. Such a circuit can be made using an LCR "acceptor," where the inductor is in series with a capacitor. At resonance the impedance of the circuit is basically that of the resistive component only, which can be made low. An example of this procedure, for rejecting the seventh harmonic of 50 Hz (350 Hz), carried out through OrCAD simulation, is shown in Figure 20.18.

20.7.3 Limits of harmonic current distortion

Under the provisions of IEEE 519, recommended limits for maximum harmonic current distortion (in percent of the maximum demand current I_L) for individual order odd harmonics are given in Table 20.1 [29]. Here, TDD is the total demand distortion defined as "the ratio of the root mean square of the harmonic content, considering harmonic components up to the 50th order and specifically excluding interharmonics, expressed as a percent of the maximum demand current" [29].

Table 20.1 **Current distortion limits (in percent of I_L) for general distribution systems (120 V to 69 kV)**

Order	$3 \leq h < 11$	$11 \leq h < 17$	$17 \leq h < 23$	$23 \leq h < 35$	$35 \leq h \leq 50$	TDD
Limit	4%	2%	1.5%	0.6%	0.3%	5%

TDD, total demand distortion.
Adapted from Ref. [29].

Table 20.2 **Planning levels of harmonic voltages in 400 V systems**

Odd harmonics (nonmultiples of three)		Odd harmonics (multiples of three)		Even harmonics	
Order "h"	Harmonic voltage (%)	Order "h"	Harmonic voltage (%)	Order "h"	Harmonic voltage (%)
5	4	3	4	2	1.6
7	4	9	1.2	4	1.0
9	3	15	0.3	6	0.5
13	2.5	21	0.2	8	0.4
17	1.6	>21	0.2	10	0.4
19	1.2			12	0.2
23	1.2			>12	0.2
25	0.7				
>25	$0.2 + 0.5(25/h)$				

Adapted from [27] Table 2, page 10.
Adapted from Ref. [27].

Even harmonics are limited to 25% of the odd harmonic limits.

Under the provisions of the UK G5/4, recommended limits for maximum voltage distortion in a 400 V system are given in Table 20.2 [30].

IEC Recommendations for voltage harmonic levels at low and medium voltage levels are given in Table 20.3, with a maximum total harmonic distortion of 8% [31].

20.8 Regulations for connection of distributed RES to the grid

When connecting a distributed RES to the grid, one needs to ensure that the system does not interfere with other users of the power network (flicker, harmonics, transients, fault in-feed). Also, ensure that the system does not present a danger to network operation (loss of mains disconnection). The owner of the distributed RES needs to ensure that the system is disconnected from the grid, properly shuts down, and protects itself under grid fault conditions. This is done if any of the following exceeds a preset level, frequency, ROCOF, or voltage level. Also, the owner needs to ensure that

Table 20.3 Compatibility levels for individual harmonic voltages in low and medium voltage networks (percent of fundamental components) [3]

Odd harmonics (nonmultiples of three)		Odd harmonics (multiples of three)		Even harmonics	
Order "h"	Harmonic voltage (%)	Order "h"	Harmonic voltage (%)	Order "h"	Harmonic voltage (%)
5	6	3	5	2	2
7	5	9	1.5	4	1.0
11	3.5	15	0.4	6	0.5
13	3	21	0.3	8	0.5
$17 \leq h \leq 49$	$2.27 \times 17/h$ $- 0.27$	$21 < h \leq 45$	0.2	$10 \leq h \leq 50$	$0.25 \times 10/h$ $+ 0.25$

Table 20.4 UK engineering recommendations that cover the connection of distributed generation to the electrical distribution network

Engineering recommendations	Generator rated power (P_{rated})	Voltage at PCC
G75	$P_{rated} \geq 5$ MW	$V_{rated} \geq 20$ kV
G59	11.1 kW $< P_{rated} < 5$ MW	400 V $\leq V_{rated} \leq 20$ kV
G83	Three-phase: $P_{rated} \leq 11$ kW	$V_{rated} = 400$ V
	Single-phase: $P_{rated} \leq 3.7$ kW	$V_{rated} = 230$ V

Adapted from Ref. [32].

the system is safe for installers, operators, and users (e.g., earthing, compliance with standards, labeling, isolation, etc.).

DGs must comply with relevant legislations in the country where it is installed, in addition to any national grid and distribution codes.

Current legislations for connection of DG in the United Kingdom are Engineering Recommendations G75, G59, and G83, which may be summarized as given in Table 20.4 [32].

In the United States, DG connection requirements may be summarized as given in Table 20.5, pursuant to FERC Order 2003a (Large Generator Interconnection) [33] and FERC Order 2006 (Small Generator Interconnection) [34]. The power limits are based upon the aggregate power output of a generating installation.

The cost of connection is dependent on several factors, including proximity to existing network connection point, level of reinforcement required to manage additional power flow, planning issues (e.g., overhead versus underground cables), capacity, and voltage level of connection in addition to any ancillary infrastructure requirements (protection, reactive power compensation, etc.).

Table 20.5 US FERC distributed generation connection requirements

Engineering recommendations	Generator rated power P_{rated}	Installation requirements
FERC order 2003a	$P_{rated} > 20\,MW$	Full engineering evaluation of impact on system
FERC order 2006	$2\,MW < P_{rated} < 20\,MW$	Full engineering evaluation of impact on system
	$P_{rated} < 2\,MW$	Fast track restricted evaluation of impact on system
	$P_{rated} \leq 10\,kW$	Application form-based permission for installation of certified inverter-based generators

Adapted from Refs [33,34].

20.9 Smart grid solutions

As described in Section 20.5, distributed RES can bring some environmental and commercial benefits as well as challenges to existing distribution networks. The impact of distributed RES is determined by the following factors, which need to be considered before connecting a system to the grid.

1. Type and capacity of the generator and the voltage level for connection to the grid (PCC).
2. Availability of power generated.
3. Variation of power generated with time and predictability of variation.
4. Reliability of the plant.
5. Technology and regulations for connection.

After consideration of these factors, if it is established that the distributed RES will negatively impact the grid, two options are possible to mitigate the impact. One is to restructure and reinforce the network but this could be prohibitively expensive. The other "smart" option is to employ new technologies such as active voltage control, demand-side management, and energy storage [35]. These are key technologies for the new concept of electricity networks of the future "smart grids" [35–37]. They will play an important role in controlling the power balance in future power networks and maximizing the energy output from distributed renewable energy generation. The smart grid requires advanced communication protocols in order to implement dynamic control and automation to allow the power network to operate closer to its capacity while maintaining system security and integrity.

As explained in Section 20.6.2, current OLTCs usually use local voltage measurements in order to bring the voltage at the transformer location towards a specified value and therefore provide an acceptable voltage profile across the feeder length. Although line compensation (measurement of current through the transformer) may help in adjusting voltage control based on load current variations, this "passive" control becomes less effective in the presence of distributed RES and may even malfunction

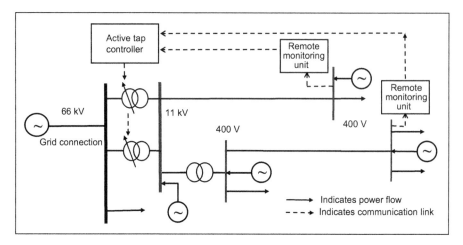

Figure 20.19 Active voltage control.

if reverse power flow occurs since line compensation is designed to measure current in one direction only.

Figure 20.19 shows an example of how the voltage rise caused by the connection of distributed generation (described in Section 20.6.2, Figure 20.15) can be mitigated by using active voltage control. This is achieved if the transformer tap changer control is based on voltage signals other than the local substation (e.g., from the remote end of a feeder). Dynamic rating may also be implemented to avoid thermal overloading of transformers and feeders.

In active network control, the voltage at a number of locations is monitored (through remote monitoring units and smart meters), and the OLTC is adjusted to maintain all of the measured locations within the desired limits. This allows the vulnerable points in the network, such as the far end of a feeder, to be monitored and kept within statutory limits. Number of remote monitoring units can be minimized by using state estimation [38].

Problems

1. Comment on the potential benefits of using distributed renewable energy generation.
2. Comment on the potential technical problems of an existing distribution power network from large-scale connection of distributed renewable energy generation. Suggest and briefly explain possible solutions using new technologies that will help to mitigate these problems.
3. According to existing legislations, how should the protection of a DG respond to a short circuit fault on the main network?
4. In grid-connected PV systems, self-commutated voltage-source inverters can be either voltage or current controlled. Briefly describe the difference between the two in terms of operating principles and main features.
5. Briefly describe, with the aid of suitable diagrams, three arrangements for connecting WECSs to the electricity grid, giving the main components and features of each arrangement.

6. Define the concept of the smart grid and show how it can help in the integration of RESs into the grid.

7. A three-phase power electronic load rated at 150 kW at 0.8 p–f lagging is fed from a 400 V supply having a fault in-feed of 2 MVA. Assuming that the source impedance is purely inductive and ignoring the effects of other loads connected to the network, calculate the percentage voltage dip at the load supply point when the load is switched direct on line.

8. A 10 km feeder is used to supply a load connected at the far end of the feeder and has a maximum capacity of 4 MVA at 0.8 p–f lagging. The feeder has a series resistance $r = 0.25$ Ω/km per phase and an inductive reactance $x = 0.1$ Ω/km per phase. The sending-end voltage is maintained at 11.8 kV.

   a. Estimate the magnitude of the load voltage at maximum loading conditions.

   b. A 2.0 MW wind turbine employing an induction generator is connected at the far end of the feeder. Assuming minimum loading conditions of 1 MVA at 0.9 p–f lagging, estimate the load voltage when the generator is operating at 80% capacity and 0.9 p–f leading.

   c. Comment on the results obtained and briefly explain how variations in wind speed can result in fluctuations in the system voltage. Suggest two possible methods to mitigate the effects of these variations.

9. A developer wishes to connect a 2 MW distributed RES to an existing distribution network. In order to do this, an underground cable is to be used to connect the generator to the nearest substation. A simplified equivalent circuit of the proposed system is shown in Figure 20.20. The existing network has a substation (PCC) with a breaker feeding other consumers, as shown in the figure. The prospective fault rating at the substation is 10 MVA at zero p–f lagging and the fault clearing capacity of the circuit breaker is 10.5 MVA. The impedance of the proposed DG (Z_G) is $j0.04$ pu and the underground cable (Z_L) is $j0.06$ pu.

   The developer needs to verify that the circuit breaker will be capable of handling the prospective fault current after DG connection. Given a per unit base of 100 kVA;

   a. calculate the per unit equivalent source impedance, Z_N.

   b. Calculate the fault power at the substation after generator connection.

   c. Comment on the implications of the results obtained in (b) for the distribution network operator and the developer. What steps might the developer take to satisfy the network operator requirement with regard to the fault level?

   d. If the impedance of the generator could be changed, what would be the minimum value required so that the fault power at the PCC does not exceed the circuit breaker capacity?

   e. Apart from fault current, list four other issues that need to be considered before connecting additional embedded generation to an existing network.

10. For the distribution system shown in Figure 20.21, the feeder is 10 km long and has a series resistance $r = 0.25$ Ω/km per phase and an inductive reactance $x = 0.1$ Ω/km per phase. The load has after-diversity minimum and maximum demand of 0.5 and 4.0 MW, respectively

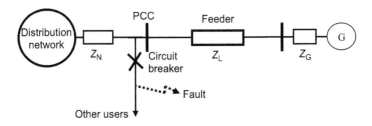

Figure 20.20 A simple distribution network with a DG.

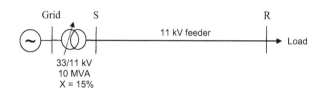

Figure 20.21 An 11 kV distribution feeder.

both at 0.9 p–f lagging. The feeder voltage is controlled by the OLTC at the 66/11 kV transformer. Assume that the grid voltage is fixed at 33 kV and that the transformer has taps of ±12.5% in steps of 2.5% (Figure 20.21).

Usually, the target voltage (control signal) for the OLTC is the voltage at busbar S and assuming that the voltage at busbar R is to be maintained within the statutory limits of 11 kV ± 5% for both minimum and maximum loading conditions. Using a base power of 10 MVA and taking the voltage at busbar R as a reference;

a. determine busbar S voltage for both minimum and maximum loading conditions.

b. Determine, with justification, the initial tap position of the OLTC (i.e., the voltage that the OLTC should maintain at busbar S) in order to keep the line voltages at busbar R within the statutory limit.

c. A 2.0 MW wind turbine employing an induction generator is connected at busbar R. Assuming minimum loading conditions and tap setting as calculated in (b) and the generator is operating at 80% capacity and 0.9 p–f leading, determine whether the OLTC is capable of maintaining the load voltage at busbar R within the statutory limits (±5%). Briefly explain why connection of distributed generation may result in a voltage rise beyond the statutory limit.

d. Active voltage control may be used to provide better control of voltage levels in a network with distributed generation. Describe how the concept of active voltage control may be implemented for the system in Figure 20.20, explaining how this can help the integration of RESs into the smart grid.

References

[1] International Energy Agency or Committee on Climate Change. The 2050 target – achieving an 80% reduction including emissions from international aviation and shipping. Available from: http://hmccc.s3.amazonaws.com/IA&S/CCC_IAS_Tech-Rep_2050Target_Interactive.pdf; 2012.

[2] Department of Energy & Climate Change Renewables statistics. Energy trends, Section 6. Available from: https://www.gov.uk/government/statistics/energy-trends-section-6-renewables. Accessed on December 22, 2014.

[3] Department of Energy & Climate Change. Digest of United Kingdom Energy Statistics 2014, a National Statistics publication, London. https://www.gov.uk/government/uploads/system/uploads/attachment_data/file/338750/DUKES_2014_printed.pdf

[4] Weedy BM, Cory BJ, Jenkins N, Ekanayake J, Strbac G. Electric power systems. Chichester, UK: John Wiley & Sons, Ltd; 2012.

[5] Julius Susanto et al., Open Electrical. Available from: http://www.openelectrical.org/wiki/index.php. Photovoltaic_Cell_Model. Accessed on January 19, 2015.

[6] Sick, F., Erge, T. Photovoltaics in buildings; a design handbook for architects and engineers. International Energy Agency, Paris, France, London: James & James; 1996

[7] Gourieres DL. Wind power plants theory and design. Oxford: Pergamon Press; 1982.

[8] Narayana M, Putrus G, Jovanovic M, Leung PS, McDonald S. Generic maximum power point controller for small scale wind turbines. Elsevier J Renew Energy 2012;44:72–9.

[9] Benassai G, Dattero M, Maffucci A. "Wave Energy Conversion Systems: Optimal Localisation Procedure". Southampton, UK: Book chapter published in "Coastal Processes" by WIT Press; 2009. pp. 129–138. ISSN 1743-3541.

[10] Tsili M, Papathanassiou S. A review of grid code technical requirements for wind farms. IET Renew Power Gen 2009;3(3):308–32.

[11] Fox B, et al. Wind power integration: connection and system operational aspects. IET Power and Energy Series 2007; London, UK.

[12] Ackermann T, Andersson G, Soder L. Distributed generation: a definition. Elec Power Syst Res 2001;57:195–204.

[13] Ingram S, Probert S, Jackson K. The impact of small scale embedded generation on the operating parameters of distribution networks, Department of Trade and Industry, UK. Report Number: K/EL/00303/04/01; 2003.

[14] Lyons PF, Taylor PC, Cipcigan LM, Trichakis P, Wilson A. Small scale energy zones and the impacts of high concentrations of small scale embedded generators, UPEC2006 Conference Proceedings. Vol. 1; September 2006, pp. 28–32.

[15] Barbier C, Maloyd A, Putrus GA. Embedded controller for LV network with distributed generation. DTI project, Contract Number: K/El/00334/00/Rep, UK; May 2007.

[16] Jenkins N, Ekanayake J, Strbac G. Distributed generation. IEE Renew Energy Series 2010; London, UK.

[17] IEEE Std 1547-2003, IEEE Standard for Interconnecting Distributed Resources with Electric Power Systems, IEEE Standards Coordinating Committee 21.

[18] IEEE Std 1547.4-2011, IEEE Guide for Design, Operation, and Integration of Distributed Resource Island Systems with Electric Power Systems, IEEE Standards Coordinating Committee 21.

[19] Lasseter R, et al. Integration of distributed energy: the CERTS microgrid concept, LBNL-50829. Berkeley, CA: Lawrence Berkeley National Laboratory; 2002.

[20] Marnay C, Robio FJ, Siddiqui AS. Shape of the microgrid, IEEE Power Engineering Society Winter Meeting, Columbus, OH; January 28–February 1, 2001.

[21] Lasseter R, Eto J. Value and technology assessment to enhance the business case for the CERTS microgrid. Madison, WI: University of Wisconsin–Madison; 2010.

[22] Asmus P, Davis B. Executive summary: microgrid deployment tracker. Boulder, CO: Pike Research; 2011.

[23] IEEE Standards 141-1993, Recommended Practice for Electric Power Distribution for Industrial Plants, IEEE Standards Board.

[24] Hingorani NG, Gyugyi L. Understanding FACTS: concepts and technology. New York, USA: IEEE Press; 2000.

[25] Bollen MHJ. Understanding power quality problems: voltage sags and interruptions. New York, USA: IEEE Press; 2000.

[26] Putrus GA, Wijayakulasooriya JV, Minns P. Power quality: overview and monitoring. Invited paper, International Conference on Industrial and Information Systems (ICIIS 2007); Peradeniya, Sri Lanka; August 8–11, 2007.

[27] IEEE Standards 1159-1995, Recommended Practice for Monitoring Electric Power Quality, IEEE Standards Board.
[28] Balda JC. Effects of harmonics on equipment. IEEE Trans Power Deliv 1993;8(2): 672–80.
[29] IEEE Std 519-2014, Recommended Practice and Requirements for Harmonic Control in Electric Power Systems, IEEE Power and Energy Society.
[30] Energy Networks Association, G5/4, February 2001.
[31] McGranaghan M, Beaulieu G. Update on IEC 61000-3-6: harmonic emission limits for customers connected to MV, HV, and EHV. Transmission and Distribution Conference and Exhibition, 2005/2006 IEEE PES. Dallas, Texas; May 2006, pp. 1158–1161.
[32] Energy Networks Association, Various Connection Engineering Recommendations. Available from: http://www.energynetworks.org/electricity/engineering/distributed-generation/distributed-generation.html [accessed on 3.18.2015].
[33] United States of America Federal Energy Regulatory Commission 18 CFR Part 35 (Docket No. RM02-1-001; Order No. 2003-A) Standardization of Generator Interconnection Agreements and Procedures [issued on 5.3.2004].
[34] United States of America Federal Energy Regulatory Commission 18 CFR Part 35 (Docket No. RM02-12-000; Order No. 2006) Standardization of Small Generator Interconnection Agreements and Procedures [issued on 12.5.2005].
[35] Ekanayake J, Liyanage K, Wu J, Yokoyama A, Jenkins N. Smart grid: technology and applications. Chichester, UK: John Wiley & Sons, Ltd; 2012.
[36] European Commission. European Smart Grids Technology Platform: Vision and Strategy for Europe's Electricity Networks of the Future, EUR22040; 2006.
[37] Putrus GA, Bentley E, Binns R, Jiang T, Johnston D. Smart grids: energising the future. Int J Environ Studies 2013;70(5):691–701.
[38] Abdelaziz AY, Ibrahim AM, Salem RH. Power system observability with minimum phasor measurement units placement. Int J Eng Sci Technol 2013;5(3):1–18.

Environmental impacts of renewable energy

21

Rosnazri Ali, Tunku Muhammad Nizar Tunku Mansur, Nor Hanisah Baharudin,
Syed Idris Syed Hassan
School of Electrical System Engineering, Universiti Malaysia Perlis (UniMAP), Arau, Perlis,
Malaysia

Chapter Outline

Electric Renewable Energy Systems
Copyright © 2016 Elsevier Inc. All rights reserved.

21.1 Introduction

Energy has become the driving force of economic growth starting from the industrial revolution in the eighteenth to the nineteenth centuries. Industrialization has shifted from human energy to create products and basic machines to a new world of coal-powered machines for mass productions. Quality of life has increased too with transportation, health, and consumer products tremendously improved. Life without electricity nowadays is unthinkable, as billions of people depend on electricity for their daily activities. An oil crisis occurred in 1973, which embarked upon the importance of energy security and renewable energy resources becoming another option to support the world's demand for electricity. Critically, the depletion and unstable supply of fossil fuel increases the cost of electricity. Furthermore, adverse effects from fossil fuel combustion to generate electricity since the industrial revolution have put further pressure on society to realize global warming issues.

It is widely known that nowadays nuclear energy is the best option to cope with the increasing demand of energy. However, the Fukushima Daiichi disaster in 2011 startled the world. The catastrophic impacts from radioactive explosion are indeed disastrous and widely damaging, and the total clean up may take up to 30–40 years, as announced by Japanese government.

Due to the nuclear disaster in Fukushima Daiichi, many countries such as Germany have taken serious actions toward nuclear power plants. In addition, European Union has committed to reduce its carbon dioxide emission less than 80%–95% compared to 1990 levels to further limit the increase of global temperature by 2°C as established by United Nations Climate Change Conference 2009 in Copenhagen. Thus, Germany has planned to phase out nuclear energy by 2022 with a program called "Energiewende," which means energy transition from fossil fuel and nuclear to 100% renewable energy. This energy-efficient program mainly improves on the renewable energy role to supply primary energy demand by up to 50%, or even 100%, by 2050 [1]. This energy transition program has projected that by 2050, 80% of electricity demand will be supplied by the country's own renewable energies and another 20% of renewable energy generated will be imported by neighboring countries such as Norway through existing hydropower reservoirs [2,3].

It is widely known that fossil fuel power plants have led to pollution and global warming issues. Thus, the idea to promote renewable energy power plants to further replace the existing fossil fuel and nuclear power plants is a better solution for a greener earth that will limit the increase in global temperatures. However, even this is quite ambiguous and ambitious. The environmental impact of cradle-to-grave renewable energy technologies will be explained in this chapter to further consider these technologies as compared with environmental impacts from fossil fuel and nuclear power plants.

21.2 Environmental concerns related to fossil fuel power plants

The reliability of fossil fuel power plants in generating electricity has successfully contributed to global economic growth and has improved quality of life for decades. However, the adverse effects of these conventional fossil fuel power plants have increased significantly toward the emissions of greenhouse gases. From 1995 to 2011, global greenhouse gas emissions increased by 38% exponentially [4]. In 2013, 31% of total US greenhouse gas emissions were from the electricity industry, which is the largest economic sector in the United States. There was a substantial rise in total greenhouse gas emissions by 11% since 1990 in conjunction with the growth of electricity demand where fossil fuels still remain the vital energy source for electricity generation [5].

The greenhouse gases allow natural heating of the earth from freezing during the winter and help plants to keep growing. However, increasing greenhouse gas concentration traps more heat inside the atmosphere and causes the earth's temperature to rise [6]. Carbon dioxide is the most abundant greenhouse gas and acts like a layer of glass surrounding the earth. Other greenhouse gases such as methane, nitrous oxide, carbon monoxide, hydrocarbons, and chlorofluorocarbons (CFCs) are also creating a transparent layer that allows high temperature sun radiation to enter the atmosphere, yet prevents heat from escaping to outer space. This creates a serious problem, which is widely known as global warming and it is mainly caused by conventional fossil fuel power plants.

United Nations Climate Change Conference 2009 in Copenhagen established to reduce greenhouse gas emissions in order to limit the increase in global temperature below 2°C. Any increment of more than 2°C of global mean surface temperature will lead to a rise in sea levels, an increase in global ocean temperature, ice expansion of the Arctic sea ice, ocean acidification, and extreme climatic events that are also called "dangerous climate change" and "catastrophic greenhouse effect" [6,7]. These calamities will be very difficult to deal with resulting in environmental disruption and enormous economic losses. In order to limit global temperature increase to 2°C, the CO_2 equivalent (CO_2e) should be stabilized within the range of 450–550 ppm CO_2e [8]. As of March 2015, the concentration of atmospheric CO_2 was 401.52 ppm, which was measured at Mauna Loa Observatory, Hawaii. It was reported that CO_2e concentration levels are increasing by more than 2 ppm yearly [9]. By delaying further necessary action, future generations will suffer even more severe detrimental effects of a changing environment when the mean surface temperature exceeds the threshold limit of 2°C.

One of the solutions to alleviate climate change is to develop ambient air quality standards at a national level in order to limit greenhouse gas concentration in the environment. Mitigating greenhouse gas concentration in the environment would also reduce CO_2 emissions and its related pollutant formation. The most important issue that can be solved here is to have a better and cleaner air quality. In 2012, 3.7 million premature deaths were reported by the World Health Organization (WHO), caused by low air quality in urban and rural areas. Improving ambient air quality will definitely improve health among the populations especially in people with cardiovascular and respiratory diseases. Since 1987, the WHO established air quality guidelines that were further revised in 1997.

Table 21.1, based on the WHO Air Quality Guidelines 2005 Global Updates, proposes a review of guidelines for the four main pollutants: sulfur dioxide (SO_2), nitrogen oxide (NO_x), ozone (O_3), and particulate matter.

21.2.1 Sulfur oxides

The great fog of December 1952 in London has become a significant effect of deteriorating ambient air quality. More than 3500 people died during this event, mostly due to

Table 21.1 WHO air quality guidelines (AQG) [10,11]

Pollutant		Averaging time	Standard level
Carbon monoxide		15 min	90 ppm
		30 min	50 ppm
		1 h	25 ppm
		8 h	10 ppm
Nitrogen oxide		1 h	200 $\mu g/m^3$
Ozone		8 h	100 $\mu g/m^3$
Sulfur dioxide		10 min	500 $\mu g/m^3$
		24 h	20 $\mu g/m^3$
Particulate matters	$PM_{2.5}$	Yearly	10 $\mu g/m^3$
		24 h	25 $\mu g/m^3$
	PM_{10}	Yearly	20 $\mu g/m^3$
		24 h	50 $\mu g/m^3$

bronchitis, emphysema, and cardiovascular disease. During the fog, concentrations of smoke and sulfur dioxide reached their highest levels of 4.46 mg/m^3 and 3.83 mg/m^3, respectively. The deadly fog was remarkable for its lengthy duration and high density. According to the Interim Report of the Committee on Air Pollution 1953, the main pollutants were from coal burning and other coal products [12].

When sulfur-containing fossil fuels such as coal and heavy oil are burned in the process of generating electricity in power plants, sulfur oxides (SO_x) are being produced when the released sulfur is combined with oxygen. Sulfur oxides refers to a group of highly reactive gases that contain sulfur and oxygen compounds such as sulfur monoxide (SO), sulfur dioxide (SO_2), sulfur trioxide (SO_3), lower sulfur oxides, higher sulfur oxides, disulfur monoxide, and disulfur dioxide [13]. The most hazardous gas among this group is SO_2, normally called black smoke [10]. It is commonly used as an indicator representing the other gases that belong to a larger gaseous group of sulfur oxides since it has the highest concentration in atmosphere than other sulfur oxides such as SO_3 and its existence may lead to another SO_x formation. Mitigating the SO_2 may also avoid the formation of fine sulfate particles, which can affect the environment and public health [14].

Fifty percent of SO_2 annual global emission is generated from coal burning and another 25–30% from oil burning [15]. Normally, concentrations of SO_2 below 0.6 ppm will not affect human beings. However, most people start recognizing SO_2 at 5 ppm since it has a nasty, sharp smell [16]. Nevertheless, even short-term exposure between 5 min and 24 h can cause respiratory illnesses such as breathing difficulty (bronchoconstriction) [17] and aggravate asthma symptoms especially in those who are actively exercising or doing outdoor activities. Scientific evidence has shown that at-risk populations such as children, the elderly, and those who have a prior history of asthma are the most affected when exposed to SO_2 even in a short term [18]. Further exposure to 10 ppm in an hour will irritate humans through breathing problems and mucus removal. The condition will worsen if the weather is stagnant, there are high ambient temperatures and humidity as well as with aerosols mixture. Furthermore, the problem will become worse if SO_2 reacts with other compounds in the environment to form fine sulfate particles and enters the digestive system [12,19].

21.2.2 Nitrogen oxides

Another group of highly reactive gases is nitrogen oxides. High-temperature combustion of fossil fuels containing nitrogen, such as coal, heavy fuel oil, and natural gas, will produce nitrogen oxides (NO_x) such as nitric oxide (NO) and nitrogen dioxide (NO_2). The generic term for both of these artificially made oxides is NO_x. Further exposure of nitrogen dioxide with volatile organic compounds, which is accelerated by photochemical effects under sunlight, will contribute to the formation of ground-level ozone (O_3) or tropospheric ozone or smog. The NO_x also react with the atmosphere to form acid rain [15].

As compared with NO, NO_2 has more adverse health effects on humans since it affects hemoglobin-oxygen affinity. The hemoglobin functions as a transporter of oxygen and carbon dioxide in our blood. However, the existence of NO_2 in our blood will interrupt the bonding of hemoglobin-oxygen and form acid in the lungs [19,20]. It has a more harmful effect than CO for the same concentration. It can be seen in urban areas as a reddish-brown layer with a sharp odor [21]. It can also affect the breathing and respiratory system from the reaction of NO_x with other compounds such as ammonia, moisture, and others to form nitric acid vapor and other particles. This will further damage lung tissue and developing fetuses. Other than this, the critical effects on respiratory diseases are caused by small particles inhaled deep into fragile lungs, which worsens respiratory diseases such as bronchitis and emphysema and intensifies existing heart disease [22] Table 21.2.

As a prominent pollutant, NO_x have major impacts on health and the environment. Besides acid rain and ground-level ozone or smog, the release of NO_x also contribute to water pollution as can be seen at Chesapeake Bay, the largest estuary in the United States. The deterioration of water quality is mainly caused by eutrophication from excessive nutrient inflows, mainly nitrogen. Water with nitrate pollution has destroyed the nutrients' ecosystem and affects aquatic plants and animals [22,23]. Eutrophication is a process by which water nutrients have been enriched to stimulate the dense growth of aquatic plant life, which eventually leads to oxygen depletion. Lacking of oxygen supply will decrease fish and shellfish populations.

Table 21.2 **Exposure effects of nitrogen dioxide on human health [19]**

Concentrations	Exposures	Effects
0.4 ppm and higher	Once	Recognized by its odor
0.06–0.1 ppm	Continuous	Respiratory illness
150–200 ppm	Few minutes	Destroys bronchioles (the smallest part of the bronchiole tube)
500 ppm	Few minutes	Acute edema (swelling due to the excessive watery liquid trapped in cellular tissue)

Besides, airborne NO_x also react with organic compounds in the atmosphere to form toxic chemicals such as nitrate radicals, nitroarenes, and nitrosamines, which may cause genetic mutation. Furthermore, particles of NO_2 can cause visibility impairment since it can inhibit the transmission of sunlight, especially in urban areas. As NO is one of the greenhouse gases, the mixture of NO with other greenhouse gases in the atmosphere also affects global warming, which can cause catastrophic climate change.

21.2.3 Ozone

Ozone can be classified as good ozone and bad ozone. Good ozone, scientifically known as stratospheric ozone, is located upwards in the atmosphere at 10–50 km altitude, whereas bad ozone or troposphere ozone is located nearer the earth, which is less than 10 km altitude [24,25]. Stratospheric ozone is a natural shield of the earth since it absorbs UVB radiation within a wavelength range of 280–315 nm. UVB radiation is emitted by the sun and is harmful to life on earth [26,27]. However, stratospheric ozone has been depleted gradually since 1970 because of synthetic chemicals, which are referred as ozone-depleting substances (ODS) including CFCs, hydrochlorofluorocarbons, halons, methyl bromide, carbon tetrachloride, and methyl chloroform. These chemicals have been used previously in refrigerators, coolants, fire extinguishers, pesticides, aerosol propellants, and solvents.

$$CFCl_3 + Sunlight \rightarrow CFCl_2 + Cl \, (Clorine \, Atom) \tag{21.1}$$

$$Cl \, (Clorine \, Atom) + O_3 \, (Ozone) \rightarrow ClO \, (Chlorine \, monoxide) + O_2 \tag{21.2}$$

$$ClO \, (Chlorine \, monoxide) + O_3 \rightarrow Cl + 2O_2 \tag{21.3}$$

These chain reactions will continuously occur and it has been estimated by scientists that 100,000 good ozone molecules can be destroyed by one chlorine atom. The most dramatic issue is that the lifetime of most CFCs is within range of 50–100 years before they are removed from the atmosphere [28]. This catastrophic effect can be seen at the Antarctic ozone hole where more than half of the ozone over the South Pole is lost during every spring due to sunlight occurrence. Due to this critical issue, the ODS usages have been controlled under the Montreal Protocol and their production has been banned by signatory countries. As a consequence, the atmospheric concentration of ODS has stabilized and declined recently [29].

Overexposure of UVB radiation to humans contributes to the risk of the most fatal skin cancer, melanoma. The risk of developing melanoma has increased more than twice since 1990. Other than this, UVB may also affect eye diseases such as cataracts and weaken the immune system. Even plants and crops are affected by this exposure, such as soybean, which leads to drops in production. Furthermore, marine life is also disturbed by UVB radiation since it leads to phytoplankton damage, which will decrease fisheries and marine food sources, as this is the base of the ocean food chain.

On the other hand, the bad or tropospheric ozone occurs when NO_2 is emitted by fossil fuels when they react with sunrays, producing NO and a free oxygen atom (O) to form O_3 at lower layers of the atmosphere [24].

$$NO_2 + \text{Sunlight} \rightarrow NO + O \qquad (21.4)$$

$$O + O_2 \rightarrow O_3 \qquad (21.5)$$

This tropospheric ozone is dangerous to human health especially in lung disease since it can worsen asthmatic patients condition and make them more sensitive to SO_2. Even healthy people have difficulty breathing when exposed to ozone. Besides, it can trigger other respiratory ailments such as chest pain, throat irritation, cough, congestion, emphysema, and bronchitis. It can also damage crops and their ecosystems, which leads to reduced agricultural production, and can destroy tree seedlings prone to diseases and pests as well as harsh weather [25]. Due to these effects, ozone has been considered one of the most common air pollutants by the WHO in addition to particulate matter, nitrogen dioxide, sulfur dioxide, and carbon monoxide.

In contrast, tropospheric ozone can be converted back to NO_2 and oxygen molecules (O_2) naturally when there is an availability of NO and lack of sunlight such as during late afternoon and night:

$$NO + O_3 \rightarrow NO_2 + O_2 \qquad (21.6)$$

Even though this natural process will help convert ozone to NO_2, this rapid chain reaction will continuously release ozone in the environment as long as NO is available in the atmosphere. Emission of hydrocarbon from transportation combustion will react with NO to release organic radicals. This competition between hydrocarbon and ozone to react with NO will also affect the continuous duration of ozone in the atmosphere [10].

21.2.4   Acid rain

Interactive reaction between SO_2 and NO_2 in the atmosphere creates another serious environmental problem: acid rain. When it is raining, water droplets run through the atmosphere mixing with the contaminant gases to form sulfuric acid (H_2SO_4), nitric acid (HNO_3), and carbonic acid (H_2CO_3). Acid rain is normally referred to as the most corrosive acid. On the other hand, H_2CO_3 is also damaging since it is produced in a larger quantity. Instead of water droplets, acid rain also can be in a form of snow, fog, mist, or dry deposition depending on the moisture content in the environment. These corrosive acid rain precipitations fall onto buildings, houses, structures, oceans, and farms. Because the increased concentration of SO_2 and NO_2 results from fossil fuel combustion it will also increase the concentration of acid rain and become more harmful.

A pure natural water pH value is 7 whereas normal rainwater is 5.6. Any pH value of less than 5 is considered to be acid rain. The acid rain will acidify lakes, ponds, or estuaries, which greatly affects their flora and fauna and ecosystems, destroying all marine life. An example of an acid rain tragedy is the acidification of 95,700 lakes in the Adirondack Mountains, USA, and Ontario, Canada, which have killed the fish population and resulted in all aquatic organisms being suppressed to survive in acidic streams [19]. Honnedaga Lake lies deep in the forest of Adirondacks and has been seen with an eerie blue color because its plankton has died due to acid rain. The acidification of water streams has a damaging effect on fisheries since it inhibits hatching and erodes the gill tissues and cellular debris between gill filaments. There are several fish species that can survive in acidic water; however, there is an increasing mercury content in their bodies. Heavy metal seeps into the rivers from soil due to acid rain. Also, this affects the ecosystem since black flies, which are a scourge during outdoor activities due to their bloody and painful bite, remain unaffected by acidic water. When most organisms die, black flies will flourish with less competition for habitat and fewer predators to control their population.

Due to the corrosive effect of acid rain, most of the hardwood forests stop growing and evergreen forests are losing their needles, which have been confirmed by the decline of red spruce forests at Camel Humps, Vermont's Green Mountain. Acid rain has drained vital nutrients from the soils and leaches aluminum from rocks, which can destroy the roots. This prevents the plants from absorbing water and other nutrients for their growth. In addition, acid rains are also damaging the structure and finish of automobiles, which causes rapid weathering. Moreover, acid rain is also leaching poisonous heavy metals, such as mercury and lead, in the water supply. Copper is also destroying the bacteria that are beneficial to septic systems [30].

21.2.5 Carbon dioxide

Carbon dioxide constitutes 0.04% of the atmosphere after oxygen, nitrogen, argon, and water vapor. A delicate balance between these mixtures and gases is really important for human beings, animals, and plants to stay healthy. However, carbon dioxide concentration has increased by 35% since the dawn of industrial revolution between the eighteenth and nineteenth centuries as measured at Mauna Loa Observatory, Hawaii. Due to intense industrialization and deforestation, carbon dioxide has been emitted in large amounts and become poisonous due to massive fossil fuel combustion for electrical energy production. This gas traps the heat in the environment as well as the sun's heat and increases the ambient temperature just like glass in a greenhouse.

As stated by United Nations Climate Change Conference 2009 in Copenhagen, a limit of 2°C of global temperature has been established due to the dangerous climate change effects [3,4]. The increasing global temperature melts the ice caps at the North and South Poles, glaciers, and snowlines. This will lead to a rise in ocean levels of 2–2.5 m by end of this century, and will cause flooding of the fertile land along coastal areas. This abrupt change disturbs the ecosystem and obviously will reduce

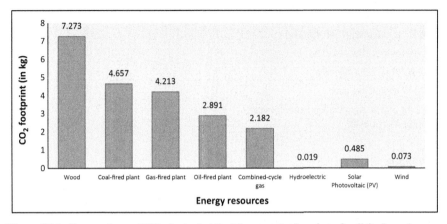

Figure 21.1 Comparison of CO_2 footprint (in kilograms) of various fossil fuels and renewable sources for production of 1 kWh of electric energy [37,38].

crop production due to limited fertile land. In addition, carbon dioxide also dissolves in the sea, which also increases the surface seawater temperature and dissipates the heat back to the atmosphere. This cycle will further intensify the effect of greenhouse effect. The increase of sea surface temperatures will also cause more destructive hurricanes and millions of people and properties will be affected. Another greenhouse effect is the heavy rainfall in some areas, which induces heavy flooding, whereas other parts will have droughts in order to balance the water cycle [16,31].

In order to minimize carbon dioxide emission, carbon footprint is a solution that can be used to investigate the summation of greenhouse gas emissions of a product or service across its lifecycle from cradle-to-grave or from the farm gate to the exit gate of the processing plants [32–34]. It is measured in units of carbon dioxide equivalent (CO_2e). It comprises two main parts, which are direct and indirect footprints. The direct or primary footprint is a direct emission of CO_2e from the fossil fuel combustion such as energy usage and transportation, whereas the indirect footprint is an indirect emission of CO_2e from the whole lifecycle of products or services [35,36]. Figure 21.1 shows comparison of carbon footprints between various energy resources.

$$\text{Carbon footprint} = \Sigma \text{ activity data } \times \text{activity emission factor} \qquad (21.7)$$

Coal is the highest carbon footprint contributor among conventional fossil fuels, which is the main reason why coal-fired power plants are the highest producer of CO_2 per kWh. Figure 21.2 shows that other conventional fossil fuels also release carbon dioxide gas to the atmosphere per unit heat energy input.

Example 21.1

Estimate the daily CO_2 footprint of a residential house with AC appliances as shown below. Consider coal to be used in generating the electrical energy with a coefficient of approximately 5.06 kg CO_2/kWh of electricity.

AC appliances	Quantity	Usage (per day)
14-ft.³ refrigerator (1080 Wh/day)	1	24 h
60-W compact fluorescents	3	8 h
70-W, 20-in. color television	1	3 h
800-W microwave oven	1	1 h
180-W, 300-ft. submersible pump for water supply from a well	1	2 h

Solution

Energy consumed can be tabulated as follows:

AC appliances	Power (W)	Usage (per day)	Wh/day
14-ft.³ refrigerator (1000 Wh/day)	1000		1000
60-W compact fluorescents	3 × 60 = 180	8 h	180 × 8 = 1440
70-W, 20-in. color television	70	3 h	70 × 3 = 210
800-W microwave oven	800	1 h	800
180-W, 300-ft. submersible pump for water supply from a well	180	2 h	360
Total energy consumed per day			3810

Carbon footprint $= (3.81\,\text{kWh/day}) \times (5.06\,\text{kgCO}_2/\text{kWh}) = 19.3\,\text{kg CO}_2/\text{day}$

21.2.6   Ashes

Series of tragedies have occurred since 1930 such as Meuse Valley, Belgium, and Donora, PA, which show that killer smog can cause death and respiratory and cardiovascular disease [39]. A total of 840,000 tons of ash are emitted from a typical 2000 MW

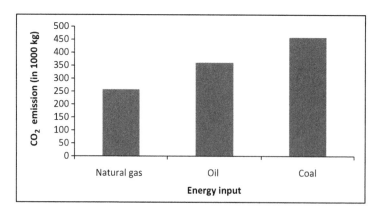

Figure 21.2 Carbon dioxide emission levels in kilograms per billion BTU of energy input [37].

conventional coal-fired power plant annually [40]. Ash is the converted inorganic impurities from coal combustion, which may discharge as bottom ash and fly ash. The bottom ash is discharged from the bottom of the furnace whereas fly ash is the discharged particles in the flue gas. Fly ash creates the primary particulate matter, which is generally referred as "PM," "PM10," or "PM2.5," and shows the value of the aerodynamic diameter of the particulate matter in micrometers [41,42]. PM10 is a coarse particle with a diameter less than 10 μm and is easily found in dusty industries and on roadways. Another fine particle with a diameter less than 2.5 μm is referred as PM2.5 and is more harmful since it can penetrate deep into the lung.

Instead of using its dynamic diameter to classify the particulate matter, classification can also be done based on its source, which are the primary and secondary particles. The primary particles are directly suspended as a result of human activity or naturally. However, secondary particles are produced by indirect reaction from chemicals emitted by primary particles that react with sunrays. Various health problems may occur due to inhalation of these particulate matters, such as coughing, irritation in airways, breathing difficulty, worsening asthmatic symptoms, bronchitis, heart attacks, and many more. In addition, there are also other adverse effects to the environment such as visibility impairment, acidification of lakes and rivers, and changes in ecosystems at coastal water and estuaries because of nutrient imbalance and leaching of nutrients from soil, which damages forests and crops, killing diversity in the ecosystems and damaging building structures [43].

In order to control suspended particle matters in the atmosphere, several technologies have been used to filter them before they are released into the environment, such as wet particulate scrubbers, fabric filters, electrostatic precipitators, mechanical collectors, and high temperature high-pressure particulate control. The wet scrubber functions by using liquid as a filter to trap the flying particulates and sulfur dioxide before they are released into the environment. On the other hand, fabric filters will trap the particulates in bags just like the vacuum cleaner before the gas is vented to the atmosphere. This method has been used widely since 1970 and its efficiency is more than 99.5%.

21.2.7   Legionnaires' disease and cooling towers

Legionnaires' disease or Legionella is a disease caused by bacterial pneumonia called *Legionella pneumophila* species [44]. The generic term "Legionellosis" or Legion fever is used for this kind of bacterial infection, which can be a mild flu-like infection of Pontiac fever up to the more severe and possibly fatal pneumonia of Legionnaires' disease [45]. It was first detected in July 1976 during an American Legion convention in Philadelphia, USA, where 29 people died out of 182 reported cases [46]. Actually *L. pneumophila* is a natural aquatic bacterium with the ability to survive in extreme environmental conditions within a range of 20–45°C [47]. The findings demonstrate that the highest infection rate occurs in summer months with samples collected monthly from thermally altered lakes. The thermally altered lakes are a cooling water lake for power plants and receive heated waste from the facility. It was revealed that the warm water and damp habitats that are suitable for algal colonization are also suitable environments for *L. pneumophila* [48].

Legionnaires' disease was an emerging disease of the twentieth century and is usually referred to as an "emerging infectious disease (UID)" is an impact of human alteration to the environment [49]. The source of legionnaires' disease is cooling towers, hot and cold water systems, spa pools, thermal pools, springs, humidifiers, domestic plumbing, and potting and compost. It can be transmitted through inhalation of the contaminated aerosol, wound infection, and aspiration. There are several contaminated places such as shopping centers, restaurants, clubs, leisure centers, sports clubs, private residences, hotels, cruise ships, camp sites, hospitals, and medical equipment [45].

Initially, cooling towers were founded to be the primary source of this disease. Cooling towers installed at public areas with dense populations, seasonal and climatic conditions, which are intermittently used, have poor engineering design, and have had little or no maintenance are the possible risks for this disease. A recent case was reported in July 2010 where a 73-year-old patient was hospitalized at Kobe University Hospital, Japan, due to alveolar hemorrhage, systemic lupus erythematosus, and antiphospholipid antibody syndrome. After 4 months in the hospital, she was then diagnosed with nosocomial Legionella pneumonia. Further treatment was given but she died because of uncontrollable pulmonary alveolar hemorrhage. Since she was diagnosed with nosocomial Legionella pneumonia, the infection control team investigated the source of the risk. The investigation revealed a link to a contaminated hospital-cooling tower; tests showed a 95% strain similarity of samples taken from the patient and the hospital cooling tower. This contaminated aerosol from the hospital-cooling tower had been inhaled by this immune-compromised patient. The actions taken by the hospital to prevent the *L. pneumophila* breeding was to increase the temperature of the hot water supply appropriately, increase the Legionella culture test frequency to three times annually, and introduce automated disinfectant insertion machines with BALSTER ST-40 N as an antiseptic reagent. After implementing the new preventive measures, no cases have been reported since from the hospital-cooling tower or the hot water system [50].

21.3 Environmental concerns related to hydroelectric power plants

Hydropower is one of the major renewable resources in the world and is considered clean energy since there is no burning of fossil fuel. Many large power plants in the world are built by using hydro technology, such as the Three Gorges Dam in China, which generates up to 22,500 MW of electricity. However, the construction of a dam to store water for hydropower could impact the environment as discussed subsequently.

21.3.1 Destruction of large area of forest and river ecosystems

There is no doubt that the construction of a dam will result in complete and irreversible destruction of a forest's ecosystem. For example, the Bakun Dam in Sarawak, which is the largest in Southeast Asia, will flood 69,640 ha of forest ecosystem, which is larger than the size of Singapore. In addition, the reservoir created by the dam will become a significant source of greenhouse gas emissions, especially carbon dioxide and methane due to microbial decomposition of the submerged forest [51].

Moreover, the construction of a dam will block the connection between the upstream and downstream of the river, which will create a sedimentation problem behind the dam. This situation will also affect some of the fish population because they cannot migrate to their spawning ground. The slow velocity of the river flow in some bays of the reservoir will influence water quality due to the high content of nutrients in water. This will cause eutrophication where excessive growth of aquatic plants will pollute the water as happened in the Three Gorges Dam [52].

21.3.2 Population resettlement

The hydropower project requires the displacement of people residing within the flooding area. The submerging area containing houses, heritage landmarks, burial grounds, crops, and ancestral land will vanish causing socioeconomic hardship. For example, thousands of indigenous people living in the surrounding Bakun Dam region will be displaced to a new settlement area. These indigenous people will lose their ancestral land and forests where they have been living independently and on which they have been relying for their agriculture, hunting, and gathering of forest products [51].

21.4 Environmental concerns related to nuclear power plants

The increasing cost of electricity generation as well as environmental impacts of conventional energy has caused many countries to shift to cleaner resources to generate electricity. Nuclear power plants are an example of a clean energy but which may cause concern to others due to their potential disastrous impact. Nuclear power has been accepted for low-cost electricity generation compared to other resources. In fact, it is cheaper than most of the conventional fossil fuel energy and renewable energy resources without being subsidized by governments. Moreover, it can generate electricity continuously with negligible greenhouse gas emissions. With increased safety features, the nuclear energy power plant is a better solution for energy security and climate change problems.

As updated by the World Nuclear Association in 2015, nuclear energy has contributed to the world's total electricity generation by 11.5% with 375,000 MW total capacities in 31 countries. Due to its advantages, many countries have become nuclear power dependent. There are several countries that depend on nuclear power for more than 30% of their electricity production, such as Belgium, Czech Republic, Finland, Hungary, Slovakia, Sweden, Switzerland, Slovenia, Ukraine, South Korea, and Bulgaria, whereas France generates 75% of its energy from nuclear power. These countries are followed with the United States, the United Kingdom, Spain, Romania, and Russia, which generate about 20% of their electricity demand by using nuclear power. Previously in 2010, Japan generated more than 25% of their electricity from nuclear power until, however, the tsunami and earthquake that hit the Fukushima Dai-ichi nuclear plant in 2011 [53].

A large nuclear power plant can save about 50,000 barrels of oil and its return of investment can be recovered within several years, which makes nuclear power a strategic plan for reliable and low-cost energy supply. An example of the energy that can be generated by complete fission of 1 g of uranium-235 (U^{235}) isotope with total energy produced per fission reaction is 200 MeV (million electron volts) that can generate 1 MW of electricity per day, which can save 3000 kg of coal or more than 2000 L of fuel per day and avoid carbon dioxide emissions of 250 kg per day [19,54].

Even though nuclear power can be considered as the future of energy to cope up with world economic growth and global warming issues, the unprecedented tsunami-induced nuclear disaster at the Fukushima Daiichi nuclear power plant on March 11, 2011 activated negative responses and damaging images toward nuclear power, which has prompted many countries to reconsider nuclear power plants and even phase out their development [55,56]. Japan is fully prepared against earthquakes and seismic activities; however, the magnitude of earthquake followed by a tsunami was one of the largest in the world and the biggest in Japan and has damaged the Pacific coastal line of northeast Japan including the Fukushima Daiichi's three nuclear reactors. Due to the nuclear spillage and tsunami, more than 90,000 local residents were evacuated from the area and 2884 people were killed. Fukushima Daiichi was the first time that three reactors were damaged on one occasion since the single reactor accident at Chernobyl in 1986 and the Three Mile Island accident in 1979.

The explosion at Fukushima Daiichi has caused extensive radioactivity emission in the atmosphere via deposited radiocesium such as ^{137}Cs, ^{134}Cs, and I^{131} [57,58]. Iodine-131 or I^{131} is the most hazardous and toxic element to humans as well as plants since it is inhaled and leads to thyroid cancer in children and adolescents [58–60]. It has been reported that thyroid cancer was the major health effect from Chernobyl disaster due to the intake of I^{131}. In addition, the long-lived deposited radiocesium ^{137}Cs and ^{134}Cs has caused relative homogeneous exposure of all organs and body tissues via its radiation even at low rates, which may also cause other cancers [61].

21.4.1 Radioactive release during normal operation

During their normal operations nuclear power plants emit radioactive particles, gases, and liquids to the environment in small amounts within the permissible limits laid down by nuclear regulatory commissions, such as the United States Nuclear Regulatory Commission, Canadian Nuclear Safety Commission, UK Nuclear Regulatory Authority, and others. Prior to being released to the environment, radioactive materials are treated, released according to the procedure, and monitored. These safety precautions are done to minimize environmental impacts on humans, animals, plants, and sea creatures. The main objective of these regulatory authorities is to make sure that the nuclear power plants are capable of controlling the nuclear reaction, cooling of radioactive materials, and containment of radioactive radiation through various ways in order to avoid radioactive materials from escaping. Table 21.3 compares several whole body radiation doses and their effect due to their exposure.

Public concern has increased due to radiation from nuclear power plant effluents as seen by vented steam and its ejection outlets at the water discharged areas. There are definitely permissible effluents that are released to the environment. However, excessive radiation will increase the risk of damage to organ and body tissues, genetic mutations, birth defects, leukemia, cancers, immune system disorders, and many more. Human body tissues are made of compounds of carbon in which isotope $_6C^{14}$ can be found in trace quantities with radioactive potassium isotope $_{19}K^{40}$ to balance cells' fluid quantities in the human body. Once radioactive gas is inhaled or consumed, this radiation will result in biological damage. As an example, radiation can penetrate and damage cells inside the respiratory tract passages, which induces lung cancer [63]. Human beings are also exposed to background radiation and also impose radiation. It has been reported that one Brazilian beach reached 800 mSv. The legal authorities have to consider the background radiation that the public have been exposed to, which is 2.5 mSv per year by normal activities in order to limit public exposure to radiation.

The discharged nuclear power plants' effluents depend on the type of reactors, which are pressurized water reactor (PWR) and boiling water reactor (BWR). The reactors that have been used in Europe are PWR, which is safer than BWR because the reactor vessels are separated from the steam generators. However, the steam in BWR is directly produced in the reactor vessel and routed to the steam turbine generator to produce electricity. The steam also brings absorbed radionuclides in gaseous form. The "offgas" system will remove the radioactive gases and delay the release of radionuclides such as krypton and xenon until their acceptable decay, but the possibilities

Table 21.3 **Comparison of whole-body radiation doses and their effects [62]**

Whole-body radiation doses (mSv)	Effects
1 mSv/year	Allowable normal radiation for the public who are exposed by discharged and direct radiation from nuclear power plants
20 mSv/year	Current allowable limit for nuclear plants workers and uranium miners
50 mSv	Allowable short-term dose for emergency workers
100 mSv	Lowest annual level that significantly increase the risk of cancer
250 mSv	Allowable short-term exposure for emergency limit such as for workers controlling the 2011 Fukushima accident
1000 mSv short term	May cause acute radiation syndrome such as nausea and decreased white blood cell count, but not death
10,000 mSv short term	May cause death with continuous exposure in a few weeks

of leakage and accidental release are the main concern of the public [19,24]. Leakage in nuclear power plants is due to mechanical failure and human error will increase because nuclear power plants are getting older, are not maintained properly, and lack monitoring from legal authorities. Even the transportation of fuel from mines to the nuclear power plant also exposes the environment to radiation.

21.4.2 Loss of coolant

The uncontrolled loss of coolant accident (LOCA) in nuclear power plants may cause ecological and economic disaster as occurred at Fukushima Daiichi in 2011 where a hydrogen explosion damaged the plant's structure and loss of coolant caused meltdown of its three unit reactors [64]. The main factor in obtaining an operating license for a nuclear power plant is safe reactor behavior during LOCA. The criteria of safe cooling ability of the reactor core are to ensure that its temperature does not exceed 1204°C between loss of coolant and start-up of emergency cooling and that its oxidation level equivalent cladding reacted of the cladding material does not exceed 17% [65–67].

If the reactor's temperature increases more than 1200°C, hydrogen molecules will be separated from the water and trapped at the ceiling of the reactor due to its light density. This may cause a fire if ignited and damage the containment structure. The explosion will emit radioactive steam into the air just like the Three Mile Island, Chernobyl, and Fukushima Daiichi accidents. Moreover, meltdown will happen if the reactor's temperature reaches 2400°C where the uranium fuel will melt down and the fuel rod will become molten wax. The molten uranium will continue to melt its way downward and penetrate the ground until the heat can be absorbed by the molten rock and soil, which is about 10 ft. deep. The molten soil surrounding the fuel will harden like glass and contain the remaining fuel. However, if the molten uranium penetrates the groundwater supplies then this will have a bigger disastrous impact, and the contamination will be spread to a larger area. In fact, very high pressure from the reactor's high temperature will explode the containment structure and release the radioactive emission into the atmosphere [24,63].

A series of tragedies have occurred due to LOCAs at nuclear power plants. As an example, severe LOCA happened in Manitoba in November 1978 where a major coolant spillage occurred due to pipe leakage of the WR-1 at Pinawa. A total of 2739 L of coolant oil leaked and discharged into the Winnipeg River. Due to the loss of coolant, the reactor reached a high temperature even though not due to meltdown temperature and damaged the three fuel elements and emitted fission products [68]. A year later, the largest nuclear disaster happened in the United States on March 28, 1979 due to loss of coolant and suffered a severe core meltdown. Fortunately, the containment structure was still intact and held almost all of the radioactive emission [69]. LOCA is the main factor that leads to meltdown and damage to the nuclear power plants. The damaging effects through radioactive release through ground soil, water supply, and atmosphere led to major health problems and even fatalities. It is not only human beings that are affected; the coolant leakage that was discharged into the river is eaten by fish and aquatic organisms. Research has been done that revealed that radiation also leads to fatalities to fish, mussels, and other aquatic species [70]. The river also provides irrigation to crops and vegetation, which then died due to the radiation; some of is eaten by humans. Livestock such as cows and sheep drink from the radiated river and provide milk and meat for human consumption. This damaging continuous food chain will surely affect humans through various carcinogenic diseases and even genetic mutation.

21.4.3 Disposal of radioactive wastes

The infinite source of nuclear energy is a better option to clean and cheap energy in recent energy security and global warming problems. However, society has to accept that this energy has disastrous damaging capabilities over a long time and takes more than a thousand years to decay. This scenario was once called a Faustian bargain with the society due to its catastrophic effects [71]. The critical issue after generating energy from nuclear power plants is management of its radioactive wastes. Since fuel rods contain uranium $_{92}U^{235}$ in the reactor they will deplete and no longer be able to sustain fission reaction, therefore it should be removed and replaced with new fuel rods. The remaining fuel normally will be stored in temporary storage, which is housed on site in lead-lined concrete pools of water. This pool will cool and contain radiation from the remaining fuel until a permanent disposal decision is made [63]. The radioactive wastes are classified into three main types according to their amounts, radioactive levels, and isotopes half-lives; these are low-level, medium-level, and high-level wastes. These radioactive wastes should be managed using the delay-and-decay method, whereby they should be stored properly until they decay their radioisotope naturally into stable, nonradioactive forms (Table 21.4).

Table 21.4 **Radioactive waste management for different types of radioactive wastes [72]**

Types of radioactive waste	Sources	Waste management methods
Low-level waste	Hospitals, laboratories, industry, nuclear fuel cycle such as paper, rags, tools, clothing, filters, and others	Incinerated in a closed container to reduce its volume and buried in shallow landfill sites
Medium-level waste	Resins, chemical sludge, reactor components, and contaminated materials from reactor decommissioning	Solidified with concrete or bitumen and deeply buried underground
High-level waste	The remaining fuel and the principal waste from reprocessing the remaining fuel	Vitrified by combining into borosilicate glass (Pyrex), sealed inside stainless steel canisters, and buried deep underground

Nuclear power plants have radioactive wastes that should be disposed of properly in a location that is a remote and geographically stable area, such as Yucca Mountain, Nevada, in United States or the Waste Isolation Pilot Plant in the Chihuahuan Desert, New Mexico. The disposal area should be built with proper security to ensure that the highly radioactive waste is stored safely within its half-life, which can be more than a thousand years [73]. Natural disposal of radioactive waste was shown at Oklo in Gabon, West Africa two billion years ago where the radioactive materials were safely contained in that area and eventually decayed into nonradioactive elements [68]. This shows that it is possible to contain radioactive wastes in secured long-term storage until they reach a safe level in unspecified time.

The safety nuclear regulatory bodies for each country that have nuclear power plants must update their policies if necessary with recent issues in order to maintain the safety of the spent fuel storage. Critical risks such as terrorist attacks, plane crashes, tsunamis, floods, tornados, and hazards from nearby activities should be taken into account in enhancing emergency responses during these conditions. These emergency responses must be in place for quite long time, which may take billions of years, and is the major challenge for radioactive waste storage.

21.5 Environmental concerns related to renewable energy

Renewable energies are all low-carbon energies and have less impact on the environment compared with fossil-fuel energies, from a global perspective. However, the impact of its implementation should not be ignored. Hence, it is important to recognize this impact and mitigation steps to limit the effect.

21.5.1 Solar energy

Solar photovoltaic (PV) energy uses unlimited energy from the sun where the main advantages are that it is environmentally friendly and emission free during its operation. However, the major concerns regarding solar PV are at the manufacturing and disposal stages.

21.5.1.1 Toxic chemicals used in manufacturing solar photovoltaic panels

Different chemicals are used in manufacturing solar PV panels, particularly during extraction of solar cells. For example, cadmium (Cd) is used in cadmium telluride (CdTe)-based thin film solar cells as a semiconductor material to convert solar energy into electrical energy, which is a highly toxic substance. The National Institute of Occupational Safety and Health considered cadmium dust and vapors as potential carcinogenic matters that can cause cancer if exposed directly to the workers [74].

21.5.1.2   Disposal and recycling of a solar panel

At the decommissioning stage of solar PV panels after completion of their expected operation period of around 25 years, their disposal on an ordinary landfill site is quite a challenge for local authorities due to the presence of hazardous materials contained in them. With the rapid growth of solar market demand nowadays, it is expected that large quantities of PV panels will be disposed of at the end of their life. Hence, proper planning is needed to handle future solar PV wastes, such as providing a dedicated landfill, special municipal incinerators, or an effective recycling management system. Otherwise, it will be very difficult to handle and will face similar problems as existing electronics product wastes [75,76].

21.5.1.3   Use of large land area

The construction of solar farms on a large scale needs land clearing, which adversely affects the natural vegetation, wildlife, and their habitats [74]. In order to achieve optimum power generation, any shading to the PV panels caused by tall trees and bushes needs to be removed and maintained. The removal of trees will reduce the absorption of carbon dioxide from the atmosphere; hence, this will defeat the purpose of reducing the greenhouse gases [77].

21.5.2   Wave energy

Most environmental concern associated with wave energy is related to marine life. The construction work and laying of submarine power cables during installation of the wave energy generator could disturb the seafloor sediments and result in the loss of habitats for some marine life. The noise produced during construction and operation of the generator may also potentially harm the aquatic life. The wave energy device may alter the sea current and affect certain fish populations in relation to their feeding and breeding grounds. There are also risks of marine life colliding with tidal turbine blades. The operation of a wave energy generator and the power cables may produce electromagnetic field (EMFs) that can directly affect marine life, such as decreases in fertility. Moreover, the EMFs may cause interference with migration and navigation, detection of prey, or escape from predators [78,79].

21.5.3   Wind energy

Wind energy is recognized as one of the cleanest energy sources and has shown rapid growth recently. The technology used has considerably matured with comparatively low costs, which has made people build more wind farms. However, development of large wind farms has caused adverse environmental issues. Although the magnitude of the impact is considered small, it should not be ignored. Some of the potential impacts are discussed as follows.

21.5.3.1 Effect on wildlife

Wind turbines could impose a danger to birds when they collide with rotating blades or its structure, such as tower, nacelles, or guy cables, which may cause severe injury or fatality. Migratory birds are potentially at high risk of hitting the rotor blades if the arrangement of the wind farm is within the fight path of their migration routes. . It is understood that birds have an ability to detect obstacles and change their direction immediately to avoid collision. However, for these migratory birds, extra effort to deviate from the route consumes their limited energy, and reduces their survival rates [80,81].

Moreover, the wind farm could also disturb local birds' fauna. Construction of a wind farm could destroy their natural habitat and may create a physical barrier between their natural breeding and feeding behaviors. The spinning of the blades during operation may frighten the birds, hence limiting their natural territories. To reduce the effects on birds, authorities need to investigate migrating birds' routes and nesting areas before deciding to build wind farms. The use of aviation radar to detect flocks of migrating birds and temporarily stop the operation may reduce the danger to birds from spinning turbine blades.

21.5.3.2 Noise

Noise produced by the wind turbine system may disturb surrounding residents especially at night where conditions are tranquil. There are two sources of noise produced by wind turbines, which are mechanical and aerodynamic noise. Mechanical noise comes from turbines' internal gear, generator, and other auxiliary components but is not affected by the size of turbine blades. Proper insulation during manufacturing and installation could reduce this noise level. In contrast, the aerodynamic noise comes from blades passing through the air and is proportional to the blades' swept area, wind speed, and speed of rotation. For example, a bigger size wind turbine is noisier than a smaller one. To mitigate the effect of noise, there should be a minimum distance between the residents and the wind farms; this practice varies between countries and regions.

21.5.3.3 Visual impact on landscaping

Erection of wind turbines could disturb the natural scenery. In addition, flickers caused by the turbine blades' movement could disturb residents. However, this problem is highly subjective to public perception and their personal feelings. Nevertheless, there should be a minimum distance between the residents and the wind farms to minimize the effect.

21.5.4 Fuel cells

A hydrogen fuel cell requires hydrogen and oxygen to generate electricity through an electrochemical process. It is expected in the near future that fuel cells will be extensively used to generate electricity due to their clean by-product, which is water.

The development of fuel cell technology especially for electric vehicles will significantly reduce greenhouse gases and further mitigate global warming issues. The main obstacle that hinders the application of the hydrogen fuel cell is the hydrogen supply itself. Even though hydrogen is available abundantly as compounds on earth such as water and hydrocarbons, the processes of producing, storing, and transportation of hydrogen are costly and may release about 10−20% of hydrogen gas during operation. If all fossil fuel plants and transportation shifted to hydrogen in the future, the large amounts of hydrogen and water vapors will be released into the atmosphere, which will triple the amount of hydrogen that is released at present. This condition will deplete the ozone layer through excessive cooling of the stratosphere, enrich heterogeneous chemistry, create more the noctilucent clouds, as well as disturb tropospheric chemistry and atmosphere−biosphere interactions. The detrimental effects of these reactions will cause the ozone hole to become larger and last longer. Thus, the idea of shifting to a cleaner energy may cause another damaging consequence and even accelerate the global warming effect [82,83].

Furthermore, hydrogen is the raw material required for the electrochemical process in the fuel cell, which can be obtained from water, biomass, and fossil fuels. However, hydrogen supplies nowadays are produced from natural gas that releases carbon dioxide during the process. Thus, the greenhouse gases are still emitted in order to achieve this clean energy [84]. Moreover, the technology of fuel cells should be improved further as it is still in a growth stage; its technological maturity is predicted to be 2018 [85].

21.5.5 Geothermal energy

Geothermal energy is thermal energy generated and stored in the earth's mantle layer. The geothermal power plants take hot fluid or steam from deep in the earth and then converted into electricity. This renewable energy resource offers many advantages over fossil fuels and is sustainable. However, several environmental concerns associated with geothermal energy are discussed as below.

21.5.5.1 Land disturbance

Active geothermal areas are normally located remotely or nearby national parks with sparse populations. The alteration and manipulation of land during construction and operation will change the landscape, natural features, and beautiful scenery. Deforestation to make way for geothermal energy development and its transmission lines may impact local flora and fauna. In addition, the geothermal plant power will become a potential source for noise during construction and drilling processes.

21.5.5.2 Atmospheric emission

Geothermal gases such as hydrogen sulfide, carbon dioxide, and methane are often discharged to the atmosphere and are harmful to the environment. Hydrogen sulfide is the most dominant noncondensable gas in geothermal fluids. It is of concern to the environment due to its smell and toxicity. When dissolved in water aerosols, it will react with oxygen to form sulfur dioxide that leads to acid rain, while methane emission is a concern for its global warming potential. In addition, there are also trace amounts of mercury, ammonia, and boron that threaten the soil and surface water near the power plant [86].

21.5.6 Biomass

Biomass in the form of charcoal and wood were used as primary fuel for energy until the nineteenth century when fossil fuel-based energy such as coal and gasoline had been used on a large scale. Today, the use of biomass as a source of energy has become attractive due to its carbon neutral nature, unlike the carbon emitting fossil fuels that cause global warming. Its carbon neutral nature is the process of releasing CO_2 to the atmosphere from burning biomass, which has been captured from the atmosphere during photosynthesis [87]. Therefore the carbon cycle between absorption by plant and emission to the atmosphere is at equilibrium. An example of biomass sources to produce biofuel for energy generation is food crops such as sugarcane, corn, soybean, and palm oil. Wastes from agriculture, sawmill factories, foods, and municipal solid waste are also useful as sources of biomass energy.

The extensive usage of biomass still has detrimental environmental concerns that must be taken care of. For example, biomass energy cultivation requires large areas of land and substantial amounts of water. The intensive harvesting of biomass will increase soil erosion, water degradation, and removal of nutrients. In addition, the use of pesticide and fertilizer could pollute water resources. Replacement of primary forests and natural ecosystems with massive energy crops plantations will significantly change natural habitats and food source for wildlife.

Another important issue is that the demand for biomass energy has created competition between crops used as food and biofuel supplies. This issue increases food prices in some regions because farmers have shifted to biofuel crops due to better earnings. It is considered inappropriate to convert food into fuel in areas where there are still many people suffering from starvation and malnutrition in order to mitigate greenhouse gas emissions.

21.6 Summary

Energy is a very important driving force to improve the standard of living and develop a country. Most of the developing countries are struggling to meet their energy demands for their populations and economic growth. With limited resources, they are unable to obtain expensive technology to filter out greenhouse gases before emitting them to the environment. As an example the installation cost of scrubbers on existing coal power plants will raise up to 30% of the capital cost [88]. Furthermore, these efficient technologies will also increase the existing cost of electricity. Efforts toward mitigating global warming issues are certainly complex since these efforts require full commitment from developed and developing countries. The rich and developed countries are keen to maintain their standard of living. However, the developing countries are struggling to increase their standard of living; their citizens also have the right to sustainable development because parts of their population are still in poverty [89].

In order to decarbonize our economic growth, nuclear power is another better option to supply electricity. It was once claimed by Lewis Strauss, the first chairman of the Atomic Energy Commission, that it was "too cheap to meter." In 1956, the Union Carbide and Carbon Corporation showed an advertisement that claimed that a pound of uranium can supply electricity for the whole of Chicago in one full day as compared to 3 million pounds of coal. Its huge and enormous energy although cheap has an interesting future and has caused many countries to depend on nuclear power, such as France with 75% of its electricity generation from nuclear energy. However, this enormous and cheap energy has catastrophic effects. Society has to live with daily normal operating radioactive emissions from nuclear power plants even if it is monitored by legal authorities. Another critical issue that has encouraged several countries to phase out their nuclear power plants is the risk of explosion of nuclear reactors, which happened at Fukushima Daiichi nuclear power plant in 2011. Although the advancement of nuclear technology has already matured there are still unforeseen factors such as natural disasters that can damage the plant. The radioactive wastes from a nuclear disaster are barely removed and proper safety precautions need to be followed in order to clean up the mess for the next thousand years. These are the expensive cradle-to-grave nuclear energy technology impacts that should be faced by our society for energy security.

The idea of transforming energy supply from fossil fuel and nuclear to totally relying on renewable energy is a serious effort that has been done, for example, by Germany. Although it is quite ambitious and ambiguous, the effort should be praised since it may envisage the future of sustainable energy that offers energy security and reduces the cost of electricity in the long run. Furthermore, energy efficient technology is also emphasized to reduce energy consumption for optimum usage and still supports their industrial baselines [90]. The idea of implementing low carbon cities by many countries, such as Bristol, Leeds, and Manchester in United Kingdom, Jalisco and Tabasco in Mexico, and Petaling Jaya in Malaysia, can create awareness and visualize sustainable energy as a pioneer project for the future, which may not only require diversifying energy supply to renewable energy but also efficiently using the electrical energy [91]. Besides, an incentive of using 100% renewable energies has been widely accepted and is being implemented in many countries such as New Zealand, Australia, Cameroon, Ghana, and other countries for their cities, residences, and business centers [92]. With advancement of renewable energy technology, a greener future will sustain the fair development of developed and developing countries.

Table 21.5 Carbon footprint of renewable energy for production of 1 kWh of electricity [37]

Fuel type	CO_2 footprint
PV	0.2204
Wind	0.03306
Hydroelectric	0.0088

Problems

1. Explain the natural greenhouse effects and their relations to the global warming issue.
2. What will happen if the global temperature keeps increasing above the limit established by United Nations Climate Change Conference 2009 in Copenhagen?
3. What is sulfur oxide and determine the most hazardous gas among the sulfur oxides.
4. Explain the health problems that may occur if humans are exposed excessively to sulfur dioxide.
5. Describe the health effects of nitrogen dioxide.
6. What happens if nitrogen oxides are released into rivers and streams?
7. Differentiate between good ozone and bad ozone.
8. Identify a serious environmental problem from a chain reaction between sulfur dioxide and nitrogen oxide in the atmosphere.
9. Discuss acid rain environmental effects with necessary examples.
10. How can carbon dioxide be a poisonous gas instead of an essential gas in the atmosphere.
11. Compute Problem 1 if oil and natural gas are used in generating electrical energy with 117,000 pounds of CO_2 per billion BTU of energy input for natural gas and 164,000 pounds of CO_2 per billion BTU of energy input for oil.
12. Estimate the carbon footprint for a 70-W LED television if it is turned on for (a) 24 h, (b) 8 h, or (c) 4 h if coal is used to produce the electricity. Assume the coefficient of carbon footprint for coal electricity generation is 1 kg CO_2/kWh.
13. Compare the carbon footprint calculated in Problem 12 if an LED TV is turned on for 24 h by using solar PV, wind, and hydroelectric (see Table 21.5).
14. Differentiate the particulate matter general term of PM10 and PM2.5.
15. Propose several methods to control suspended particle matters in the atmosphere.
16. Define Legionnaires' disease and describe the main sources of this disease.
17. Justify the advantage of nuclear energy in terms of resources and carbon footprint.
18. Identify the damaging effects of excessive radioactive emissions to the environment.
19. Explain the meaning of LOCA and why it is so important in a nuclear power plant.
20. Provide several examples of tragedies that have occurred due to LOCA in nuclear power plants.
21. Give recommendations on radioactive waste management for different types of radioactive wastes.
22. Discuss the disposal of permanent radioactive waste worldwide and its effect to the environment.
23. Compare the environmental impacts caused by impoundment type and run-of-river type hydroelectric power plants.
24. Identify recommended distances and noise limits between a wind farm and residents implemented by various countries and regions.
25. Determine hazardous material associated with the manufacturing of a silicon-based PV module.

References

[1] Scholz R, Beckmann M, Pieper C, Muster M, Weber R. Considerations on providing the energy needs using exclusively renewable sources: Energiewende in Germany. Renew Sustain Energy Rev 2014;35:109–25.

[2] Research Cooperation Renewable Energies (FVEE), Energiekonzept 2050, 2010.

[3] Smart Energy for Europe Platform, Joint Norwegian-German declaration: for a long term collaboration to promote renewables and climate protection. 2012.

[4] WRI (World Resources Institute), Climate Analysis Indicators Tool (CAIT) 2.0: WRI's climate data explorer, 2014. [Online]. Available from: http://cait2.wri.org/.

[5] U.S. Environmental Protection Agency. Inventory of U.S. Greenhouse Gas Emissions and Sinks: 1990–2013, http://www.epa.gov/climatechange/emissions/usinventoryreport. html; 2015.

[6] Nag PK. Power plant engineering. 3rd ed. New Delhi: Tata McGraw-Hill Publishing Company Limited; 2008.

[7] Richardson K, Steffen W, Liverman D. Climate change: global risks, challenges & decisions. United Kingdom: Cambridge University Press; 2011.

[8] Kriegler E, Weyant JP, Blanford GJ, Krey V, Clarke L, Edmonds J, Fawcett A, Luderer G, Riahi K, Richels R, Rose SK, Tavoni M, van Vuuren DP. The role of technology for achieving climate policy objectives: overview of the EMF 27 study on global technology and climate policy strategies. Clim Change 2014;123:353–67.

[9] Stern N. What is the economics of climate change? World Econ 2006;7(2):1–10.

[10] World Health Organization (WHO), WHO air quality guidelines: Global Update 2005. 2005, pp. 1–21.

[11] World Health Organization (WHO), Air quality guidelines for Europe, 2nd ed., No. 91. 2000.

[12] Greater London Authority, 50 years on: the struggle for air quality in London since the great smog of December 1952. 2002, pp. 1–40.

[13] Larøi V, Karlsen H, Skinner R. Generations. ABB Marine and Cranes 2012;77.

[14] United States Environmental Protection Agency (EPA), Sulfur dioxide: health. [Online]. Available from: http://www.epa.gov/airquality/sulfurdioxide/health.html.

[15] Flynn D. Thermal power plant simulation and control. London, United Kingdom: The Institution of Electrical Engineers; 2003.

[16] Australian Government Department of the Environment and Heritage. Air quality fact sheet: sulfur dioxide, https://www.environment.gov.au/protection/publications/factsheet-sulfur-dioxide-so2; 2005.

[17] Balmes JR, Fine JM, Sheppard D. Symptomatic bronchoconstriction after short term inhalation of sulfur dioxide. Am Rev Respir Dis 1987;136(5):1117–21.

[18] United States Environmental Protection Agency (EPA). Fact sheet revisions to the primary national ambient air quality standard, monitoring network and data reporting requirements for sulfur dioxide. pp. 1–6; 2010.

[19] El-Wakil MM. Powerplant technology. Singapore: McGraw-Hill, Inc; 1985.

[20] Stepuro TL, Zinchuk VV. Nitric oxide effect on the hemoglobin-oxygen affinity. J Physiol Pharmacol 2006;57(1):29–38.

[21] United States Environmental Protection Agency (EPA), Nitrogen Dioxide. [Online]. Available from: http://www.epa.gov/airquality/nitrogenoxides/.

[22] United States Environmental Protection Agency (EPA). NOx: how nitrogen oxides affect the way we live and breathe. Office of Air Quality Planning and Standards, pp. 2–3; 1998.

[23] Krupnick A, McConnell V, Austin D, Cannon M, Stoessell T, Morton B, The Chesapeake Bay and the control of NOx emissions: a policy analysis, 1998.

[24] El-Sharkawi MA. Electric energy an introduction. 3rd ed. CRC Press; 2013.

[25] United States Environmental Protection Agency (EPA). Ozone: good up high bad nearby. Office of Air and Radiation, pp. 1–2; 2003.

[26] Björn LO. Stratospheric ozone, ultraviolet radiation, and cryptogams. Biol Conserv 2007;135:326–33.

[27] Bjorn LO. Photobiology the science of light and life. Netherlands: Springer; 2002.

[28] de Jager D, Manning M, Kuijpers L. IPCC/TEAP Special Report: safeguarding the ozone layer and the global climate system: issues related to hydrofluorocarbons and perfluoro-carbons, Technical Summary; 2007.

[29] Horneman A, Stute M, Schlosser P, Smethie W, Santella N, Ho DT, Mailloux B, Gorman E, Zheng Y, van Geen A. Degradation rates of CFC-11, CFC-12 and CFC-113 in anoxic shallow aquifers of Araihazar, Bangladesh. J Contam Hydrol 2008;97:27–41.

[30] Sheehan JF. Acid rain: a continuing national tragedy. Elizabethtown, NY: Adirondack Council; 1998.

[31] Nag PK. Power plant engineering. 3rd ed. New Delhi: Tata McGraw-Hill Publishing Company Limited; 2008.

[32] Johnson E. Charcoal versus LPG grilling: a carbon-footprint comparison. Environ Impact Assess Rev 2009;29(6):370–8.

[33] Dormer A, Finn DP, Ward P, Cullen J. Carbon footprint analysis in plastics manufacturing. J Clean Prod 2013;51:133–41.

[34] Vergé XPC, Maxime D, Dyer JA, Desjardins RL, Arcand Y, Vanderzaag A. Carbon footprint of Canadian dairy products: calculations and issues. J. Dairy Sci. 2013;96(9):6091–104.

[35] Tukker A, Jansen B. Environmental impacts of products: a detailed review of studies. J Ind Ecol 2006;10(3):159–82.

[36] Kenny T, Gray NF. Comparative performance of six carbon footprint models for use in Ireland. Environ Impact Assess Rev 2009;29(1):1–6.

[37] Keyhani A. Design of smart power grid renewable energy systems. New Jersey: John Wiley & Sons, Inc; 2011.

[38] Energy Information Administration, Natural Gas 1998 Issues and Trends, 1999.

[39] Pope CA. Health effects of particulate matter air pollution. In: EPA Wood Smoke Health Effects Webinar; 2011.

[40] Steen M. Greenhouse gas emissions from fossil fuel fired power generation systems, http://publications.jrc.ec.europa.eu/repository/handle/JRC21207; 2001.

[41] IEA Clean Coal Centre, Particulate emission control technologies. [Online]. Available from: http://www.iea-coal.org.uk/site/ieacoal/databases/ccts/particulate-emissions-control-technologies.

[42] Klingspor JS, Vernon JL. Particulate control for coal combustion. London, United Kingdom: IEA Coal Research; 1988.

[43] United States Environmental Protection Agency (EPA), Particulate matter (PM) research. [Online]. Available from: http://www.epa.gov/airscience/air-particulatematter.htm.

[44] Atlas RM. Legionella: from environmental habitats to disease pathology, detection and control. Environ Microbiol 1999;1:283–93.

[45] Bartram J, Chartier Y, Lee JV, Pond K, Surman-Lee S. Legionella and the prevention of legionellosis, vol. 14. Geneva, Switzerland: World Health Organization Press; 2008.

[46] Fraser DW, Tsai TR, Orenstein W, Parkin WE, Beecham HJ, Sharrar RG, Harris J, Mallison GF, Martin SM, McDade JE, Shepard CC, Brachman PS. Legionnaires' disease − description of an epidemic of pneumonia. N Engl J Med 1977;(297):1189–97.

[47] Legionella management. [Online]. Available from: http://www.ges-water.co.uk/legionel-la-management/.

[48] Fliermans CB, Cherry WB, Orrison LH, Smith SJ, Tison DL, Pope DH. Ecological distribution of Legionella pneumophila. Appl Environ Microbiol 1981;41(1):9–16.

[49] MacFarlane JT, Worboys M. Showers, sweating and suing: Legionnaires' disease and "new" infections in Britain, 1977—90. Med Hist 2012;56:72–93.

[50] Osawa K, Shigemura K, Abe Y, Jikimoto T, Yoshida H, Fujisawa M, Arakawa S. A case of nosocomial Legionella pneumonia associated with a contaminated hospital cooling tower. J Infect Chemother 2014;20(1):68–70.

[51] Keong CY. Energy demand, economic growth, and energy efficiency — the Bakun dam-induced sustainable energy policy revisited. Energy Policy 2005;33:679–89.

[52] Xu X, Tan Y, Yang G. Environmental impact assessments of the Three Gorges Project in China: issues and interventions. Earth-Science Rev 2013;124:115–25.

[53] World Nuclear Association, Nuclear power in the world today, 2015. [Online]. Available from: www.world-nuclear.org/info/info8.html.

[54] The science of nuclear power. [Online]. Available from: http://nuclearinfo.net/Nuclear-power/TheScienceOfNuclearPower.

[55] Hatamura Y, Abe S, Fuchigami M, Kasahara N, Iino K. The 2011 Fukushima Daiichi Nuclear Power Plant Accident. United Kingdom: Woodhead Publishing/Elsevier Ltd.; 2015.

[56] Bird DK, Haynes K, van den Honert R, McAneney J, Poortinga W. Nuclear power in Australia: a comparative analysis of public opinion regarding climate change and the Fukushima disaster. Energy Policy 2014;65:644–53.

[57] Hirose K. 2011 Fukushima Dai-ichi nuclear power plant accident: summary of regional radioactive deposition monitoring results. J Environ Radioact 2012;111:13–7.

[58] Onda Y, Kato H, Hoshi M, Takahashi K, Nguyen ML. Soil sampling and analytical strategies for mapping fallout in nuclear emergencies based on the Fukushima Dai-ichi nuclear power plant accident. J Environ Radioact 2014;139:300–7.

[59] Miyake Y, Matsuzaki H, Fujiwara T, Saito T, Yamagata T, Honda M, Muramatsu Y. Isotopic ratio of radioactive iodine ($^{129}I/^{131}I$) released from Fukushima Daiichi NPP accident. Geochem J 2012;46:327–33.

[60] Hatch M, Ostroumova E, Brenner A, Federenko Z, Gorokh Y, Zvinchuk O, Shpak V, Tereschenko V, Tronko M, Mabuchi K. Non-thyroid cancer in Northern Ukraine in the post-Chernobyl period: short report. Cancer Epidemiol 2015;39(3):279–83.

[61] Bouville A, Likhtarev IA, Kovgan LN, Minenko VF, Shinkarev SM, Drozdovitch VV. Radiation dosimetry for highly contaminated Belarusian, Russian and Ukrainian populations, and for less contaminated populations in Europe. Health Phys 2007;93(5):487–501.

[62] World Nuclear Association, Nuclear radiation and health effects, 2015. [Online]. Available from: http://www.world-nuclear.org/info/Safety-and-Security/Radiation-and-Health/Nuclear-Radiation-and-Health-Effects/.

[63] Schobert HH. Energy and society: an introduction. New York: Taylor & Francis; 2002.

[64] Mahmoodi R, Shahriari M, Zolfaghari A, Minuchehr A. An advanced method for determination of loss of coolant accident in nuclear power plants. Nucl Eng Des 2011;241(6):2013–9.

[65] Hache G, Chung HM, The history of LOCA embrittlement criteria, in Conference: 28th Water Reactor Safety Information Meeting, 2001, pp. 1–32.

[66] Grosse MK, Stuckert J, Steinbrück M, Kaestner AP, Hartmann S. Neutron radiography and tomography investigations of the secondary hydriding of zircaloy-4 during simulated loss of coolant nuclear accidents. Phys Procedia 2013;43:294–306.

[67] Chung HEEM. Fuel behavior under loss-of-coolant accident situations. Nucl Eng Technol 2005;37:327–62.

[68] Taylor D. Manitoba's forgotten nuclear accident. Winnipeg Free Press; 2011. Available from: http://www.winnipegfreepress.com/opinion/analysis/manitobas-forgotten-nuclear-accident-118563039.html.

[69] United States Nuclear Regulatory Commission. Three mile island accident, http://www.nrc.gov/reading-rm/doc-collections/fact-sheets/3mile-isle.pdf; 2013.

[70]   Dempsey CH. Ichthyoplankton entrainment. J Fish Biol 1988;33:93–102.
[71]   Weinberg AM. Social institution and nuclear energy. Science 1972;177:27–34.
[72]   World Nuclear Association, Waste management: overview. [Online]. Available from: http://www.world-nuclear.org/info/Nuclear-Fuel-Cycle/Nuclear-Wastes/Waste-Management-Overview/.
[73]   Fanchi JR, Fanchi CJ. Energy in the 21st century. 2nd ed. Singapore: World Scientific Publishing Co. Pte. Ltd; 2011.
[74]   Aman MM, Solangi KH, Hossain MS, Badarudin A, Jasmon GB, Mokhlis H, Bakar AHA, Kazi S. A review of safety, health and environmental (SHE) issues of solar energy system. Renew Sustain Energy Rev 2015;41:1190–204.
[75]   Cyrs WD, Avens HJ, Capshaw ZA, Kingsbury RA, Sahmel J, Tvermoes BE. Landfill waste and recycling: use of a screening-level risk assessment tool for end-of-life cadmium telluride (CdTe) thin-film photovoltaic (PV) panels. Energy Policy 2014;68:524–33.
[76]   Bakhiyi B, Labrèche F, Zayed J. The photovoltaic industry on the path to a sustainable future − environmental and occupational health issues. Environ Int 2014;73:224–34.
[77]   Dessouky MO. The environmental impact of large scale solar energy projects on the MENA deserts: best practices for the DESERTEC initiative in IEEE EuroCon 2013, 2013, no. July, pp. 784–788.
[78]   Lin L, Yu H. Offshore wave energy generation devices: impacts on ocean bio-environment. Acta Ecol Sin 2012;32(3):117–22.
[79]   Frid C, Andonegi E, Depestele J, Judd A, Rihan D, Rogers SI, Kenchington E. The environmental interactions of tidal and wave energy generation devices. Environ Impact Assess Rev 2012;32(1):133–9.
[80]   Dai K, Bergot A, Liang C, Xiang W-N, Huang Z. Environmental issues associated with wind energy – a review. Renew Energy 2015;75:911–21.
[81]   Leung DYC, Yang Y. Wind energy development and its environmental impact: a review. Renew Sustain Energy Rev 2012;16(1):1031–9.
[82]   Cartlidge E, Fuel cells: environmental friend or foe? 2003. [Online]. Available from: http://physicsworld.com/cws/article/news/2003/jun/13/fuel-cells-environmental-friend-or-foe.
[83]   Tromp TK, Shia R-L, Allen M, Eiler JM, Yung YL. Potential environmental impact of a hydrogen economy on the stratosphere. Science 2003;300(2003):1740–2.
[84]   Environmental and Energy Study Institute, Hydrogen fuel cell. [Online]. Available from: http://www.eesi.org/topics/hydrogen-fuel-cells/description.
[85]   Ho JC, Saw EC, Lu LYY, Liu JS. Technological barriers and research trends in fuel cell technologies: a citation network analysis. Technol Forecast Soc Change 2014;82:66–79.
[86]   Bayer P, Rybach L, Blum P, Brauchler R. Review on life cycle environmental effects of geothermal power generation. Renew Sustain Energy Rev 2013;26:446–63.
[87]   Abbasi T, Abbasi SA. Biomass energy and the environmental impacts associated with its production and utilization. Renew Sustain Energy Rev 2010;14:919–37.
[88]   United States Environmental Protection Agency (EPA). Air pollution control technology fact sheet, http://www3.epa.gov/ttncatc1/dir1/fcyclon.pdf; 2002.
[89]   Hodgson PE. Energy, the environment and climate change. London, United Kingdom: Imperial College Press; 2010.
[90]   Morris C, Pehnt M. Energy transition the German Energiewende. [Online]. Available from: http://energytransition.de/.
[91]   The Carbon Trust of the UK, Low carbon cities, 2014. [Online]. Available from: http://www.lowcarboncities.co.uk/cms/.
[92]   Renewable 100 Policy Institute, Go 100% renewable energy. [Online]. Available from: http://www.go100percent.org/cms/index.php?id=4.

Author Index

Subject Index

Printed in the United States
By Bookmasters